Geostatistics Oslo 2012

Quantitative Geology and Geostatistics

VOLUME 17

For further volumes:
http://www.springer.com/series/6466

Petter Abrahamsen • Ragnar Hauge •
Odd Kolbjørnsen
Editors

Geostatistics
Oslo 2012

Editors
Petter Abrahamsen
Norwegian Computing Center
Oslo
Norway

Odd Kolbjørnsen
Norwegian Computing Center
Oslo
Norway

Ragnar Hauge
Norwegian Computing Center
Oslo
Norway

ISSN 0924-1973 Quantitative Geology and Geostatistics
ISBN 978-94-007-4152-2 ISBN 978-94-007-4153-9 (eBook)
DOI 10.1007/978-94-007-4153-9
Springer Dordrecht Heidelberg New York London

Library of Congress Control Number: 2012938533

Printed on acid-free paper

Springer is part of Springer Science+Business Media (www.springer.com)

Foreword

The series of International Geostatistics Congresses was initiated in Frascati in 1975 and has been arranged every fourth year since the Avignon congress in 1988:

Frascati, Italy in 1975
Lake Tahoe, USA in 1983
Avignon, France in 1988
Tróia, Portugal in 1992
Wollongong, Australia in 1996
Cape Town, South Africa in 2000
Banff, Canada in 2004
Santiago, Chile in 2008
Oslo, Norway in 2012

The International Geostatistics Congress has been an important arena for sharing and discussing the development of geostatistics in all these years. We are proud to welcome the congress back to Europe after 20 years of traveling the world.

The Ninth International Geostatistics Congress was arranged by the Norwegian Computing Center, University of Oslo, and the Norwegian University of Science and Technology.

Local organizing committee:

Petter Abrahamsen (Chairman)	Norwegian Computing Center
Ragnar Hauge	Norwegian Computing Center
Odd Kolbjørnsen	Norwegian Computing Center
Ørnulf Borgan	University of Oslo
Arnoldo Frigessie	University of Oslo
Henning Omre	Norwegian University of Science and Technology
Richard Sinding-Larsen	Norwegian University of Science and Technology
Øivind R. Lie	Congress-Conference AS

Preface

These conference proceedings consist of 44 technical papers presented at the *Ninth International Geostatistics Congress* held in Oslo, Norway June 11. to 15. 2012. The papers have been reviewed by a panel of specialists in Geostatistics. There is an additional set of papers that is published electronically and is available at the website of the International Association for Mathematical Geosciences: iamg.org. These papers have not been reviewed prior to the publication.

The proceedings are divided into four main sections: Theory, Petroleum, Mining, and Environmental, Climate and Hydrology. The first section focuses on new ideas of general interest to many fields of applications. The next sections are more focused on the particular needs of the particular industry or activity.

Geostatistics is vital to any industry dependent on natural resources. Methods from geostatistics are used for estimating reserves, quantifying economical risk and planning of future industrial operations. Geostatistics is also an important tool for mapping environmental hazard and integrating climate data.

Oslo, Norway

Petter Abrahamsen
Ragnar Hauge
Odd Kolbjørnsen

Preface

Acknowledgements

We would like to thank all the contributors, reviewers, chairmen and -women, and delegates who have helped making the Ninth International Geostatistics Congress a success. We would in particular like to thank the following colleagues for comments and suggestions:

Executive committee:

Amílcar Soares, Portugal
Christian Lantuejoul, France
Clayton V. Deutsch, Canada
Jaime Gómez-Hernández, Spain
Jef Caers, USA
Julián M. Ortiz, Chile
Peter Dowd, Australia
Roussos Dimitrakopoulos, Canada
Sean Mc Kenna, USA

Consultative committee:

Christina Dohm, South Africa
Hiroshi Okabe, Japan
Håkon Tjelmeland, Norway
Ian Glacken, Australia
João Felipe Coimbra Leite Costa, Brazil
Jose Almeida, Portugal
Oy Leuangthong, Canada
Sanjay Srinivasan, USA
Winfred Assibey-Bonsu, South Africa
Xavier Emery, Chile

Last but not least we are grateful to the company sponsors who contributed generously to the success of the congress:

Norwegian Computing Center
Statoil
Geovariances
Roxar

Contents

Part III Mining

Part I
Theory

Sequential Simulation with Iterative Methods

Daisy Arroyo, Xavier Emery, and María Peláez

Abstract The sequential Gaussian algorithm is widespread to simulate Gaussian random fields. In practice, the determination of the successive conditional distributions only accounts for the information available in a moving neighborhood centered on the target location, which provokes a loss of accuracy with respect to a unique neighborhood implementation. In order to reduce this loss of accuracy, iterative methods for solving large kriging systems of equations are used to improve the determination of the conditional distributions, taking the results obtained in a moving neighborhood as a first approximation. Numerical experiments are presented to show the proposed strategies and the improvements in the reproduction of the correlation structure of the simulated random field.

1 Introduction

This paper deals with the non-conditional sequential simulation of a Gaussian random vector $Y = (Y_1, Y_2, \ldots, Y_n)^T$, with zero mean and given variance-covariance matrix C_Y. This vector may be the restriction to a set of n spatial locations $x_1, x_2, \ldots, x_n$ of a Gaussian random field whose correlation structure has been inferred from regionalized data. In the following, we will denote by $\tilde{Y} = (\tilde{Y}_1, \tilde{Y}_2, \ldots, \tilde{Y}_n)^T$ the actually simulated random vector and by $C_{\tilde{Y}}$ its variance-covariance matrix. Ideally, this matrix should match C_Y, but this is often not the case due to an approximate implementation of the simulation algorithm [7, 8, 10, 15].

D. Arroyo (✉) · M. Peláez
Northern Catholic University, Av. Angamos 0610, Antofagasta, Chile
e-mail: darroyo@ucn.cl

M. Peláez
e-mail: mpelaez@ucn.cl

X. Emery
University of Chile, Av. Tupper 2069, Santiago, Chile
e-mail: xemery@ing.uchile.cl

P. Abrahamsen et al. (eds.), *Geostatistics Oslo 2012*,
Quantitative Geology and Geostatistics 17,
DOI 10.1007/978-94-007-4153-9_1,

2 Traditional Sequential Gaussian Simulation

The random vector Y can be simulated sequentially, by putting [6, 10]:

$$\tilde{Y}_p = \sum_{i=1}^{p-1} \lambda_{i,p} \tilde{Y}_i + \sigma_p U_p, \quad \forall p = 1, \ldots, n, \tag{1}$$

where $\Lambda_p = (\lambda_{1,p}, \ldots, \lambda_{p-1,p})^T$ are the simple kriging weights assigned to $Y_1, \ldots, Y_{p-1}$ when predicting Y_p, σ_p is the associated kriging standard deviation, and U_p is a standard Gaussian random variable independent of $U_1, \ldots, U_{p-1}$. The kriging weights are found by solving a system of linear equations of size $p-1$, in which the coefficient matrix is the variance-covariance matrix of the conditioning variables. As for the kriging standard deviation, it can be calculated as [4]:

$$\sigma_p = \sqrt{C_Y(x_p, x_p) - 2\sum_{i=1}^{p-1} \lambda_{i,p} C_Y(x_i, x_p) + \sum_{i=1}^{p-1}\sum_{j=1}^{p-1} \lambda_{i,p}\lambda_{j,p} C_Y(x_i, x_j)}. \tag{2}$$

Accounting for the simple kriging equations, this standard deviation simplifies into:

$$\sigma_p = \sqrt{C_Y(x_p, x_p) - \sum_{i=1}^{p-1} \lambda_{i,p} C_Y(x_i, x_p)}. \tag{3}$$

When the number of components of the random vector to simulate is large, the process of calculating the kriging weights involves much computational work. In practice, the variables considered for kriging are the nearest to the target location, i.e., kriging is performed in a moving neighborhood instead of a unique neighborhood, based on the screening-effect approximation [6, 10]. This moving neighborhood strategy generally involves some inaccuracies that tend to propagate as simulation progresses, so that the variance-covariance matrix $C_{\tilde{Y}}$ of the simulated random vector does no longer match the desired matrix C_Y [7, 8, 15]. The question therefore arises of how to improve the solution of sequential simulation within a moving neighborhood at relatively low costs.

3 Iterative Methods for Solving Linear Systems of Equations

Iterative methods can be used to solve large systems of kriging equations. They consist in considering an initial approximation $\Lambda_p^{(0)} = (\lambda_{1,p}^{(0)}, \ldots, \lambda_{p-1,p}^{(0)})^T$ and constructing a sequence of vectors $\{\Lambda_p^{(m)} : m = 1, 2, \ldots\}$ that converge to the exact solution: $\lim_{m\to\infty} \Lambda_p^{(m)} = \Lambda_p$. Iterative methods can be classified into stationary and non-stationary. The former are simple and easier to implement, but their rate of convergence is usually low, so their use is justified only with some specific types of matrix systems. The main stationary methods are the Jacobi, Gauss-Seidel, Successive Overrelaxation and Symmetric Successive Overrelaxation methods [1, 2, 5].

Non-stationary methods, also known as projection methods, differ from the previous ones in the use of information, which allows obtaining the solution dynamically. The Conjugate Gradient [11, 13] is the most prominent non-stationary iterative method, which solves linear systems under the restriction that the coefficient matrix is symmetric and positive definite. Among the existing non-stationary iterative methods for non-symmetric matrices, one finds the Generalized Minimal Residual method [12] and the Biconjugate Gradient Stabilized method [16].

For the numerical experiments presented in Sect. 5, two specific iterative methods have been chosen: the stationary Gauss-Seidel (GS) method for its easy implementation and the non-stationary Conjugate Gradient (CG) method for its effectiveness from the standpoint of accuracy and computational cost. These two methods converge, irrespective of the choice of the initial approximation, when the coefficient matrix of the linear systems of equations is a symmetric positive definite matrix [5, 9]. The Gauss-Seidel method consists in splitting the left-hand side matrix of the system into three components: a matrix of diagonal terms, a matrix of lower triangular terms and a matrix of upper triangular terms, so that the linear system can be rewritten and solved by the fixed point iteration method [14]. The Conjugate Gradient method proceeds by generating sequences of iterates (i.e., successive approximations to the solution) and residuals, and search directions used to update the iterates and residuals. Although the length of these sequences can become large, only a small number of vectors needs to be kept in memory. In every iteration of the method, two inner products are performed in order to compute update scalars that are defined to make the sequences satisfy certain orthogonality conditions. On a symmetric positive definite linear system, these conditions imply that the distance to the true solution is minimized in some norm.

4 Proposed Methodology

It can be shown [8] that the variance-covariance matrix of the random vector simulated by the sequential algorithm, as per (1), is

$$C_{\tilde{Y}} = \Lambda^{-1}\left(\Lambda^{-1}\right)^T, \tag{4}$$

with

$$\Lambda = \begin{pmatrix} \frac{1}{\sigma_1} & 0 & \cdots & \cdots & 0 \\ -\frac{\lambda_{1,2}}{\sigma_2} & \frac{1}{\sigma_2} & 0 & \cdots & 0 \\ \vdots & & \ddots & 0 & \vdots \\ -\frac{\lambda_{1,n-1}}{\sigma_{n-1}} & \cdots & -\frac{\lambda_{n-2,n-1}}{\sigma_{n-1}} & \frac{1}{\sigma_{n-1}} & 0 \\ -\frac{\lambda_{1,n}}{\sigma_n} & \cdots & \cdots & -\frac{\lambda_{n-1,n}}{\sigma_n} & \frac{1}{\sigma_n} \end{pmatrix}.$$

When considering a unique neighborhood to solve the kriging system, the target variance-covariance matrix is perfectly reproduced: $C_{\tilde{Y}} = C_Y$. In contrast, this may no longer be the case when considering a moving neighborhood. To mitigate the loss of accuracy, we will incorporate an iterative method at each step of the algorithm in order to better approximate the kriging weights that would be obtained under a unique neighborhood implementation.

At step p of the sequential algorithm, we first construct an initial approximation of the simple kriging weights and then apply one of the chosen iterative methods for solving the simple kriging system. Specifically, we first use the classical implementation with a moving neighborhood and obtain k weights

$$\lambda_{\tau(1),p}, \lambda_{\tau(2),p}, \ldots, \lambda_{\tau(k),p},$$

where $\tau(1), \ldots, \tau(k)$ are the indices of the k variables selected in the neighborhood. Provided with these k weights, we construct the initial approximation of size $p-1$:

$$\Lambda_p^{(0)} = (0, \ldots, 0, \lambda_{\tau(1),p}, 0, \ldots, 0, \lambda_{\tau(2),p}, 0, \ldots, 0, \lambda_{\tau(k),p}, 0, \ldots, 0)^T, \tag{5}$$

where the variables outside the moving neighborhood are assigned zero weights. We then apply an iterative method to solve the kriging system in a unique neighborhood, using $\Lambda_p^{(0)}$ as an initial approximation. After m iterations, the components of vector $\Lambda_p^{(m)} = (\lambda_{1,p}^{(m)}, \ldots, \lambda_{p-1,p}^{(m)})^T$ are used in (1) and (2), instead of the components $\lambda_{1,p}, \ldots, \lambda_{p-1,p}$ of the exact solution. In practice, because the number of iterations is finite, one does not obtain the exact solution, so the proposed algorithm will provide a simulated vector $\tilde{Y}$ with a variance-covariance matrix $C_{\tilde{Y}}$ different from the desired matrix C_Y. This matrix is given by (4), except that the triangular matrix Λ has to be replaced by

$$\tilde{\Lambda} = \begin{pmatrix} \frac{1}{\tilde{\sigma}_1} & 0 & \cdots & \cdots & 0 \\ -\frac{\lambda_{1,2}^{(m)}}{\tilde{\sigma}_2} & \frac{1}{\tilde{\sigma}_2} & 0 & \cdots & 0 \\ \vdots & & \ddots & 0 & \vdots \\ -\frac{\lambda_{1,n-1}^{(m)}}{\tilde{\sigma}_{n-1}} & \cdots & -\frac{\lambda_{n-2,n-1}^{(m)}}{\tilde{\sigma}_{n-1}} & \frac{1}{\tilde{\sigma}_{n-1}} & 0 \\ -\frac{\lambda_{1,n}^{(m)}}{\tilde{\sigma}_n} & \cdots & -\frac{\lambda_{n-2,n}^{(m)}}{\tilde{\sigma}_n} & -\frac{\lambda_{n-1,n}^{(m)}}{\tilde{\sigma}_n} & \frac{1}{\tilde{\sigma}_n} \end{pmatrix}, \tag{6}$$

where the standard deviations $\{\tilde{\sigma}_1, \ldots, \tilde{\sigma}_n\}$ are calculated by using the approximate kriging weights in (2). In the following section, numerical experiments are presented in order to demonstrate the performance of the proposed algorithm.

5 Numerical Experiments

5.1 Implementation Details

In this section, we consider the simulation of a second-order stationary Gaussian random field Y on a regular $2D$ grid with 65×65 nodes, with the following implementation:

- The grid nodes are visited through a midpoint displacement sequence: the corners of the grid are visited first, then, at each subsequent step, one visits the midpoints of the already visited points, until all the grid nodes are visited [8].
- Four isotropic covariance models are considered for the Gaussian random field, each with a unit sill:

  1. spherical covariance with range 10
  2. spherical covariance with range 40
  3. spherical covariance with range 10 and 30 % nugget effect
  4. spherical covariance with range 40 and 30 % nugget effect

- Two moving neighborhoods are considered for kriging, containing up to $k = 20$ and $k = 50$ conditioning variables, respectively.
- When solving linear systems by an iterative method, the method stops after $s = 50$ iterations.

Following Emery and Peláez [8], the accuracy of sequential simulation with the proposed methodology will be assessed by calculating the Frobenius norm of the difference between the model covariance matrix (C_Y) and the covariance matrix of the simulated vector ($C_{\tilde{Y}}$), standardized by the Frobenius norm of the model covariance matrix:

$$\eta = \frac{\|C_{\tilde{Y}} - C_Y\|_F}{\|C_Y\|_F}. \tag{7}$$

The parameter η so defined has the meaning of a relative error and will be referred to as the *standardized Frobenius norm* of $C_{\tilde{Y}} - C_Y$. For each of the above-mentioned covariance models and moving neighborhoods, this standardized norm is indicated in Table 1, for classical sequential simulation using a moving neighborhood and for sequential simulation combined with the Gauss-Seidel and Conjugate Gradient methods. All experiments were performed in MATLAB on a Pentium $i5$ with 64 bits. By incorporating an iterative method into sequential simulation, the reproduction of the target covariance turns out to be more accurate for all the assumed covariance models, especially when using the Conjugate Gradient method. Conjugate Gradient also compares favorably with Gauss-Seidel in terms of CPU times, as explained in Appendix.

5.2 Selection of Conditioning Variables

As the number of components of the Gaussian random vector to simulate is large, solving the kriging equations with an iterative method is still CPU-expensive since the method works with large weights vectors $\Lambda_p^{(0)}, \Lambda_p^{(1)}, \ldots, \Lambda_p^{(m)}$. These vectors can however be reduced without much loss of accuracy, insofar as the variables located far from the target node are usually assigned very small kriging weights [4, 10]. To this end, one can define a second moving neighborhood containing no more than q previously simulated variables and restrict the application of the iterative

Table 1 Standardized Frobenius norms of $C_{\tilde{Y}} - C_Y$ for sequential Gaussian simulation (SGS) with and without Gauss-Seidel (GS) method and Conjugate Gradient (CG) method. Smallest norms are indicated in bold type, largest norms in italic type.

Covariance model		SGS		SGS + GS (50 iterations)		SGS + CG (50 iterations)	
Range	Nugget	Moving neighborhood					
		$k=20$	$k=50$	$k=20$	$k=50$	$k=20$	$k=50$
10	0 %	*0.0599*	0.0286	0.0014	0.0014	0.000299	**0.000281**
10	30 %	*0.1616*	0.0598	0.0016	0.0016	**0.000001**	**0.000001**
40	0 %	*0.0342*	0.0104	0.0007	0.0006	0.000110	**0.000084**
40	30 %	*0.1334*	0.0786	0.0022	0.0005	0.000005	**0.000002**

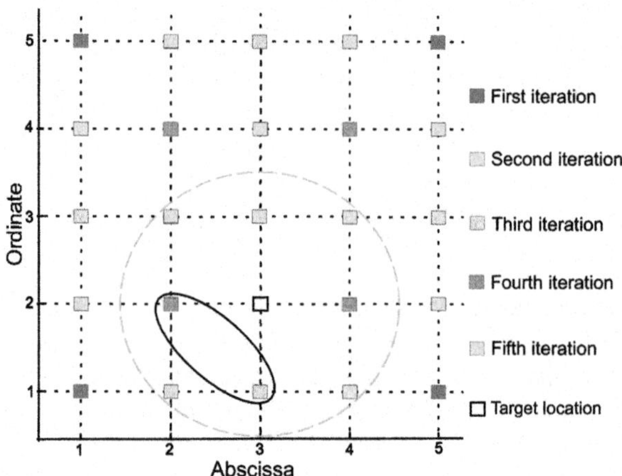

Fig. 1 Simulation grid with 5 × 5 nodes, midpoint displacement sequence, first moving neighborhood containing $k = 2$ variables (*black*) and second moving neighborhood (*blue*) containing $q = 8$ variables

methods to these q variables (Fig. 1). When the number of previously simulated variables $p - 1$ is greater than q, we construct an initial approximation of size q

$$\Lambda_q^{(0)} = (0, \ldots, 0, \lambda_{\tau(1),p}, 0, \ldots, 0, \lambda_{\tau(2),p}, 0, \ldots, 0, \lambda_{\tau(k),p}, 0, \ldots, 0)^T, \tag{8}$$

where the variables outside the moving neighborhood of size k are assigned zero weights. With the iterative methods, we solve the following kriging system of size q (instead of a system of a size $p - 1$), with vector $\Lambda_q^{(0)}$ as the initial approximation.

To assess the accuracy of this alternative approach, the same numerical experiments as in Sect. 5.1 have been realized, with two options for the second moving neighborhood ($q = 200$ and $q = 400$) (Table 2). Although the reproduction of the target covariance matrix is not as good as with the initial proposal (Table 1), it is still much better than with the classical implementation of sequential simulation: the standardized Frobenius norm of $C_{\tilde{Y}} - C_Y$ is between 1 and 2 orders of magnitude smaller when using the iterative methods.

Some additional comments are worth being made:

Table 2 Standardized Frobenius norms of $C_{\tilde{Y}} - C_Y$ for sequential Gaussian simulation (SGS) with and without Gauss-Seidel (GS) method and Conjugate Gradient (CG) method. Smallest norms are indicated in bold type, largest norms in italic type

Covariance model		SGS		SGS + iterative method (50 iterations)							
Range	Nugget	Moving neighborhood									
		$k=20$	$k=50$	$k=20$				$k=50$			
				Second neighborhood							
				$q=200$		$q=400$		$q=200$		$q=400$	
				GS	CG	GS	CG	GS	CG	GS	CG
10	0 %	*0.0599*	0.0286	0.0093	0.0092	0.0026	**0.0022**	0.0093	0.0093	0.0025	**0.0022**
10	30 %	*0.1616*	0.0598	0.0244	0.0244	0.0093	**0.0092**	0.0244	0.0244	0.0093	**0.0092**
40	0 %	*0.0342*	0.0104	0.0021	0.0021	0.0010	**0.0009**	0.0021	0.0021	0.0010	**0.0009**
40	30 %	*0.1334*	0.0786	0.0138	0.0138	0.0068	**0.0067**	0.0138	0.0138	0.0067	**0.0067**

Table 3 Standardized Frobenius norms of $C_{\tilde{Y}} - C_Y$ for sequential Gaussian simulation (SGS) with and without iterative method (initial vector of zeros for $\Lambda_p^{(0)}$). Smallest norms are indicated in bold type, largest norms in italic type

Covariance model		SGS	SGS + iterative method	
Range	Nugget	($k = 20$)	GS (50 iterations)	CG (50 iterations)
10	0 %	*0.0599*	0.0137	**0.002031**
10	30 %	*0.1616*	0.0086	**0.000011**
40	0 %	0.0342	*0.3293*	**0.002096**
40	30 %	*0.1334*	0.1072	**0.000028**

1. A second neighborhood containing $q = 400$ variables yields a better reproduction of the target covariance matrix than a neighborhood with $q = 200$ variables. In the case of a simulation grid much larger than the one considered here, the number of variables for the second neighborhood could be increased to several thousands (e.g., $q = 10000$), while the direct resolution of kriging systems would be impractical.
2. The reproduction of the target covariance matrix is almost as accurate by using the Gauss-Seidel or the Conjugate Gradient method. Both methods are therefore valid options for the proposal, although the latter is faster from the computational time standpoint.
3. The experiments have been realized with different numbers of iterations for Gauss-Seidel and Conjugate Gradient (up to $s = 100$), but no significant decrease of the standardized Frobenius norm of $C_{\tilde{Y}} - C_Y$ has been observed.

5.3 *Choice of the Initial Vector in the Iterative Methods*

Because the iterative methods converge to the solution irrespective of the initial vector, it is also interesting to run the algorithm with a vector of zeros as an initial approximation of the weights vector, instead of a vector of the form (5), in order to reduce CPU time. The results of this variant (Table 3) indicate that the reproduction of the target covariance matrix is poorer when using the Gauss-Seidel method, especially with the larger range models, but remains equally good when using the Conjugate Gradient method.

One may also combine this alternative with the first one, by restricting the application of the iterative methods to q neighboring variables. The results show that the Conjugate Gradient method still provides a good reproduction of the target covariance matrix (Table 4).

Table 4 Standardized Frobenius norms of $C_{\tilde{Y}} - C_Y$ for sequential Gaussian simulation (SGS) with and without iterative method (initial vector of zeros for $\Lambda_q^{(0)}$). Smallest norms are indicated in bold type, largest norms in italic type

Covariance model		SGS	SGS + iterative method			
Range	Nugget	($k = 20$)	GS (50 iterations)		CG (50 iterations)	
			Second neighborhood			
			$q = 200$	$q = 400$	$q = 200$	$q = 400$
10	0 %	*0.0599*	0.0162	0.0132	0.0092	**0.0022**
10	30 %	*0.1616*	0.0265	0.0116	0.0245	**0.0092**
40	0 %	0.0342	0.0510	*0.1094*	0.0021	**0.0009**
40	30 %	*0.1334*	0.0203	0.0262	0.0138	**0.0067**

6 Conclusions

We proposed a method to simulate a Gaussian random vector, based on the incorporation of iterative methods into the sequential algorithm, in order to improve the determination of the kriging weights required for determining the successive conditional distributions. The initial weight vector is taken as that obtained with the classical implementation based on a moving neighborhood.

Two iterative methods have been put to the test: Gauss-Seidel and Conjugate Gradient, showing that the latter yields a better reproduction of the target covariance matrix than the former, with a smaller computational cost. This cost is considerably reduced, with little loss of accuracy, if one further selects the neighboring variables that are deemed relevant for kriging (typically, a few hundreds or thousands variables around the target location), or if one considers an initial vector of zeros in the iterative methods.

Acknowledgements This research was funded by the Chilean program MECESUP UCN0711 and the FONDECYT project 11100029.

Appendix: Analysis of Computational Cost

In this appendix we assess the computational cost of sequential simulation using a moving neighborhood of fixed size k, and the additional cost of solving the kriging system with an iterative method in each step of sequential simulation.

A.1 Solving Kriging Systems with a Moving Neighborhood

If we sequentially simulate a Gaussian random vector with n components using a moving neighborhood containing k variables, one has to solve one kriging system of size 1×1, one system of size $2 \times 2, \dots$, one system of size $(k-1) \times (k-1)$ and $(n-k)$ systems of size $k \times k$. To solve a linear system of equations of size $p \times p$ by means of the Gaussian elimination (direct method), the number of floating point operations is $2p^3/3$ [3]. Accordingly, the computational cost in floating point operations to solve all the kriging systems in the sequential simulation with a neighborhood of k variables is $O(nk^3)$:

$$\begin{aligned} C_{\text{neighborhood}\,k} &= \frac{2}{3} \times 1^3 + \frac{2}{3} \times 2^3 + \dots + \frac{2}{3}(k-1)^3 + \frac{2}{3}(n-k)k^3 \\ &= \frac{2}{3}\left(1^3 + 2^3 + \dots + k^3\right) + \frac{2}{3}(n-k-1)k^3 \\ &= \frac{1}{6}k^2\left[1 + (4n-2)k - 3k^2\right]. \end{aligned} \tag{9}$$

If we consider the ideal case of a unique neighborhood implementation ($k = n - 1$), the number of floating point operations is $O(n^4)$:

$$C_{\text{unique neighborhood}} = \frac{1}{6}n^2(n-1)^2. \tag{10}$$

A.2 Solving Kriging Systems with Iterative Methods

The computational cost to calculate the kriging weights and that are used to construct initial approximations (to apply the iterative methods), using a moving neighborhood of size k and Gaussian elimination, is given by (9). To this cost, one has to add the cost of the iterative method from component $(k+2)$ to component n of the simulated vector.

A.2.1 Gauss-Seidel Method

To solve a linear system of equations of size $p \times p$ with the Gauss-Seidel iterative method, the number of floating point operations is $2sp^2$, where s is the number of iterations performed [3]. Therefore, the number of floating point operations for the $(n-k-1)$ systems where we apply the Gauss-Seidel method is:

$$\begin{aligned} &2s\left[(k+1)^2 + (k+2)^2 + \dots + (n-1)^2\right] \\ &\quad = \frac{1}{3}s\left[n(n-1)(2n-1) - k(k+1)(2k+1)\right]. \end{aligned} \tag{11}$$

With respect to the classical implementation (9), the additional cost is $O(sn^3)$.

Table 5 Standardized Frobenius norms of $C_{\tilde{Y}} - C_Y$ (η) and CPU times (in seconds) for sequential Gaussian simulation (SGS) and sequential Gaussian simulation with Conjugate Gradient method (SGS + CG) when the first and second neighborhoods are equal (a circle of radius 30). Smallest CPU times are indicated in bold type, largest CPU times in italic type

Covariance model			SGS		SGS + CG	
Model	Range	Nugget	η	CPU time	η	CPU time
Spherical	10	0 %	0.0011	*5642*	0.0018	**4600**
Spherical	10	30 %	0.0005	*5813*	0.0005	**4481**
Spherical	50	0 %	0.0615	*7004*	0.0615	**5767**
Spherical	50	30 %	0.1043	*6965*	0.1043	**5786**

A.2.2 Conjugate Gradient Method

The cost to solve a linear system of equations of size $p \times p$ with Conjugate Gradient method is $s(6p+3+2u_p)$, where u_p is the number of nonzero elements of the left-hand side matrix of the system of equations [9]. Accordingly, the number of floating point operations for solving the last $(n-k-1)$ kriging systems with the Conjugate Gradient method is:

$$s \sum_{p=k+1}^{n-1} (6p+3+2u_p) = 3s\left(n^2 - k^2 - 2k - 1\right) + 2s \sum_{i=k+1}^{n-1} u_i.$$

The computational order of this additional is $O(sn^2)$ when the matrix of the system has many zero entries.

A.3 CPU Time on a Numerical Example

To illustrate the gain in CPU time when using an iterative method (here, the Conjugate Gradient) rather than the direct method (Gaussian elimination) to solve the successive kriging systems, we have run an experiment on a regular 2D grid with 100×100 nodes, which are visited through a random sequence using multiple grids [15]. Table 5 shows the standardized Frobenius norm of $C_{\tilde{Y}} - C_Y$ and the CPU time for direct sequential Gaussian simulation (SGS) and for the proposed algorithm (SGS + CG), using the same set of conditioning variables to construct the successive conditional distributions (specifically, the variables located within a circle of radius 30 centered on the target node). The initial approximation for the Conjugate Gradient method is a vector of zeros. It is seen that the accuracy is comparable for both approaches (SGS and SGS + CG), insofar as the standardized Frobenius norms are practically the same, but there are considerable differences in CPU times. These differences increase for larger simulation grids and for larger sizes of the neighborhood used to select the conditioning variables.

References

1. Axelsson O (1985) A survey of preconditioned iterative methods for linear systems of algebraic equations. BIT Numer Math 25:166–187
2. Barret R (1994) Templates for the solution of linear systems: building blocks for iterative methods. SIAM publications, vol 117. SIAM, Philadelphia
3. Burden R, Faires D (2010) Numerical analysis, 9th edn. Brooks/Cole, Boston, 872 pp
4. Chilès JP, Delfiner P (1999) Geostatistics: modeling spatial uncertainty. Wiley, New York, 695 pp
5. Datta BN (2010) Numerical linear algebra and applications, 2nd edn. Society for Industrial and Applied Mathematics, Philadelphia, 530 pp
6. Deutsch CV, Journel AG (1998) GSLIB: geostatistical software library and user's guide, 2nd edn. Oxford University Press, New York, 369 pp
7. Emery X (2004) Testing the correctness of the sequential algorithm for simulating Gaussian random fields. Stoch Environ Res Risk Assess 18(6):401–413
8. Emery X, Peláez M (2011) Assessing the accuracy of sequential Gaussian simulation and cosimulation. Comput Geosci 15(4):673–689
9. Golub GH, Van Loan CF (1996) Matrix computations, 3rd edn. Johns Hopkins University Press, Baltimore, 694 pp
10. Gómez-Hernández JJ, Cassiraga EF (1994) Theory and practice of sequential simulation. In: Armstrong M, Dowd PA (eds) Geostatistical simulations. Kluwer Academic, Dordrecht, pp 111–124
11. Hestenes MR, Stiefel E (1952) Methods of conjugate gradients for solving linear systems. J Res Natl Bur Stand 49(6):409–436
12. Saad Y (1981) Krylov subspace methods for solving large unsymmetric linear systems. Math Comput 37:105–126
13. Saad Y (2003) Iterative methods for sparse linear systems, 2nd edn. SIAM, Philadelphia, 567 pp
14. Stoer J, Burlisch R (2002) Introduction to numerical analysis, 3rd edn. Springer, New York, 732 pp
15. Tran TT (1994) Improving variogram reproduction on dense simulation grids. Comput Geosci 20(7–8):1161–1168
16. Van der Vorst HA (1992) Bi-CGSTAB: a fast a smoothly converging variant of Bi-CG for the solution of nonsymmetric linear systems. SIAM J Sci Comput 13:631–644

Applications of Randomized Methods for Decomposing and Simulating from Large Covariance Matrices

Vahid Dehdari and Clayton V. Deutsch

Abstract Geostatistical modeling involves many variables and many locations. LU simulation is a popular method for generating realizations, but the covariance matrices that describe the relationships between all of the variables and locations are large and not necessarily amenable to direct decomposition, inversion or manipulation. This paper shows a method similar to LU simulation based on singular value decomposition of large covariance matrices for generating unconditional or conditional realizations using randomized methods. The application of randomized methods in generating realizations, by finding eigenvalues and eigenvectors of large covariance matrices is developed with examples. These methods use random sampling to identify a subspace that captures most of the information in a matrix by considering the dominant eigenvalues. Usually, not all eigenvalues have to be calculated; the fluctuations can be described almost completely by a few eigenvalues. The first k eigenvalues corresponds to a large amount of energy of the random field with the size of $n \times n$. For a dense input matrix, randomized algorithms require $O(nn\log(k))$ floating-point operations (flops) in contrast with $O(nnk)$ for classical algorithms. Usually the rank of the matrix is not known in advance. Error estimators and the adaptive randomized range finder make it possible to find a very good approximation of the exact SVD decomposition. Using this method, the approximate rank of the matrix can be estimated. The accuracy of the approximation can be estimated with no additional computational cost. When singular values decay slowly, power method can be used for increasing efficiency of the randomized method. Comparing to the original algorithm, the power method can significantly increase the accuracy of approximation.

V. Dehdari (✉) · C.V. Deutsch
Center for Computational Geostatistics, University of Alberta, 3-133 NREF Building, Edmonton, Alberta, Canada, T6G 2W2
e-mail: dehdari@ualberta.ca

C.V. Deutsch
e-mail: cdeutsch@ualberta.ca

P. Abrahamsen et al. (eds.), *Geostatistics Oslo 2012*,
Quantitative Geology and Geostatistics 17,
DOI 10.1007/978-94-007-4153-9_2,

1 Introduction

Sequential Gaussian simulation is a popular method for generating realizations. This method is based on recursive application of Bayes law. There are two problems related to this method. The cost of generating n realization is n times of cost of generating one realization. Generating one realization can be reasonably fast, but if 100 realizations are to be generated, it can be time consuming. Second, reproducing long range variogram structures using this method is sometimes difficult. LU simulation is another method that can be used for generating unconditional or conditional realizations [1, 2]. This method is based on decomposing the covariance matrix. This method reproduces covariance matrix and once covariance matrix decomposed, generating large number of realization can be done without significant cost. For this reason, covariance matrix should be defined which contains covariance between data to data C_{11}, data to unsampled nodes C_{12} and unsampled nodes to unsampled nodes C_{22}. This covariance matrix has the following form:

$$C = \begin{bmatrix} C_{11} & C_{12} \\ C_{21} & C_{22} \end{bmatrix}. \tag{1}$$

In this matrix $C_{12} = C_{21}^T$. If there are n data and N unsampled nodes, C_{11} is $n \times n$, C_{12} is $n \times N$ and C_{22} is $N \times N$. For generating realizations, the covariance matrix can be decomposed using Cholesky decomposition. Cholesky decomposition is a popular method when we have a symmetric and positive definite matrix. Using this method covariance matrix can be decomposed to the lower and upper triangular matrices such that $C = LU = LL^T$ where L and U are lower and upper triangular matrices respectively. Another popular method is eigenvalue-eigenvector decomposition. In Cholesky decomposition, unconditional realization can be generated by decomposing C_{22} and using $m_i = L_{22}z_i$ formula. In this formula, z_i is random vector from Gaussian distribution with zero mean and unit variance and L_{22} is lower triangular matrix in Cholesky decomposition. Also kriging can be used for generating conditional realizations [2]. There are two problems related to the LU decomposition:

1. Accumulated roundoff errors can cause imaginary values for poorly conditioned covariance matrix [8].
2. When the covariance matrix is very large, full decomposition needs a very large additional storage and CPU time. In this case, decomposition cannot be done.

But as an advantage, this method can reproduce the mean and covariance matrices very well.

As another approach, Davis proposed the following method for generating conditional realization in just one step [2]. In this method by decomposing the original covariance matrix which contains information about covariance among all of grids and data, conditional realizations can be generated easily. Assume

$$\begin{bmatrix} y_1 \\ y_2 \end{bmatrix} = \begin{bmatrix} L_{11} & 0 \\ L_{21} & L_{22} \end{bmatrix} \begin{bmatrix} z_1 \\ z_2 \end{bmatrix}, \tag{2}$$

where y_1 is $n \times 1$ column vector of normal score of conditioning data, and y_2 is $N \times 1$ column vector of required simulation values. From this system of equations $z_1 = L_{11}^{-1} y_1$ can be found. Also z_2 is $N \times 1$ vector of independent normal score data from Gaussian distribution. By replacing this equation into the second equation vector of required simulation values can be found from the following equation:

$$y_2 = L_{21} L_{11}^{-1} y_1 + L_{22} z_2. \tag{3}$$

The eigenvalue-eigenvector decomposition (SVD or singular value decomposition), can be used to decompose the covariance matrix to the form

$$C = U \Lambda U^T = U \Lambda^{1/2} \Lambda^{1/2} U^T = \left(U \Lambda^{1/2} U^T\right)\left(U \Lambda^{1/2} U^T\right)^T, \tag{4}$$

where U is the matrix of eigenvectors—each column of U is one eigenvector of covariance matrix. Λ is a diagonal matrix of eigenvalues; each element on the diagonal is one of the eigenvalues. These eigenvalues are ordered so that the first one is the largest one and they decrease to zero. The number of non-zero eigenvalues is the rank of matrix. In contrast to LU decomposition, in this case $U \Lambda^{1/2} U^T$ is not triangular and previous approach (3) cannot be used for generating conditional realizations. Assume that:

$$U \Lambda^{1/2} U^T = \begin{bmatrix} A & B \\ C & D \end{bmatrix}. \tag{5}$$

In this case, if there are n data and N unsampled node, A is $n \times n$, B is $n \times N$, C is $N \times n$ and D is $N \times N$. Conditional simulation can be found from the following formula

$$y = \begin{bmatrix} y_1 \\ y_2 \end{bmatrix} = \begin{bmatrix} A & B \\ C & D \end{bmatrix} \begin{bmatrix} z_1 \\ z_2 \end{bmatrix}. \tag{6}$$

The elements in z_1 can be found by the relation $A z_1 + B z_2 = y_1$. From this formula we obtain:

$$z_1 = A^{-1}(y_1 - B z_2). \tag{7}$$

Again z_2 is $N \times 1$ vector of independent normal score data from Gaussian distribution. As a result, conditional simulation can be found from the following formula:

$$y = \begin{bmatrix} y_1 \\ C A^{-1}(y_1 - B z_2) + D z_2 \end{bmatrix}. \tag{8}$$

In this case, the matrix A usually is much smaller than the size of covariance matrix and its inverse can be found easily. Usually Cholesky decomposition is faster than SVD decomposition, but as we mentioned before, if the dimension of covariance matrix is large, decomposing the covariance matrix would be impossible. To prevent this problem, this paper introduces a randomized low rank approximation method for decomposing the covariance matrix. The most important reason for using SVD decomposition instead of LU decomposition is to permit the use of randomized low rank approximation methods. Using this method, the SVD decomposition or inversion of large matrices can be done very fast. In the next section, the application of randomized methods for decomposing large matrices will be presented.

2 Randomized Low Rank Approximation

Randomized method allows designing provably accurate algorithms for very large or computationally expensive problems. This method has been investigated by many authors [5–7, 9, 10]. This method is very popular method in computer science for image processing. In this method, by randomized sampling of rows or columns of a matrix, a new matrix can be found that is much smaller in size than the original matrix and captures most of the action of that matrix in a subspace. Using this method, the original matrix can be restricted to the subspace and then decomposition can be done in a subspace using standard factorization methods such as QR or SVD decomposition. Assume that rank of a matrix is R. In this case this matrix has R eigenvalues greater than zero. Using the randomized low rank approximation method, the approximated SVD decomposition of this matrix can be found by considering the first k dominant eigenvalues. This is called rank-k approximation of this matrix. The energy of a system can be defined as a summation of the first k largest eigenvalues divided by summation of all of eigenvalues. Usually by considering 90–95 % of the energy of a system, a reasonable approximation of the matrix decomposition can be done without missing major variabilities. Random sampling from a matrix A with size $m \times m$ can be done by multiplying this matrix by another matrix Ω which has the size of $m \times k$ and has been drawn randomly from the Gaussian distribution. Then, instead of decomposing the original matrix, this matrix which has much smaller size than the original matrix should be decomposed. If $Y = A\Omega$, in this case matrix Y can be decomposed to the Q and R:

$$Y = QR = [\,Q_1 \quad Q_2\,]\begin{bmatrix} R_1 \\ 0 \end{bmatrix} = Q_1 R_1. \tag{9}$$

If Y has dimension of $m \times k$, R_1 is a $k \times k$ upper triangular matrix, Q_1 is $m \times k$, Q_2 is $m \times (m-k)$ and Q_1 and Q_2 both have orthogonal columns. Q_2 matrix is related to the null space of matrix Y. QR factorization of Y is equal to the QR factorization of A in the subspace which is approximate decomposition of matrix A in the full space. If SVD decomposition of A is needed, matrix B can be found such that $B = Q^T A$. Finally SVD decomposition of matrix A can be found from full SVD decomposition of matrix B which has much smaller size than the original matrix. The randomized method for decomposing matrix A with size of $m \times n$ as described in [5] is:

1. Draw an $m \times k$ Gaussian random matrix Ω.
2. Form the $m \times k$ sample matrix $Y = A\Omega$.
3. Form an $m \times k$ orthonormal matrix Q such that $Y = QR$.
4. Form the $k \times m$ matrix $B = Q^T A$.
5. Compute the SVD of the small matrix B: $B = \widehat{U}\Sigma V^T$.
6. Form the matrix $U = Q\widehat{U}$.

In terms of computation cost, assume that the first k eigenvalues of a dense matrix with the size of $m \times m$ should be approximated. In this case, randomized algorithms require $O(m\,m \log k)$ floating-point operations (flops) in contrast with $O(m\,m\,k)$ for classical algorithms. Also randomized methods usually required a constant number of passes over the original matrix instead of $O(k)$ passes for classical algorithms [5].

When singular values decay slowly, the algorithm should be modified to increase efficiency of the randomized method. For this purpose, the power method can be used. Assume that matrix A should be decomposed and its eigenvalues decay slowly. Comparing to the original algorithm, the power method can significantly increase the accuracy of approximation. For solving this problem, instead of decomposing matrix A, matrix B can be decomposed:

$$B = \left(AA^*\right)^q A, \tag{10}$$

where A^* is the conjugate transpose of matrix A and q is an integer which is called the power. Both matrices A and B have the same singular vectors, but singular values of matrix B decay much faster than matrix A [7, 10]. If the ith singular value shows with the notation σ_i then:

$$\sigma_i(B) = \left(\sigma_i(A)\right)^{2q+1}. \tag{11}$$

In some cases, if the first 10 000 eigenvalues represent 95 % of the energy of the random field with size of $10^6 \times 10^6$, using the power method entails that approximately the first 1 000 eigenvalues could represent 95 % of the energy of the random field. Due to the large size of the matrix A, finding $(AA^*)^q A$ is almost impossible, this expression typically can be evaluated via alternating application of A and A^*. In this case, after finding $Y = A\Omega$, QR decomposition of Y without significant computation cost can be found. Matrix Q is orthogonal and it is approximated QR decomposition of matrix A. Therefore $QQ^* = I$ and as a result $A \approx QQ^*A$. Matrix Q has much smaller size than matrix A. Then in an iterative procedure without any significant cost $(AA^*)^q A$ can be found.

The algorithm of randomized subspace power method as found in [5] is summarized by:

1. Draw an $n \times k$ Gaussian random matrix Ω.
2. Form $Y_0 = A\Omega$ and compute its QR factorization $Y_0 = Q_0 R_0$
3. For $j = 1, 2, \ldots, q$ where q is integer power
   Form $\widetilde{Y}_j = A^T Q_{j-1}$ and compute its QR factorization $\widetilde{Y}_j = \widetilde{Q}_j \widetilde{R}_j$
   Form $Y_j = A\widetilde{Q}_j$ and compute its QR factorization $Y_j = Q_j R_j$
   End
4. $Q = Q_q$
5. Form the $k \times n$ matrix $B = Q^T A$.
6. Compute the SVD of the small matrix B: $B = \widehat{U} \Sigma V^T$.
7. Form the matrix $U = Q\widehat{U}$.

This method is much more efficient than the original method and increases accuracy of approximation significantly. It works very well for problems where the singular values decay slowly. Usually $q = 2$ or 3 is a good choice. Larger q will not significantly improve estimation.

There are different challenges in using this method. First, before full decomposition of a matrix, the eigenvalues are unknown. As a result finding the total energy of

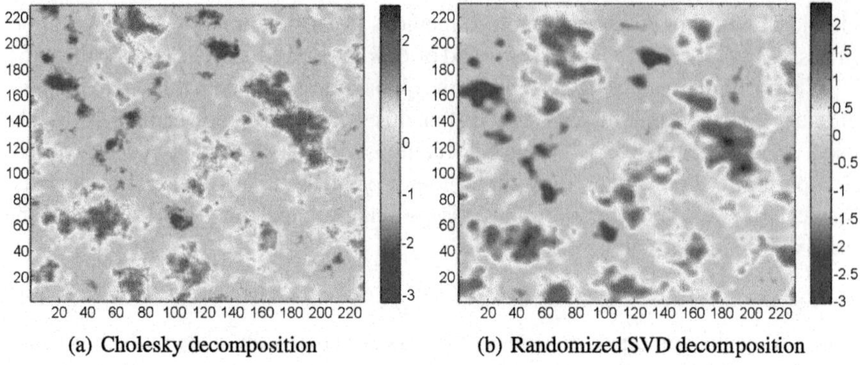

(a) Cholesky decomposition

(b) Randomized SVD decomposition

Fig. 1 Unconditional realization using exponential covariance

the system for computing the fraction of energy in the approximation is a problem. Second, the rank of the matrix is unknown in advance, so the number of eigenvalues for a good estimation is not known in advance. In the case of decomposing the covariance matrix, a solution of the first problem can be found easily. In this case, as an advantage of the covariance matrix, summation of eigenvalues is equal to the trace of covariance matrix, which is the summation of the elements on the diagonal. Because these values are equal to the variance, even before finding the covariance matrix, the total energy of the system can be easily found. Assume that the dimension of a covariance matrix is $50\,000 \times 50\,000$ and variance of the standard normal data is 1, in this case total energy of the system is equal to 50 000. After finding approximate decomposition, the summation of the eigenvalues can be found easily. In order to find the fraction of energy we used for simulation, this value can be divided by the total energy of the system.

Regarding the unknown rank of the matrix, the randomized range finder can be used. This method starts with an initial guess. Usually for a matrix with the size of $50\,000 \times 50\,000$, the first 1 000 eigenvalues are enough for approximating at least 90 % energy of the system. We can start with $k = 1\,000$ and by finding fraction energy of the system, if energy was not enough, the number of eigenvalues can be increased gradually to find a reasonable approximation. As another test, the accuracy of the approximation can be estimated with no additional computational cost [5]. This can be done by finding $l2$ norm of a matrix which is difference of original matrix and approximate matrix:

$$\left\| A - U \Sigma V^T \right\| = \left\| A - Q Q^T A \right\| < \varepsilon. \tag{12}$$

The spectral norm of a matrix is the largest eigenvalue of that matrix. This is similar to the rank-1 approximation of that matrix which can be found very fast. For finding a good approximation, ε should be a small number. Using these methods, the accuracy of estimation can be estimated easily.

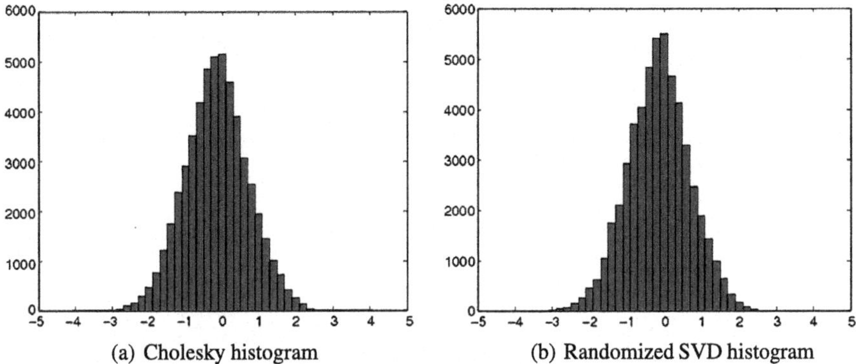

(a) Cholesky histogram (b) Randomized SVD histogram

Fig. 2 Histogram comparison using exponential covariance

3 Results and Discussion

The goal of this paper is to compare the efficiency of a randomized method with Cholesky decomposition, different synthetic examples with reasonable size have been considered. The dimension of the domain in these examples is 230×230 blocks. In this case, size of covariance matrix would be $52\,900 \times 52\,900$ and it has about 2.8 billion elements. This was the largest matrix that I could decompose with the LU simulation. Larger matrix can be created and decomposed using randomized methods, but because the goal of this paper is comparing the efficiency of these methods, this example is large enough for this reason. For the first example, a random field has been generated with variance 1 and also exponential covariance matrix. Using unconditional simulation and just by decomposing the C_{22} in the covariance matrix, one realization for both of LU and randomized SVD has been generated. In this example, correlation range is 60 blocks. This range is equal to practical range of covariance matrix (95 % of the sill). SVD used the first 2 000 eigenvalues for generating the result. In this case, full decomposition has 52 900 eigenvalues, but the first 2 000 eigenvalues equal to about 93 % energy of the system. In this case, the spectral norm is less than 0.01 which shows a good precision. Figure 1 shows result of each method. As Fig. 1(b) shows, randomized algorithm eliminated very short variabilities. It means that if correlation range is very small, very small eigenvalues should not be eliminated. In this case, more eigenvalues should be considered for storing 95 % energy of that system. In the limiting case, when correlation range is zero, none of eigenvalues are zero. In this case, covariance matrix would be a diagonal matrix and its decomposition using Cholesky, even if the matrix is very large, would be very fast. As a result, if range is very small (which is rare in the real cases) using Cholesky decomposition could be a better choice. What is important in the simulation is histogram and variogram reproduction. Figure 2 and Fig. 3(a) show histogram and variogram comparisons using both of methods.

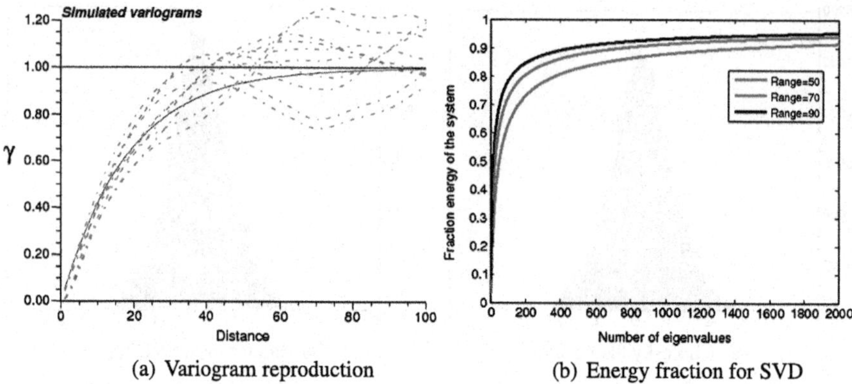

(a) Variogram reproduction (b) Energy fraction for SVD

Fig. 3 Variogram reproduction and energy fraction using exponential covariance. In (**a**) *red dashed lines* are variograms of simulated realizations using randomized SVD, *blue dashed lines* are variograms of simulated realizations using Cholesky decomposition and *gray solid line* is theoretical variogram. In (**b**) *colors* indicates correlation range in the covariance function

Usually histogram reproduction is not a big problem in the simulation. As you can see, Fig. 2 shows comparison between reproduced histograms using both methods. In this case both of histograms are normal and they are pretty the same.

In Fig. 3(a), five realizations generated for each method. Red dashed lines are variograms of simulated realizations using randomized SVD, blue dashed lines are variograms of simulated realizations using Cholesky decomposition and gray solid line is theoretical variogram. It seems that average of variograms in both methods can reproduce theoretical variogram. A small correlation range means the eigenvalues will decay slowly and by increasing the range, they will decay very fast. For smaller ranges, the power method can be used for increasing the decay rate. Using power method may increase computation time a little bit which is not significant, but it can solve problem related to the slow decay rate. As another solution, the number of desired eigenvalues can be increased. For example if 2 000 eigenvalues can give 0.95 energy of a system for range 70, 3 000 eigenvalues can give this amount of energy for range 40. The best results can be found by increasing the number of desired eigenvalues and also power simultaneously. Instead of selecting 3 000 eigenvalues with power equal to 2, 2 500 eigenvalues can be selected with power equal to 3. Figure 3(b) shows changing in the fraction energy of the system with changing the correlation range. Power of 3 has been selected for all of these figures, for this reason decreasing the range does not have a significant effect on the fraction of energy of the system using the first 2 000 eigenvalues. The first 200 eigenvalues with a range 90 gives about 85 % energy of the system, but with a range of 50 gives only about 70 % energy of the system. This method is very efficient in terms of matrix decomposition. Even for this small example, the run time of the Cholesky decomposition was more than 1.5 hours, but randomized SVD method decomposed the covariance matrix in 20 minutes for finding the first 2 000 eigenvalues. A small change in the number of desired eigenvalues does not change the computation time significantly. Unconditional realizations were generated for a random field with the

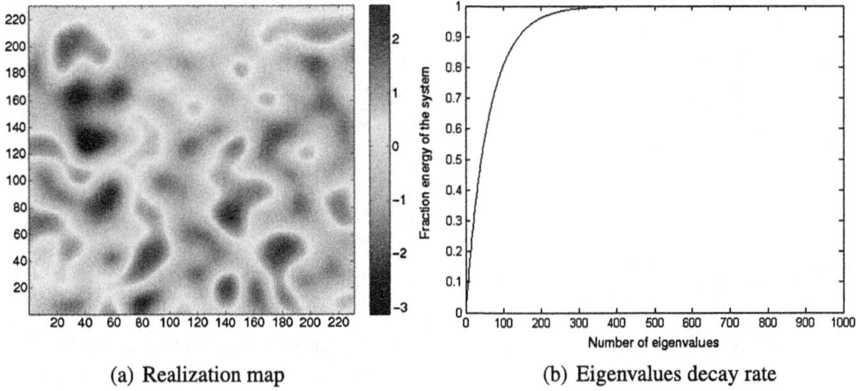

(a) Realization map (b) Eigenvalues decay rate

Fig. 4 Realization generated using Gaussian covariance

size of 230 × 230 blocks with the correlation range of 30 blocks using a Gaussian covariance matrix. Figure 4 shows the generated realization and the decay rate of the eigenvalues.

As you can see in Fig. 4(b), although a small range was selected for this case, the rate of eigenvalues decay is very fast. Only with the first 400 eigenvalues 99.99 % of the energy of the system has been stored. All other eigenvalues have very small values. They are smaller than the order of 10^{-4} and they do not have any significant effect on the estimations. In this case, run time was about 9 minutes compare to the 1.5 hours for Cholesky decomposition. This example shows that rate of eigenvalues decay in the Gaussian covariance is too fast. This finding confirms the results of [3] and [4] about rate of eigenvalues decay in different covariance matrices. They showed that eigenvalues of a Gaussian covariance matrix decay exponentially fast in contrast to the eigenvalues of an exponential covariance matrix which may decay slowly. Even by selecting a larger range, e.g. 65 blocks, the first 150 eigenvalues can show 99.99 % energy of the system. Full decomposition can give exactly the same realization. Suppose that there is a very large field with very large covariance matrix. In this case, using the first 1 000 eigenvalues, large number of realizations can be generated very fast. Nice thing about the LU or SVD simulation methods is that once covariance matrix decomposed, large number of realizations can be generated with a negligible computation time.

The last covariance function that we considered in this paper is spherical covariance function. As we considered different examples, we noticed that rate of eigenvalues decay for spherical covariance is not as fast as the Gaussian covariance, but it is faster than exponential covariance function. Again, same as the first example, one unconditional realization for a random field with the size of 230 × 230 blocks with the range of 50 blocks using the spherical covariance matrix generated. Figure 5 shows comparison between generated realizations using Cholesky and randomized SVD methods. In this case, run time was about 13 minutes compare to the 1.5 hours for Cholesky decomposition.

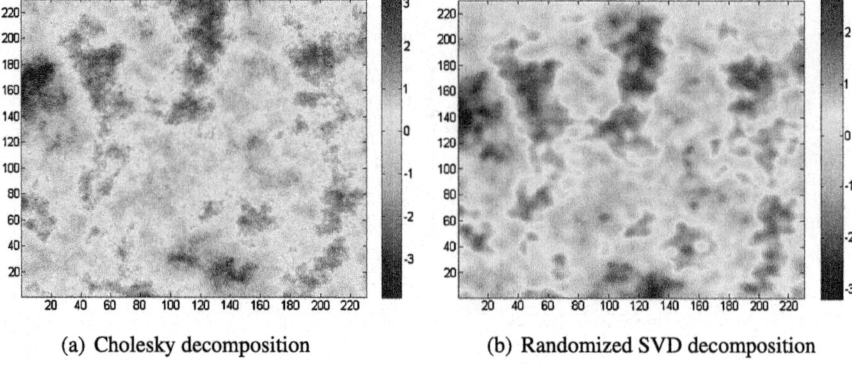

(a) Cholesky decomposition

(b) Randomized SVD decomposition

Fig. 5 Realization generated using spherical covariance

The histogram and variogram reproduction for both the Gaussian and spherical covariance functions was satisfactory. The most difficult case is related to the exponential covariance function which causes eigenvalues to decay slower than other types. For the last example, a conditional simulation example has been considered. For this purpose, a synthetic model with the size of 230×230 blocks with the range of 50 blocks using the exponential covariance matrix generated. As we showed before, due to the small decay rate of eigenvalues, exponential covariance is the most challenging type of covariance function. Figure 6 shows location map of data, generated realizations using LU and SVD and also variogram reproduction.

Based on the Cholesky and SVD formulations, all of the input data can be reproduced after simulation. Also as you can see, for five different realizations the average of the variograms reproduce the input variogram, so covariance reproduction has been done satisfactorily. Again red dashed lines are variograms of simulated realizations using randomized SVD, blue dashed lines are variograms of simulated realizations using Cholesky decomposition and gray solid line is theoretical variogram. In this problem, SVD decomposition has been done using the first 3 000 eigenvalues and $q = 3$. In this case more than 95 % energy of the system has been found. Also spectral norm after decomposition was about 0.01.

The efficiency of this method has been considered with several examples. This method is very efficient when the size of the covariance matrix is very large. Once a matrix can be generated, decomposing it with this method is possible and the accuracy of the method can be made acceptable. The only problem is related to the storage of large matrices. Full decomposition of that matrix is another important problem. If the size of the matrix is very large, full decomposition of that matrix is almost impossible or at least takes several hours. In the case of using Cholesky decomposition, using double precision variables is necessary. As we mentioned before, roundoff error is one of the problems of this method, but in the case of SVD decomposition, there is no problem with roundoff error and the matrix does not need to be positive definite. Even for storing a larger matrix, single precision variables can be used to decrease the memory usage.

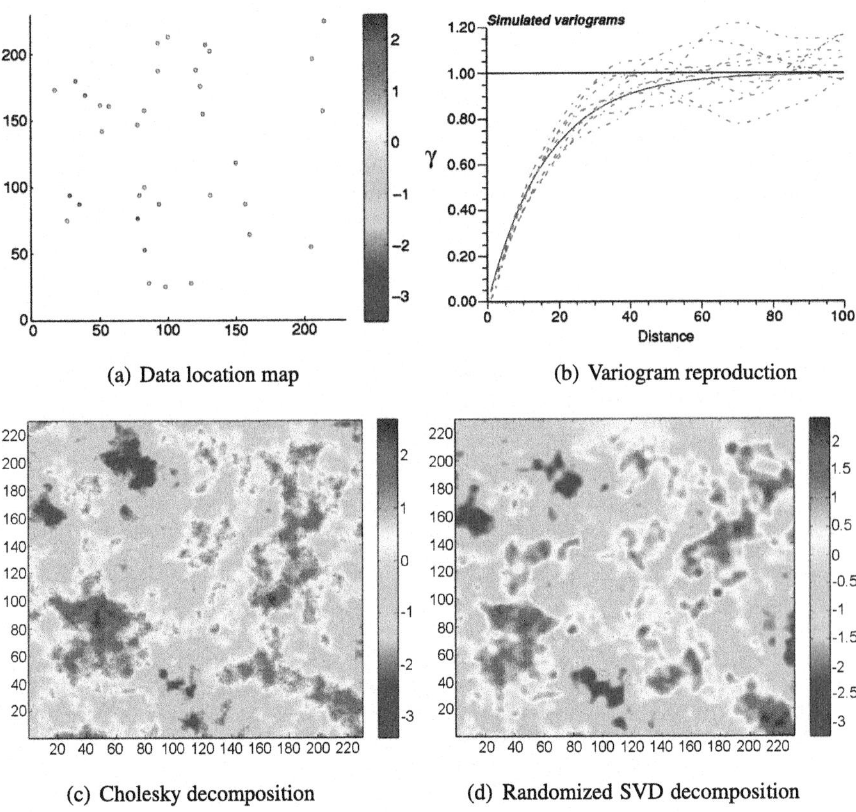

(a) Data location map

(b) Variogram reproduction

(c) Cholesky decomposition

(d) Randomized SVD decomposition

Fig. 6 Generating conditional simulation realization

4 Conclusions

In this paper, randomized SVD decomposition has been considered as an efficient method for decomposing covariance matrix and generating simulation realizations. This method is based on the randomized sampling of covariance matrix for finding a subspace which has much smaller size than the original matrix and captures most of the action of that matrix. Using this randomized low rank approximation method, approximated SVD decomposition of covariance matrix can be found by considering the first largest k eigenvalues. For testing efficiency of this method, different examples for different types of covariance functions have been considered. This method works very well for the cases that eigenvalues decay very fast. An example of this case is when the field correlation can be defined using Gaussian covariance function. For other cases like the spherical or exponential covariance functions that eigenvalues decay slower, using power method is a very good choice for solving this problem. The accuracy of results can be estimated by computing energy fraction or spectral norm. A good estimation should use about 95 % energy of the system and gives a small spectral norm. For assessing efficiency of this method, examples compared with Cholesky decomposition results. Comparisons show good efficiency of this method. Even for small examples, this method is much faster than Cholesky decomposition method. Using this method, very large covariance matrices can be decomposed which is impossible using Cholesky method. The only limitation of this method is related to storing a very large covariance matrix in the memory; the covariance matrix must be stored in machine memory.

References

1. Alabert F (1987) The practice of fast conditional simulations through the LU decomposition of the covariance matrix. Math Geol 19(5):369–386
2. Davis M (1987) Production of conditional simulations via the LU triangular decomposition of the covariance matrix. Math Geol 19(2):91–98
3. Efendiev Y, Hou T, Luo W (2007) Preconditioning Markov chain Monte Carlo simulations using coarse-scale models. SIAM J Sci Comput, 28(2):776
4. Frauenfelder P, Schwab C, Todor R (2005) Finite elements for elliptic problems with stochastic coefficients. Comput Methods Appl Mech Eng 194(2–5):205–228
5. Halko N, Martinsson P, Tropp J (2011) Finding structure with randomness: probabilistic algorithms for constructing approximate matrix decompositions. SIAM Rev 53:217
6. Liberty E, Woolfe F, Martinsson P, Rokhlin V, Tygert M (2007) Randomized algorithms for the low-rank approximation of matrices. Proc Natl Acad Sci USA 104(51):20167
7. Martinsson P, Szlam A, Tygert M (2010) Normalized power iterations for the computation of SVD. In: NIPS workshop on low-rank methods for large-scale machine learning, Vancouver
8. Oliver D, Reynolds A, Liu N (2008) Inverse theory for petroleum reservoir characterization and history matching. Cambridge University Press, Cambridge
9. Papadimitriou C, Raghavan P, Tamaki H, Vempala S (2000) Latent semantic indexing: a probabilistic analysis. J Comput Syst Sci 61(2):217–235
10. Rokhlin V, Szlam A, Tygert M (2009) A randomized algorithm for principal component analysis. SIAM J Matrix Anal Appl 31(3):1100–1124

Event-Based Geostatistical Modeling: Description and Applications

Michael J. Pyrcz, Timothy McHargue, Julian Clark, Morgan Sullivan, and Sebastien Strebelle

Abstract Event-based methods provide unique opportunities to improve the integration of geologic concepts into reservoir models. This may be accomplished over a continuum of rule complexity from very simple geometric models to complicated dynamics. Even the application of simple rules, including few conceptual interactions based on an understanding of stratigraphic relationships and parametric geometries for event scale depositional and erosion features, have been shown to efficiently produce complicated and realistic reservoir heterogeneities. In more complicated applications, initial and boundary conditions from analysis of paleobathymetry and external controls on sediment supply and the event rules may be informed by process models. These models have interesting features that depart from typical geostatistical model; they demonstrate emergent behaviors and preserve all information at all scales during their construction. These models may be utilized to produce very realistic reservoir models and their unique properties allow for novel applications. These modeling applications include; impact of model scale, seismic resolvability, value of information, flow relevance of advanced architecture, iterative and rule-based conditioning to sparse well and seismic data, numerical analogs

M.J. Pyrcz (✉) · M. Sullivan
Chevron Energy Technology Company, Houston, TX, USA
e-mail: mpyrcz@chevron.com

M. Sullivan
e-mail: morgan.sullivan@chevron.com

T. McHargue
Stanford University, Stanford, CA, USA
e-mail: timmchar@stanford.edu

J. Clark · S. Strebelle
Chevron Energy Technology Company, San Ramon, CA, USA

J. Clark
e-mail: julian.clark@chevron.com

S. Strebelle
e-mail: stsb@chevron.com

P. Abrahamsen et al. (eds.), *Geostatistics Oslo 2012*,
Quantitative Geology and Geostatistics 17,
DOI 10.1007/978-94-007-4153-9_3,

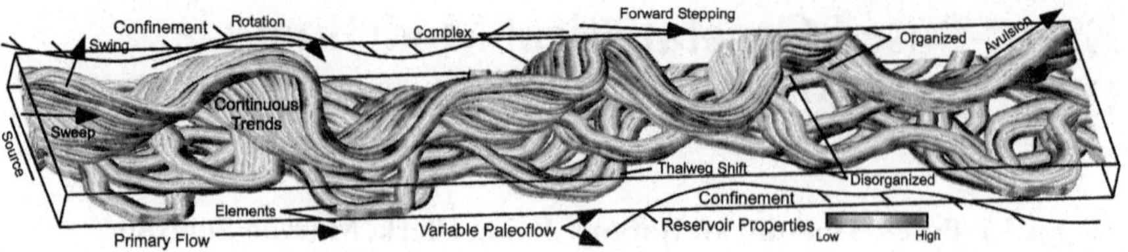

Fig. 1 Complicated deepwater channelized architecture generated with event-based modeling

for architectural concepts, statistical analysis and classification of architectures, unstructured grid construction and utilization as training and visualization tools.

1 Introduction

Throughout this paper we describe event-based methods [16, 19, 20, 24]. We recognize and reference previous and parallel research that use terms such as process-mimicking, process-oriented or hybrid models to name a few, including fluvial and deepwater channels models [3, 10, 30], deepwater lobes [1], deepwater slope valleys [27] and methods to further integrate process into reservoirs [18]. Space is not available here to describe implementation details, although the reader will find many of these implementations details for various event-based approaches in other publications [19, 20, 22, 23].

The efforts to further improve geologic features are inseparable from the theoretical and practical development of geostatistics. The developments of universal kriging, sequential indicator simulation, Gaussian truncated simulation, object-based modeling, simulated annealing/optimization-based simulation and multiple-point simulation have all been motivated by the need for improved reproduction of geologic concepts. While acknowledging the value and practical success of this previous work, we see a niche for even more improved geologic realism. In this context we describe event-based modeling as an effort to improve the reproduction of geologic concepts by integrating geologic rules into a geostatistical framework.

Current geostatistical algorithms, pixel-based or object-based, using semivariograms, training images or geometric parameters, enable the reproduction of spatial statistics inferred from available conditioning data and analogues, but rarely integrate information related to depositional processes. Indeed, because conventional geostatistical models are constructed without any concept of time or depositional sequence, their ability to incorporate sedimentological rules, which explain facies geobodies interactions and intra-body porosity/permeability heterogeneity, is quite limited. Consider the architecture in Fig. 1 generated with an event-based model based on a set of expert derived deepwater channel rules. It would not be possible to reproduce these details with standard geostatistical methods.

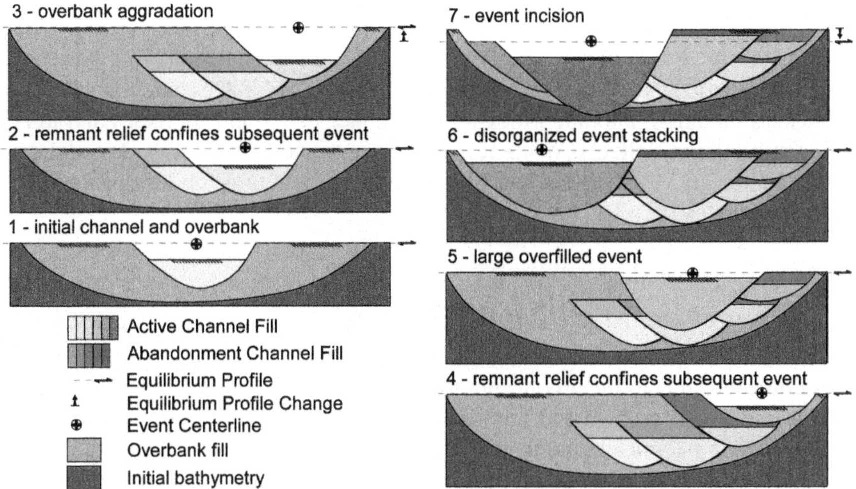

Fig. 2 Time sequence in a single strike section demonstrating the architecture resulting from seven events in response to varying allogenic parameters in event-based modeling. Note the impact of under-filled versus filled channels on stacking pattern (from [24])

2 The Fundamental Elements of Event-Based Models

The fundamental differences between event-based and typical geostatistical methods are covered briefly below. Event-based methods are proposed as a new set of tools with distinct strengths, applications and limitations to augment the other tools available in geostatistics. Specifically, event-based methods and resulting reservoir models may be immediately useful when geologic complexity is required in sparse conditioning settings.

2.1 Forward Modeling

Temporal considerations are central to geologic concepts. Yet, conventional geostatistical methods have no concept of time. With adoption of the forward modeling framework event-based models readily integrate these concepts. Time is integrated as a sequence of events. While it is possible to impose a hierarchy of scales, no effort is made to rigorously account for the exact lapse of time, instead the method accounts for the sequence of placement (and erosion) of significant reservoir elements (see a simple strike section example in Fig. 2).

Accommodation, or space available to accumulate sediments, is specified, eroded or created by the event-based model. The model, based on rules, places architectural elements in sequence that aggrade over and denude previous deposits. Model construction ceases when specified metric(s) are met, including data and trend conditioning and matching global probability density functions. This is another difference as an event-based model may respond to confinement.

Typically geostatistical methods rely on simulation along in a random order; pixel-based methods follow a random path to perform sequential simulation and object-based models randomly place objects. Random order does have valuable consequences including maximum flexibility to honor data, input statistics and trends while minimizing artifacts. While it is necessary to move to a forward model context to honor these geologic rules, we acknowledge a significant loss of flexibility to honor data and trends.

2.2 Hierarchical Modeling

It follows that hierarchical concepts may be directly integrated into the forward model utilized in event-based methods. Hierarchy may be imposed by cycles in the geometric parameters associated with the elements, or it may emerge during model construction due to aggradation, avulsion, eroding and filling of new channel complexes to build nested complexes within complex sets [17]. The event-based hierarchy has been described previously [21].

By comparison, due to the random order, conventional geostatistics honors hierarchies with nested modeling methods [4] or through the imposition of detailed nonstationary statistics models. Both of these methods require the explicit modeling of each hierarchical scale and associated boundaries (preferable if boundaries are mapable). Conversely, it is more difficult for event-based models to reproduce rigidly mapped hierarchies, as this potentially becomes a difficult inversion problem. This is further illustrated in the subsequent discussion on path routing.

2.3 Path Routing

Object-based modeling relies on a two step process, first random assignment of geometry parameters, and then random placement of this geometry in the model (including position and orientation). This methodology is very flexible for honoring data and trends, but prevents these models from honoring rule-based constraints such as source locations and types, and topography and erodability constraints.

Event-based models are initiated at a source and utilize rules-based flow routing to "grow" a path through the model. As a result, the constructed architectures appropriately conform to source, topography and erodability. In a simple case, channels conform to and may erode a slope valley. In a more complicated case an under-filled channel acts as an attractors for the next channel resulting in a distinct organized stacking patterns similar to results described by Jerolmack and Paola [11]. In the methods described in Pyrcz and others [24], this routing routine has two components, a statistical model based on preferred planform morphology (e.g. sinuosity, wavelength or even a detailed azimuth power spectrum) and rules that constrain this morphology with local features [17]. These paths may terminate outside the model or within the model as is the case for distributary lobes and splays. In addition, these paths may bifurcate and merge as is the case in braided fluvial channels.

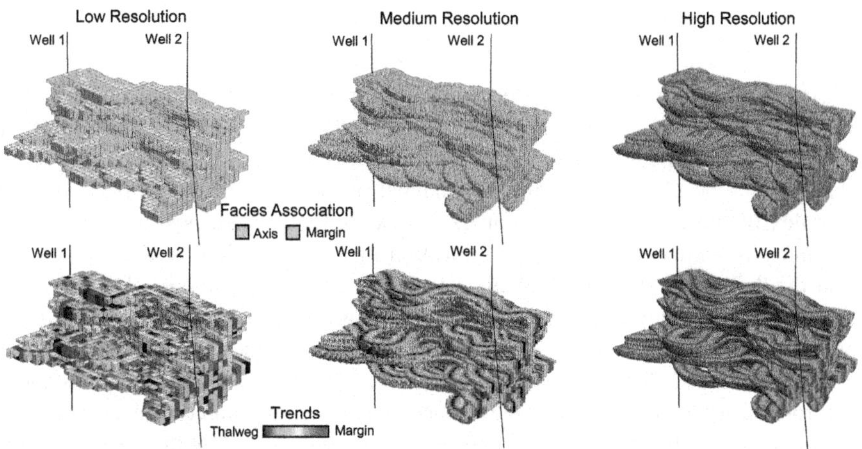

Fig. 3 Example of exact down scaling of a sector model with facies and reservoir quality

2.4 Flow-Axis-Based/Grid-Free Methods

The event-based elements are represented as a continuous mathematical model of the central event axis and geometric parameters associated with the attached architecture [6, 31]. For example, an event is routed by sequentially placing control nodes from source to terminus. Then a continuous spline function is fit to these control nodes. The path is investigated to fit a rule-based continuous model of geometric parameters. In a simple case, the channel thalweg is related to curvature, in a more complicated case the transition from erosional channels to depositional lobes is determined by the integration of a flow energy proxy along the path [20, 22].

Implicit to this method is the independence from a grid. Whereas traditional object-based model immediately rasterize their parametric geometries to the grid [5]. This rasterization in event-based modeling is only performed when a model is required. The architectural model is preserved graphically in the geomodeling software allowing for hand editing of the architecture and as an ASCII file allowing subsequent reloading to any grid. This results in the ability to interact with the models in a very flexible manner. For example, one may add a well intercept or move a channel to a well, to test the impact on flow response. In addition, exact downscaling is possible to any resolution [24]. We acknowledge the parallel research of Hassanpour and Deutsch [8] with the development of grid-free object-based models with similar properties and have adapted their term "grid-free" and also the mesh-free development [13]. An example of exact downscaling is shown in Fig. 3 with sector models between two wells of two channel complexes and associated reservoir quality models down-scaled for detailed connectivity and flow analysis.

2.5 Preservation of Information Content

During model construction all information concerning the genesis and interrelationships of all elements, complexes and complex sets is preserved. For any location in the model the specific hierarchical assignment is known, for example, a specific channel may be identified as the first channel of the second disorganized channel complex of the third channel complex set. Also, the relative position within the hierarchy is known. For example, a location may be specified as at the base and near the thalweg of the first channel, near the base and outer margin of the channel complex and so forth. Additionally, much more information is available such as the precise depositional coordinates (vertical stratal, transverse, longitudinal position and orientation [21]) at each hierarchical scale, the allogenic and input statistical controls on the model at the time of placement, geomorphometric measures such as curvature, gradient, proxies for energy, depth, grain size, flow thickness etc. and all of this is grid-free. This preservation of information allows for a variety of applications. The most practical being very detailed facies proportion and continuous property trend models.

Event-based modeling can be a numerical laboratory to explore emergent features, test the complicated interactions of expert derived rules and to learn new rules [16, 17, 22]. With very efficient methods, large models are generated in seconds on a typical PC, and visualization and statistical summaries allow for large experiments to be conducted rapidly and new insights to be developed, such as the controls on channel architecture [16, 17] and the formation of composite slope valley surfaces [27].

2.6 Reproducing Statistics vs. Emergent Behaviors

Geostatistical algorithms honor input statistics and they adhere to decisions of property stationarity. In general it is expected for input statistics to be honored. MPS represents a departure from this paradigm as there is no guarantee of reproduction of the associated model of spatial continuity (i.e. the training image). Event-based models take this even further as they are capable of producing emergent features that result from the interaction between rules and between rules and local constraints.

As a result, it may be difficult to honor very specific heterogeneity concepts informed from local data and precise trend information. A practical work around is to simplify the rule set for practical conditional reservoir models to enable the reproduction of the basic required architectural concepts. Then for a numerical laboratory approach, advanced rules are enabled resulting in complicated emergent behaviors. Two components of emergent behaviors are new heterogeneities that are not imposed directly and drift from initial seed architectures. The second is a concern as it may be considered an artifact of mismatch of the seed form and the rules. This may be dealt with through inversion for the rules parameters given the seed form or vice versa or by running the rules until the form stabilizes and removing the transitional products [10]. The emergence of low entropy features not explicitly constrained in the model inputs is considered as the *predictive nature* of these models. This is a challenging area of research and admittedly the extent of the predictive value and reasonable applications need to be explored. Our experience has been that the efficient reproduction of emergent reservoir scale features has resulted in new concepts and directions for investigation.

3 Applications

The following discussion covers applications with event-based modeling supporting conventional geostatistical workflows and applications with event-based providing numerical analog models.

3.1 Conventional Geostatistics

The ability to efficiently and easily build a large suite of scale refine-able, information rich, detailed numerical representations of reservoir architecture is useful in conventional reservoir geostatistics. The potential applications include training images and non-stationary statistics [24, 26].

Exhaustive training images are required by MPS to calculate the large number of required conditional probabilities for specified data events. It is natural to consider the use of event-based methods to furnish such images. Yet, experience has shown that non-stationarity and complexity in training images typically degrades the quality of MPS realizations. In fact, the best practice is to construct greatly simplified training images and impose nonstationarity in the MPS simulation [7, 26]. Given this consideration, we do not recommend event-based models for MPS, instead we recommend more simplified methods such as the training image libraries and generators [2, 15]. Yet, there may be opportunities to utilize event-based models to extract traditional spatial continuity models such as semivariogram models and indicator semivariogram models and transitional probabilities.

Another application is to calculate non-stationary statistics such as trends and locally variable azimuths from event-based models and apply them to guide conventional geostatistical models. For example, event-based methods may be useful for inferring consistent, detailed locally variable azimuth models, given local constraints. This is straightforward, given the preservation of information in these models. The design of unstructured grids from these detailed architectures could also be useful.

3.2 Numerical Analog Models

In a more advanced application, the event-based models are treated as numerical analog models for the reservoir of interest. Once this decision is made a variety

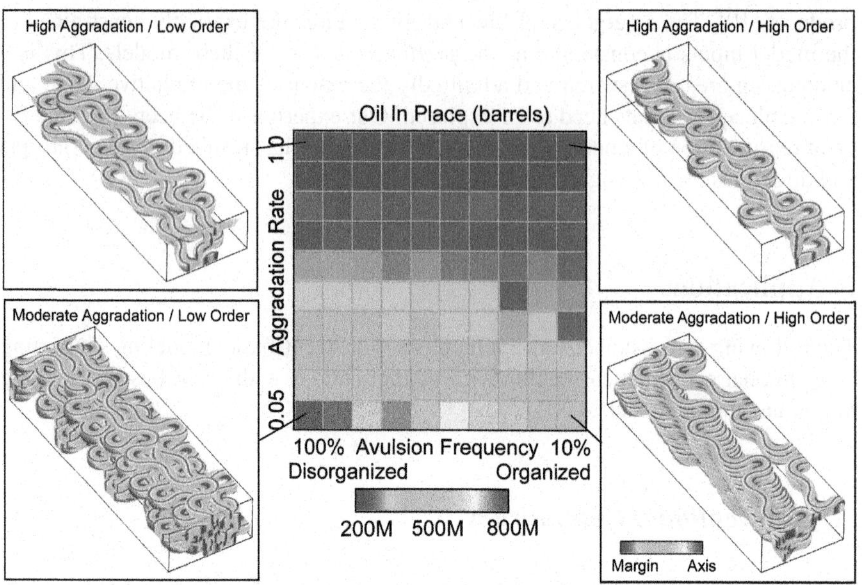

Fig. 4 Response surface for oil in place from a spectrum of event-based channelized models with variable aggradation rate and degree of channel organization as constrained by the frequency of avulsions. Assuming that the sediment-fill composition of channel elements remains constant, the volume of oil in place is constrained primarily by high aggradation rates and secondly by disorganized stacking of channel elements (from [24])

of applications are possible, including: architectural relationships, well risk analysis and value of well data. Admittedly the following applications rely on strong assumptions of the chosen models being analogous and characterizing appropriate models of uncertainty.

It may be important to understand the relationships between model architectural parameters and/or between these and reservoir parameters such as fluid volumes, connectivity and flow response. With a detailed set of realistic architectural models and their emergent behaviors this is possible. For example, in a simple channel setting, the influence of the channel stacking and system aggradation rate can be directly related to reservoir volume. Event-base methods enable this experiment, as this volume is not the sum of the volume of individual channels, nor constrained by the model as an input, but the result of a complicated preservation operator inherent to the rules (see Fig. 4). In another case, the preservation potential of components of the element fill may be quantified, for example the fraction of axial channel lag preserved in an aggrading and meandering channel model. There are much more complicated experiments possible, such as quantifying reservoir connectivity for various types and frequencies of channel avulsion.

Journel and Bitnavo [12] and Maharaja [14] developed methodologies for exploring NTG uncertainty through spatial bootstrap from reservoir models. In a similar manner, a proposed well design may be applied to a suite of event-based models to calculate the resulting probability distribution of any associated well result of inter-

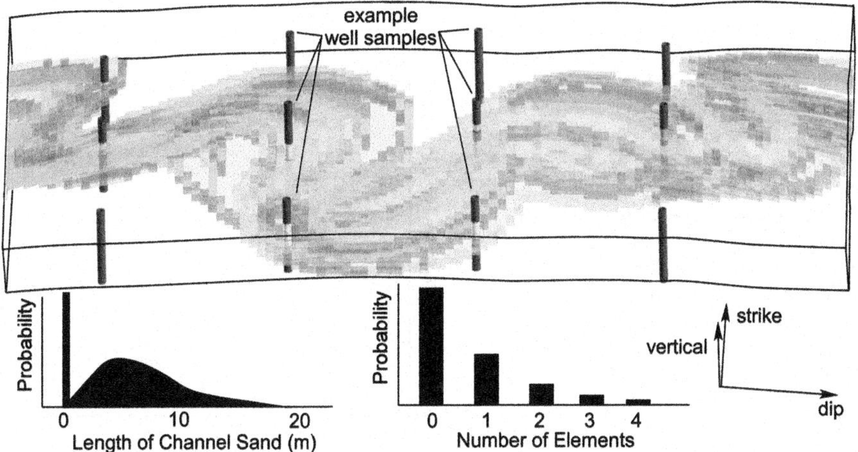

Fig. 5 Simple organized channel complex with channel axis (*yellow*), off-axis (*orange*) and margin (*green*) and a subset of regularly spaced spatial bootstrap samples and two example well-based statistics from exhaustive sampling

est, such as the net pay length, average NTG, proportion the well of above specific thresholds, number of isolated units etc. If information is available concerning well site selectivity in the analog models, this may be imposed with a selectivity bias surface that adjusts spatial bootstrap sampling rates (see a suite a well samples and well-based statistics from an event-based model in Fig. 5).

This workflow is useful, as it may be difficult to transfer 3D concepts of architecture to 1D concepts of a well outcome (along the trajectory). In addition, this method allows for the comparison of multiple well plans and their associated risks. For example, the distribution of possible NTG can be compared for a single well and averaged over a multiple well template, providing a direct indication of the mitigation of well risk through multiple wells. Also, vertical, horizontal and deviated wells may be compared to assess the value of the added cost of directional drilling to mitigating well risk.

As an extension of well risking, the value of well information may be analyzed. It is straightforward to sample well templates through a suite of event-based numerical analogs to assess the probability of a well result, given the architectural model. Bayesian inversion allows for the assessment of the probability of a specific architecture given a well result. This concept of inferring architecture from well statistics is similar to parallel work [9].

We propose a simple numerical approach to this inversion. A large suite of models are generated that are deemed to represent reservoir uncertainty. Then a variety of well configurations are sampled from all of these models and statistics of interest are binned and tabulated. A simple example for a several well template and from 4,000 architectural models is shown in Fig. 6. Note the difference in the probability contours between the 1 well vs. 7 well case. Changes in the number of wells, well rules and type of wells also impact these conditional distributions and provides

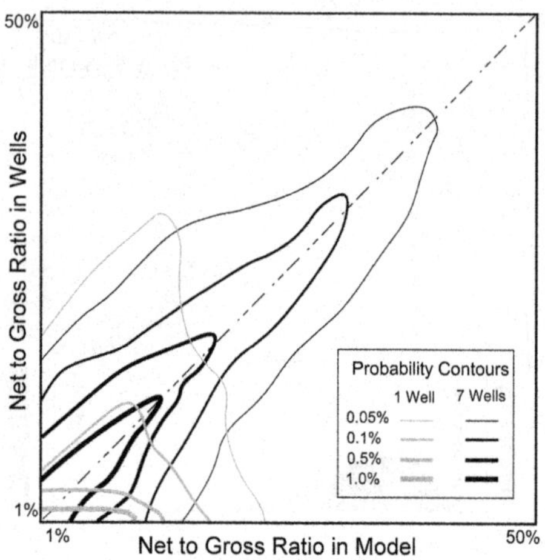

Fig. 6 Net to gross in model vs. net to gross observed in one or several wells bivariate relationship calculated by scanning 4,000 channel models with both well templates, allowing for inversion for the probability distribution for NTG given well NTG. The degree of separation in the conditional distributions (net to gross distribution in model given net to gross observed in wells) indicates information content for wells to inform reservoir

information on the value of a well plan (measured by separation of conditional probabilities $P\{\text{model}|\text{well}\}$).

3.3 Conditioning to Reservoir Models

Methods to condition object- or event-based models include; (1) dynamically constrain model parameters during model construction to improve data match [1, 3], (2) pseudo-reverse modeling [28], (3) apply as a trends/training image for multiple-point geostatistics [24] and (4) direct fitting of geometries to data [25], (5) model perturbation/updating [5, 19] and (6) extracting features from the model and associated statistics [18, 24]. Each of these techniques has significant limitations either in efficiency, robustness or the ability to retain complicated geometries and interrelationships.

A new conditioning method is proposed with the caveat that its robustness is not guaranteed in all settings and it is known to only be effective when conditioning data is sparse and trend models are smooth. For each event, conditioning data is coded into the rules constraints. For example, the erodability map is adjusted to account for conditioning and the geometric rules are adjusted to allow for the appropriate element intercepts and trend results (similar to the attraction approach by Abrahamsen and others [1]). Once approximate conditioning is achieved then the event is perturbed for exact conditioning. In the case of amalgamated beds, conditioning applies stochastic rules that allow for a variety of amalgamated element configurations accounting for uncertainty in the well interpretation.

Advanced geologic quantification, that is new statistics that describe heterogeneity, presents new conditioning challenges [29]. New modeling methods are required to honor these statistics.

4 Conclusions

Event-based modeling utilizes a new framework to readily integrate process and expert derived rules. This results in very complicated and realistic architectures and various new applications, such as improving the traditional geostatistics models with nonstationary statistics and analysis of architectural and reservoir relationships, well risking and value of well information. Conditioning through rules has been successful in sparse data settings, but still remains a challenge in dense data settings.

References

1. Abrahamsen P, Fjellvoll B, Hauge R, Howell J, Aas T (2007) Process based stochastic modeling of deep marine reservoirs. In: EAGE petroleum geostatistics, Cascias, Portugal
2. Boisvert JB, Leuangthong O, Ortiz JM, Deutsch CV (2008) A methodology to construct training images for Vein type deposit. Comput Geosci 34(5):491–502
3. Cojan I, Fouche O, Lopez S (2005) Process-based reservoir modelling in the example meandering channel. In: Leuangthong O, Deutsch CV (eds) Geostatistics Banff 2004. Springer, New York, pp 611–620
4. Deutsch CV, Wang L (1996) Hierarchical object-based stochastic modeling of fluvial reservoirs. Math Geol 28(7):857–880
5. Deutsch CV, Tran TT (2002) FLUVSIM: a program for object-based stochastic modelling of fluvial depositional systems. Comput Geosci 28:525–535
6. Georgsen F, Omre H (1997) Combining fibre processes and Gaussian random functions for modeling fluvial reservoirs. In: Soares A (ed) Geostatistics Troia 1993. Kluwer Academic, Dordrecht, pp 425–439
7. Harding A, Strebelle S, Levy M, Thorne J, Xie D, Leigh S, Preece R, Scamman R (2004) Reservoir facies modeling: new advances in MPS. In: Leunangthong O, Deutsch C (eds) Geostatistics Banff 2004. Springer, New York, pp 559–568
8. Hassanpour R, Deutsch CV (2010) An introduction to grid-free object-based facies modeling. In: Deutsch CV (ed) Centre for computational geostatistics report 12, paper 107
9. Hong G, Deutsch CV (2010) Fluvial channel size determination with indicator variograms. Pet Geosci 16:161–169
10. Howard AD (1992) Modeling channel migration and floodplain sedimentation in meandering streams. In: Carling PA, Petts GE (eds) Lowland floodplain rivers. Wiley, Chichester, pp 1–41
11. Jerolmack DJ, Paola C (2007) Complexity in a cellular model of river avulsion. Geomorphology 91:259–270
12. Journel AG, Bitanov A (2004) Uncertainty in NG ratio in early reservoir development. J Pet Sci Eng 44(1–2):115–130
13. Liu GR (2002) Mesh free methods: moving beyond the finite element method. CRC Press, New York
14. Maharaja A (2007) Global net-to-gross uncertainty assessment at reservoir appraisal stage. PhD dissertation, Stanford University, 144 pp
15. Maharaja A (2008) TiGenerator: object-based training image generator. Comput Geosci 34(12):1753–1761
16. McHargue T, Pyrcz MJ, Sullivan MD, Clark JD, Fildani A, Romans BW, Covault JA, Levy M, Posamentier HW, Drinkwater NJ (2010) Architecture of turbidite channel systems on the continental slope: patterns and predictions. Mar Petroleum Geol 28(3):728–743
17. McHargue T, Pyrcz MJ, Sullivan M, Clark J, Fildani A, Levy M, Drinkwater N, Posamentier H, Romans B, Covault J (2011) Event-based modeling of tubidite channel fill, channel stacking pattern and net sand volume. In: Martinsen OJ, Pulham AJ, Haughton PD, Sullivan MD

(eds) Outcrops revitalized: tools, techniques and applications. SEPM concepts in sedimentology and paleontology, vol 10, pp 163–174

18. Miller J, Sun T, Li H, Stewart J, Genty C, Li D, Lyttle C (2008) Direct modeling of reservoirs through forward process-based models: can we get there. In: International petroleum technology conference, Kuala Lumpur, Malaysia, Dec 3–5
19. Pyrcz MJ (2004) Integration of geologic information into geostatistical models. PhD dissertation, University of Alberta, 250 pp
20. Pyrcz MJ, Catuneanu O, Deutsch CV (2005) Stochastic surface-based modeling of turbidite lobes. Am Assoc Pet Geol Bull 89:177–191
21. Pyrcz MJ, Leuangthong O, Deutsch CV (2005) Hierarchical trend modeling for improved reservoir characterization. In: International association of mathematical geology annual conference 2005, IAMG, Toronto, Canada
22. Pyrcz MJ, Sullivan M, Drinkwater N, Clark J, Fildani A, Sullivan M (2006) Event-based models as a numerical laboratory for testing sedimentological rules associated with deepwater sheets. In: Slatt RM, Rosen NC, Bowman M, Castagna J, Good T, Loucks R, Latimer R, Scheihing R Smith R (eds) GCSSEPM 26th Bob F. Perkins research conference. GCSSEPM, Houston, pp 923–950
23. Pyrcz MJ, Boisvert J, Deutsch CV (2009) Alluvsim: a conditional event-based fluvial model. Comput Geosci 35:1671–1685
24. Pyrcz MJ, McHargue T, Sullivan M, Clark J, Drinkwater N, Fildani A, Posamentier H, Romans B, Levy M (2011) Numerical modeling of channel stacking from outcrop. In: Martinsen OJ, Pulham AJ, Haughton PD, Sullivan MD (eds) Outcrops revitalized: tools, techniques and applications. SEPM concepts in sedimentology and paleontology, vol 10, pp 149–162
25. Shmaryan L, Deutsch CV (1999) Object-based modeling of deep-water reservoirs with fast data conditioning: methodology and case studies. In: SPE annual technical conference and exhibition
26. Strebelle S (2002) Conditional simulation of complex geological structures using multiple-point statistics. Math Geol 34(1):1–21
27. Sylvester Z, Pirmez C, Cantelli A (2010) A model of submarine channel-levee evolution based on channel trajectories: implications for stratigraphic architecture. Mar Petroleum Geol 28(3):716–727
28. Tetzlaff DM (1990) Limits to the predictive ability of dynamic models the simulation clastic sedimentation. In: Cross TA (ed) Quantitative dynamic stratigraphy. Prentice Hall, New York, pp 55–65
29. Wang Y, Straub KM, Hajek EA (2011) Scale dependant compensational stacking: an estimate of autogenic timescales in channelized sedimentary deposits. Geology 39(9):811–814
30. Wen R (2005) SBED studio: an integrated workflow solution for multi-scale geo-modelling. In: EAGE 67th conference, Madrid, Spain
31. Wietzerbin LJ, Mallet JL (1993) Parameterization of complex 3D heterogeneities: a new CAD approach. In: SPE annual technical conference and exhibition

A Plurigaussian Model for Simulating Regionalized Compositions

Xavier Emery and Ignacio Gálvez

Abstract Regionalized compositions correspond to coregionalized variables that are non-negative and sum to a constant. In geostatistical applications, such variables are often modeled via log-ratio transformations, which allow converting the composition into a set of unconstrained variables on which estimation or simulation techniques can be applied. A different approach is explored in this work, extending the plurigaussian model used to represent categorical variables with mutually exclusive categories. The proposed model depends on a set of scalar parameters and independent stationary Gaussian random fields. In order to determine the model parameters, four steps are considered, which consist in ordering the components of the composition and in fitting the expected values, second-order moments and direct and cross variograms of these components. The conditional simulation of the composition can be performed by recourse to the Gibbs sampler and to classical Gaussian simulation algorithms. The applicability and versatility of the proposed model are illustrated with a case study in ore body evaluation, in which the composition of interest consists of the relative proportions of sulfide minerals in rock samples.

1 Introduction

A regionalized composition is a set of non-negative coregionalized variables that sum to a constant value at every location. Many examples of compositions are found in geochemistry, mining, petroleum engineering and environmental sciences, where one measures the relative amounts of constituents of sediment, soil or rock samples, e.g., the percentages of sulfide minerals in an ore deposit or the proportions of water, oil and rock in a reservoir. Another example is given by the indicators of a categorical variable with mutually exclusive categories: at each location, one

X. Emery (✉) · I. Gálvez
University of Chile, Avenida Tupper 2069, Santiago, Chile
e-mail: xemery@ing.uchile.cl

I. Gálvez
e-mail: igalvez@ing.uchile.cl

P. Abrahamsen et al. (eds.), *Geostatistics Oslo 2012*,
Quantitative Geology and Geostatistics 17,
DOI 10.1007/978-94-007-4153-9_4,

indicator is equal to one and the others are equal to zero, so that the indicators sum to one.

In this context, it is of interest to design interpolation methods that reproduce the compositional nature of the coregionalization, i.e., that satisfy the constant sum and the non-negativity constraints. Examples of such methods are compositional kriging [11] or cokriging of a transformation of the composition, such as centered log-ratio and additive log-ratio transformations [1, 8, 9]. In this paper, rather than spatial prediction, we are interested in the simulation of regionalized compositions. To this end, a specific, yet versatile, model will be designed, which generalizes the plurigaussian model for simulating categorical variables.

2 Model Presentation

2.1 A Preliminary Lemma

Our idea is to construct a compositional vector random field by using a set of basic scalar random fields with outcomes in the interval (0, 1) and fixed mean values. In the traditional plurigaussian model, these basic random fields are the indicators of independent Gaussian random fields. To generalize this model, here we will work with the conditional expectations of such indicators, based on the following lemma.

Lemma 1 *Let U be a standard Gaussian random variable and G its cumulative distribution function. Then, the following equality holds*:

$$E\left\{G\left(\frac{y-sU}{\sqrt{1-s^2}}\right)\right\} = G(y), \quad \forall s \in [0,1),\ \forall y \in \mathbb{R}. \tag{1}$$

Proof In the d-dimensional Euclidean space $\mathbb{R}^d$, consider a standard Gaussian random field $\{Y(x) : x \in \mathbb{R}^d\}$. Given a set of data on this random field and a real number y, the conditional expectation of the indicator function $1_{Y(x)\leq y}$ (equal to 1 if $Y(x) \leq y$, 0 otherwise) at location x is

$$[1_{Y(x)\leq y}]^* = G\left(\frac{y - Y^*}{\sigma^*}\right), \tag{2}$$

with Y^* the simple kriging predictor of $Y(x)$ and σ^* the standard deviation of the associated kriging error, see e.g. [10]. On the one hand, as Y^* is a Gaussian random variable with zero mean and variance $s^2 = 1 - \sigma^{*2}$, one has

$$Y^* = sU, \tag{3}$$

with U a standard Gaussian random variable. On the other hand, the expected value of $[1_{Y(x)\leq y}]^*$ is the same as that of $[1_{Y(x)\leq y}]$, i.e., $G(y)$. It follows that

$$G(y) = E\left\{G\left(\frac{y-Y^*}{\sigma^*}\right)\right\} = E\left\{G\left(\frac{y-sU}{\sqrt{1-s^2}}\right)\right\}, \tag{4}$$

which completes the proof of Lemma 1, insofar as s can take any value in [0, 1). □

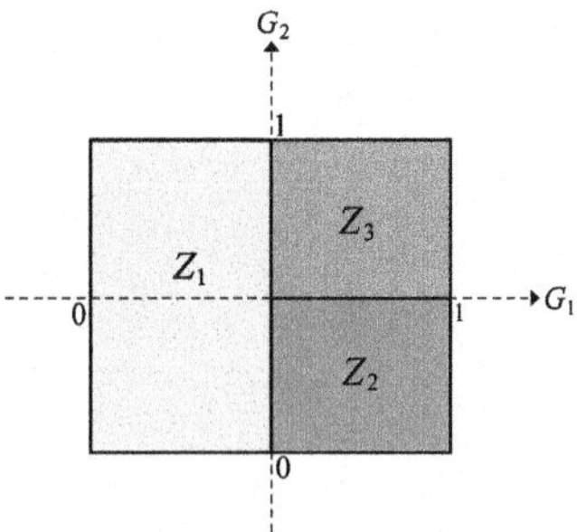

Fig. 1 Schematic representation of the generalized plurigaussian model defined in (6)

2.2 A Three-Component Compositional Model

Consider two real numbers (y_1 and y_2) and four mutually independent random fields defined in $\mathbb{R}^d$ (U_1, U_2, S_1 and S_2), the first two ones with standard Gaussian univariate distributions and the last two ones with outcomes in the interval $(0, 1)$. Based on Lemma 1, let us define two independent basic random fields with outcomes in $(0, 1)$ and mean values $G(y_1)$ and $G(y_2)$, respectively, by posing:

$$G_i(x) = G\left(\frac{y_i - S_i(x)U_i(x)}{\sqrt{1 - S_i(x)^2}}\right), \quad \forall i \in \{1, 2\}, \ \forall x \in \mathbb{R}^d. \tag{5}$$

A compositional vector random field with three components, $Z = (Z_1, Z_2, Z_3)$, is then obtained as follows:

$$\begin{aligned} Z_1(x) &= G_1(x), \\ Z_2(x) &= \left[1 - G_1(x)\right]G_2(x), \\ Z_3(x) &= \left[1 - G_1(x)\right]\left[1 - G_2(x)\right]. \end{aligned} \tag{6}$$

As a particular case, if $S_1(x)$ and $S_2(x)$ are close to 1, the outcomes of $G_1(x)$ and $G_2(x)$ are close to 0 or 1, and one finds the traditional plurigaussian model in which $Z_1(x)$, $Z_2(x)$ and $Z_3(x)$ are the indicators associated with non-overlapping categories [2, 6]. In the general case when the outcomes of $S_1(x)$ and $S_2(x)$ pertain to the interval $(0, 1)$, so do the outcomes of $Z_1(x)$, $Z_2(x)$ and $Z_3(x)$. As for the traditional plurigaussian model, one can represent the model in (6) through a two-dimensional flag, in which the axes are associated with the random fields G_1 and G_2 (Fig. 1). In the next sections, we make the following simplifying assumptions:

1. The random fields S_1 and S_2 have univariate beta distributions and can be transformed into random fields Y_1 and Y_2 with standard Gaussian univariate distributions:

$$S_i(x) = F_{a_i,b_i}^{-1}\left(G\left(Y_i(x)\right)\right), \quad \forall i \in \{1, 2\}, \ \forall x \in \mathbb{R}^d, \tag{7}$$

   where $F_{a,b}$ is the cumulative distribution function of a beta random variable with parameters $a > 0$ and $b > 0$. The beta distribution has been chosen because of its varied shapes in the interval $(0, 1)$, depending on the selected parameters, which will allow a flexible fitting of the univariate distributions of the composition components.

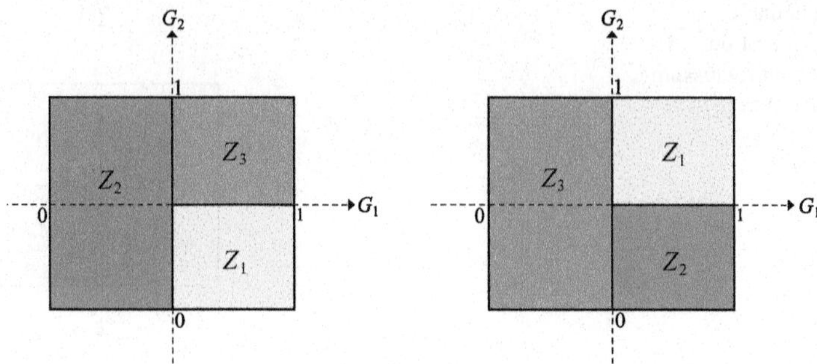

Fig. 2 Representations of alternative plurigaussian models

2. U_1, U_2, Y_1 and Y_2 are stationary Gaussian random fields. The motivation for this assumption is the availability of many algorithms for conditional simulation.

3 Inference of Model Parameters

The above-defined model depends on the ordering of the components (Z_1, Z_2, Z_3) of the compositional vector, the scalar thresholds y_1 and y_2, the scalar parameters a_1, b_1, a_2, b_2 of the beta distributions, and the auto-covariance functions (equivalently, the variograms) of the Gaussian random fields U_1, U_2, Y_1, Y_2. In the following subsections, we examine how these parameters can be determined from a set of data on the vector random field Z.

3.1 Ordering of Vector Components

In the model proposed in (6), the ordering of the components (Z_1, Z_2, Z_3) is arbitrary, and two alternative models can be designed (Fig. 2). A simple way to discriminate between the models represented in Fig. 1 and Fig. 2 is to examine the direct and cross variograms of Z_1, Z_2 and Z_3. For instance, let us consider the model in (6) and calculate the cross variogram γ_{12} between Z_1 and Z_2 for a given lag separation vector h. According to definitions we get

$$\begin{aligned}
2\gamma_{12}(h) &= E\{[Z_1(x+h) - Z_1(x)][Z_2(x+h) - Z_2(x)]\} \\
&= E\{[G_1(x+h) - G_1(x)][1 - G_1(x+h)]G_2(x+h)\} \\
&\quad - E\{[G_1(x+h) - G_1(x)][1 - G_1(x)]G_2(x)\} \\
&= E\{[G_1(x+h) - G_1(x)][1 - G_1(x+h)]\}G(y_2) \\
&\quad - E\{[G_1(x+h) - G_1(x)][1 - G_1(x)]\}G(y_2) \\
&= -2G(y_2)\gamma_1(h),
\end{aligned} \tag{8}$$

with γ_1 the direct variogram of Z_1. Similarly, one finds $\gamma_{13}(h) = [G(y_2) - 1]\gamma_1(h)$. That is, the variogram of Z_1 is proportional to the cross variograms between Z_1 and Z_2 and between Z_1 and Z_3. In contrast, in the model presented in the left-hand side of Fig. 2, the variogram of Z_2 is proportional to the cross variograms between Z_1 and Z_2 and between Z_2 and Z_3, whereas in the model presented in the right-hand side of Fig. 2, the variogram of Z_3 is proportional to the cross variograms between Z_1 and Z_3 and between Z_2 and Z_3. The comparison of the sample variograms of Z_1, Z_2 and Z_3 is therefore helpful to determine which of the three models is best suited to the available data.

3.2 Determination of Thresholds

Considering the model in (6) and based on Lemma 1, one has:

$$\begin{aligned} E[Z_1(x)] &= G(y_1), \\ E[Z_2(x)] &= [1 - G(y_1)]G(y_2), \\ E[Z_3(x)] &= [1 - G(y_1)][1 - G(y_2)]. \end{aligned} \tag{9}$$

Accordingly, there is a one-to-one correspondence between the thresholds y_1 and y_2 and the expectations of the vector components (Z_1, Z_2, Z_3). The knowledge of the latter therefore allows determining the former.

3.3 Determination of Parameters of Beta Distributions

The parameters a_1, b_1, a_2, b_2 have an impact on the distribution of the random vector $Z(x)$, in particular, on its second-order moments:

$$\mu_{ij} = E[Z_i(x)Z_j(x)], \quad \forall i, j \in \{1, 2, 3\}, \ \forall x \in \mathbb{R}^d. \tag{10}$$

Accounting for the definition of the vector component (6) and for Lemma 1, one has:

$$\begin{aligned} \mu_{11} &= E[G_1(x)^2], \\ \mu_{22} &= \{1 - 2G(y_1) + E[G_1(x)^2]\}E[G_2(x)^2], \\ \mu_{33} &= \{1 - 2G(y_1) + E[G_1(x)^2]\}\{1 - 2G(y_2) + E[G_2(x)^2]\}, \\ \mu_{12} &= \{G(y_1) - E[G_1(x)^2]\}G(y_2), \\ \mu_{13} &= \{G(y_1) - E[G_1(x)^2]\}\{1 - G(y_2)\}, \\ \mu_{23} &= \{1 - 2G(y_1) + E[G_1(x)^2]\}\{G(y_2) + E[G_2(x)^2]\}. \end{aligned} \tag{11}$$

The following lemma is helpful to express the expectations of $G_1(x)^2$ and $G_2(x)^2$.

Lemma 2 *For $i \in \{1,2\}$, the random field G_i defined in* (5) *can be expanded in the following fashion*:

$$G_i(x) = G(y_i) + g(y_i) \sum_{p=1}^{\infty} \frac{1}{\sqrt{p}} H_{p-1}(y) S_i(x)^p H_p\big(U_i(x)\big), \tag{12}$$

where g denotes the standard Gaussian probability density function and $\{H_p, p \in \mathbb{N}\}$ the normalized Hermite polynomials, defined as [10]:

$$H_p(y) = \frac{1}{\sqrt{p!}\, g(y)} \frac{\mathrm{d}^p g(y)}{\mathrm{d}y^p}, \quad \forall p \in \mathbb{N},\ \forall y \in \mathbb{R}. \tag{13}$$

Proof Let us consider a real number y and a standard Gaussian random field $\{Y(x) : x \in \mathbb{R}\}$. Adopting the same notations as in (2), the following expansion holds [4]:

$$[1_{Y(x) \le y}]^* = G(y) + g(y) \sum_{p=1}^{\infty} \frac{1}{\sqrt{p}} H_{p-1}(y) \big(1 - \sigma^{*2}\big)^{\frac{p}{2}} H_p\left(\frac{Y^*}{\sqrt{1 - \sigma^{*2}}}\right). \tag{14}$$

Lemma 2 can therefore be established by using (2) and (3). □

Accounting for (12) and for the orthonormality of the Hermite polynomials with respect to the standard Gaussian distribution, the expectation of $G_i(x)^2$ for $i \in \{1,2\}$ is found to be:

$$E\big[G_i(x)^2\big] = G(y_i)^2 + g(y_i)^2 \sum_{p=1}^{\infty} \frac{1}{p} H_{p-1}(y)^2 E\big[S_i(x)^{2p}\big]. \tag{15}$$

Furthermore, given that $S_i(x)$ has a beta distribution with parameters (a_i, b_i), one finally obtains:

$$E\big[G_i(x)^2\big] = G(y_i)^2 + g(y_i)^2 \sum_{p=1}^{\infty} \frac{1}{p} H_{p-1}(y)^2 \frac{\Gamma(a_i + 2p)\Gamma(a_i + b_i)}{\Gamma(a_i + b_i + 2p)\Gamma(a_i)}. \tag{16}$$

Having specified the thresholds y_1 and y_2 from the previous stage, the parameters of the beta distributions (a_1, b_1, a_2 and b_2) can be chosen in order to fit the second-order moments of the vector components (Z_1, Z_2, Z_3), based on (11) and (16).

3.4 Determination of Variograms

To complete the specification of the model, it remains to determine the variograms of the Gaussian random fields U_1, U_2, Y_1, Y_2. Such variograms define the spatial correlation structure of the compositional random vector, therefore they have an impact on the direct and cross variograms of Z_1, Z_2 and Z_3. Since the analytical calculation of the direct and cross variograms of Z_1, Z_2 and Z_3 as a function of the variograms of U_1, U_2, Y_1 and Y_2 is quite complex, a numerical calculation is proposed instead. It consists of the following steps:

Algorithm 1 Conditional simulation

1. Simulate $U_1(x_A)$, $U_2(x_A)$, $Y_1(x_A)$ and $Y_2(x_A)$ conditionally to $Z(x_A) = z_A$.
2. Simulate $U_1(x_I)$ conditionally to $U_1(x_A)$.
3. Simulate $U_2(x_I)$ conditionally to $U_2(x_A)$.
4. Simulate $Y_1(x_I)$ conditionally to $Y_1(x_A)$. Derive $S_1(x_I)$, as per (7).
5. Simulate $Y_2(x_I)$ conditionally to $Y_2(x_A)$. Derive $S_2(x_I)$, as per (7).
6. Derive $Z(x_I) = (Z_1(x_I), Z_2(x_I), Z_3(x_I))$, as per (5), (6).

1. Calculate the sample direct and cross variograms of Z_1, Z_2 and Z_3 for a given set of lag separation vectors $\{h_k : k = 1 \dots K\}$.
2. Propose prior variogram models for U_1, U_2, Y_1 and Y_2.
3. For $k = 1 \dots K$:
   a. Construct a large number of realizations of the bivariate Gaussian pairs $(U_1(0), U_1(h_k))$, $(U_2(0), U_2(h_k))$, $(Y_1(0), Y_1(h_k))$, $(Y_2(0), Y_2(h_k))$.
   b. Using (5) to (7), derive realizations of the pair $(Z(0), Z(h_k))$.
   c. From the realizations, calculate the direct and cross variograms of the Z-components for lag h_k.
4. Compare the sample variograms calculated at Step 1 with those obtained at Step 3c. If the fit is not satisfactory, go back to Step 2.

4 Conditional Simulation

We now tackle the problem of simulating the vector random field Z conditionally to a set of data on this random field. For brevity, let us denote by $z_A = \{z_\alpha : \alpha \in A\}$ the conditioning data values, $x_A = \{x_\alpha : \alpha \in A\}$ the conditioning data locations, and $x_I = \{x_i : i \in I\}$ the set of locations targeted for simulation. Algorithm 1 is proposed for conditional simulation. Steps 2–5 in Algorithm 1 boil down to the conditional simulation of stationary Gaussian random fields, which can be realized by using one of the many multivariate Gaussian simulation algorithms proposed in the literature [3, 6]. Step 1 in Algorithm 1 can be solved by an iterative algorithm (Gibbs sampler), as explained in Algorithm 2. To avoid dealing with infinite Gaussian values at the initialization stage (Step c) or the iteration stage (Step e), it is convenient to slightly modify the conditioning data values that are equal to 0 or 1, e.g., by setting these values to 0.0001 or 0.9999. For practical applications, this modification is inconsequential.

5 Application

In this section, the proposed approach is applied to a data set from a porphyry copper deposit that has been recognized through a set of exploration diamond drill holes.

Algorithm 2 Gibbs sampler for Step 1 of conditional simulation

Initialization. For each data location x_α with $\alpha \in A$:

(a) Simulate $Y_1(x_\alpha)$ and $Y_2(x_\alpha)$ as two independent standard Gaussian random variables.
(b) By using (7), derive $S_1(x_\alpha)$ and $S_2(x_\alpha)$.
(c) By using (5), (6) and the conditioning values of $Z_1(x_\alpha)$, $Z_2(x_\alpha)$ and $Z_3(x_\alpha)$, derive $U_1(x_\alpha)$ and $U_2(x_\alpha)$.

Iteration.

(a) Select one index α at random (uniformly) in A. Let us denote by $B = A - \{\alpha\}$ the indices of the remaining data locations.
(b) Determine the distribution of $U_1(x_\alpha)$ conditional to $U_1(x_B)$. This conditional distribution is Gaussian, with mean equal to the simple kriging prediction of $U_1(x_\alpha)$ from $U_1(x_B)$ and variance equal to the simple kriging variance. Likewise, determine the distributions of $U_2(x_\alpha)$, $Y_1(x_\alpha)$ and $Y_2(x_\alpha)$ conditional to $U_2(x_B)$, $Y_1(x_B)$ and $Y_2(x_B)$.
(c) From the corresponding conditional distributions, simulate candidate values for $Y_1(x_\alpha)$ and $Y_2(x_\alpha)$.
(d) By using (7), derive candidate values for $S_1(x_\alpha)$ and $S_2(x_\alpha)$.
(e) By using (5), (6) and the conditioning values of $Z_1(x_\alpha)$, $Z_2(x_\alpha)$ and $Z_3(x_\alpha)$, derive candidate values for $U_1(x_\alpha)$ and $U_2(x_\alpha)$.
(f) Calculate the conditional joint density (denoted by p) for the current set of values of $U_1(x_\alpha)$ and $U_2(x_\alpha)$, and the conditional joint density (denoted by p') for the candidate set of values. Since U_1 and U_2 are independent, the joint densities are the products of the marginal conditional densities of $U_1(x_\alpha)$ and $U_2(x_\alpha)$.
(g) Generate a value v uniformly on [0, 1].
(h) If $v < \frac{p'}{p}$, replace the current values of $U_1(x_\alpha)$, $U_2(x_\alpha)$, $Y_1(x_\alpha)$ and $Y_2(x_\alpha)$ by the candidate values, in accordance with Metropolis-Hastings acceptance criterion [5, 7].
(i) Go back to Step (a) until a large number of iterations have been achieved.

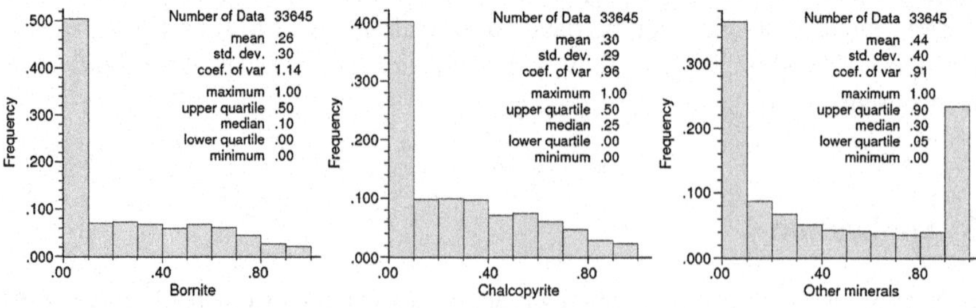

Fig. 3 Histograms of compositional variables (original data set)

Table 1 Sample univariate statistics of compositional variables

Variable	Number of data	Minimum	Maximum	Mean	Variance
Bornite	33,645	0	1	0.263	0.0899
Chalcopyrite	33,645	0	1	0.301	0.0841
Other minerals	33,645	0	1	0.436	0.1568

Table 2 Sample variance-covariance matrix of compositional variables

Variable	Bornite	Chalcopyrite	Other minerals
Bornite	0.158	−0.076	−0.082
Chalcopyrite	−0.076	0.085	−0.009
Other minerals	−0.082	−0.009	0.090

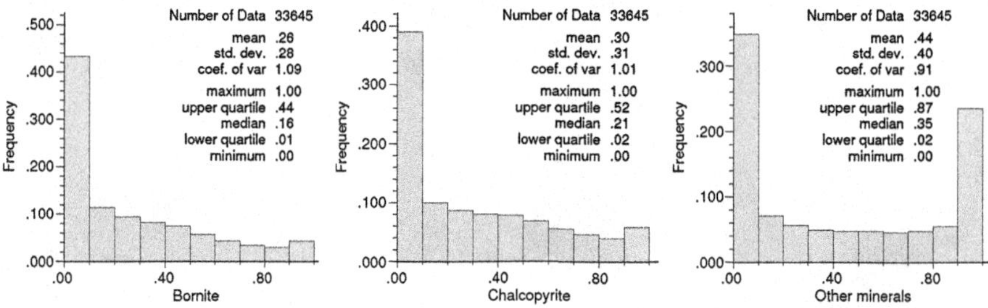

Fig. 4 Histograms of compositional variables (non-conditional realizations)

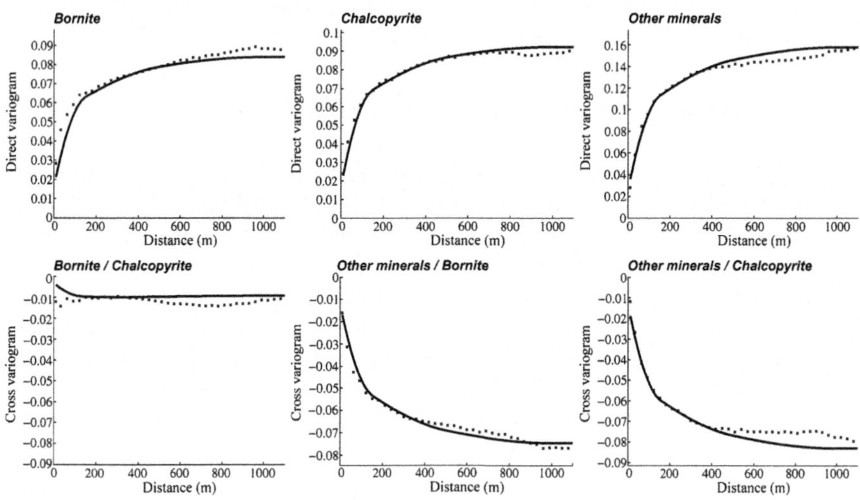

Fig. 5 Sample (*crosses*) and modeled (*solid lines*) direct and cross variograms of compositional variables

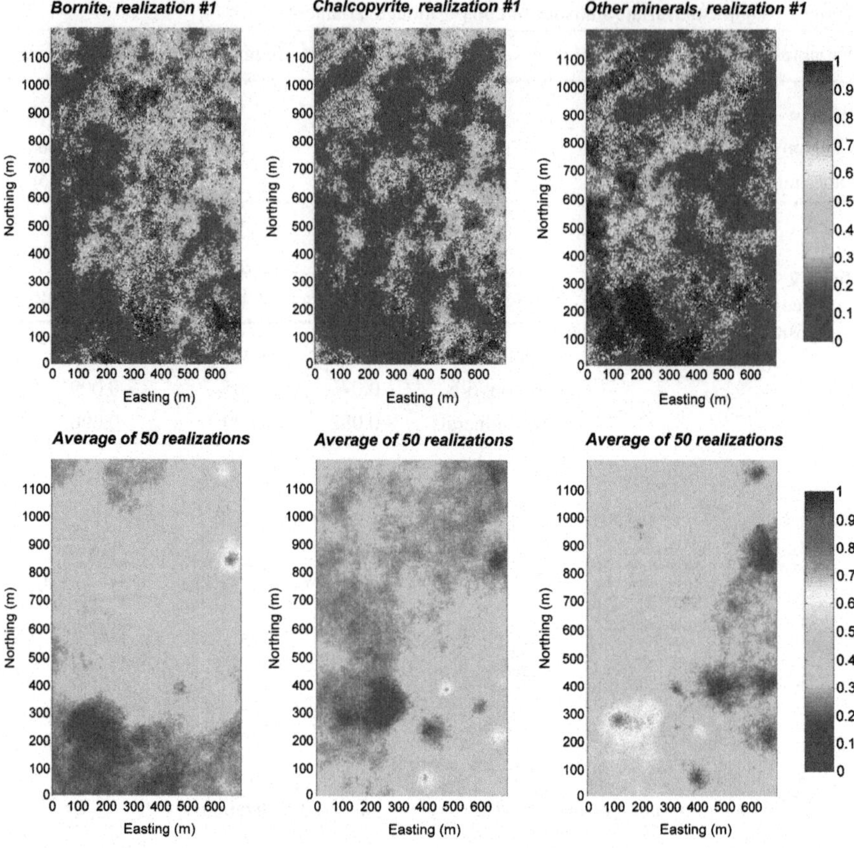

Fig. 6 Conditional realization (*top*) and average of 50 conditional realizations (*bottom*) of the relative proportions of bornite, chalcopyrite and other minerals

For each drill hole sample, the relative proportions of ten sulfide minerals (bornite, chalcopyrite, chalcocite, covellite, digenite, enargite, galena, pyrite, tennantite and tetrahedrite) have been measured. For the sake of simplicity, only the two main minerals (bornite and chalcopyrite) are considered, together with the sum of the other minerals. The histograms and basic statistics of the variables under study are presented in Fig. 3 and Tables 1 and 2. After examining the direct and cross variograms of the compositional variables, it is decided to associate the variable "other minerals" with the random field Z_1, insofar as its direct variogram has a shape similar to its cross variograms with the other two variables (chalcopyrite and bornite, which will be associated with the random fields Z_2 and Z_3, respectively).

Accounting for the sample means in Table 1 and for (9), the thresholds are found to be $y_1 = -0.165$ and $y_2 = 0.084$. The parameters of the beta distributions are then chosen in order to reproduce the second-order moments of the compositional variables (Table 2), based on (11) and (16), as well as their univariate distributions, based on a graphical comparison of the sample distributions (Fig. 3) with the distri-

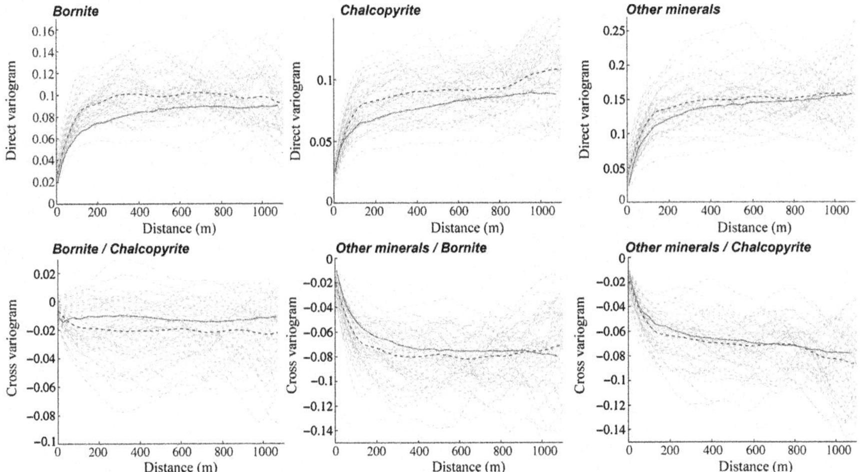

Fig. 7 Sample variograms (*grey dashed lines*) and average of sample variograms (*black dashed lines*) of 50 conditional realizations, calculated along the north-south direction, and omnidirectional sample variograms of original data (*solid lines*)

butions of a set of non-conditional realizations of the vector random field Z (Fig. 4). After trials and errors, one finds $a_1 = 9.20$, $b_1 = 1.04$, $a_2 = 2.57$ and $b_2 = 1.03$. The last step in the determination of the model parameters is the fitting of the direct and cross variograms of the compositional variables. Because no obvious anisotropy has been detected, only omnidirectional sample variograms are calculated. The variograms of the Gaussian random fields (U_1, U_2, Y_1, Y_2) are modeled by nested nugget effects and isotropic spherical structures with ranges 150 m, 500 m and 1,000 m. The resulting variogram models for the compositional variables are presented in Fig. 5, together with the sample variograms.

Despite the apparent complexity of the model, the inference of its parameters can be performed even in the presence of sparse data, as soon as these data allow the calculation of sample histograms and sample direct and cross variograms that are deemed representative of the entire coregionalization. Once the model parameters are determined, it is possible to simulate U_1, U_2, Y_1, Y_2 at the data locations, then at any other location targeted for simulation, following the algorithm proposed in Sect. 4. As an illustration, one conditional realization of the composition and the average of 50 conditional realizations are shown in Fig. 6, for a specific elevation in the deposit. The number of iterations in the Gibbs sampler has been chosen so that each data is visited on average 1,000 times. Apart from the visual inspection of the simulated maps (Fig. 6), one can check that the realizations reproduce the spatial correlation structure of the original compositions, by comparing the sample direct and cross variograms of the realizations with those of the original data (Fig. 7).

6 Conclusions

A generalized plurigaussian model has been proposed to simulate a regionalized composition with three components. The model turns out to be versatile, insofar as its parameters can be fitted in order to reproduce the first- and second-order moments and the spatial correlation structure of the composition components, while conditional simulation can be performed by means of iterative and classical multivariate Gaussian algorithms. Variations of the proposed model to more than three components can easily be designed, by using plurigaussian flags with more categories and/or with more dimensions.

Acknowledgements This research has been funded by the Chilean Fund for Science and Technology Development, through Project FONDECYT 1090013.

References

1. Aitchison J (1986) The statistical analysis of compositional data. Monographs on statistics and applied probability. Chapman & Hall, London
2. Armstrong M, Galli A, Beucher H, Le Loc'h G, Renard D, Doligez B, Eschard R, Geffroy F (2011) Plurigaussian simulations in geosciences. Springer, Heidelberg
3. Chilès JP, Delfiner P (1999) Geostatistics: modeling spatial uncertainty. Wiley, New York
4. Emery X (2005) Simple and ordinary multigaussian kriging for estimating recoverable reserves. Math Geol 37:295–319
5. Hastings WK (1970) Monte Carlo sampling methods using Markov chains and their applications. Biometrika 57:97–109
6. Lantuéjoul C (2002) Geostatistical simulation: models and algorithms. Springer, Heidelberg
7. Metropolis N, Rosenbluth AW, Teller AH, Teller E (1953) Equation of state calculation by fast computing machines. J Chem Phys 21:1087–1092
8. Pawlowsky V, Olea RA, Davis JC (1995) Estimation of regionalized compositions: comparison of three methods. Math Geol 27:105–127
9. Pawlowsky-Glahn V, Olea RA (2004) Geostatistical analysis of compositional data. Studies in mathematical geology, vol 7. Oxford University Press, New York
10. Rivoirard J (1994) Introduction to disjunctive kriging and non-linear geostatistics. Clarendon Press, Oxford
11. Walvoort DJJ, de Gruijter JJ (2001) Compositional kriging: a spatial interpolation method for compositional data. Math Geol 33:951–966

New Flexible Non-parametric Data Transformation for Trans-Gaussian Kriging

Alexander Gribov and Konstantin Krivoruchko

Abstract This paper proposes a new flexible non-parametric data transformation to Gaussian distribution. This option is often required because kriging is the best predictor under squared-error minimization criterion only if the data follow multivariate Gaussian distribution, while environmental data are often best described by skewed distributions with non-negative values and a heavy right tail. We assume that the modeling random field is the result of some nonlinear transformation of a Gaussian random field. In this case, the researchers commonly use a certain parametric monotone (for example, power or logarithmic) or variants of normal score transformation. We discuss drawbacks of these methods and propose a new flexible non-parametric transformation. We compare the performance of simple kriging with the proposed data transformation to several other data transformation methods, including transformation based on a mixture of Gaussian kernels and multiplicative skewing with several base distributions. Our method is flexible, and it can be used for automatic data transformation, for example, in black-box kriging models in emergency situations.

1 Introduction

Transformations are used in geostatistics to bring data close to normal distribution and to satisfy stationarity assumptions. The logarithmic transformation is often used for a skewed data distribution with a small number of large values. Formulas for lognormal simple and ordinary kriging predictions and prediction standard errors can be found in [3]. Approximate formulas based on expansion of the transformation function in a second-order Taylor series are used for other functional data transformation such as power and arcsine, see [2]. A popular power transformation can be used with positive data only and has issues with transformation of small data values

A. Gribov (✉) · K. Krivoruchko
Environmental Systems Research Institute, 380 New York St., Redlands, CA 92373, USA
e-mail: agribov@esri.com

K. Krivoruchko
e-mail: kkrivoruchko@esri.com

P. Abrahamsen et al. (eds.), *Geostatistics Oslo 2012*,
Quantitative Geology and Geostatistics 17,
DOI 10.1007/978-94-007-4153-9_5, © Springer Science+Business Media Dordrecht 2012

to multivariate normal distribution. It is not flexible enough for real data transformation because it has just one parameter. The normal score transformation (NST) ranks the dataset from lowest to highest values and matches these ranks to equivalent ranks from a standard normal distribution. The result of the NST is always a univariate normal distribution. Any trend in the data should be removed before NST. An important consequence of successful NST is that kriging prediction standard error becomes a function of the observed data in addition to a function of the data configuration, so that the prediction standard error often becomes larger in the areas where the observed values are larger, see examples and a discussion about accurate data back-transformation in [5]. Successful NST requires estimation of smooth probability density function [5]. A mixture of Gaussian distributions with different means and standard deviations can be used to smooth the probability density function. Gaussian mixture produces a multi-modal distribution. This can be a problem because typically we assume that the stationary data are generated by one particular process, not a combination of several simple processes. With large number of kernels, the cumulative empirical distribution can be approximated almost exactly, but it is very unlikely that the approximate distribution density corresponds to any real physical process which can be responsible for the multi-modal data distribution. An alternative is to use multiplicative skewing of a base distribution (usually normal, lognormal, or gamma), see [6] and below.

The number of the available data transformation methods is large and it is difficult to choose an optimal method for the data at hands objectively and automatically. This motivated us to develop a new data transformation method. This method is estimating a nonlinear function which transforms the observed data to multivariate Gaussian distribution. This function has the following features: it is monotonously increasing, it has at least two derivatives, it is scalable and shift resistant. We show that our method is flexible and it can be used for automatic data transformation, for example, in the black-box kriging models in emergency situations, such as the recent radioactive accident at the Fukushima nuclear power plant in Japan.

2 Multiplicative Skewing

In practice, the sample distributions are rarely accurately described by a particular theoretical distribution. Therefore, it is natural to introduce some skewness into the appropriate theoretical distribution and retain the properties of that distribution. In this paper, we follow the idea to separate the skewing mechanism from the base theoretical distribution via inverse probability integral transformation proposed in [4]. Basically, the resulting skewed distribution $f_s(x) = f(x)p(F(x))$ is a weighted version of the selected theoretical distribution $f(x)$ with symmetric or asymmetric weights $p(F(x))$, not linked to the theoretical distribution, where $f(x)$ and $F(x)$ are respectively the probability density and the cumulative distribution of random variable x.

We follow [6], and use the Bernstein densities (a weighted sum of beta probability distribution functions) to model weights $p(F(x))$ defined on the unit interval

[0, 1]. A Bernstein density is defined as

$$p(x \mid m, \mathbf{w}) = \sum_{j=1}^{m} w_j \,\mathrm{Beta}(x \mid j, m-j+1), \quad w_j \geq 0, \sum_{j=1}^{m} w_j = 1,$$

where Beta(·) is the beta probability density function:

$$\mathrm{Beta}(x \mid j, m-j+1) = \frac{m!}{(j-1)!(m-j)!} x^{j-1}(1-x)^{m-j}.$$

Note that when $m = 1$, the base theoretical distribution remains unchanged. Authors of [6] used computational-intensive Bayesian approach for estimation of the weights w_i, but we found that they can be accurately estimated using maximum likelihood approach with the following log-likelihood function:

$$\sum_{i=1}^{n} \ln\left[f(x_i) \sum_{j=1}^{m} w_j \,\mathrm{Beta}\big(F(x_i) \mid j, m-j+1\big) \right].$$

The likelihood function can be weighted if the data declustering algorithm is used, but in this paper we assume that the data are equally weighted. A stable and efficient iterative EM algorithm proposed in [1] is used for the weights estimation:

1. The initial weights approximation is $w_j^{(1)} = \frac{1}{m}$, for all $j = 1, \ldots, m$.
2. Conditional probability of selecting component number j, $j = 1, \ldots, m$, given the data x_i, $i = 1, \ldots, n$, is:
$$w_{j|i}^{(k)} = \frac{w_j^{(k)} \,\mathrm{Beta}(F(x_i) \mid j, m-j+1)}{\sum_{j=1}^{m} w_j^{(k)} \,\mathrm{Beta}(F(x_i) \mid j, m-j+1)}.$$
3. The weights are approximated using $w_j^{(k+1)} = \frac{1}{n} \sum_{i=1}^{n} w_{j|i}^{(k)}$.
4. Stop iterations when improvement in loglikelihood function is sufficiently small, otherwise go to step 2.

Any data distribution can be used as the base distribution, and we use three theoretical distributions (Student's t, gamma and lognormal) and two empirical distributions discussed in the section empirical data transformation below. We found that in practice, combining a maximum of eight beta distribution skewers, $m \in \{1, 2, \ldots, 8\}$, is sufficient to approximate most of sample distributions. The Akaike's information criterion (AIC) is used to choose the optimal mixture of the beta skewers.

We illustrate how the skewing mechanism works with a subset of the data used for validation of the INTAMAP project, http://www.intamap.org, Fig. 1a. The data values are clearly different on the sides of the imaginary line between small (circles with values less than or equal 90) and large (black points with values larger than 90) values. The data are displayed on top of the geological map with clearly different characteristics, and this could be a reason for the abrupt changes in the data values. A histogram of the sample data values in Fig. 1b looks like a mixture of two distributions. The result of fitting the sample distribution using the lognormal base and maximum of 1, 4, 6, and 8 skewers is displayed with lines (fitting using

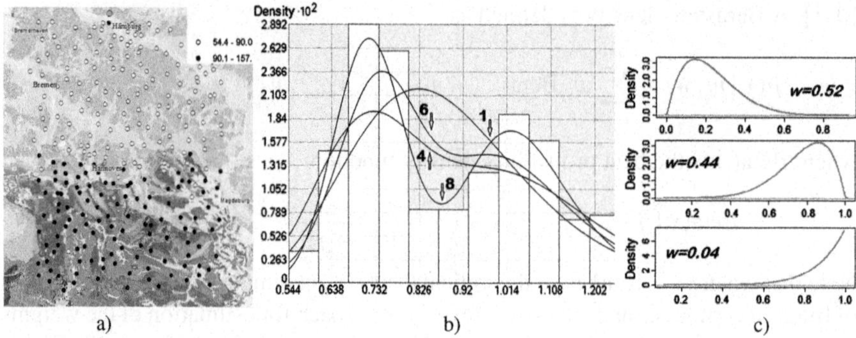

Fig. 1 (**a**) Subset of the radiation data from http://www.intamap.org/sampleRadiation.php. (**b**) Fitting the bimodal distribution using maximum of 1, 4, 6, and 8 skewers. (**c**) Three beta distributions with non-zero weights used in the best fit of the sample distribution

this base distribution produced lower AIC value than fitting with the Student's t and gamma bases). The best fit according to the AIC criterion was achieved with eight beta skewers, although only three of them have non-zero weights. These three beta distributions and the estimated weights are shown in Fig. 1c.

3 Multiplicative Skewing Using Simulated Data

In this section we show how the distribution of the simulated data can be approximated using the multiplicative skewing algorithm. We simulated two datasets from the gamma and univariate distributions using the following `R` commands:

```
G1<- rgamma(1000, shape=2, rate=5); U1<- runif(1000);
```

Then we created a third dataset `G2` using the mixture of distributions `G1` and `U1` as follows:

```
G2 <- 0.5*G1 + 0.5*U1;
```

The multiplicative skewing algorithm reconstructs theoretical distribution `G1` exactly, line in Fig. 2a. In fact, the sample mean and variance of distribution `G1` were automatically reproduced without skewing by choosing the gamma base and $m = 1$ in the Bernstein density. The normal quantile plot shown at the top right corner of Fig. 2a suggests that the approximated distribution can be successfully transformed to the univariate normal distribution. At first look a noisy version of distribution `G1`, distribution `G2` in Fig. 2b, looks simple and it seems that it can be easily approximated and then transformed to the normal distribution. However, gamma distribution with the sample `G2` data mean and variance in Fig. 2b is not satisfactory visually and from the diagnostic point of view: the normal quantile plot shows that the approximated line is clearly different from the desirable 1:1 line and the AIC value is far from the optimal one as shown in Table 1 (second row).

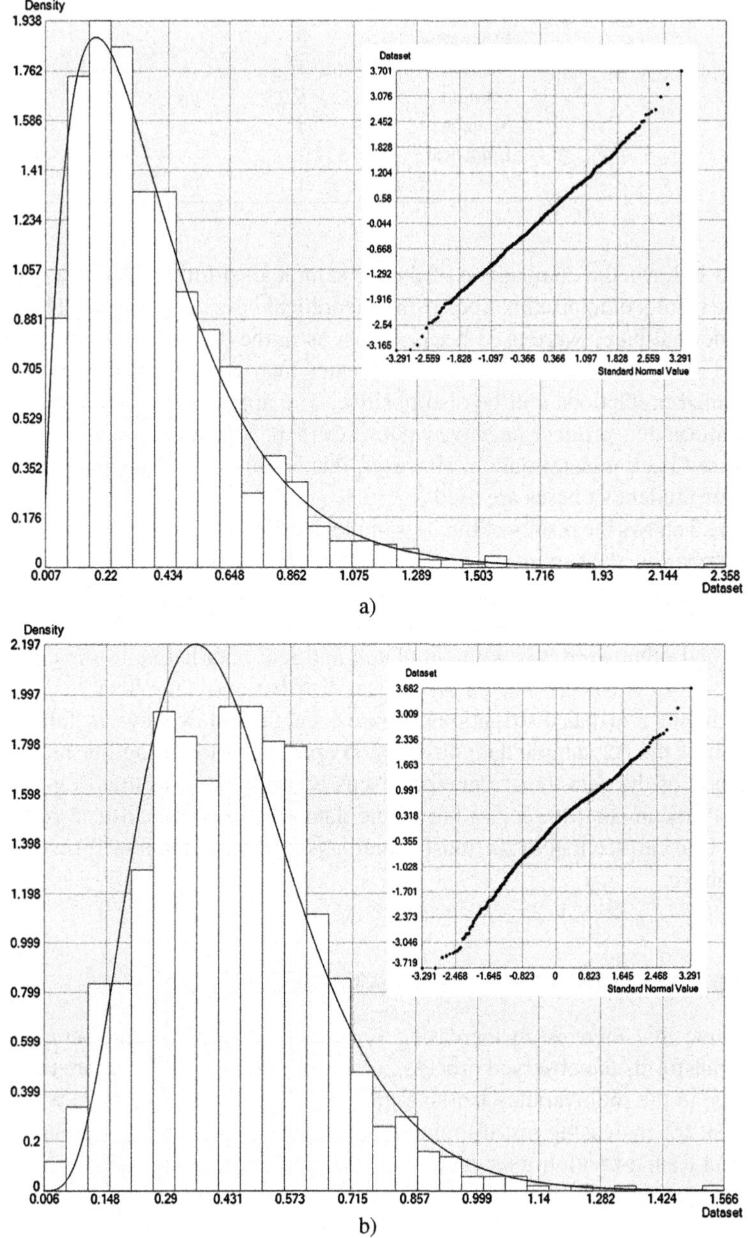

Fig. 2 (**a**) Fitting gamma distribution with parameters *shape* = 2 and *rate* = 5. (**b**) Fitting a noisy version of the same gamma distribution

Table 1 Fitting of the G2 sample data distribution comparison

Base distribution	Number of components in the Bernstein density	AIC
Lognormal	8	−216.32
Gamma	1	−399.43
Gamma	8	−428.32
Student's t	8	−440.66
Empirical	1	−449.84
Log-empirical	1	−454.28

Table 1 shows the comparison of the G2 sample data fitting. The first four rows show the results of approximations using theoretical base distributions. The gamma and lognormal bases were used because, just as in the case of most environmental data, we know that the data values are greater than zero. In this case, the back-transformed predictions will be also positive. The approximated back transformation formulas may produce negative values, see [5, p. 340] for one possible approach of unbiased back transformation. However, this feature is not guaranteed when the normal or Student's t bases are used.

Figure 3 shows the results of the G2 sample distribution approximation with eight beta distribution skewers using gamma, lognormal and Student's t bases. The last two approximations are bimodal although the sample data were simulated from the unimodal distribution contaminated by noise.

This and other exercises with simulated and real data led us to the conclusion that the set of the bases based on theoretical distributions is not flexible enough for approximating real data distributions. Figure 4 and the last two rows in Table 1 show the result of the G2 sample data distribution approximation using our new empirical method of the data transformation discussed in the next section (log-empirical in Fig. 4b means that the logarithm of the data values is used instead of the data values). One can see that these transformations outperform the transformations discussed above.

4 Empirical Data Transformation

We assume that there is an increasing function $f(z(s_i) \mid \mathbf{w})$ with parameters $\mathbf{w}$, which transforms the observed process $z(s_i)$, where s_i, $i = 1, \ldots, n$, are the sample locations, to the multivariate Gaussian process $y(s_i)$, $y(s_i) = f(z(s_i) \mid \mathbf{w})$. The requirement for increasing transformation function is necessary to guarantee unique back data transformation after modeling and predicting in the transformed space. We also require everywhere smooth function because finding unique non-smooth function based on the finite number of observations is very difficult and because we do not know good justification for the non-smoothness property. Next, we assume that the transformation function has at least first and second derivatives. Then the first derivative $f_1(z(s_i) \mid \mathbf{w})$ is positive because $f(z(s_i) \mid \mathbf{w})$ is an increasing function. Finding a positive function is an easier task than finding an increasing

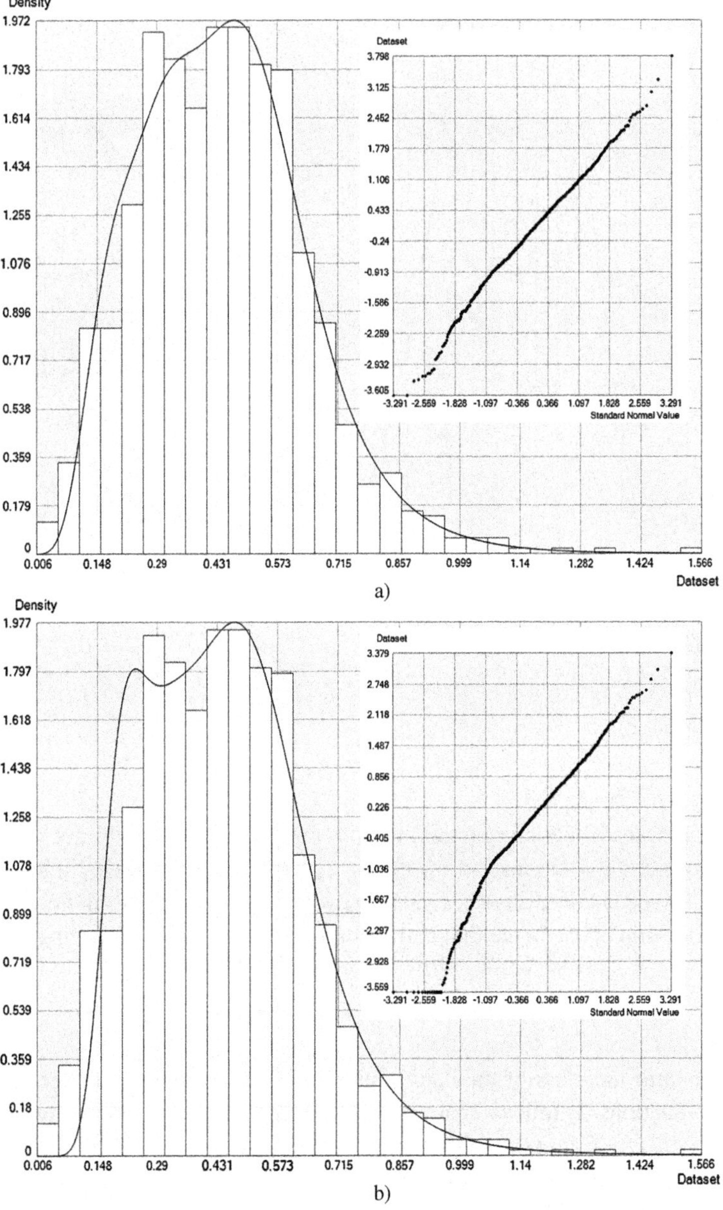

Fig. 3 Fitting G2 sample distribution with 8 skewers using (**a**) gamma base; (**b**) lognormal base; and (**c**) Student's t base

function. Taking into account that adding a constant to the function $f(z(s_i) \mid \mathbf{w})$ does not change the result, we have:

$$f\big(z(s_i) \mid f_1(\cdot \mid \mathbf{w})\big) = \int_0^z f_1\big(z' \mid \mathbf{w}\big)\, \mathrm{d}z'. \tag{1}$$

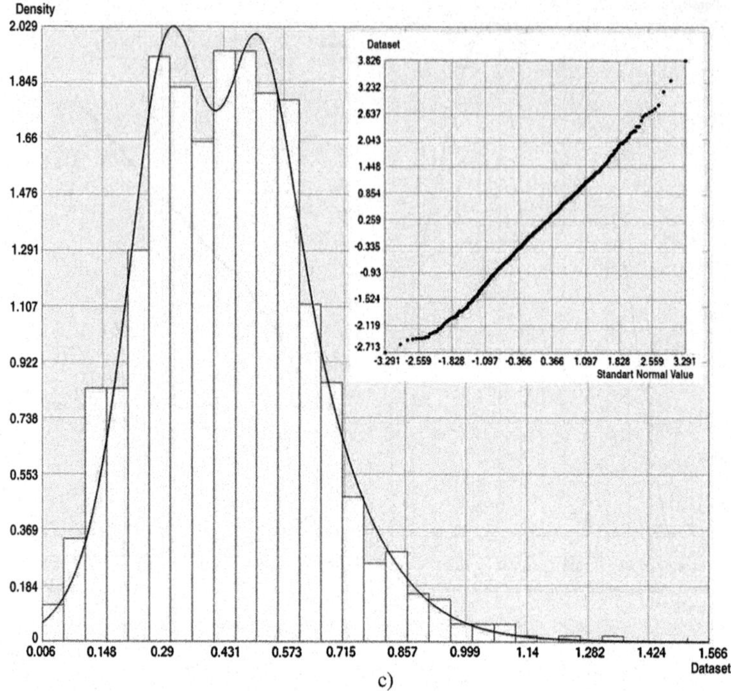

c)

Fig. 3 (continued)

Stable data transformation requires that the first derivative should not be close to zero or to infinity. Then instead of finding $f_1(z(s_i) \mid \mathbf{w})$ with these limitations, we look for the second derivative $f_2(z(s_i) \mid \mathbf{w})$ with constraints for its definite integral. The relationship between the first and second derivatives is the following:

$$f_1\big(z \mid f_2(\cdot \mid \mathbf{w})\big) = \int_{-\infty}^{z} f_2\big(z' \mid \mathbf{w}\big)\,\mathrm{d}z' + C. \tag{2}$$

Next, we limit the class of the data transformations by allowing either convex or concave functions. It follows that one of the following inequalities should be satisfied: $\forall z,\ f''(z \mid f_2(\cdot \mid \mathbf{w})) \geq 0$ or $\forall z,\ f''(z \mid f_2(\cdot \mid \mathbf{w})) \leq 0$. The result of the data transformation is not sensitive to multiplication by a constant and we can write the following expression instead of (2):

$$f_1\big(z \mid f_2(\cdot \mid \mathbf{w})\big) = k\int_{-\infty}^{z} f_2\big(z' \mid \mathbf{w}\big)\,\mathrm{d}z' + kC, \quad k > 0.$$

We select k in such a way that deviation of the first derivate from the value of 1 is the same at positive and negative infinities, that is $f_1(-\infty \mid \mathbf{w}) - 1 = 1 - f_1(+\infty \mid \mathbf{w})$ which is the same as $f_1(-\infty \mid \mathbf{w}) + f_1(+\infty \mid \mathbf{w}) = 2$. If this condition is satisfied,

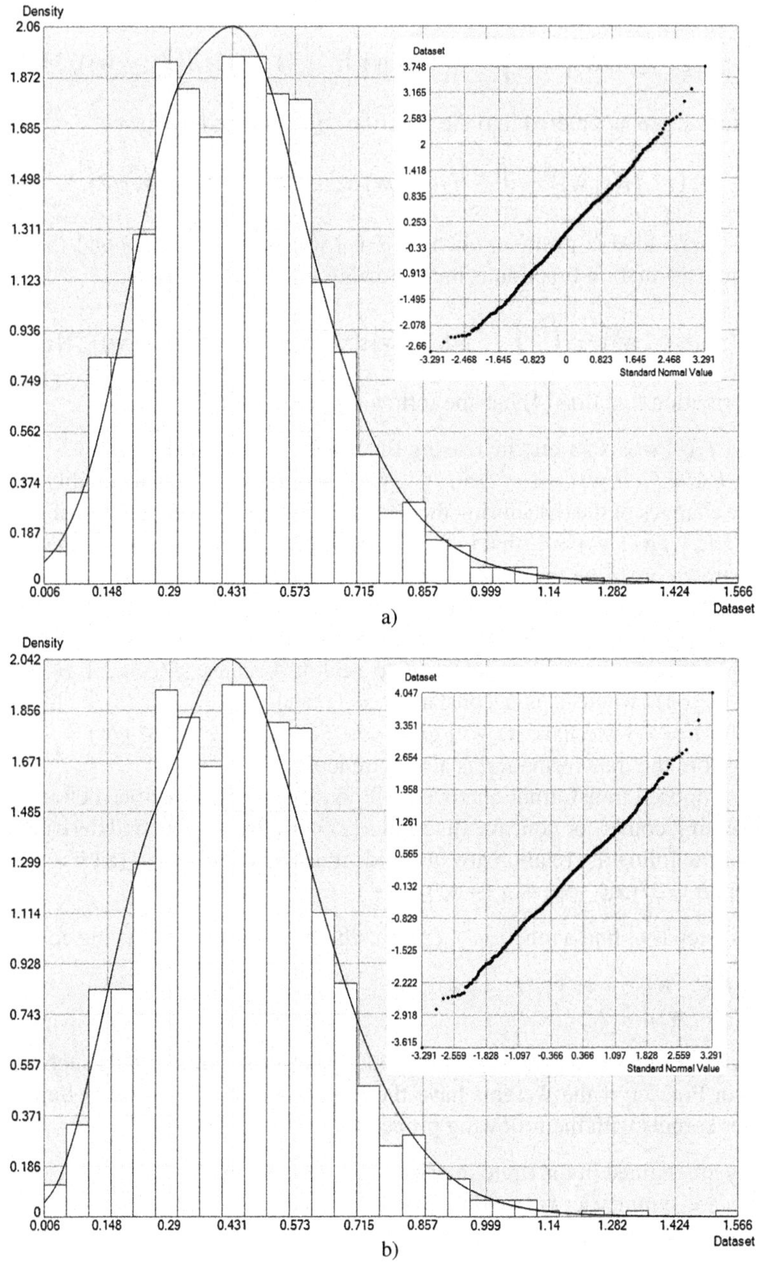

Fig. 4 Fitting the G2 sample data distribution using (**a**) empirical and (**b**) log-empirical data transformations

the second derivative is limited: $kC = 1 - k \times a(f_2(\cdot \mid \mathbf{w}))/2$, where $a(f_2(\cdot \mid \mathbf{w})) = \int_{-\infty}^{+\infty} f_2(z' \mid \mathbf{w})\,dz'$. Which gives:

$$f_1\big(z \mid f_2(\cdot \mid \mathbf{w})\big) = k \int_{-\infty}^{z} f_2\big(z' \mid \mathbf{w}\big)\,dz' + \big(1 - k \times a\big(f_2(\cdot \mid \mathbf{w})\big)/2\big).$$

A constant k can be entered into the function $f_2(\cdot \mid \mathbf{w})$, and we have:

$$f_1\big(z \mid f_2(\cdot \mid \mathbf{w})\big) = \int_{-\infty}^{z} f_2\big(z' \mid \mathbf{w}\big)\,dz' + \big(1 - a\big(f_2(\cdot \mid \mathbf{w})\big)/2\big). \tag{3}$$

The first derivative is positive when $|a(f_2(\cdot \mid \mathbf{w}))| < 2$. Using (1) and (3), the required transformation function is the following:

$$f\big(z \mid f_2(\cdot \mid \mathbf{w})\big) = \int_{0}^{z} \int_{-\infty}^{z'} f_2\big(z'' \mid \mathbf{w}\big)\,dz''\,dz' + \big(1 - a\big(f_2(\cdot \mid \mathbf{w})\big)/2\big)z. \tag{4}$$

Transformation function (4) has the following properties:

1. $f(z \mid f_2(\cdot \mid \mathbf{w}))$ is strictly increasing function.
2. $\forall z, |f'(z \mid f_2(\cdot \mid \mathbf{w})) - 1| \le |a(f''(\cdot \mid f_2(\cdot \mid \mathbf{w})))/2| < 1$. This inequality prevents large changes of the data during the data transformation and back transformation.
3. $\exists z, f'(z \mid f_2(\cdot \mid \mathbf{w})) = 1$, that is there is a location where the transformation does not stretch or shrink the data.
4. The transformation is not sensitive to the shift because $f(z + c \mid g(\cdot)) = f(z \mid f_2(\cdot)) + b$, where c and b are constants and $g(z) = f_2(z - c)$.
5. The transformation is not sensitive to scaling because $f(c \times z \mid g(\cdot)) = c \times f(z \mid f_2(\cdot))$, where c is a constant, $c \ne 0$, and $g(z) = f_2(z/c)/c$. In particular, for $c = -1$ we have $f(-z \mid g(\cdot)) = -f(z \mid f_2(\cdot))$, where $g(z) = -f_2(-z)$. Therefore the transformation is also symmetric.
6. The proposed transformation covers all possible transformations defined by the increasing convex or concave function $u(z)$ with first and second derivatives because the following relation always holds true: $u(z) = u(0) + f(z \mid b \times u''(\cdot))/b$, where $b = 2/(u'(-\infty) + u'(+\infty))$.

The next step is to find a function $f_2(z \mid \mathbf{w})$ which satisfies the following conditions:

1. $\forall z, f_2(z \mid \mathbf{w}) \ge 0$ or $\forall z, f_2(z \mid \mathbf{w}) \le 0$.
2. $|a(f_2(\cdot \mid \mathbf{w}))| < 2$.

These conditions are satisfied for a weighted sum of several kernels such as those shown in Fig. 5a, if the weights have the same sign and their sum is limited. We consider kernels with the following properties:

1. They are defined in the finite interval: $\forall |x| \ge 1, k(x) = 0$.
2. They are symmetric: $k(x) = k(-x)$.
3. They are positive: $k(x) \ge 0$.
4. They are continuous.
5. They are monotonic, $\forall |x| \le |y|, k(y) \le k(x)$.
6. Mixture of two kernels reconstructs a constant function: $\forall 0 \le x \le 1, k(x) + k(1 - x) = 1$. It follows that the following relationship holds: $\int_{-\infty}^{+\infty} k(x)\,dx = 1$.

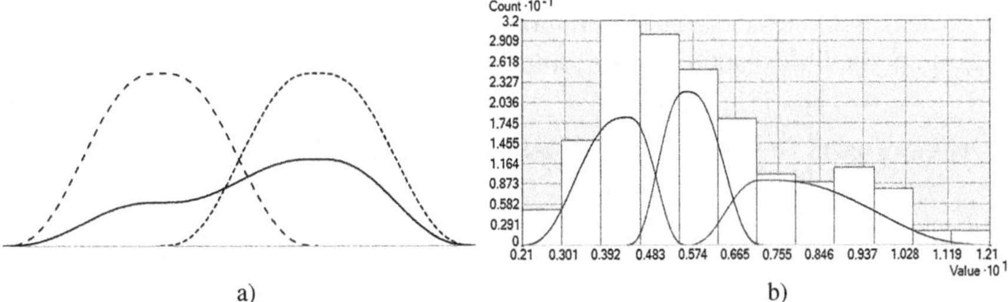

Fig. 5 (**a**) A weighted sum, *solid line*, of two kernels with weights 0.25 and 0.5, *dashed lines*, for approximation of $f_2(z \mid \mathbf{w})$. (**b**) The centers of three kernels are located at the first quartile, median, and third quartile of the sample data distribution

One kernel which has all properties discussed above is the following:

$$k(x) = \begin{cases} 1 - |x|^3(10 - |x|(15 - 6|x|)), & \text{if } |x| < 1; \\ 0, & \text{otherwise.} \end{cases}$$

This kernel has continuous first and second derivatives, including points $x = \{-1, 0, 1\}$ and the transformation function based on this kernel has four continuous derivatives.

In practice, transformation functions can be estimated only for intervals with sufficiently large number of measurements. Therefore we place centers of the kernels at particular quantile values, for example, centers of three kernels can be located at the first quartile, median, and third quartile values. Figure 5b shows such kernels' configuration for the tutorial Geostatistical Analyst data, ozone measurements collected in California. For m kernels, the transformation function for the $\{z_i\}$ can be estimated as follows. Lets x_0 is equal to the minimum value of z_i, x_i, $i = 1, \ldots, m$, are equal to data $(m+1)$-quantiles, and x_{m+1} is equal to the maximum value of z_i. Then the second derivative is

$$f_2(x \mid \mathbf{w}) = 2\sum_{i=0}^{m-1}\left[\frac{w_i}{x_{i+2} - x_i} \times k\left(\frac{x - x_{i+1}}{(x_{i+1} - x_i)\delta_{x<x_{i+1}} + (x_{i+2} - x_{i+1})\delta_{x_{i+1}\le x}}\right)\right],$$

where w_i, $i = 0, \ldots, m-1$, are the weights with the following properties:

1. All weights have the same sign: $w_i \ge 0$, $\forall i = 0, \ldots, m-1$ or $w_i \le 0$, $\forall i = 0, \ldots, m-1$.
2. $|\sum_{i=0}^{m-1} w_i| < 1$.

We define the function

$$\psi\big(z(s_i), \mathbf{w}\big) = \ln\big[\phi\big(f\big(z(s_i) \mid f_2(\cdot \mid \mathbf{w})\big); m_n(\mathbf{w}), \sigma_n^2(\mathbf{w})\big)\big],$$

where

$$\phi\big(x; \mu, \sigma^2\big) = \frac{1}{\sqrt{2\pi\sigma^2}} \exp\left\{-\frac{(x-\mu)^2}{2\sigma^2}\right\},$$

is the density of the Gaussian distribution with mean μ and variance σ^2, and use the maximum likelihood approach for estimating the optimal weights:

$$\mathbf{w} = \arg\max_{\mathbf{w}} \left\{ \max_{m_n(\mathbf{w}),\sigma_n^2(\mathbf{w})} \left\{ \sum_{i=1}^{n} \left(\psi\left(z(s_i),\mathbf{w}\right) + \ln\left[f'\left(z(s_i) \mid f_2(\cdot \mid \mathbf{w})\right)\right] \right) \right\} \right\}. \quad (5)$$

Optimal values for $m_n(\mathbf{w})$ and $\sigma_n^2(\mathbf{w})$ can be found from the following formulas:

$$\hat{m}_n(\mathbf{w}) = \frac{1}{n} \sum_{i=1}^{n} f\left(z(s_i) \mid f_2(\cdot \mid \mathbf{w})\right),$$

$$\hat{\sigma}_n^2(\mathbf{w}) = \frac{1}{n} \sum_{i=1}^{n} \left(f\left(z(s_i) \mid f_2(\cdot \mid \mathbf{w})\right) - \hat{m}_n(\mathbf{w}) \right)^2.$$

Define thus the corresponding

$$\psi_{\text{opt}}\left(z(s_i),\mathbf{w}\right) = \ln\left[\phi\left(f\left(z(s_i) \mid f_2(\cdot \mid \mathbf{w})\right);\, \hat{m}_n(\mathbf{w}),\hat{\sigma}_n^2(\mathbf{w})\right)\right].$$

Substituting this into (5) gives

$$\mathbf{w} = \arg\max_{\mathbf{w}} \left\{ \sum_{i=1}^{n} \left(\psi_{\text{opt}}\left(z(s_i),\mathbf{w}\right) + \ln\left[f'\left(z(s_i) \mid f_2(\cdot \mid \mathbf{w})\right)\right] \right) \right\}$$
$$= \arg\max_{\mathbf{w}} \left\{ -\frac{n}{2} \ln\left(\hat{\sigma}_n^2(\mathbf{w})\right) + \sum_{i=1}^{n} \ln\left[f'\left(z(s_i) \mid f_2(\cdot \mid \mathbf{w})\right)\right] \right\}.$$

Note that the usage of the logarithm of the data values in the case of log-empirical data transformation leads to changing (5) by a constant value which can be omitted.

5 Simple Kriging with Empirical Data Transformation

The multiplicative skewing data transformation with theoretical or empirical bases can be used as the first step of the NST algorithm. The NST requires specification of the data mean value or mean surface and in practice this means that this transformation can be used with simple or disjunctive kriging. In this section we show the empirical data transformation usage for data interpolation with simple kriging. Simple kriging is the basis for linear mixed models and for various Bayesian kriging models which become dominating in modern statistical research and applications.

Figure 6a shows a histogram of a subset of the ^{137}Cs soil contamination data collected in Belarus in 1992 (the complete dataset is provided in [5]). The data distribution does not appear Gaussian or lognormal. Also, the data variance is larger than the data mean. Consequently, the data distribution approximation based on a particular continuous theoretical distribution is problematic, and the log-empirical transformation gives the best AIC and cross-validation diagnostic statistics. Another good transformation from the prediction diagnostics point of view is a mixture of Gaussian kernels shown in Fig. 6b and described in detail in [5]. Indeed, using a

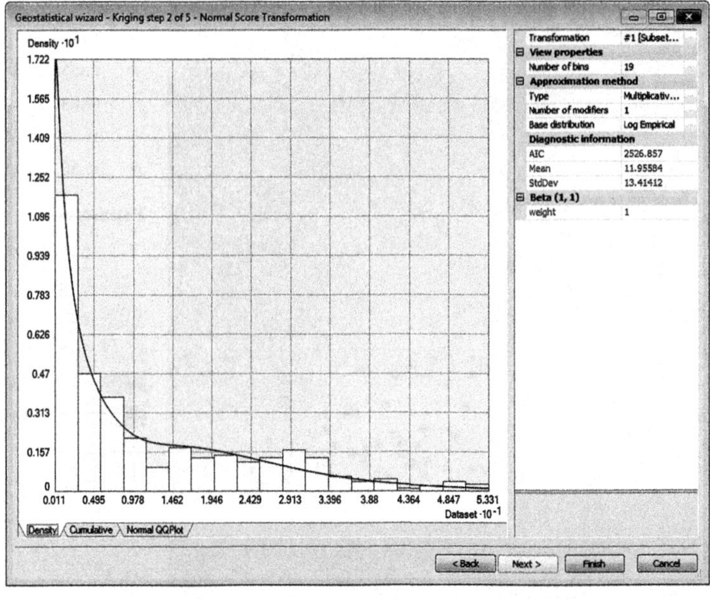

a)

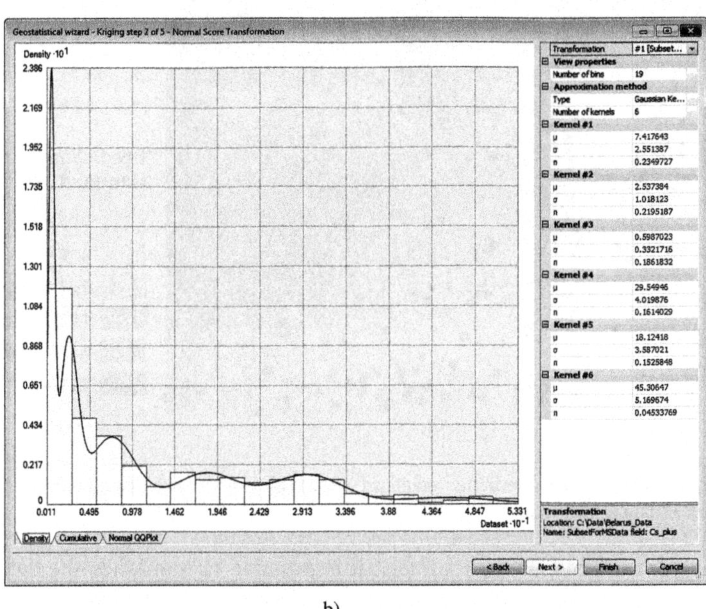

b)

Fig. 6 (**a**) The distribution of the ^{137}Cs soil contamination data approximated using log-empirical transformation. (**b**) The data distribution is approximated by the formula $\sum_{i=1}^{6} p_i \times \phi(x \mid \mu_i, \sigma_i^2)$ with the following constraint to the weights: $\sum_{i=1}^{6} p_i = 1$. The values for p_i, μ_i, and σ_i are provided at the *right side* of the Geostatistical Analyst's dialog

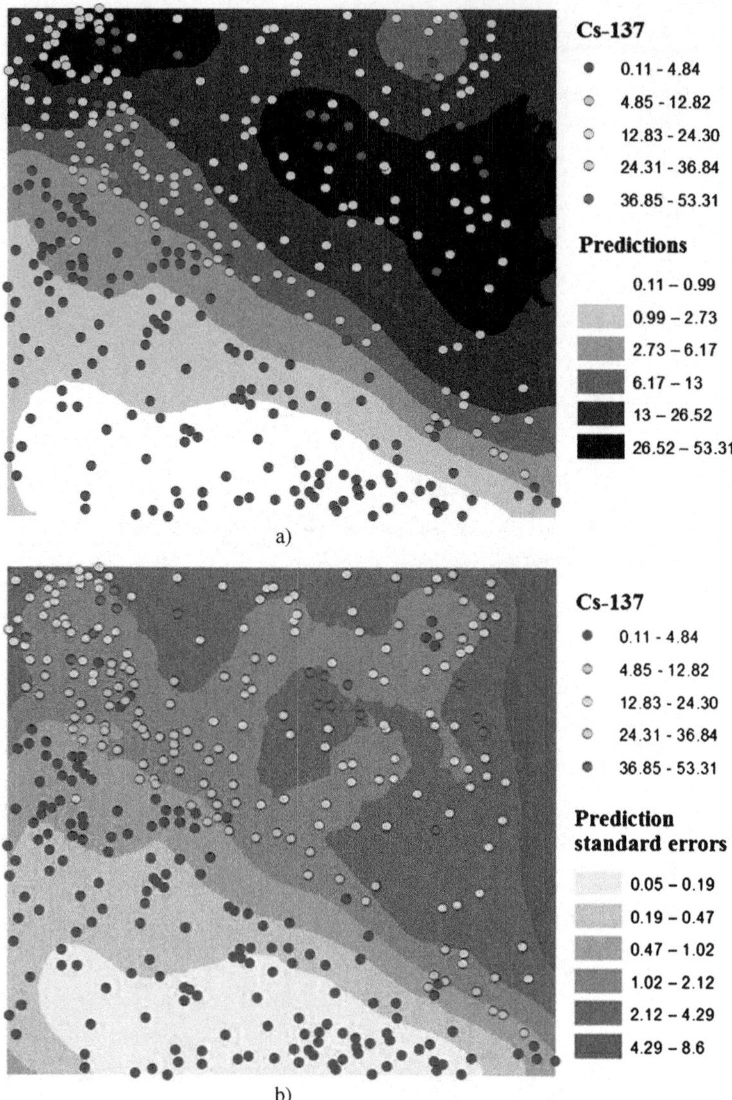

Fig. 7 Predictions (**a**) and prediction standard errors (**b**) maps created by simple kriging with log-empirical data transformation

weighted sum of six Gaussian kernels, it is possible to reproduce the data distribution accurately, and for these data, the cross-validation diagnostic is similar for simple kriging with log-empirical and with a mixture of Gaussian kernels transformations. However, it is very unlikely that the radioactive cesium soil contamination in relatively small area was resulted from six different processes and it is best described by the distribution approximation with six modes. Therefore, we prefer the unimodal approximation in Fig. 6a. Figure 7 shows the prediction and prediction standard error maps created by simple kriging with log-empirical data transformation. Note that these maps are related: areas with large predictions have large prediction standard errors as well. This is a desired and important property of the radiation data interpolation because radioactive decay is a Poisson process and an essential property of a Poisson process is that its mean is equal to its variance.

6 Conclusion

We developed a new flexible non-parametric data transformation which successfully transforms the observed data to the multivariate normal distribution. This option allows using simple kriging as a black-box automatic interpolator of complex data in emergency situations.

Our transformation is also used as an option for the non-stationary empirical Bayesian kriging (EBK). In this case, the transformation function becomes a random variable, and a large number of transformations are used to account for the uncertainty of the estimated sampling distribution. An important feature of the EBK with empirical data transformation is that the data distribution is modeled without an assumption about data independence as it is implicitly assumed in this paper. It becomes possible because our EBK model subsets the data, and the maximum number of samples in the subset is limited to 500 samples. Then for the relatively small dataset, the transformation and the spatial covariance model estimation can be done simultaneously. The resulting data transformation takes into account the data dependency. The EBK with and without the empirical data transformation option will be discussed in the forthcoming paper by the same authors.

A new non-parametric data transformation to Gaussian distribution is implemented in the ArcGIS Geostatistical Analyst, version 10.1, for both classical kriging models and EBK. This version of the software will be released in the second quarter of 2012.

References

1. Celeux G, Chrétien S, Forbes F, Mkhadri A (2001) A component-wise EM algorithm for mixtures. J Comput Graph Stat 10(4):697–712
2. Cressie N (1993) Statistics for spatial data, revised edn. Wiley, New York
3. Cressie N (2006) Block kriging for lognormal spatial processes. Math Geol 38:413–443
4. Ferreira JTAS, Steel MFJ (2006) A constructive representation of univariate skewed distributions. J Am Stat Assoc 101(474):823–829
5. Krivoruchko K (2011) Spatial statistical data analysis for GIS users. ESRI Press, Redlands, 928 pp
6. Quintana FA, Steel MFJ, Ferreira JTAS (2009) Flexible univariate continuous distributions. Bayesian Anal 4(3):497–522

Revisiting the Linear Model of Coregionalization

Denis Marcotte

Abstract The linear coregionalization model (LCM) is the model most commonly used in practice for cokriging. The LCM model amounts to represent the observed variables as linear combinations of sets of independent (or at least uncorrelated) underlying variables. The popularity of the LCM stems mostly from the ease of modeling and verification of the admissibility of the model. However, LCM has strong limitations as it implies symmetrical cross-covariances for the variables under study. This symmetry could explain why the cokriging is generally unsuccessful to improve significantly over kriging when the observed variables are collocated. One possible generalization to the LCM (GLCM) is to consider the observed variables as linear combinations of underlying independent variables and of variables representing deterministic functions of some of these underlying variables. Candidate functions are for example regularization, spatial shift and derivatives, the last two functions enabling to introduce asymmetry (and anisotropy) in the cross-covariances when needed. We note that GLCM comprises the deterministic cokriging of a variable and a given function of it (partial derivative, regularization, ...) as a particular case. To ensure admissibility of the model, the modeling is done directly on the coefficients of the underlying variables, and functions of it, using an iterative non-linear optimization approach that compares the theoretical covariances computed from the GLCM to the experimental ones. Various synthetic cases are presented illustrating the flexibility of the GLCM. The well-known Gslib data file is also used to compare the relative performances of the LCM and of the GLCM. GLCM is shown to provide more precise estimations than the LCM model for the test case.

1 Introduction

The multiplication of remote sensing devices and the increasing usage of geophysical methods has brought numerous new sources of data to geoscience studies. The problem of data integration therefore poses an important challenge to multivariate

D. Marcotte (✉)
Département des génies civil, géologique et des mines, École Polytechnique, C.P. 6079, Succ. Centre-ville, Montréal, Québec, Canada, H3C 3A7
e-mail: denis.marcotte@polymtl.ca

P. Abrahamsen et al. (eds.), *Geostatistics Oslo 2012*, Quantitative Geology and Geostatistics 17,
DOI 10.1007/978-94-007-4153-9_6,

geostatistics. In current practice, the class of available multivariate models is limited. Except for very specific cases where a deterministic relation between variables links the covariance models with equivalent deterministic relations, most of the time one has to resort to the sole linear coregionalization model (LCM) [7]. Many practitioners even simplifies further the LCM by using the simplified version designed as collocated cokriging even though [9] and [10] has shown that there is no LCM that reduces to collocated cokriging. Therefore one incurs a loss of information when using collocated cokriging rather than the full LCM.

The limitations of LCM are well known. First, it produces only symmetric models of cross-covariance. Second, the cross-correlation is maximum at lag $h = 0$. Third, when a structure appears on the cross-covariance, it must also be present on the simple structures of the corresponding variables. Forth, the admissibility condition implies that the eigenvalues of the individual coefficient matrices of the LCM all be positive (sufficient but not necessary condition). This could call for an iterative procedure [6] to enforce the condition. Alternatively, one can use the perturbed LCM obtained by putting all negative eigenvalues equal to zero and reconstructing the coefficient matrices. Reference [8] proposed a more general and flexible model than the LCM they call spectral additive components. However, in the multivariate case, the number of parameters in the model and the computational requirement increases quickly. Moreover, ensuring an admissible model is not immediate.

We propose a new generalization of the LCM model (GLCM) that enables to alleviate the above mentioned limitations. The basic idea is simple: rather than expressing the random vector $\mathbf{Z}$ as the sum of a number of independent variables, we express it as the sum of a few independent variables and well chosen functions of some of these independent variables. In this study, for illustration, the functions considered are: 1—the partial first derivatives, 2—the shift operator in direction of a unit vector u, and 3—the regularization operator. Together, these functions are able to produce many kinds of asymmetric cross-covariances. Of course, alternative functions more closely related to the variables under study might also be used. The three proposed functions have a clear physical interpretation. For example, the derivative naturally occurs when mapping the elevation of a geological unit using elevation and lithology attitude in the borehole (e.g. with a dip-meter). Reference [1] used the fact that the aquifer's log(transmissivity) field is proportional to the derivative of the hydraulic head field. Shifts can occur when, for example, a given mineral precipitates under different physicochemical conditions than a second one [7, p. 41]. It is also observed with some geophysical properties like electrical conductivity and chargeability which often appear spatially shifted relatively to the location of the core of a mineralized body [11]. Finally, the regularization operator could be useful with variables defined on different physical supports (e.g. field data and remote sensing data). It could also be useful for the study of diffusion process, either for variables measured at different times or when measured at different locations in the presence of an advective component.

We do not try to fit directly the model parameters of the GLCM to a series of simple and cross-covariances as is usually done with the LCM. Instead, we proceed indirectly by estimating all coefficients linking the underlying variables or underlying functions of variables to the observed random vector much as in [2]. All the cross-covariances can then be computed from these coefficients. This ensures automatically the whole model is admissible for all choice of coefficients of the underlying variables and functions.

The paper is structured as follows. We first recall the LCM, then we describe in details the functions used to define the GLCM and their main properties. We describe the estimation approach. We show with a synthetic example the applicability of the estimation method. Finally we apply it to a set of data for which the regular LCM cannot be modeled adequately and does not describe well the various simple and cross-covariances. We compare on this example the results obtained with the LCM and the GLCM.

2 The Linear Coregionalization Model

Let $\mathbf{Z}$ be the $p \times 1$ random vector under study. In the LCM model, $\mathbf{Z}$ is related linearly to an underlying (therefore unobserved) $m \times 1$ random vector $\mathbf{Y}$ with unit variance and orthogonal components:

$$\mathbf{Z} = \mathbf{A}\mathbf{Y}, \tag{1}$$

where $\mathbf{A}$ is the $p \times m$ matrix of unknown coefficients. In common practice, the $\mathbf{A}$ matrix coefficients is not estimated directly [2], in fact it is not estimated at all. The variograms and cross-variograms (or covariances and cross-covariances) are computed, and a LCM is adjusted. The LCM is expressed as

$$\mathbf{C}(\mathbf{h}) = \mathbf{B}_1\mathbf{C}_1(\mathbf{h}) + \mathbf{B}_2\mathbf{C}_2(\mathbf{h}) + \cdots + \mathbf{B}_k\mathbf{C}_k(\mathbf{h}), \tag{2}$$

where each $\mathbf{B}_i$ matrix of size $p \times p$ is semi-positive definite and component structure $\mathbf{C}_i(\mathbf{h})$ is one admissible model. Note that the LCM implicitly assumes that $m = p \times k$ underlying orthogonal variables Y exist (p variables for each of the k identified structure) and are combined to make up the vector $\mathbf{Z}$. Each $\mathbf{B}_i$ matrix is the ith block of size $p \times p$ found along the diagonal of the matrix $\mathbf{A}\mathbf{A}'$. Despite this large number of underlying variables, the LCM is plagued with the limitations described in the introduction. Therefore, it lacks flexibility to model different types of coregionalization.

3 A Generalized Linear Coregionalization Model

To simplify, we consider the 2D case but the approach applies equally to 3D. Consider the following simple functions of underlying variables Y_i:

$$Y_i^x = \frac{\partial Y_i}{\partial x}, \tag{3}$$

$$Y_i^y = \frac{\partial Y_i}{\partial y}, \tag{4}$$

$$Y_i^v = \frac{1}{v}\int_v Y_i(\mathbf{u})\,d\mathbf{u}, \tag{5}$$

$$Y_i^s = Y_i(\mathbf{x}+\mathbf{s}). \tag{6}$$

The first two functions are the partial derivatives along two orthogonal directions, the third one is the regularization over a support v, the last one is simply the shift operator. With these functions, we can now write the GLCM as

$$\mathbf{Z} = \mathbf{AY} + \mathbf{A}^x\mathbf{Y}^x + \mathbf{A}^y\mathbf{Y}^y + \mathbf{A}^v\mathbf{Y}^v + \mathbf{A}^s\mathbf{Y}^s. \tag{7}$$

All $\mathbf{A}$ matrices are of size $p \times k$, where p is the number of observed variables and k is the number of underlying variables. Note that the underlying variables are assumed spatially orthogonal, i.e. $\mathrm{Cov}(Y_i(\mathbf{x}), Y_j(\mathbf{x}+\mathbf{h})) = 0$ $(\forall i \neq j, \forall \mathbf{h})$. A few underlying variables might share the same spatial covariance model, or each can have its own structure. We assume for simplicity that each underlying variable has a single basic structure. The $\mathbf{A}$, $\mathbf{A}^x$, $\mathbf{A}^y$ and $\mathbf{A}^s$ coefficient matrices can have any real coefficients. In particular, each matrix column j can be filled with zeros if the function of the corresponding underlying variable j is not used in the model. Each matrix can be filled entirely of zeros if the function is not used at all in the model.

For example, with two Z-variables, two Y-variables and only the two partial derivatives of Y_1 used as functions, the Z-covariance model can be expressed more compactly as

$$\mathbf{C_Z}(\mathbf{h}) = \begin{pmatrix}\mathbf{A}_1 & \mathbf{A}_1^x & \mathbf{A}_1^y & \mathbf{A}_2\end{pmatrix} \begin{pmatrix} C_{Y_1}(\mathbf{h}) & \frac{\partial C_{Y_1}(\mathbf{h})}{\partial x} & \frac{\partial C_{Y_1}(\mathbf{h})}{\partial y} & 0 \\ -\frac{\partial C_{Y_1}(\mathbf{h})}{\partial x} & -\frac{\partial^2 C_{Y_1}(\mathbf{h})}{(\partial x)^2} & -\frac{\partial^2 C_{Y_1}(\mathbf{h})}{\partial x \partial y} & 0 \\ -\frac{\partial C_{Y_1}(\mathbf{h})}{\partial y} & -\frac{\partial^2 C_{Y_1}(\mathbf{h})}{\partial x \partial y} & -\frac{\partial^2 C_{Y_1}(\mathbf{h})}{\partial y^2} & 0 \\ 0 & 0 & 0 & C_{Y_2}(\mathbf{h}) \end{pmatrix} \times \begin{pmatrix}\mathbf{A}_1' \\ \mathbf{A}_1^{x\prime} \\ \mathbf{A}_1^{y\prime} \\ \mathbf{A}_2'\end{pmatrix}, \tag{8}$$

where $\mathbf{A}_1$, $\mathbf{A}_1^x$, $\mathbf{A}_1^y$ and $\mathbf{A}_2$ are each 2×1 matrices. Because the intermediate matrix in (8) is non-symmetric, clearly $\mathbf{C_Z}$ is also non-symmetric. It is also anisotropic as the derivatives of a stationary variable are anisotropic by definition. Figure 1 illustrates the corresponding GLCM with two variables and a single common structure. Parameters were chosen, more or less arbitrarily to show asymmetry, as $\mathbf{A}_1 = [1,\ 2]'$, $\mathbf{A}_1^x = [0,\ 30]'$, $\mathbf{A}^s = [0,\ 3]'$, the shift being 25 units along x. Covariance of Y_1 is Gaussian with $a = 20$ and unit sill. Thus we get $Z_1 = Y_1$ and $Z_2 = 2\,Y_1 + 30\,Y_1^x + 3\,Y_1^s$. Which gives:

$$\begin{aligned}\mathrm{Var}(Z_2) = {} & 4\,\mathrm{Var}(Y_1) + 900\,\mathrm{Var}(Y_1^x) + 9\,\mathrm{Var}(Y_1^s) + 120\,\mathrm{Cov}(Y_1, Y_1^x) \\ & + 12\,\mathrm{Cov}(Y_1, Y_1^s) + 90\,\mathrm{Cov}(Y_1^x, Y_1^s) + 90\,\mathrm{Cov}(Y_1^s, Y_1^x)\end{aligned}$$

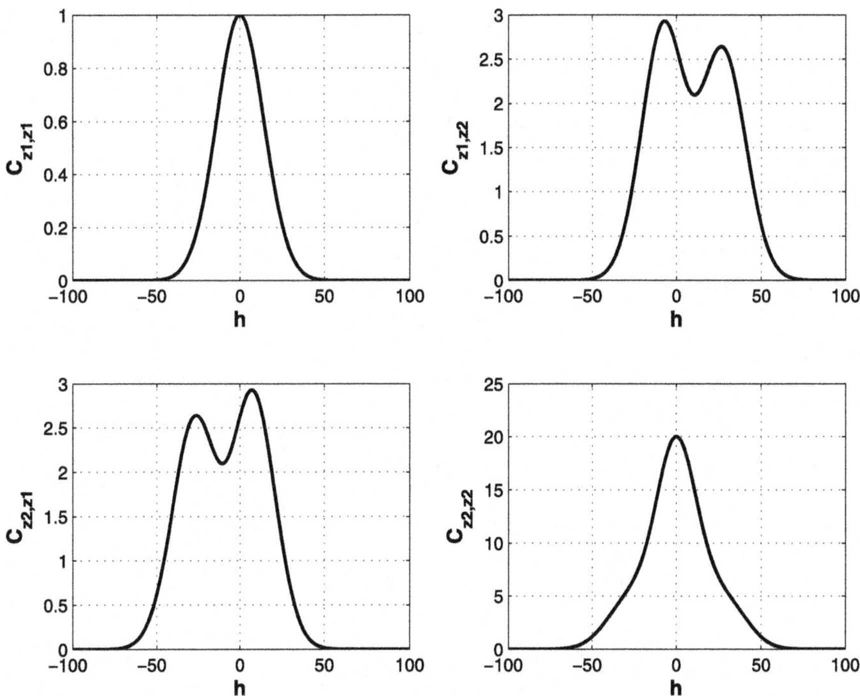

Fig. 1 Example of simple and cross covariances for a model with one structure, its partial derivative along x, and a shift of 25 along x

$$\begin{aligned}&= 4 + 900(1/200) + 9 + 120(0) + 12(0.2096)\\&\quad + 90(0.0262) + 90(-0.0262)\\&= 20.015.\end{aligned}$$

Figure 2 illustrates the model with $\mathbf{A}_1 = [1,\ 2]'$, $\mathbf{A}_1^x = [-2,\ 3]'$, $\mathbf{A}_1^y = [2,\ -1]'$, $\mathbf{A}^s = [0,\ 3]'$, the shift being 25 units along x, and $\mathbf{A}_2 = [2,\ -3]'$. Y_1 covariance is Gaussian with effective range 10 and unit sill, Y_2 covariance is cubic with range 20 and unit sill.

4 Estimating the Model Parameters

The model specification requires to select the number of underlying variables and the functions associated to it. For each underlying variable, the basic structure must be decided. This is done, as usual, by visual inspection of the simple and cross covariances or variograms. At this step, it should be possible to decide, at least tentatively, of the functions that will be selected for each underlying variable. For example, a shift between two variables is evident on the experimental cross-covariances. The presence of a component on the cross-covariances with different signs on each

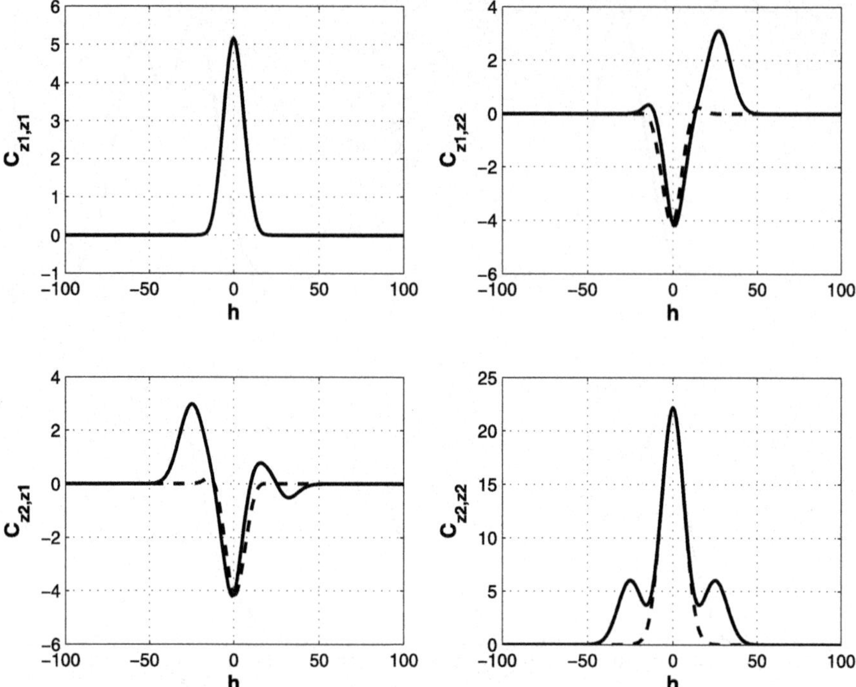

Fig. 2 Example of simple and cross covariances for a model with two structures, partial derivatives along x and y on the first structure, and a shift of 25 along x on the 2nd structure. Covariances along x direction (*solid line*) and y direction (*dashed line*)

side of distance lag 0 suggests the inclusion of the derivative operator. Different observed variables showing similar ranges but linear and parabolic behavior at the origin suggest the inclusion of the regularization function, etc. Then, the shape and scale parameters, and eventually the shift, of each basic structure are estimated efficiently by unconstrained non-linear optimization. Care must be taken to select underlying variables, corresponding functions and basic structures that are compatible. For example, to use the derivative function Y_1^x, the underlying variable Y_1 must have no nugget effect and have a twice derivable covariance at $h = 0$. Another approach, when the covariance of the underlying variable is not differentiable, would be to apply the partial derivative operator on the regularized underlying variable rather than directly on the underlying variable. This can be seen as a new (composite) function of the underlying variable.

The total number of scale parameters in the GLCM has upper bound given by

$$n_{\max} = n_f \times k \times p, \tag{9}$$

where n_f is the number of functions applied (including the identity function), k is the number of underlying Y variables retained, p is the number of observed variables. This upper bound is to compare to the number of parameters adjusted in the

LCM which is $p(p+1)/2 \times n_k$, where n_k is the number of elementary structures (each structure defines implicitly p underlying variables).

To keep the number of parameters as low as possible one idea is to select best underlying variables and functions using some kind of stepwise procedure. The objective function is then the square differences between experimental and model covariances. The underlying variables and functions that decreases most the objective function are selected first. We keep including additional functions and variables up to the point where the addition of any further variable gives only a marginal improvement to the objective function. Another approach would be to use maximum likelihood (if one is willing to assume multi-Gaussian field). Significance of one or many added underlying variable, or of function(s), can then be tested by log-likelihood ratio. Here, neither of these approaches were applied for the simple examples presented in the next sections.

4.1 An Example of Model Fitting

The Cholesky method is used to simulate in 1D a variable Z and its derivative $\mathrm{d}Z/\mathrm{d}x$ at 1500 points. The simulated covariance model is cubic with range $a = 50$. As expected, the experimental variogram shows a range around 50. The scale parameters are adjusted automatically with the range fixed at 50. The adjusted scale coefficients are: $\mathbf{A} = [A_{Z_1}, A_{Z_2}]' = [1.07, 0.002]'$ for the underlying variable, and $\mathbf{A}^x = [A^x_{Z_1}, A^x_{Z_2}]' = [-0.003, 1.077]$ for the derivative. These values are to compare to the known model values of $\mathbf{A} = [A_{Z_1}, A_{Z_2}]' = [1, 0]$ and $\mathbf{A}^x = [A^x_{Z_1}, A^x_{Z_2}]' = [0, 1]$. The means of Z and its derivative were also correctly estimated close to 0 by the fitting program. The resulting fit is shown on the non-centered experimental covariances in Fig. 3.

5 The Gslib (Walker Lake) Data

The Walker Lake data subset included in Gslib [5] is used as a case study. The original data file has two variables computed from a numerical terrain model on a 50×50 regular grid. The first variable (Z_1) is studied. In addition, two new variables are defined: the gradient vector components along x and y, identified as Z_2 and Z_3 respectively. Figure 4 shows the experimental covariances and cross-covariances for these three variables. Clearly, the $Z_1 - Z_2$ and $Z_1 - Z_3$ covariances show a non-symmetrical behavior incompatible with a LCM. Instead, the GLCM should be used.

The chosen GLCM has a single structure identified from inspection of Z_1 experimental covariances as a cubic covariance with range 12. To this structure corresponds the underlying variable Y. The x and y derivatives of Y component, Y^x and Y^y were also included in the GLCM as the experimental $Z_1 - Z_2$ and $Z_1 - Z_3$ cross-covariances show typical asymmetries found between a variable and its derivative.

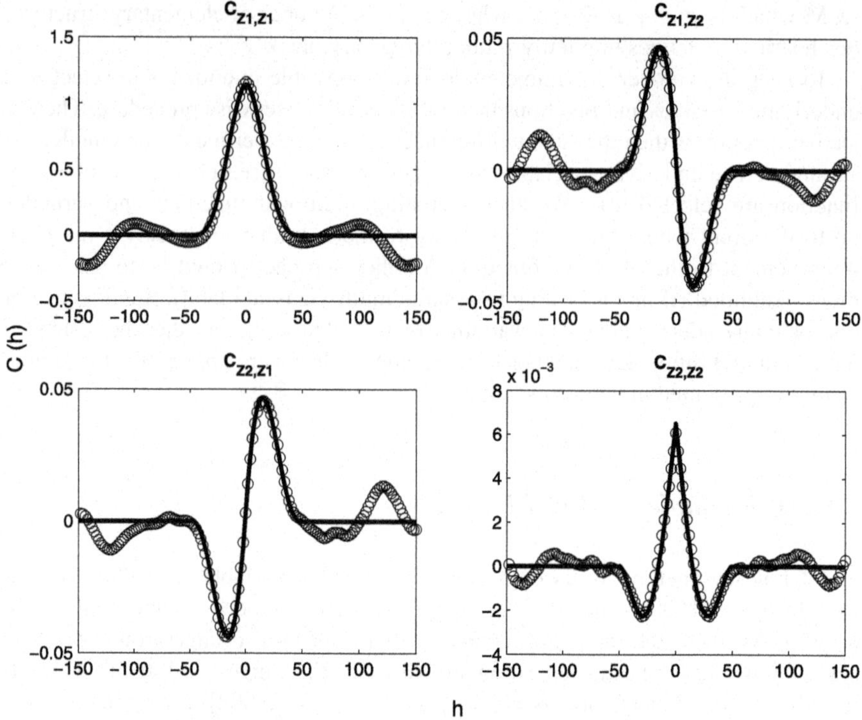

Fig. 3 Example of experimental covariance fitting for a 1D model and two variables Z_1 and Z_2. The GLCM has one underlying variable Y_1 and its derivative Y_1^x. Experimental covariances (*circles*) and adjusted model (*solid line*). See text for parameters

Table 1 Estimated scale parameters of the model

Observed variables	Underlying variables		
	Y	Y^x	Y^y
Z_1	2.58	1.41	−1.82
Z_2	0.00	2.33	−0.30
Z_3	0.00	−0.45	2.29

In all, 7 scale components, out of 9, are estimated to describe the GLCM model. The two remaining scale parameters being simply set to 0. The scale parameters are presented in Table 1. Figure 5 shows the corresponding experimental and model covariances and cross-covariances. With this GLCM, one has

$$\begin{aligned} Z_1 &= 2.58Y + 1.41Y^x - 1.82Y^y, \\ Z_2 &= 2.33Y^x - 0.30Y^y, \\ Z_3 &= -0.45Y^x + 2.29Y^y. \end{aligned} \tag{10}$$

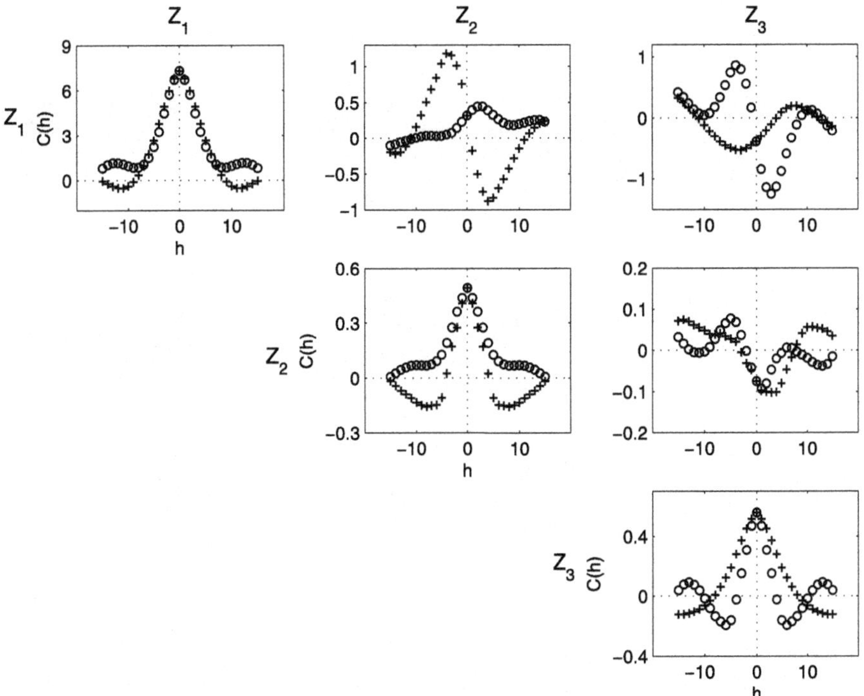

Fig. 4 Experimental covariances and cross-covariances along x (+) and y (o) directions; h measured in pixels

5.1 Kriging Comparison

The model adjusted in the previous section is used for kriging using Z_1 alone and for ordinary cokriging using Z_1, Z_2 and Z_3. In the cokriging case, it is assumed that Z_2 and Z_3 are known at each data location and at the point currently estimated. That is, a multi-collocated neighborhood is used. The sample sizes 50, 100 and 200 are considered for Z_1. The estimated values are compared to the true Z_1 values over the 2500 grid points and the Root Mean Square (RMS) of the errors is computed. Table 2 presents the RMS and the Pearson correlation coefficients with the true surface data for the different scenarios studied. Note that the RMS with the LCM are larger than for kriging with the same number of points. For the LCM, the correlation between Z_1 and Z_2 and Z_3 were only 0.17 and −0.19, respectively. Therefore, the auxiliary variables could not contribute to improve the estimation. In fact, they rather seem to introduce a noise detrimental to the precision of the estimation. On the contrary, the GLCM improves noticeably over the kriging results. Therefore, a better modeling of the simple and cross-covariances of the variables proved to be rewarding. Figure 6 shows the true data surface, the kriging surface and the GLCM cokriging surface for the case with $n = 200$ samples. One can appreciate the better estimates provided by GLCM cokriging.

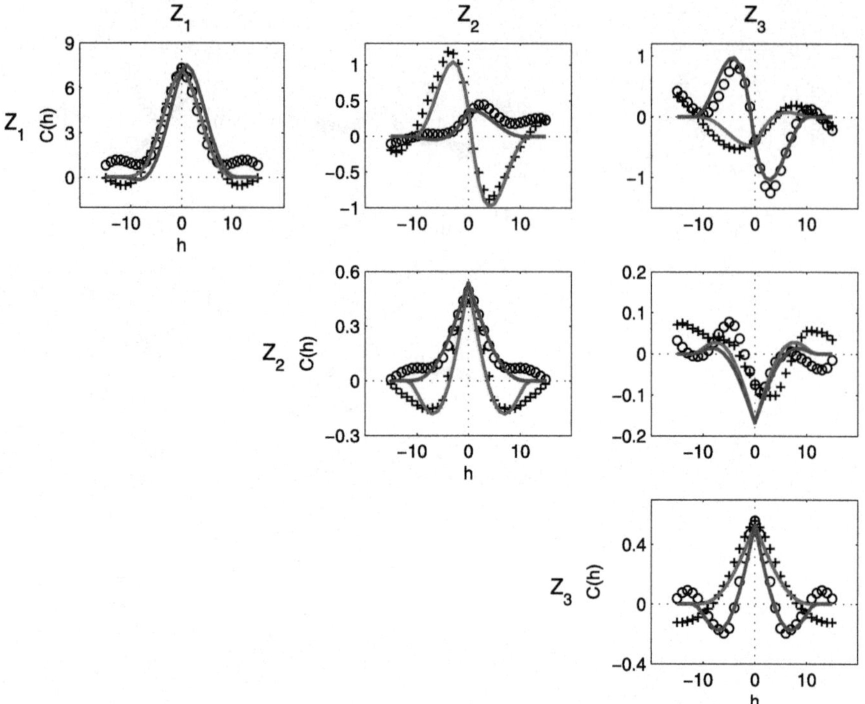

Fig. 5 Experimental and model covariances and cross-covariances along x direction (+, and *red line*) and y direction (o and *blue line*); h measured in pixels

Table 2 Kriging and cokriging RMS for different sample sizes; Pearson correlation coefficient with true surface in parentheses

Sample size	Kriging	Cokriging (LCM)	Cokriging (GLCM)
$n = 50$	2.06 (0.67)	2.09 (0.65)	1.34 (0.87)
$n = 100$	1.37 (0.86)	1.64 (0.80)	0.89 (0.94)
$n = 200$	0.75 (0.96)	1.04 (0.93)	0.60 (0.98)

6 Discussion

The model fit to the experimental covariances and cross-covariances of the Gslib data (Fig. 5), although not perfect, is deemed acceptable. It is obtained automatically thanks to an unconstrained non-linear optimization procedure (Nelder-Mead algorithm programmed in function *fminsearch* of Matlab). However, as often with non-linear optimization, it is advisable to monitor closely the fit obtained. For example, in the objective function, weights were introduced for each different covariances so as to favor a better simultaneous adjustment of the different experimental covariances. This enables to account, in the objective function, that the scales of Z_1, Z_2

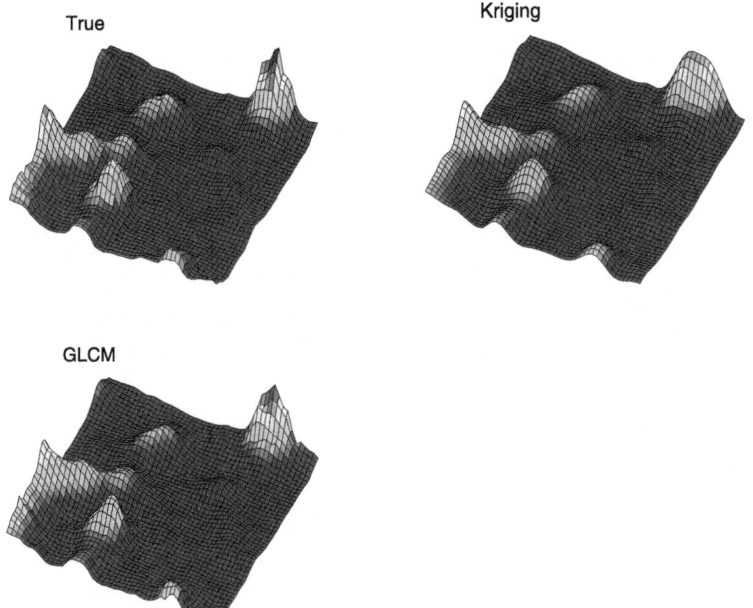

Fig. 6 Result for example. True (*top left*), kriged (*top right*) and GLCM-cokriged (*bottom left*) surface for the Gslib data

and Z_3 are different. Moreover, weights could be also added to give more importance to the small distance lags or to consider the number of pairs in each lag [4]. Here, the shape (and range) of the underlying basic function has been chosen by mere visual inspection of the experimental covariances.

The GLCM is built on the same framework as the LCM. The main differences is that linear combination of underlying variables is done with some variables that are functionally linked to others. This dependence enables to introduce asymmetries in the model. The functions considered here are simple: regularization, shift and first order derivatives. This set can be extended provided the function is stationary and that its theoretical covariance and cross-covariance with the underlying variable, and other functions of the underlying variable, can be computed. Note that the resulting model is necessarily admissible as it is simply a linear combination of dependent and independent stationary random variables, each with its own structure.

In multivariate geostatistics, the use of experimental cross-covariances is not so common. This is a bit surprising as it is well known [3, 12] that the cross-covariances carries more structural information than the cross-variograms which implicitly assumes the cross-covariances are symmetrical. One of the reason for this apparent lack of interest could be the unavailability of a general model to account for non-symmetrical covariances. The GLCM provides such a model.

There is another practical reason to favor the cross-covariances over the cross-variograms: it does not require that the different variables share a sufficient number of common locations, that is, the sampling pattern can be completely heterotopic.

This kind of sampling pattern is very common in practice. For example, one could merge data from different sources like remote sensing images and point sampling, airborne geophysics with geochemical borehole data, numerical terrain model with hydraulic head, and so on. In all these cases, the cross-variograms cannot be computed directly. Interpolating one variable to the locations of the other variable, in order to be able to compute the cross-variograms, necessarily distorts the true spatial link between these variables. On the other hand, when using cross-covariances, one has to estimate the means of the different variables, so the resulting experimental cross-covariances are biased. To avoid this bias, one idea is to use instead non-centered covariances and to estimate simultaneously the means as additional parameters in the automatic fitting procedure.

7 Conclusion

A new multivariate covariance model, the GLCM, is proposed to extend the modeling of the LCM to cases with asymmetric cross-covariances. The model considers that the observed variables are linear combinations of underlying variables and functionals of some of the underlying variables. The model is admissible by construction and its parameters can be estimated by non-linear optimization. Applied on the Gslib data set, it provided better cokriging results than the classical LCM or kriging.

References

1. Ahmed S, de Marsily G (1989) Co-kriged estimates of transmissivities using jointly water level data. In: Armstrong M (ed) Geostatistics, pp 615–628
2. Bourgault G, Marcotte D (1991) The multivariable variogram and its application to the linear coregionalization model. Math Geol 23:899–928
3. Chilès J-P, Delfiner P (1999) Geostatistics: modeling spatial uncertainty, 1st edn. Wiley-Interscience, New York
4. Cressie N (1991) Statistics for spatial data. Wiley-Interscience, New York
5. Deutsch C, Journel AG (1992) GSLIB: geostatistical software library and user's guide. Oxford University Press, London
6. Goulard M, Voltz M (1992) Linear coregionalization: tools for estimation and choice of cross-variogram matrix. Math Geol 24:269–286
7. Journel AG, Huijbregts CJ (1978) Mining geostatistics. Academic Press, San Diego
8. Marcotte D, Powojowski M (2005) Covariance models with spectral additive components. In: Leuangthong O, Deutsch CV (eds) Geostatistics Banff 2004, pp 115–124
9. Rivoirard J (2001) Which models for collocated cokriging. Math Geol 33:117–131
10. Rivoirard J (2004) On some simplifications of cokriging neighbourhood. Math Geol 40:425–443
11. Telford WM, Geldart LP, Sheriff RE (1995) Applied geophysics, 2nd edn. Cambridge University Press, Cambridge
12. Wackernagel H (2003) Multivariate geostatistics: an introduction with applications, 3rd edn. Springer, Berlin

Modeling Nonlinear Beta Probability Fields

K. Daniel Khan and J.A. Vargas-Guzman

Abstract Experiments show how proportions correspond to nonlinear conditional Beta probability distributions. The encountered Beta probability (Beta p-) fields are not classic p-fields. Instead, the shape of the local probability density function (pdf) for the proportion $p(x)$ random variable at each location x changes with the Beta distribution parameters, α and β which are in turn functions of the local proportions themselves. In addition, Beta random variables with different shapes of pdf's have nonlinear pairwise relations; therefore, Beta p-fields are nonlinear. A novel approach was devised to transform these complex Beta random variables to the Gaussian domain which can be modeled with linear geostatistics. A property of the proposed numerical approach is that kriging estimates, when back transformed to the Beta domain using Riemann's integral produce unbiased Beta parameters. This contribution completely redefines the p-field concept with a theoretical framework enabling the simulation of Beta p-fields of proportions. This work represents new findings in geostatistics and motivates further studies of probability or proportion as a random variable.

1 Introduction

One of the most powerful tools of stochastic modeling is the use of nonlinear transformations on simulated Gaussian variables. As an example, stochastic flow models are transformations which are unavoidable because the nonlinear outputs (e.g., a pressure or fluid saturation field) cannot be directly estimated by linear geostatistics (e.g., [2]). In addition, data transformations are routinely applied to data for fitting to the Gaussian model. For the spatial estimation of conditional probability density function (pdf) parameters, one could argue that linear kriging is defined for

K.D. Khan (✉)
5-709 Luscombe Pl., Victoria, BC, Canada, V9A 7L6
e-mail: k.danielkhan@gmail.com

J.A. Vargas-Guzman
P.O. Box 13734, Dhahran, Saudi Arabia
e-mail: anton_varguz@hotmail.com

P. Abrahamsen et al. (eds.), *Geostatistics Oslo 2012*,
Quantitative Geology and Geostatistics 17,
DOI 10.1007/978-94-007-4153-9_7,

second order variables only. Direct Sequential Simulation (DSS) [10, 14] was proposed with the idea that the conditional distributions could be estimated from data directly, without a Gaussian transformation and simulated to produce a realization of a random field. Practicality and the need to reproduce heteroscedastic [11] and nonlinear features require linearizing data transformations which may or may not yield Gaussian variables. Nonlinear relations among variables can be treated with a methodology whereby the attributes can be represented as power transformations of Gaussian variables with parameters which can be estimated with linear geostatistics. The approach, based on Riemann integration was also applied to the lognormal distribution [17].

Any stochastic random variable can also be seen as a transformation of a uniformly distributed probability $U[0, 1]$ to the domain of the random variable. This transformation is equivalent to the classic simulation approach, where the probability scores $p(x)$ are projected to generate the $z(x)$ scores using a conditional cumulative distribution function. A realization of a correlated conditional probability field $p(x) = F(z; x|n)$ locally conditioned on nearby data events, n underlies a realization of the spatially-distributed attribute, $z(x)$, at each location x. The original classic probability field (i.e., p-field) concept proposed a field of correlated probabilities generated with stationary linear models [6, 15]. Subsequent analyses [12, 16] questioned the assumption of linear correlation between random variables representing conditional probabilities. The likelihood distributions of conditional probabilities comprising a correlated field of random variates, $p(x)$ and their covariances have remained unexplored in geostatistical theory.

The objective of this paper is not to revisit the p-field concept per se, but to examine the nature of spatial conditional probability, as observed from measuring proportions in natural phenomena. First, empirical evidence of the shape of the probability density functions (pdf's) for proportion random variables is gathered to define adequate probability models for conditional proportions. The pdf's must then be used to gain insights on the spatial relations among proportion random variables $p(x)$. If the proportions correspond to nonlinear relations; the final goal is to find a linearization approach, and implement it to enable the modeling of the variables $p(x)$.

2 Experimental Design

Are stochastic proportions uniformly distributed? Recall that an attribute Z (e.g., a rock property) at a location x can be modeled as a random variable $z(x)$ by estimating its conditional cumulative probability distribution function (ccdf) $F_c(z(x)|n)$ where n is an influential vector of data. The abscissa of $F_c(z|\cdot)$ is the random variable z and the cumulative probability, p on the interval [0, 1], is plotted on the ordinate of the cdf. Now consider that the conditional probability $p = F_c(z|\cdot)$ of the random variable $z(x)$ is itself a proportion random variable with an unknown conditional pdf. This pdf may be seen from $p = F_g(z|\cdot)$, where $F_g(.)$ is the global uniform [0, 1] pdf for a stationary field $Z(x)$. Proportions are random variables because they cannot be exactly measured from sampling or counting everywhere. The following experiments clarify the concept of conditional random proportions. Note that probabilities are indeed estimated as proportion data obtained by counting events in practice.

2.1 Experiment 1

Consider a standard normally-distributed variable z in which we are interested in the realizations of its conditional proportions $F_c(p)$, instead of its known conditional distribution values $F_c(z \mid \cdot)$. We project the z values corresponding to a set of ccdf's, $F_c(z \mid \cdot)$ onto the axis of uniform [0, 1] global (i.e., unconditional) cumulative probability of the z cdf, $F_g(z)$. For the unconditional or global variate $z(x)$, the operation simply returns a variable that corresponds to the uniform [0, 1] distribution; however, for any conditional distribution, $F_c(z \mid \cdot)$ it yields the conditional random variable of the conditional probability itself, $p_\beta(x) = F_g(z(x) \mid \cdot)$. The result is a set of conditional histograms of the [0, 1] valued probability random variate $p_\beta(x)$, one for each conditional random variable projected (Fig. 1). These density functions have nonuniform shapes ranging systematically from a symmetric function about the median to extreme skewed distributions toward the extremes. This is an important result which seems to have eluded numerous studies on spatial proportion distributions, as far as we are aware. We can see evidence that conditional distributions of proportions or probability random variables are not uniform. This result is not confined to the Gaussian distribution model; the same result is observed when the global and conditional z-cdf's are lognormally-distributed or follow the F-distribution. This result could be general, as it is physically impossible that proportions (i.e., counted indicator events) could be all uniform in frequency, in natural phenomena. The conditional distributions in Fig. 1 can all be fitted to Beta density functions, which have the form

$$f(x; \alpha, \beta) = \frac{\Gamma(\alpha + \beta)}{\Gamma(\alpha)\Gamma(\beta)} p^{\alpha-1}(1 - p)^{\beta-1} \quad \text{for } p \in [0, 1], \tag{1}$$

where $\Gamma(\cdot)$ is the gamma function, and p is the proportion attribute or probability random variable in a Beta probability field. For clarity, we will call the attribute $p_\beta(x)$ a proportion random variable for the rest of the paper.

The Beta distribution, has been presented in the geostatistical literature as a means for modeling the uncertainty in categorical facies proportions [1, 8] and for modeling change of support effects in categorical facies proportions [4].

2.2 Experiment 2

Consider the univariate distributions of conditional proportions on an exhaustive image of a real complex spatial attribute. The well-known Walker Lake data [9] was

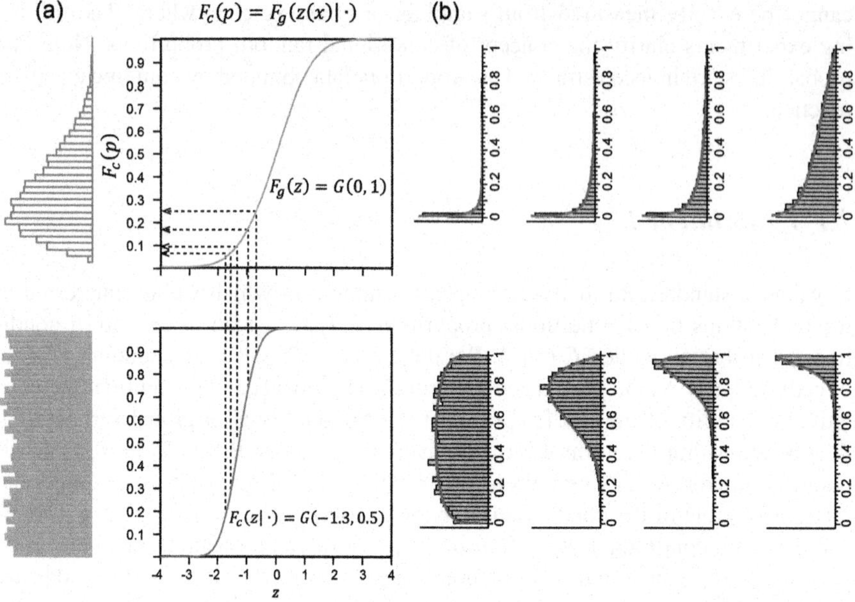

Fig. 1 (**a**) Distributions of conditional probability are generated by projection onto the unconditional uniform [0, 1] axis of the standard normal (i.e., unconditional) Gaussian CDF; (**b**) Distribution of proportions from eight conditional Gaussian distributions with mean values $(-2.5, -1.64, -1.27, -0.67, 0, 0.67, 1.27, 1.64)$ and variance $(0.9, 0.9, 0.9, 0.9, 0.9, 0.5, 0.5, 0.5)$ read from *top left*

selected for its accessibility to readers interested in reproducing this experiment. In order to sample the local distributions of an exhaustive attribute, a viable approach is to gently perturb the data in such a way that simulated realizations are generated with ergodic fluctuations about the original reference image.

The program SASIM [3] was used to perturb the p-field corresponding to the Walker Lake *cumulative probability* field. This is the attribute constructed through the global cumulative probability distribution function, $p(\mathbf{x}) = F_g[z(\mathbf{x})]$ of this exhaustive image, which can be considered a realization of a random field of proportions, where $z(x)$ is the original Walker Lake attribute. The Walker Lake p-field was perturbed subject to the following constraints: reproduction of the global histogram and variogram of the $U[0,1]$ transform of the data; reproduction of indicator variograms modeling the probability thresholds 0.1, 0.2, 0.3, and 0.9, each with a separate model inferred from the reference image; and an empirical correlation coefficient and scatterplot between the reference image and a smoothed version of the same image. A single annealing realization and the E-type map of the ensemble of 500 realizations demonstrate that the results of the exercise fairly resampled the reference image (Fig. 2).

Four pairs of nodes were arbitrarily selected in regions of low, moderate, and high values of the resulting proportions (i.e., p-field) at a lag separation distance of 15 units along the approximate direction of maximum continuity. Each node in

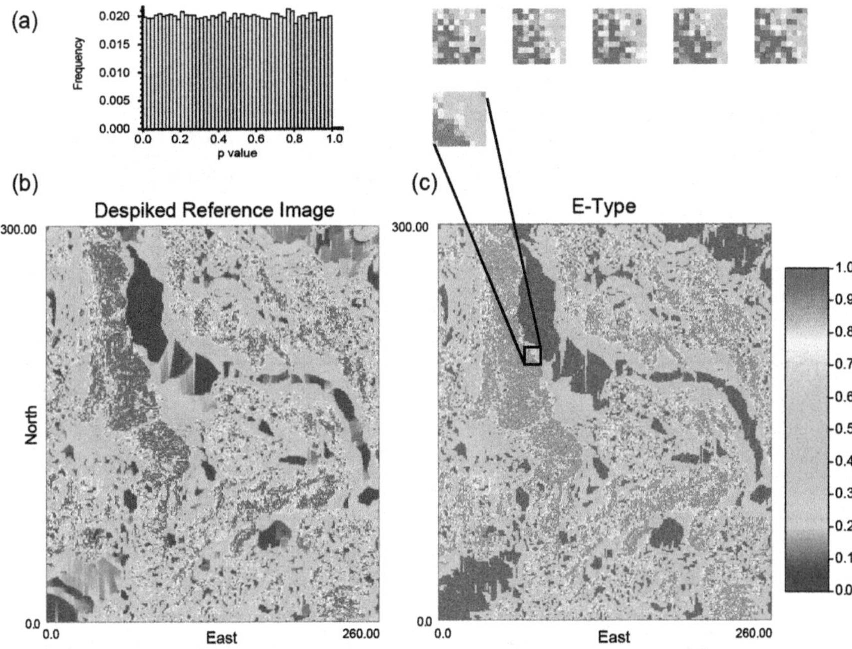

Fig. 2 (**a**) Histogram of (**b**) despiked Walker Lake p-field; (**c**) E-type ensemble average of annealing realizations showing fluctuations for a 10×10 cutout about the average

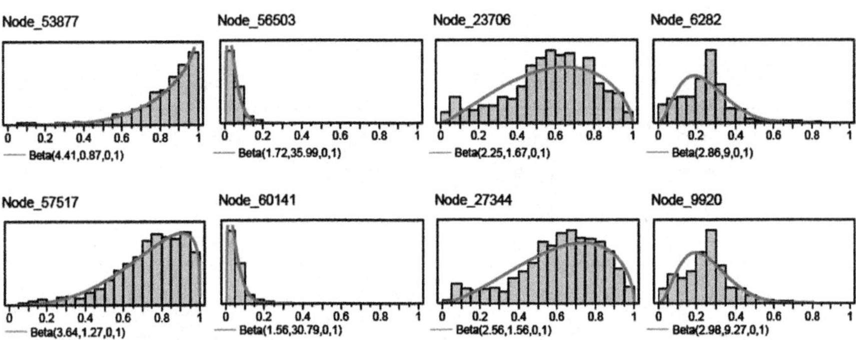

Fig. 3 Empirical distributions derived from annealing sampling experiment on Walker Lake p-field. Node pairs are arranged vertically, for example: Nodes 53877 and 57517 are separated by the specified constant lag of 15 units and taken from a like-valued region in the field. In all cases the distributions are fittable by Beta distribution functions (*solid curves*)

the image has a resultant histogram of conditional proportions given by the realizations of the 500 perturbed probability fields. The sample distributions at the node pairs were fit by modeled Beta distributions whose shape parameters were allowed to float to obtain a best fit (Fig. 3). The parameters of the fitted Beta distributions are close to those of the empirical univariate proportions (Table 1). In cases where

Table 1 Statistics of the conditional p-ccdf's sampled from the annealing exercise shown in Fig. 3

Node 53877			Node 56503		
RV	Empirical	Predicted	RV	Empirical	Predicted
E(·|n)	**0.835**	**0.831**	**E(·|n)**	**0.046**	**0.046**
Var(·|n)	**0.022**	**0.020**	**Var(·|n)**	**0.0011**	**0.0013**
α	4.412	4.900	α	1.716	1.557
β	0.873	1.000	β	35.985	30.792
Node 23706			Node 6282		
RV	Empirical	Predicted	RV	Empirical	Predicted
$E(\cdot\|n)$	**0.574**	**0.561**	$E(\cdot\|n)$	**0.241**	**0.241**
Var$(\cdot\|n)$	**0.0497**	**0.0544**	Var$(\cdot\|n)$	**0.0142**	**0.0183**
α	2.251	1.980	α	2.862	2.160
β	1.670	1.550	β	9.002	6.810
Node 57517			Node 60141		
RV	Empirical	Predicted	RV	Empirical	Predicted
$E(\cdot\|n)$	**0.741**	**0.738**	$E(\cdot\|n)$	**0.048**	**0.049**
Var$(\cdot\|n)$	**0.032**	**0.032**	Var$(\cdot\|n)$	**0.0014**	**0.0016**
α	3.648	3.720	α	1.557	1.370
β	1.273	1.320	β	26.700	26.700
Node 27344			Node 9920		
RV	Empirical	Predicted	RV	Empirical	Predicted
$E(\cdot\|n)$	**0.621**	**0.612**	$E(\cdot\|n)$	**0.243**	**0.243**
Var$(\cdot\|n)$	**0.0459**	**0.0506**	Var$(\cdot\|n)$	**0.0139**	**0.0175**
α	2.562	2.260	α	2.982	2.310
β	1.561	1.430	β	9.273	7.200

the empirical distributions do not look exactly Beta, we attribute this to sampling artifacts of the annealing such that information from adjacent locations is mixed with the local distribution which should be from a single Beta distribution. The key result is that fitted Beta distributions quite closely identify the local conditional mean and variances of $p_\beta(x)$ sampled from the annealing perturbations (Table 1). A basic observation is that other pdf models such as the uniform, Gaussian, lognormal, and skewed power pdfs are not suitable to fit all the variable shapes encountered in a single image. Therefore, the Beta "family" was found as the best choice for modeling these data.

Transforming the local Beta distributions $p_\beta(x)$ through the global cdf $F_g(z(x) \mid \cdot)$ of the Walker Lake "U" variable yields a realization of the attribute

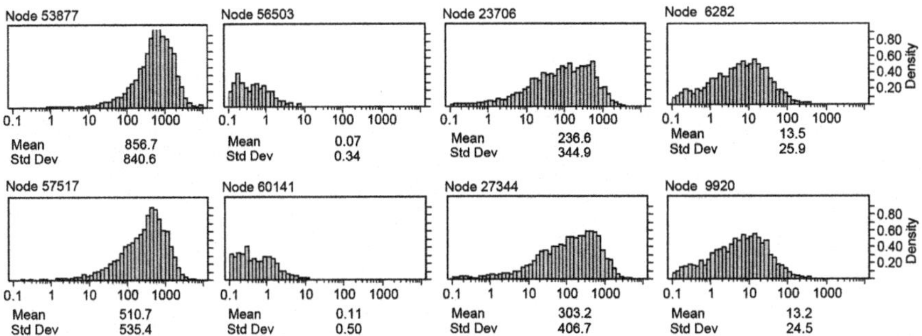

Fig. 4 Direct Z-conditional distributions of the Walker Lake "U"-variable obtained by quantile transformation $z(p) = F_Z^{-1}(p_\beta)$ of Beta-models (*solid curves* of Fig. 3)

directly. This is an important result because we have in fact characterized the local conditional $z(x \mid \cdot)$ random variables' ccdf's through modeling the Beta distributions of conditional proportions (Fig. 4).

3 Nonlinearity and Nonstationarity of Beta p-Fields

Correlated Beta processes or random fields have not been successfully constructed yet [7]. The univariate Beta distribution of conditional proportions $p_\beta(x|n)$ at a location x can take any number of shapes, and will only be proportional to the distribution at any neighboring location if the mean and variance, and therefore their Beta shapes, are very similar. The heterogeneous shapes of pdf's comprising a conditional Beta p-field should most often result in bivariate distributions with nonlinear relations between the random variables. The greater the dissimilarity in the shapes of the pdf's of the local random variables, the less likely it is to get any significant linear correlation between them. However, various nonlinear relations may hold simultaneously with separate neighboring locations.

It is interesting to visualize the functional relations between pairs of Beta distributed proportions (Fig. 5). Such pairwise functions illustrate the possible nonlinearity between the Beta variables for two spatial locations $(x, x+h)$. An interpretation of these functions is the bivariate copulas between two correlated conditional Gaussian variables (Experiment 1). Recall that a bivariate Gaussian copula is the joint cumulative distribution function of the probability in $U[0,1]$. That is, $c(U_1, U_2) = (F_1(z_1), F_2(z_2))$ of the joint cdf of a random vector (z_1, z_2). Replacing the uniform by the Beta conditional distributions for any two locations is the bivariate copula.

Experiment 1 suggested that a second order stationary Gaussian field becomes a Beta field with nonstationary nonlinear relations among variables in the probability or proportion domain, and vice-versa. Figure 5 implies that modeling stationary variograms or covariances for Beta proportions is inadequate. Therefore we need a way to linearize the Beta p-field to a Gaussian domain to enable the use a stationary

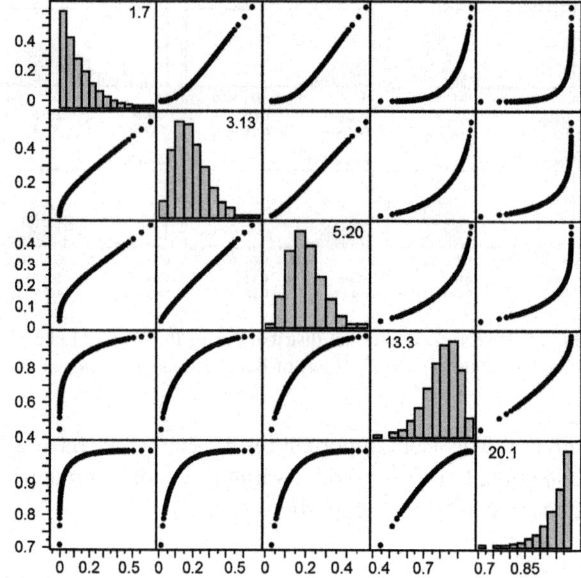

Fig. 5 Matrix of deterministic correlation functions between selected Beta distributions of conditional probability $p_{\beta}(x)$. The distribution shape parameters α, β are shown alongside the density function on the diagonal of the matrix. These functions may be interpreted as bivariate copulas

covariance model and kriging. Linearization into the Gaussian space would eliminate nonlinear correlations, but some cases may remain nonstationary. The practical methods to treat such nonstationary Gaussian fields with universal kriging and/or generalized covariances are well known in Geostatistics, and they are not repeated here.

4 Geostatistical Modeling of Beta Probability Fields

Utilizing the preceding concepts in geostatistical modeling requires estimating the conditional mean proportion, $\hat{p}$ and its associated estimation variance $\mathrm{Var}(\hat{p})$ from surrounding samples to obtain the $p_{\beta}(x)$ random variables.

The Beta distribution on [0, 1] describes the uncertainty about the proportion, and is completely defined by a conditional mean and variance. The well-known relations are:

$$\alpha = \hat{p}\left[\frac{\hat{p}(1-\hat{p})}{\mathrm{Var}(\hat{p})} - 1\right] \quad \text{and} \quad \beta = (1-\hat{p})\left[\frac{\hat{p}(1-\hat{p})}{\mathrm{Var}(\hat{p})} - 1\right], \tag{2}$$

where $\hat{p} = E[p(x)]$ and $\mathrm{Var}(\hat{p}) = E[p(x)^2] - (E[p(x)])^2$. Because the Beta-distributed proportions $p_{\beta}(x)$ are nonlinear and nonstationary in their correlations a direct kriging of proportions is not applied.

4.1 A Beta to Gaussian Transformation Function

An apparent possibility for the transform function is the logistic transformation of the proportions [13]. The problem with the logistic transformation is the resultant transformed variable is far from being Gaussian or symmetric for asymmetric Beta distributed variables.

An appropriate transformation for our purposes is one that will transform any Beta-distributed variable into something that is very close to a Gaussian-distribution. A simple and novel transformation was devised. The logic leading to the proposed transform is based on basic principles of indicators. The development of the transform will be presented separately due to its length and the considerations of the scope of this paper. In summary it was found that the logarithm of the variance of indicators, $\ln(\frac{p-p^2}{0.25})$ is a squared Gaussian because $p(1-p)$ is indeed a second order variable. The linearization transformation $\varphi(p)$ is

$$y = \varphi(p) = \begin{cases} -\sqrt{|\ln(\frac{p-p^2}{0.25})|} & \text{when } p < 0.5, \\ \sqrt{|\ln(\frac{p-p^2}{0.25})|} & \text{when } p \geq 0.5. \end{cases} \tag{3}$$

4.1.1 The Gaussian to Beta Back Transformation Function

From (3) a back-transformation is constructed after solving the quadratic relation between the proportion and the Gaussian random variable. This is:

$$p = \varphi^{-1}(y) = \frac{1}{2} \pm \frac{1}{2}\sqrt{1 - \exp(-y^2)}. \tag{4}$$

Note that p is the target Beta-distributed proportion random variable, and $q = (1-p)$ is the proportion random variable for the complementary event. Therefore, (4) obeys the closure composition condition $p + q = 1$.

4.2 Estimating Moments of Proportions

It is well known that nonlinear transformation of random variables cannot be applied to expected values because a direct back transformation of a Gaussian expected value would result in a strongly biased expected value for the nonlinear random variable—the Beta proportion in this case. So the expected values for moments must be properly evaluated using Riemann's integral for power random variables. The general integral expression is in [17].

Substituting (4) into the integral yields the correct back transformation of the Gaussian kriging estimates $\hat{\mu}_y$ and $\hat{\sigma}_y^2$ into any moment of order $\{1, 2, 3, \ldots, m\}$.

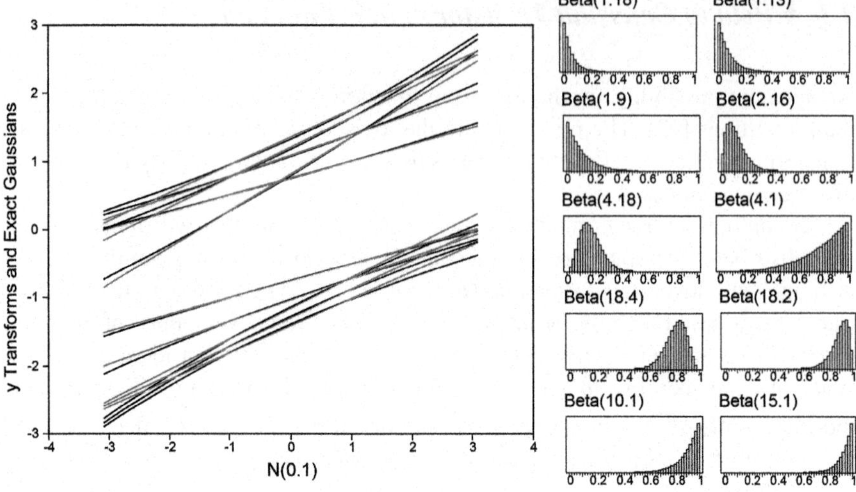

Fig. 6 Proposed pseudo-Gaussian Y transformations (*black curves*) compared to perfect Gaussian-distributed Z variables with Y mean and variance, $G_z(\hat{\mu}_y, \hat{\sigma}_y^2)$ (*red lines*) all plotted against the standard normal distribution for an array of asymmetric Beta distributions (*bottom panel*)

Therefore, $(\varphi^{-1}(y))^m$ is a power random variable of a Gaussian. The resulting expectation is:

$$E[p^m] \cong \int_{-\infty}^{\infty} \left(\frac{1}{2} \pm \frac{1}{2}\sqrt{1-\exp(-y^2)}\right)^m \frac{1}{\sigma\sqrt{2\pi}} \exp\left(-\frac{(y-\hat{\mu}_y)^2}{2\hat{\sigma}_y^2}\right) \mathrm{d}y. \quad (5)$$

The corrected mean, $\hat{\mu}_{p_\beta}(x) = E[p]$ and variance, $\hat{\sigma}^2_{p_\beta}(x) = E[p^2] - E[p]^2$ are used to compute the Beta parameters as indicated in (2). The application of (5) to kriging estimates is straightforward, following the general recommendations and workflow for power random variables in Vargas-Guzman [17]. However, instead of searching for an analytical integration, (5) is numerically integrated here, and that allows adjusting the input pseudo-Gaussian into a perfect Gaussian random variable.

The transformation of (3) is exact for any symmetric Beta distribution ($\alpha = \beta$), for integer Beta shape parameters. For asymmetric Beta distributions ($\alpha \neq \beta$), the transformation is approximate. The approximation is evident from the curvature when plotted against the standard normal $N(0, 1)$ reference (Fig. 6). Because kriging returns the Gaussian conditional mean and variance for a truly Gaussian variable z, we would need to introduce the slight shift of the Gaussian variable to the pseudo-Gaussian before applying the back transform, as $p \sim \varphi^{-1}(z + \varepsilon)$, for exact results.

Differences between the modeled Beta random variables and the array of asymmetric reference Beta RV's (Fig. 6) are small. The shape parameters between the reference and modeled Beta RV's are tabulated for comparison (Table 2). The results indicate the proposed transform and numerical back transformation are quite sufficient for practical purposes.

Table 2 Beta shape parameters of reference (see Fig. 6) and modeled Beta distributions

α, β										
Reference	1, 18	1, 13	1, 9	2, 16	4, 18	4, 1	18, 4	18, 2	10, 1	15, 1
Model	0.8, 14.4	0.8, 10.8	0.9, 7.8	1.9, 15.1	4.0, 17.9	3.8, 0.9	17.9, 4.0	16.8, 1.9	8.6, 0.9	12.3, 0.8

4.3 Geostatistical Conditional Simulation of Beta Probability Fields

Simulation of conditional Beta probability fields may now proceed via the conventional sequential Gaussian simulation approach, with the embedded Riemann back transformation of parameters, and forward Gaussian transformation of simulated scores as the following additional embedded steps:

- Back transform the estimates $\hat{\mu}_y(x)$ and $\hat{\sigma}_y^2(x)$ with the Riemann integration (5), and use these back transformed parameters to obtain the Beta ccdf parameters α and β (2).
- Draw a random number, u_p [0, 1] from the uniform pdf for the realization at the location x, and obtain the inverse of the local Beta distribution as a simulated realization, $p(x)$ of the random variable $p_\beta(x)$.
- Transform the new simulated value $p(x)$ to $y(x)$ in the Gaussian domain using (3), and iterate on the estimation of parameters for the next location.
- Repeat until all nodes have been visited.
- An additional step for continuous (nonproportion) attributes is the straightforward quantile mapping of the Beta simulated proportions to the variable through the global CDF.

Notes: The inverse of the local Beta ccdf's are obtained by implementing the algorithm by [5]. The last step amounts to generating a realization of the Z-attribute itself as $z^*(\mathbf{x}) = F_Z^{-1}(p_\beta(\mathbf{x}))$. After transforming the proportion data into the pseudo-Gaussian y variable, nonlinear relations have vanished, but nonstationarity may still be present and need proper handling with known tools.

5 Discussion and Conclusions

The hypothesis that proportions may be Beta distributed was tested with two experiments. The first experiment was based on conditional simulated Gaussian fields, and the second experiment recurred to real phenomena as described by the Walker Lake geostatistical data set, and so was effectively free from any *a priori* model assumptions or simplifications. Both experiments showed that conditional proportions are not uniform, Gaussian, lognormal or any other common pdf. They corresponded to the Beta distribution. These results may be dependent on scale of an unknown indicator averaging incorporated in the data proportion. This means that the Beta distribution may be valid only within a range of scales, and if the samples represent very large volumes in relation to the individual indicator events, the proportion may tend to be Gaussian, due to the Central Limit Theorem.

This paper has endeavored to provide a workflow for the unbiased estimation of correlated Beta pdf conditional parameters and simulation of a Beta p-field, or equivalently, of the associated continuous attribute. The effort led to the use of Riemann expectation, which required a transform from the Beta domain to the Gaussian domain and vice-versa. Such a transform was devised by a methodology which is reported separately due to its length and the scope of this paper. The validity of the transform has been thoroughly tested and presented with examples. We expect this contribution will enable not only geostatistical simulation based on Beta p-fields of proportions but will inspire new studies of probability or proportion as a random variable.

References

1. Biver P, Haas A, Bacquet C (2002) Uncertainties in facies proportion estimation II: application to geostatistical simulation of facies and assessment of volumetric uncertainties. Math Geol 34(6):703–714
2. Carrera J, Alcolea A, Medina A, Hidalgo J, Slooten LJ (2005) Inverse problem in hydrogeology. Hydrogeol J 13(1):206–222
3. Deutsch CV, Journel AG (1998) GSLIB, geostatistical software library and users guide, 2nd edn. Oxford University Press, New York, Oxford, 369 pp
4. Deutsch CV, Lan Z (2008) The Beta distribution for categorical variables at different support. In: Graeme Bonham-Carter QC (ed) Progress in geomathematics. Springer, Berlin, pp 445–456
5. DiDinato AR, Morris A (1993) Algorithm 708: significant digit computation of the incomplete Beta function ratios. ACM Trans Math Softw 18:360–373
6. Froidevaux R (1993) Probability field simulation. In: Soares A (ed) Geostatistics Troia. Kluwer Academic, Norwell, pp 73–84
7. Goetschalckx R, Pascal P, Hoey J (2011) Continuous correlated Beta processes. In: Walsh T (ed.) International joint conference on artificial intelligence (IJCAI). Barcelona, Spain
8. Haas A, Formery P (2002) Uncertainties in facies proportion estimation I. Theoretical framework: the Dirichlet distribution. Math Geol 34(6):679–702
9. Isaacs EH, Srivastava RM (1989) An introduction to applied geostatistics. Oxford University Press, New York, 561 pp
10. Journel AG (1994) Modeling uncertainty: some conceptual thoughts. In: Dimitrakopoulos R (ed) Geostatistics for the next century. Kluwer Academic, Dordrecht
11. Leuangthong OL (2004) The promises and pitfalls of direct simulation. In: Leuangthong OL, Deutsch CV (eds) Geostatistics Banff 2004. Springer, Banff
12. Pyrcz M, Deutsch CV (2001) Two artifacts of probability field simulation. Math Geol 33(7):775–799
13. Rasmussen CE, Williams CKI (2006) Gaussian processes for machine learning. MIT Press, Cambridge
14. Soares A (2001) Direct sequential simulation and cosimulation. Math Geol 33(8):911–926
15. Srivastava RM (1992) Reservoir characterization with probability field simulation. SPE Form Eval 7(4):927–937
16. Srivastava RM, Froidevaux R (2004) Probability field simulation: a retrospective. In: Leuangthong OL, Deutsch CV (eds) Geostatistics Banff 2004. Springer, Banff, pp 55–64
17. Vargas-Guzman JA (2004) Geostatistics for power models of Gaussian fields. Math Geol 36:307–322

Approximations of High-Order Spatial Statistics Through Decomposition

Ryan Goodfellow, Hussein Mustapha, and Roussos Dimitrakopoulos

Abstract Many of the existing multiple-point and high-order geostatistical algorithms depend on training images as a source of patterns or statistics. Generating these training images, particularly for continuous variables, can be a labor-intensive process without any guarantee that the true high-order statistics of the deposit are accurately represented. This work proposes a decomposition of a high-order statistic (moment) into a set of weighted sums that can be used for approximating spatial statistics on sparse data sets, which could lead to new data-driven simulation algorithms that forgo the use of training images. Using this decomposition, it is possible to approximate the n-point moment by searching for pairs of points and combining the pairs in the various directions at a later step, rather than searching for replicates of the n-point template, which is often unreliable for sparse data sets. Experimental results on sparse data sets indicate that the approximations perform much better than using the actual n-point moments on the same data. Additionally, the quality of the approximation does not appear to degrade significantly for higher-orders because it is able to use more information from pairs of data, which is generally not true for the moments from the sample data, which requires all n points in a template.

1 Introduction

Over the past several decades, it has been assumed that the first two orders of statistics (histogram and variogram, respectively) were sufficient to characterize most of the problems encountered when estimating or simulating spatial random fields [2, 3, 7, 9]. Over the past two decades, it has become apparent that these traditional

R. Goodfellow (✉) · R. Dimitrakopoulos
McGill University, 3450 University St., Montreal, QC, Canada, H3A 2A7
e-mail: ryan.goodfellow@mail.mcgill.ca

R. Dimitrakopoulos
e-mail: roussos.dimitrakopoulos@mcgill.ca

H. Mustapha
Schlumberger Abingdon Technology Center, Abingdon, UK
e-mail: hmustapha@slb.com

P. Abrahamsen et al. (eds.), *Geostatistics Oslo 2012*,
Quantitative Geology and Geostatistics 17,
DOI 10.1007/978-94-007-4153-9_8,

two-point methods are unable to characterize curvilinear features and complex geometries that are common in geological environments and are unable to reproduce the spatial connectivity of extreme values that are of key interest in most applications.

Dimitrakopoulos et al. [4], Mustapha and Dimitrakopoulos [12] and Mustapha et al. [15] introduce the concept of spatial cumulants to model complex non-Gaussian and non-linear geological phenomena, where spatial cumulants are simply combinations of lower- or equal-order spatial moments. This method is different from existing multiple-point geostatistics algorithms [1, 8, 16, 17] because it attempts to quantify spatial interactions using maps of high-order statistics. Mustapha and Dimitrakopoulos [13, 14] propose a high-order sequential simulation algorithm for continuous variables, HOSIM, which has been shown to accurately reproduce many orders of spatial statistics on sparse data sets. The method uses Legendre polynomials [10] and high-order spatial statistics to reconstruct the joint probability density function (*jpdf*) for z_0 using neighboring nodes $z_1, \ldots, z_N$, as follows:

$$\begin{aligned} &f(z_0, z_1, \ldots, z_N) \\ &\quad \approx \sum_{i_0=1}^{\omega} \cdots \sum_{i_{N-1}=0}^{i_{N-2}} \sum_{i_N=0}^{i_{N-1}} L_{\bar{i}_0,\ldots,\bar{i}_{N-1},i_N} P_{\bar{i}_0}(z_0) \ldots P_{\bar{i}_{N-1}}(z_{N-1}) P_{\bar{i}_N}(z_N), \end{aligned} \tag{1}$$

where $f(z_0, z_1, \ldots, z_N)$ is a multivariate *jpdf*; ω is a specified maximum order of moment or cumulant, $L_{\bar{i}_0,\ldots,\bar{i}_{N-1},i_N}$ is the Legendre coefficient for the template derived from the moments or cumulants in a training image and data set, $\bar{i}_k = i_k - i_{k+1}$ for $k < N$; and finally $P_i(z_i)$ is a Legendre polynomial function. The conditional distribution used to simulate z_0, $f(z_0|z_1, \ldots, z_N)$, can be obtained by applying Bayes' Theorem, i.e. by dividing (1) by its marginal distribution, $\int f(z_0, z_1, \ldots, z_N)\,\mathrm{d}z_0$. Given that it is difficult to obtain experimental spatial cumulants (or moments) for higher orders from sparse data sets to define the Legendre coefficients, $L_{\bar{i}_0,\ldots,\bar{i}_{N-1},i_N}$, the method relies on a training image to supplement the statistics that cannot be inferred from the data alone. For a more detailed discussion on the HOSIM algorithm, the reader is referred to [13] and [14].

While training images have helped to use more information than a data set has to offer, some practical implementation issues include: (i) generating of the training image on a point scale; (ii) the geologists' experience, interpretation; and (iii) the ability to incorporate various levels of spatial statistics into the image [5]. There is interest in being able to approximate or model high-order spatial statistics using only the available data, thus avoiding many of the problems associated with training images. Approximations for high-order spatial statistics could then be integrated into existing and new simulation algorithms that make full use of available data and would lead to higher-quality and representative models of spatial uncertainty.

In the following sections, the fundamental concepts of spatial moments are reviewed. The method for decomposing the high-order spatial moments into a set of weighted sums is presented, followed by a discussion on how this can be used to separate a high-order spatial moment centered at a node into pairs, where the points at the unknown locations are approximated. The approximations are then tested on various amounts of data, and the performance of the approximations is measured. Finally, conclusions and recommendations for future work are discussed.

2 Decomposition of High-Order Spatial Statistics

2.1 Spatial Moments

A spatial random field $Z(\mathbf{x}), \mathbf{x} \in \Re^n$ is a group of random variables $\{Z(\mathbf{x}_i),\ i = 1, \ldots, n\}$ at locations $\mathbf{x}_1, \mathbf{x}_2, \ldots, \mathbf{x}_i$ where each random variable is defined on a probability space $(\omega, \Im, P)$ and takes values in a measurable space $(\Re, \beta(\Re))$. Assuming $Z(\mathbf{x})$ is a stationary and ergodic random field indexed in $\Re^n$, the rth-order spatial moment, $m_r^z(\mathbf{h}_1, \mathbf{h}_2, \ldots, \mathbf{h}_{r-1})$, is defined as

$$m_r^z(\mathbf{h}_1, \mathbf{h}_2, \ldots, \mathbf{h}_{r-1}) = \mathrm{Mom}\big[Z(\mathbf{x}), Z(\mathbf{x}+\mathbf{h}_1), Z(\mathbf{x}+\mathbf{h}_2), \ldots, Z(\mathbf{x}+\mathbf{h}_{r-1})\big]$$
$$= E\left\{Z(\mathbf{x}) \prod_{i=1}^{r-1} Z(\mathbf{x}+\mathbf{h}_i)\right\}, \tag{2}$$

where the terms $\mathbf{h}_i = h_i \mathbf{e}_i$ represent vectors oriented from the point $\mathbf{x}$, where h_i represents the distance along the unit vector $\mathbf{e}_i$.

2.2 Proposed Decomposition

The following decomposition is generalized for any r^{th}-order spatial moments; for more specific details on the decomposition, the reader is referred to [6]. Consider an rth-order spatial moment, centered at a point $Z(\mathbf{x})$, and is defined by an r-point template. The underlying principle of the decomposition is to represent the product of the random variables as a sum of weighted components, as

$$E\left\{\prod_{i=0}^{r-1} Z(\mathbf{x}+\mathbf{h}_i)\right\} = \underbrace{\sum_{i=1}^{r-1} E\left\{\alpha_{2i}^{h_1,\ldots,h_{r-1}} Z(\mathbf{x}+\mathbf{h}_i)\right\}}_{\text{Weighted external nodes}}$$
$$+ \underbrace{\sum_{i=1}^{r-1} E\left\{\alpha_{2i-1}^{h_1,\ldots,h_{r-1}} Z(\mathbf{x})\right\}}_{\text{Weighted central node}}, \tag{3}$$

where $\alpha_{2i}^{h_1,\ldots,h_{r-1}} = 1, \forall i = \{1, \ldots, r-1\}$ are the weights of the external $(r-1)$ nodes separated by vectors $\mathbf{h}_i = h_i \mathbf{e}_i, \forall \{i = 1, \ldots, r-1\}$. The weights of the central node, $\alpha_{2i-1}^{h_1,\ldots,h_{r-1}}$, have been decomposed into $(r-1)$ components, and are unknown and are assumed equal for $i = \{1, \ldots, r-1\}$. The weights for the central node

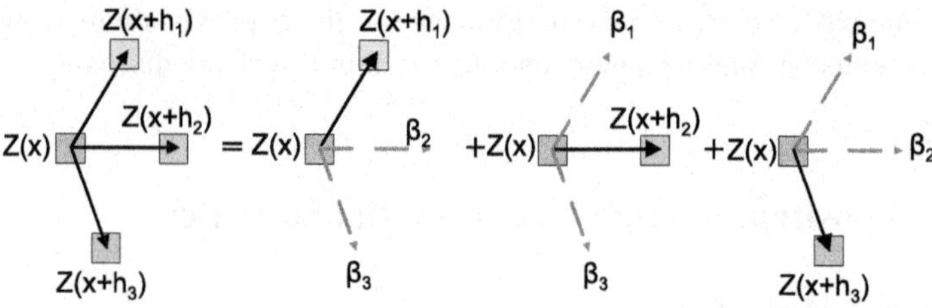

Fig. 1 Decomposition of a fourth-order template into sets of pairs. Approximations are performed by approximating the remaining unknown points in the four-point template

$\alpha_{2i-1}^{h_1,\ldots,h_{r-1}}$, $i = \{1, \ldots, r-1\}$ can be calculated for a set of given data points by rearranging (3) as follows

$$\alpha_{2i-1}^{h_1,\ldots,h_{r-1}} = \frac{Z(\mathbf{x})Z(\mathbf{x}+\mathbf{h}_i)\prod_{j=1,j\neq i}^{r-1} Z(\mathbf{x}+\mathbf{h}_j) - Z(\mathbf{x}+\mathbf{h}_i) - \sum_{j=1,j\neq i}^{r-1} Z(\mathbf{x}+\mathbf{h}_j)}{(r-1)Z(\mathbf{x})}. \tag{4}$$

If one considers only a pair of points in the high-order template and the other points are unknown, it is possible to state that $\prod_{j=1,j\neq i}^{r-1} Z(\mathbf{x}+\mathbf{h}_j) = \prod_{j=1,\,j\neq i}^{r-1} \beta_j$ and $\sum_{j=1,\,j\neq i}^{r-1} Z(\mathbf{x}+\mathbf{h}_j) = \sum_{j=1,\,j\neq i}^{r-1} \beta_j$, where the values at the unknown locations, β_j, are to be approximated. By separating the r points that define the rth-order spatial moment into pairs and substituting (4) into (3), it is possible to approximate the rth-order spatial moment as follows:

$$\begin{aligned} E\left\{\prod_{i=0}^{r-1} Z(\mathbf{x}+\mathbf{h}_i)\right\} &\approx \sum_{i=1}^{r-1} E\{\alpha_{2i}^{h_1,\ldots,h_r-1} Z(\mathbf{x}^{(i)}+\mathbf{h}_i)\} \\ &+ \frac{1}{(r-1)}\sum_{i=1}^{r-1} E\left\{Z(\mathbf{x}^{(i)})Z(\mathbf{x}^{(i)}+\mathbf{h}_i)\prod_{j=1,j\neq i}^{r-1}\beta_j\right\} \\ &- \frac{1}{(r-1)}\sum_{i=1}^{r-1} E\left\{Z(\mathbf{x}^{(i)}) + \prod_{j=1,j\neq i}^{r-1}\beta_j\right\}. \end{aligned} \tag{5}$$

2.3 Approximations of Decomposed High-Order Spatial Statistics

By applying the decomposition derived in (5), approximating the rth-order spatial moment $E\{\prod_{i=0}^{r-1} Z(\mathbf{x}+\mathbf{h}_i)\}$ is a matter of finding pairs of points $\{z(\mathbf{x}), z(\mathbf{x}+\mathbf{h}_i)\}$ from a sample data set for any vector $\mathbf{h}_i$, performing the approximation for the remaining unknown points $\beta_j = Z(\mathbf{x}+\mathbf{h}_j)$, $j = 1, \ldots, r-1$, $j \neq i$, and combining the results for all vectors $\mathbf{h}_j$ at a later step. For example, Fig. 1 shows how a fourth-order spatial moment can be decomposed into pairs, where each of the unknown

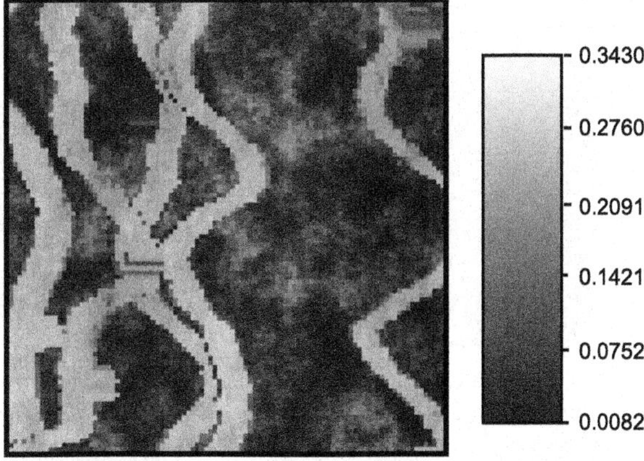

Fig. 2 2-D section of the Stanford V porosity data set (100×100 points)

points for each of the components of the decomposition can be approximated independently.

There are two critical components that influence the quality of the approximation: the range of the data values and the method used to generate the unknown points. For the approximations outlined in this paper, the data has been rescaled to lie between a small interval, [0.9, 1], and the unknown points in a template are obtained by randomly sampling from the univariate distribution of the sample data. A detailed analysis and discussion of the data scaling and distribution sampling methods can be found in [6].

3 Experimental Approximations Through Decomposition on Sparse Data Sets

The following experiments are performed on a 2-D section of the Stanford V [11] porosity data set (Fig. 2), which contains 10 000 data points (100×100 regular grid, where each pixel in the image will herein be assumed to separated by 1 m in the x- and y-directions). To generate the sparse data sets, a specified percentage of the original 2-D section of the Stanford V porosity data set is randomly retained as sample data. The unit vectors defining the template follows the coordinate axis. In order to generate the approximations the algorithm searches along the unit vectors at length h_1 and h_2 to find pairs of sample data points, and performs the approximation for the unknown point.

3.1 Third-Order Spatial Moment Approximations

Figure 3 shows the approximated and true L-shape (vectors pointing to 90° and 0° azimuth) spatial moment maps that are calculated experimentally from 100 %

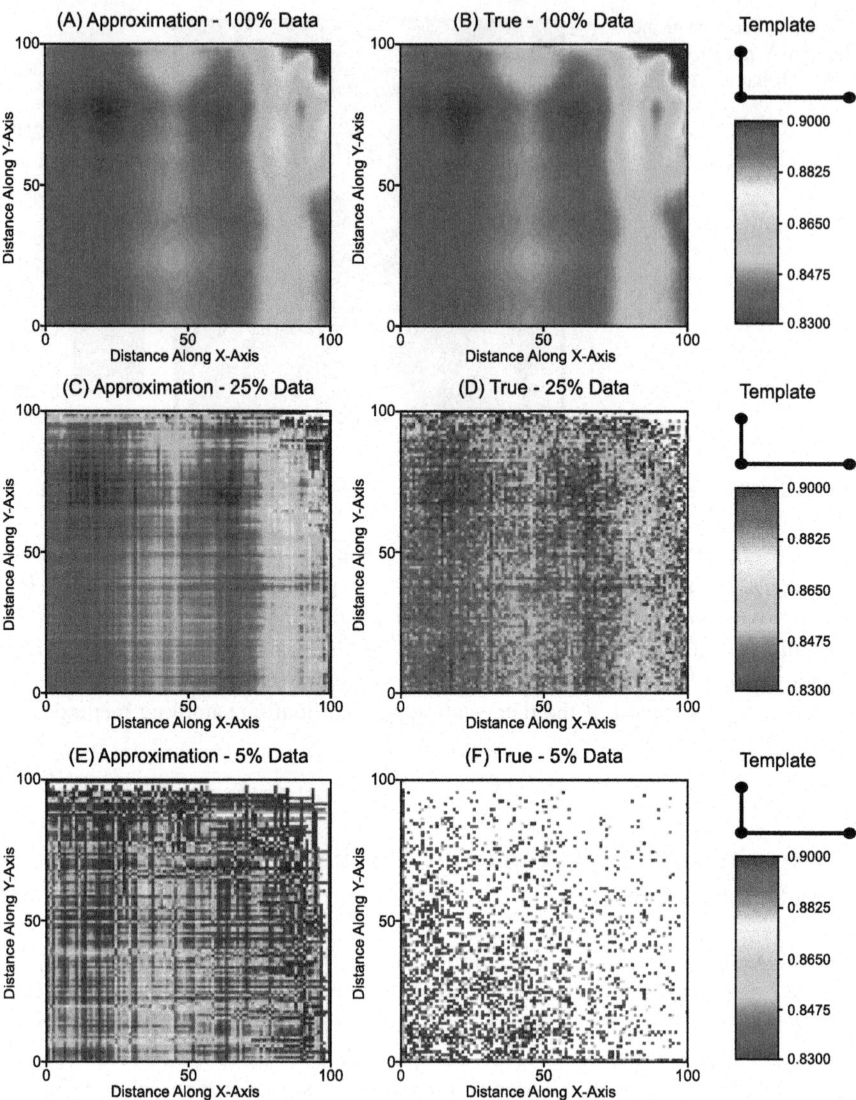

Fig. 3 Comparison between approximations and experimental spatial moments for sparse grids for 100 %, 25 % and 5 % of the original data set. The calculations are based on 1 m lag distances and 0 m lag tolerance

(10 000 points), 25 % (2 500 points) and 5 % (500 points) of the original data set. It is noted that this approximation is performed on 1 m lag spacing without any lag tolerance. It is apparent that as a smaller percentage of the data set is retained, the quality of both the approximation and the true (non-approximated) spatial moment maps deteriorate when compared to the original L-shape spatial moment map (Fig. 3 (B)); this is to be expected given that there are fewer points in the data set

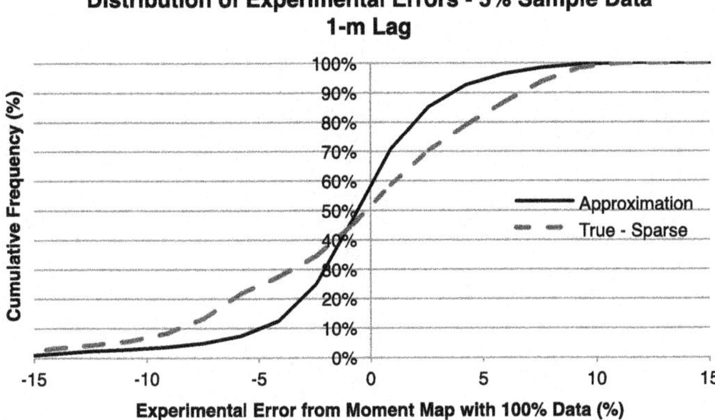

Fig. 4 Distribution of errors for the approximated and true experimental spatial moment maps with 5 % of the sample data on a 1 m lag distance and 0 m lag tolerance

to describe the complex geometries of the channels. It is noted that in all cases, regardless of the percentage of the data retained as sample data, the approximation performs much better in terms of replicating the main features of the original spatial moment map, including the two zones of high values at $[25 \leq x \leq 50, 0 \leq y \leq 100]$ and $[80 \leq x \leq 90, 0 \leq y \leq 100]$. When the amount of data is reduced, there are substantially fewer spatial moments calculated from the true experimental map, and the structures in the map deteriorate rapidly which results in to a highly variable map. This is to be expected, given that there are very few triplets of points to be able to infer the statistic. The spatial moment maps for lower percentages of sample data are not presented, given that the quality of both the approximation and true spatial moment maps are poor, which is caused by the fact that the lag distance is very short and a lag tolerance is not used.

Figure 4 shows the distribution of experimental errors for the approximated and true spatial moment maps using 5 % of the sample data at 1 m lag spacing and 0 m lag tolerance (Figs. 3 (E) and (F)). The experimental errors are calculated using the true spatial moment maps from 100 % of the data (Fig. 3 (B)) as follows:

$$\%\ \text{Error} = \frac{(\text{Experimental Moment}_{x\,\%\,\text{Data}} - \text{True Moment}_{100\,\%\,\text{Data}})}{\text{True Moment}_{100\,\%\,\text{Data}}}. \tag{6}$$

It is noted that Fig. 4 is a standardized cumulative frequency distribution, where the number of occurrences has been standardized according to the number of points available; for example, the approximated spatial moment map for 5 % of the data contains 9 356 values (out of the 10 000 spatial moments that the full data set provides), whereas the actual sample spatial moment map only contains 2526 spatial moments. Not only does the proposed approximation provide more moments for sparse data sets, but there is also substantially less error with approximation when compared to the actual sample spatial moments.

Figure 5 shows the approximations and the actual third-order spatial moment maps for an L-shape template using 100 %, 5 % and 1 % of the sample data, with

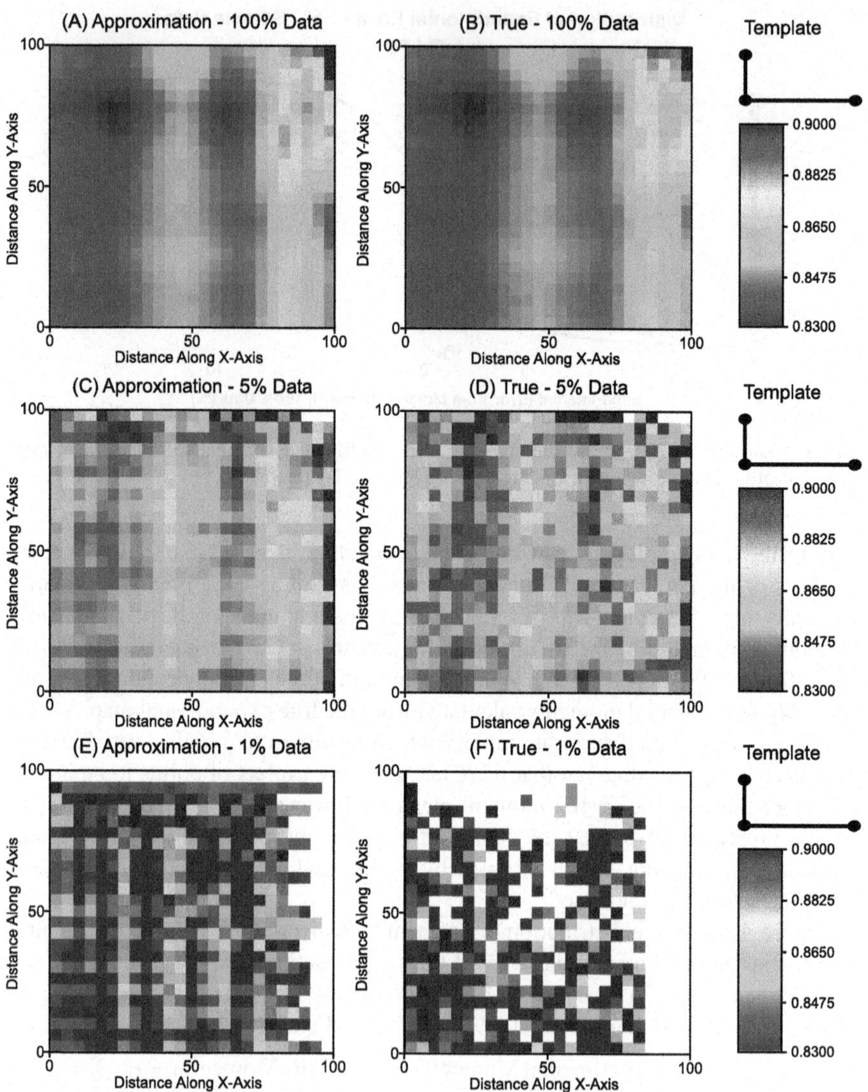

Fig. 5 Comparison between approximations and experimental spatial moments for sparse grids for 100 %, 5 % and 1 % of the original data set. The calculations are based on 4 m lag distances and 2 m lag tolerance

4 m lag spacing and a 2 m lag tolerance. A 4 m spacing was chosen in order to have a high likelihood of finding pairs for each lag spacing. It is apparent that Fig. 5 (A) and Fig. 5 (B) are almost identical, which also indicates that the quality of the approximation is unaffected by the lag spacing and tolerance used. The quality of the approximation using 5 % of the data (Fig. 5 (A)) appears to be substantially better than the actual experimental spatial moments (Fig. 5 (B)). Similar to Fig. 3 (E),

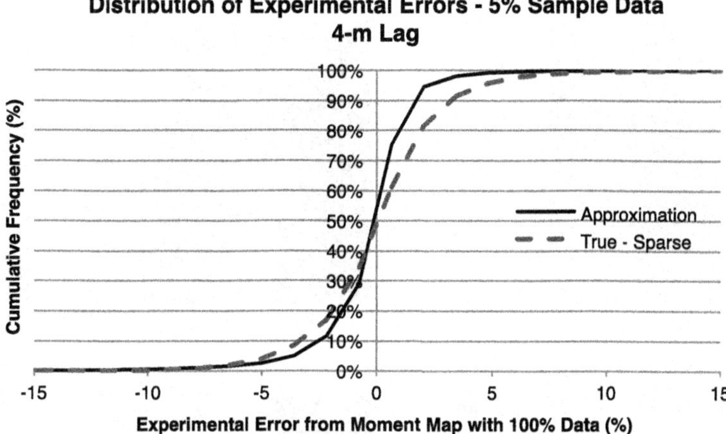

Fig. 6 Distribution of errors for the third-order spatial moment maps, with an L-shaped template, and 4 % of the sample data on a 4 m lag distance

even with 5 % of the data, it is possible to see high-value relationships between the channels, which is difficult to interpret on the actual sample spatial moment map, given the high variability in spatial moment values from one location to the next. At 1 % of the data, the approximation (Fig. 5 (E)) once again provides more spatial moments (given that it is more difficult to find any replicates for a template configuration), and it is still possible to interpret the high-value relationships between the channels better than the actual sample map (Fig. 5 (F)). It is important to keep in mind that only 100 out of the 10 000 points are used, and the lag spacing of 4 ms is relatively small compared to the density of the samples on the grid.

Figure 6 shows the distribution of experimental errors for the approximation and the actual sample spatial moments with an L-shape template when using 5 % of the data on a 4 m lag spacing with 2 m lag tolerance, and is calculated using (6). The range of experimental errors drastically decreases when compared to the errors measured at 1 m lag spacing without lag tolerance, and is a result of smoothing and having more pairs of sample data available. The variability of the experimental errors for the approximated spatial moment map is less than that of the actual sample spatial moments, indicating that the use of the approximated moments is better suited for sparse data sets than using the sample spatial moments directly. It is noted, however, that since the lag spacing and tolerance are increasing, it is easier to find 3-point replicates for the true (non-approximated) spatial moments, so the differences between the distributions for the experimental error decreases.

3.2 Fifth-Order Spatial Moment Approximations

Approximating the fifth-order spatial moment requires approximating the values at three unknown locations for each pair of points in a vector h_j. Figure 7 shows a

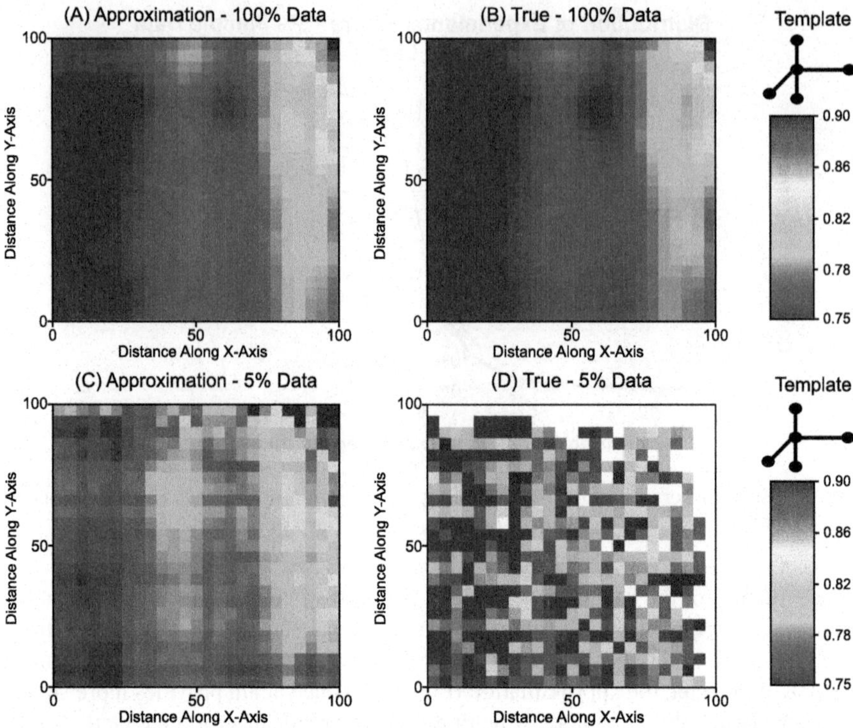

Fig. 7 Comparison between fifth-order spatial moment maps $[h_0, h_{90}, h_{180}, h_{225} = 5\text{ m}]$ for sparse data sets based on 4 m lag distances

2-D section of the fifth-order spatial moment map $(\{h_{90^\circ}, h_{0^\circ}, h_{180^\circ}, h_{225^\circ}\})$ that compares the true and approximated spatial moments when using 100 % and 5 % of the available data, where the 180° and 225° azimuth directions are fixed at a lag distance of five meters. Even for higher orders, the approximation and true spatial moment maps using all data (Figs. 7 (A) and (B), respectively) are very similar. With only 5 % of the available data, the approximation outperforms the actual sample moments; the approximation, Fig. 7 (C), reproduces many of the dominant features from the full spatial moment map. The discrepancies between the approximation with 5 % of the data and the true map are generally caused by the fact that there are fewer points that describe the channels, and are not a result of the degradation of the approximation. For example, Fig. 7 (C) shows an area of high-values at $[25 \le x \le 50, 50 \le y \le 75]$, and is not present in Fig. 7 (A); this high-valued region is also present in the third-order spatial moment shown in Fig. 5 (C), which indicates that it is not an artifact of degradation.

Figure 8 shows the cumulative distribution of experimental errors for the approximated and actual sample spatial moments when using 5 % of the data and a 4 m lag spacing, and is calculated using (6). Evidently, the approximation has significantly less variability in approximation error than the actual sample spatial

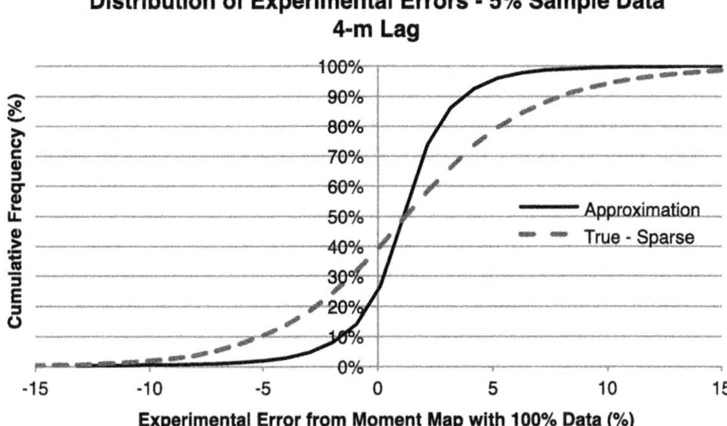

Fig. 8 Cumulative frequency distribution of errors for fifth-order spatial moment maps $[h_0, h_{90}, h_{180}, h_{225} = 5\text{ m}]$ with 5 % of the sample data on a 4 m lag distance

moments, which is a result of a lower number of replicates for the given template to infer the true fifth-order statistic from the sample data set. This indicates that the approximation is more suitable for providing spatial moments of higher-orders than using the actual sample moments.

4 Conclusions

This paper proposes a method for approximating high-order spatial statistics on sparse data sets for a stationary and ergodic random field through decomposing a high-order spatial moment into a set of weighted sums. Using the proposed decomposition, it is only necessary to search for pairs of data in each of the vectors that define the high-order template, which is more practical on sparse data sets than traditional methods that require a large number of replicates for a given template.

Experimental results with sparse data sets indicate that the proposed approximation method generates spatial moment maps that better-resemble the true spatial moment map (calculated with all data) than using the actual sparse spatial moments. Additionally, the error for the approximations on sparse data sets is lower than the actual sparse moments. Results up to the fifth-order spatial moments indicate that the approximation error only slightly degrades, and the experimental error for higher-order spatial moments is lower than the actual sample moments calculated from the sparse data set. For these reasons, it may be possible to integrate the approximated spatial moments into stochastic simulation algorithms and forgo the use of training images. Future work will seek to integrate the approximations into the high-order simulation (HOSIM) algorithm for simulation without using training images.

References

1. Arpat GB, Caers J (2007) Conditional simulation with patterns. Math Geosci. doi:10.1007/s11004-006-9075-3
2. Chilès JP, Delfiner P (1999) Geostatistics: modeling spatial uncertainty. Wiley, New York
3. David M (1988) Handbook of applied advanced geostatistical ore reserve estimation. Elsevier, Amsterdam
4. Dimitrakopoulos R, Mustapha H, Gloaguen E (2010) High-order statistics of spatial random fields: exploring spatial cumulants for modelling complex, non-Gaussian and non-linear phenomena. Math Geosci. doi:10.1007/s11004-009-9258-9
5. Goodfellow R, Albor F, Dimitrakopoulos R, Lloyd T (2010) Quantifying multi-element and volumetric uncertainty in a mineral deposit. In: MININ 2010: proceedings of the 4th international conference on mining innovation, Santiago, Chile, pp 149–158
6. Goodfellow R, Mustapha H, Dimitrakopoulos R (2011) Approximations of high-order spatial statistics through decomposition. COSMO research report, No 5, vol 2, Montreal, QC
7. Goovaerts P (1998) Geostatistics for natural resources evaluation. Cambridge University Press, Cambridge
8. Guardiano J, Srivastava RM (1993) Multivariate geostatistics: beyond bivariate moments. In: Soares A (ed) Geostatistics Tróia '92, vol 1. Kluwer, Dordrecht, pp 133–144
9. Journel AG, Huijbregts ChJ (1978) Mining geostatistics. Academic Press, San Diego
10. Lebedev NN, Silverman R (1972) Special functions and their applications. Dover, New York
11. Mao S, Journel AG (1999) Generation of a reference petrophysical/seismic data set: the Stanford V reservoir. Report 12, Stanford Center for Reservoir Forecasting, Stanford, CA
12. Mustapha H, Dimitrakopoulos R (2010) A new approach for geological pattern recognition using high-order spatial cumulants. Comput Geosci. doi:10.1016/j.cageo.2009.04.015
13. Mustapha H, Dimitrakopoulos R (2010) High-order stochastic simulations for complex non-Gaussian and non-linear geological patterns. Math Geosci. doi:10.1007/s11004-010-9291-8
14. Mustapha H, Dimitrakopoulos R (2011) HOSIM: a high-order stochastic simulation algorithm for generating three-dimensional complex geological patterns. Comput Geosci. doi:10.1016/j.cageo.2010.09.007
15. Mustapha H, Dimitrakopoulos R, Chatterjee S (2011) Geologic heterogeneity representation using high-order spatial cumulants for subsurface flow and transport simulations. Water Resour Res. doi:10.1029/2010WR009515
16. Strebelle S (2002) Conditional simulation of complex geological structures using multiple point statistics. Math Geosci. doi:10.1023/A:1014009426274
17. Zhang T, Switzer P, Journel AG (2006) Filter-based classification of training image patterns for spatial simulation. Math Geosci. doi:10.1007/s11004-005-9004-x

Multiple-Point Geostatistical Simulation Based on Genetic Algorithms Implemented in a Shared-Memory Supercomputer

Oscar Peredo and Julián M. Ortiz

Abstract Multiple-point geostatistical simulation aims at generating realizations that reproduce pattern statistics inferred from some training source, usually a training image. The most widely used algorithm is based on solving a single normal equation at each location using the conditional probabilities inferred during the training process. Simulated annealing offers an alternative implementation that, in addition, allows to incorporate additional statistics to be matched and imposing constraints based, for example, on secondary information. Another class of stochastic simulation algorithms, called genetic algorithms (GA), allows to incorporate additional statistics in the same way as simulated annealing. This paper focuses on a sequential implementation of a genetic algorithm to simulate categorical variables to reproduce multiple-point statistics, and also the details concerning its parallelization and execution in a shared-memory supercomputer. Examples are provided to show the simulated images with their objective functions and running times.

1 Introduction

In geostatistical multiple-point simulation, the most popular method is a sequential approach based on Bayes postulate to infer the conditional distribution from the frequencies of multiple-point arrangements obtained from a training image. This method, originally proposed by Guardiano and Srivastava [13], and later efficiently implemented by Strebelle and Journel [28], is called single normal equation simulation (snesim) (see also [27]). This method has been the foundation for many variants such as simulating directly full patterns [1, 11] and using filters to approximate the

O. Peredo (✉)
Barcelona Supercomputing Center (BSC-CNS), Department of Computer Applications in Science and Engineering, Edificio NEXUS I, Campus Nord UPC, Gran Capitán 2-4, Barcelona 08034, Spain
e-mail: oscar.peredo@bsc.es

J.M. Ortiz
ALGES Lab, Advanced Mining Technology Center, Department of Mining Engineering, University of Chile, Av. Tupper 2069, Santiago 837-0451, Chile
e-mail: jortiz@ing.uchile.cl

P. Abrahamsen et al. (eds.), *Geostatistics Oslo 2012*,
Quantitative Geology and Geostatistics 17,
DOI 10.1007/978-94-007-4153-9_9, © Springer Science+Business Media Dordrecht 2012

patterns [29]. The use of a Gibbs Sampling algorithm to account directly for patterns has also been proposed [2, 17]. A sequential method using a fixed search pattern and a 'unilateral path' also provides good results [6, 7, 23]. Other approaches available consider the use of neural networks [3, 4], updating conditional distributions with multiple-point statistics as auxiliary information [20–22] or secondary variable [16], and simulated annealing [9].

Simulated annealing provides a very powerful framework to integrate different types of statistics, and potentially, generate models subject to constraints that cannot be handled by other methods. Its stochastic nature is closely related with other kind of methods, the genetic algorithms [15], that uses some features of the annealing process but bases its strength in a set of genetic operations iteratively applied to a population of possible realizations of the simulation on course.

Like the simulated annealing algorithm, the execution time (and memory space) of the genetic algorithms can be prohibitive for large images or large pattern templates, but the recent increase in availability of powerful multiple-processor computers and multiple-core central processing units (CPUs), as well as the use of graphics processing units (GPUs) for parallelizing the calculations required for some heavy computing tasks, motivates researching new applications in this framework and lets us revisit algorithms that were too demanding for the technology existing a few years ago. Some examples can be reviewed in [24] and [18]. In order to develop an efficient geostatistical multiple-point simulator based in a parallel genetic algorithm, we need to develop an efficient sequential genetic algorithm and that is the central topic of this paper. Details about the parallel implementation and its execution using a shared-memory supercomputer are also provided, together with several examples speedup results. Where speedup is defined in terms of execution time:

$$\text{speedup} = \text{exec. time}_{\text{sequential}} / \text{exec. time}_{\text{parallel}}.$$

2 Genetic Algorithms

Genetic algorithms (GA) were developed in the 1970s with the work of Holland [15] and in subsequent decades with De Jong [8] and Goldberg [12]. Initially used to find good feasible solutions for combinatorial optimization problems, today they are used in various industrial applications, and recent advances in parallel computing have allowed their development and continuing expansion.

In the canonical approach of GA, typically there is an initial population of individuals, where each individual is represented by a string of bits, as $\text{indiv}_k = 000110101$, and a fitness function $f(\text{indiv}_k)$ which represents the performance of each individual. The fitness function, or objective function, is the objective that must be minimized through the generations over all the individuals. A termination criteria must be defined in order to achieve the desired level of decrement in the fitness function. The main steps and operations performed in a canonical GA can be viewed in Algorithm 1. Theoretical details on the convergence of this algorithms can be reviewed in [15] and an approach where Markov chains are used as theoretical framework to demonstrate the convergence is described in [10] and [25].

Algorithm 1 Canonical genetic algorithm

```
Initial population: N random individuals
Evaluate a fitness function f in each individual
while termination criteria is not achieved do
    Selection: select the best individuals based in its fitness functions
    Crossover: breed new individuals crossing bits of individuals from the
    selection
    Mutation: breed new individuals mutating some bits of individuals from
    the selection
    Replace old individuals by new ones
    Evaluate a fitness function f in each individual
end while
```

3 Shared-Memory Supercomputing

Supercomputers are tools for the simulation of complex processes that require a very large number of operations to be solved in a reasonable time. The first models were based on very powerful single-processors and had specific hardware that gave them the additional speed to execute numerical applications. Step by step, machines that contained several processors were constructed. Those processors worked together with the execution of the same application. This trend, that started almost 30 years ago, with very few processors, has been continued and today the most powerful supercomputer has more than 500000 processors. This technological advance has forced the application experts to adapt themselves to the new way of designing computers and have developed parallel versions of their algorithms using specific well-proven programming models.

The two main models of parallel machines are: distributed-memory and shared-memory machines, with their respective best known programming models MPI [26] and OpenMP [5]. In the first model, each processor has its own private memory and the data interchanged between processors travels through a network in chunks of messages. The speed of this communication depends on the speed of the interconnection network. In the second model, each processor has access to a common memory so the data does not need to travel through a network, instead it can be reached at system bus speed.

The advantage of the shared-memory model is that it offers a unified memory space in which all data can be found. The speed of communication of this model exceeds the speed of the distributed-memory model but its main disadvantage is its non-scaling property, in the sense that two shared-memory machines cannot cooperate under the shared-memory model, they need to be communicated through a network and then a hybrid shared-distributed-memory model applies.

A third model, which has had rapid growth in recent years, is called General-Purpose computing in GPUs and uses the power of the graphical processing units to perform computations in applications traditionally handled by the CPU. This new technology combined with new programming models such as Compute Unified Device Architecture (CUDA) [19], allows to increase the speed of a shared-memory machine in a scalable way without requiring an external network to perform the communication. Several shared-memory processors combined with several graphics processing units seems to be the short-term future of the hardware framework of supercomputers.

With those ideas in mind, our aim is to explore algorithms that can exploit the features of the current technologies and be able to adapt to future trends. Genetic algorithms receive the classification of *embarrassingly parallel* technique to solve problems. This classification comes from the fact that separating the workload of the problem into several parallel tasks is trivial. This property motivates its investigation and application in the field of geostatistics, and particularly in multiple-point simulation. An efficient implementation developed in a shared-memory supercomputer can be a first step in its implementation for the next generation of desktop and laptop low-cost processors, that probably will adopt the trends of the current and future supercomputing.

4 Implementation

In this work, we implemented a GA to generate realizations associated with the problem of multi-point geostatistical simulation using data patterns. The basic idea is to start with a population of realizations that does not honor the statistics and perform the genetic operators over them generating a new population until the statistics are close enough to the target. This is done by defining a fitness function that corresponds to a weighted sum of component objective functions. Each one of these components corresponds to a measure of mismatch between the target statistics and the current statistics for each individual, which are expressed as a mathematical expression. In general, the fitness function can be written as

$$f(\text{indiv}_k) = \sum_{i=1}^{N_c} w_i O_i(\text{indiv}_k), \tag{1}$$

where N_c is the number of components in the objective function, w_i are the weights assigned to each one of the components, and O_i is the mismatch value for component i between the individual indiv_k and the training image.

For example, this function could be composed by the mismatch in histogram reproduction, defined as the squared difference in the cumulative frequencies measured at some quantiles for the model simulated versus the target histogram, a mismatch in variogram reproduction, composed by squared differences between the target variogram model and the variogram calculated from the realization being perturbed, for a number of lag distances, and a mismatch in the reproduction of multiple-point statistics.

Regarding the genetic operators, in addition to the canonical steps (selection, crossover and mutation), in the multiple-point version of GA, we use a restart step, in which several individuals of the population are restarted or *reborn* in the sense

Algorithm 2 Multiple-point statistics genetic algorithm

INPUT:

- Training image of size $n_{\text{row}} \times n_{\text{col}}$
- Template of size $m_{\text{row}} \times m_{\text{col}}$
- Population size N_{popul}
- Maximum number of generations $\text{max}_{\text{generations}}$
- Fitness function f from (1)
- Percentage of selection $p_{\text{selection}}$
- Percentage of mutation p_{mutation}
- Percentage of cut-points $p_{\text{cut-points}}$
- Percentage of restart p_{restart}

Initial population: N_{popul} random individuals of size $n_{\text{row}} \times n_{\text{col}}$
Evaluate fitness function f in each individual
generations $\leftarrow 0$
while generations $\leq \text{max}_{\text{generations}}$ **do**
  Sort the individuals in ascending order using the fitness function f
  if $\frac{\max_k f(\text{indiv}_k) - \min_k f(\text{indiv}_k)}{\min_k f(\text{indiv}_k)} \leq$ tolerance **then**
    Restart: select $p_{\text{restart}} \times N_{\text{popul}}$ individuals and set restart their bits
  end if
  Selection: select the first $p_{\text{selection}} \times N_{\text{popul}}$ individuals from the sorted list
  Crossover: breed new individuals performing a multiple cut point crossover between individuals from the selection, using $p_{\text{cut-points}} \times n_{\text{row}} \times n_{\text{col}}$ cut-points.
  Mutation: mutate each bit of the new individuals with probability p_{mutation}
  Replace old individuals by new ones
  Evaluate fitness function f in each individual
  generations $\leftarrow$ generations $+ 1$
end while
OUTPUT: indiv_k with best fitness function

that all their bits are generated randomly again, as in the first population. This step is crucial in the convergence of the algorithm because it allows to jump out from local optima. The key aspect of the restart step is to store the best individual [25] for the next generation although the majority of the population has been altered. In subsequent generations, those new individuals will introduce variability to the population and in particular to the previously stored best individual, contributing to the minimization of the objective function.

The steps of the implemented genetic algorithm can be viewed in Algorithm 2. The language used was `Fortran 90` using a test-driven methodology for encapsulated routines.

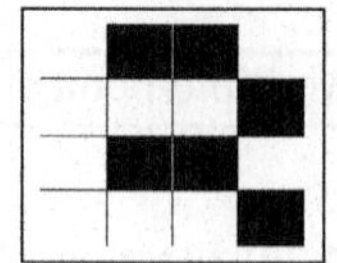 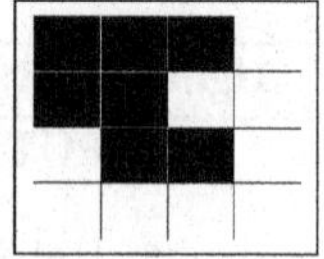

Fig. 1 Training image example of size 4 × 4 (*left*) and an example individual of size 4 × 4 (*right*)

4.1 Construction of the Histogram of Pattern Frequencies

An interesting feature of the implementation concerns the construction of the histogram of pattern frequencies for the training image and individuals in each generation. If we store the frequencies of the patterns that appear in the training image in a k-ary tree, with k the number of categories and depth equal to $m_{\text{row}} \times m_{\text{col}}$ the amount of space used depends directly on the number of different patterns found. For example, for the training image on the left of Fig. 1, the histogram is stored in the binary tree described in Fig. 2 (black nodes have code "1" and white nodes "0", and the patterns are calculated from top-left to bottom-right). The frequencies of the associated patterns are stored in the leaves of the tree.

4.2 Calculation of Fitness Function

A fitness function calculation is performed for each individual in each generation, so it is important that this procedure is as efficient as possible. To calculate the fitness function we compare the histograms of the individual and the training image. A considerable bottle-neck for this calculation is the access to the k-ary tree where lives the training image histogram, because for each extracted pattern from an individual, a search must be performed over the tree in order to see if this pattern belongs to the histogram of the training image or not. A proposed solution to this problem is to store an index of the lexicographical order of the patterns in the k-ary tree associated to the training image. This index allows to search the existence of a pattern in the tree with a worst case performance of order $O(\log n)$ with n the number of leaves with data in the tree.

For example, let us suppose that we want to calculate the fitness function of the individual depicted in the right of Fig. 1, with respect to the training image in the left of Fig. 1 with a 2 × 2 template. The first pattern extracted from the individual, from top-left to bottom-right is 0100. We need to search this pattern in the binary tree associated to the histogram of the training image described in Fig. 2. The lexicographical order can be viewed in the lowest level of rounded boxes of the tree in Fig. 2. From left to right, the patterns are stored in ascending order. Using this order, a binary search is performed. In this case there is a match (the second box from left to right) and the pattern must be compared with the frequency present in the corresponding leaf.

The fitness function corresponds to a modified version of the L^2-norm or Euclidean distance between two histograms. If only the patterns that appear in the

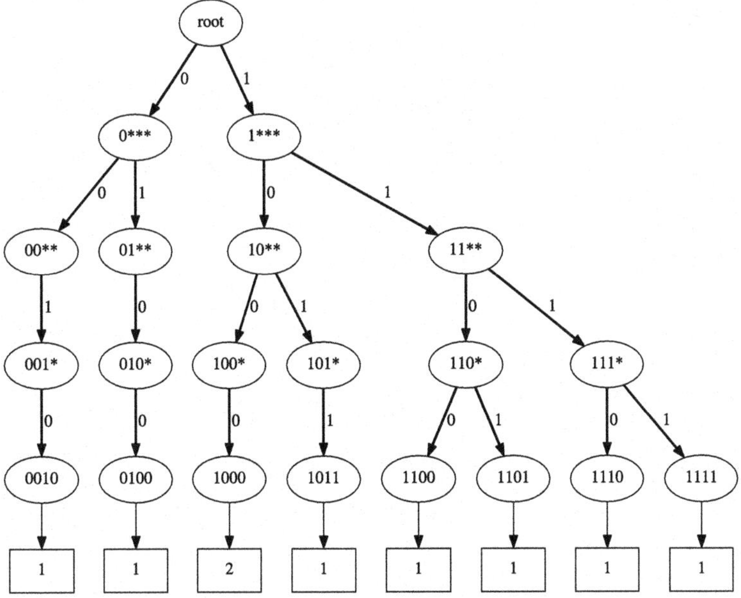

Fig. 2 Data structure to store the histogram of 2 × 2 template of training image from the left of Fig. 1

training image are counted in this sum, we obtain a weighted fitness function as

$$f_{2'}(\text{indiv}_k) = \sum_{i=1}^{N_{\text{MPE}}} \lambda_i' \left(f_i^{\text{TI}} - f_i^{\text{indiv}_k}\right)^2, \quad \text{where } \lambda_i' = \begin{cases} 1 & \text{if } f_i^{\text{TI}} \neq 0, \\ 0 & \text{if } f_i^{\text{TI}} = 0. \end{cases} \tag{2}$$

Other possible fitness functions can be viewed in [24].

4.3 *Parallelization*

The parallelization of the sequential Algorithm 2 essentially consists in distributing the workload carried out in each generation using OpenMP 3.0 [5] directives in the main loops of the code. In this work we only focus our research in the aspects of parallelization concerning a shared-memory machine. In steps 3 and 14 of Algorithm 2, the evaluation of the fitness function for each individual can be performed in parallel by multiple processes using the `DO` directive of OpenMP. Using this scheme, if P processes are working together and the loop goes from 1 to N_{popul}, each process will calculate the fitness function of $\lfloor \frac{N_{\text{popul}}}{P} \rfloor$ different individuals. Data races are not allowed because the automatic distribution of workload performed by the `DO` directive ensures that different individuals are given to different processes. The same approach can be used in the step 10 of the Algorithm 2 to copy the best $p_{\text{selection}} \times N_{\text{popul}}$ individuals and store them in a new memory location. In the steps

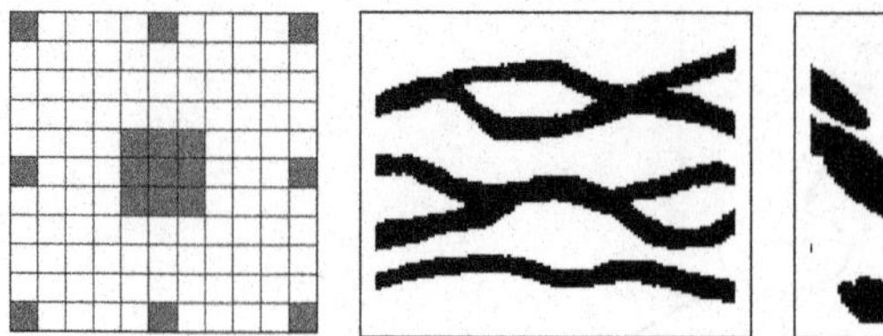

Fig. 3 Template of 17 non-connected nodes (*left*) and training images: channels (*center*) and ellipses 45 (*right*)

11 and 12 of the Algorithm 2, we have a loop over the parents to breed the new generation using crossover and mutation operators. Initially, two parents are *randomly* selected from the copies stored in the selection step. Using the `DO` directive, if two processes are selecting the same parent, no problem is generated, because only reading operations are performed over the memory location that stores this parent. The problem arises when two processes choose the same pair of parents in each iteration. This behavior occurs when the processes have the same series of pseudo-random numbers. In that case, their selections of parents will be the same and no variability will be introduced because the same offspring is generated. In order to avoid this problem, we use different seed numbers for the pseudo-random generator for each process. This generators are identified by the process id and use an algorithm to reproduce a *uniform* (0, 1) process.

5 Examples

Some examples are presented to illustrate the implementation and results over two training images of size 100 × 100. The tests were carried out considering multiple-point statistics obtained for a 17-node non-connected template (Fig. 3-left).

No conditioning data was used in the tests. The fitness function used in the tests corresponds to (2). The parameter set was the following: population size $N_{\text{popul}} = 1000$, maximum number of generations $\max_{\text{generations}} = 12000$, percentage of selection $p_{\text{selection}} = 0.5$, percentage of mutation $p_{\text{mutation}} = 0.2$, percentage of cut-points $p_{\text{cut-points}} = 0.01$ and percentage of restart $p_{\text{restart}} = 0.9$. The implementation was performed using `Fortran 90` and the Intel© Fortran compiler `ifort`, and the execution was tested on a SGI Altix 4700 shared-memory machine with 128 Intel© Itanium© 9030 1.6 GHz processors (2 cores each one). The average, maximum and minimum speedup achieved in 10 runs for each template and each training image is presented in Table 1, considering runs with 2, 4, 8, 16, 32 and 64 processors.

The first test consists of simulating a geological setting of sinuous channels in a background, using the training image presented in Fig. 3-center [27] with size 100 × 100. The second example consists in simulating a setting of ellipses in a background, each of them rotated in 45° with respect to the horizontal axis, using the training image presented in Fig. 3-right. The simulated images with their convergence and pattern frequency plots are presented in Fig. 4 and Fig. 5.

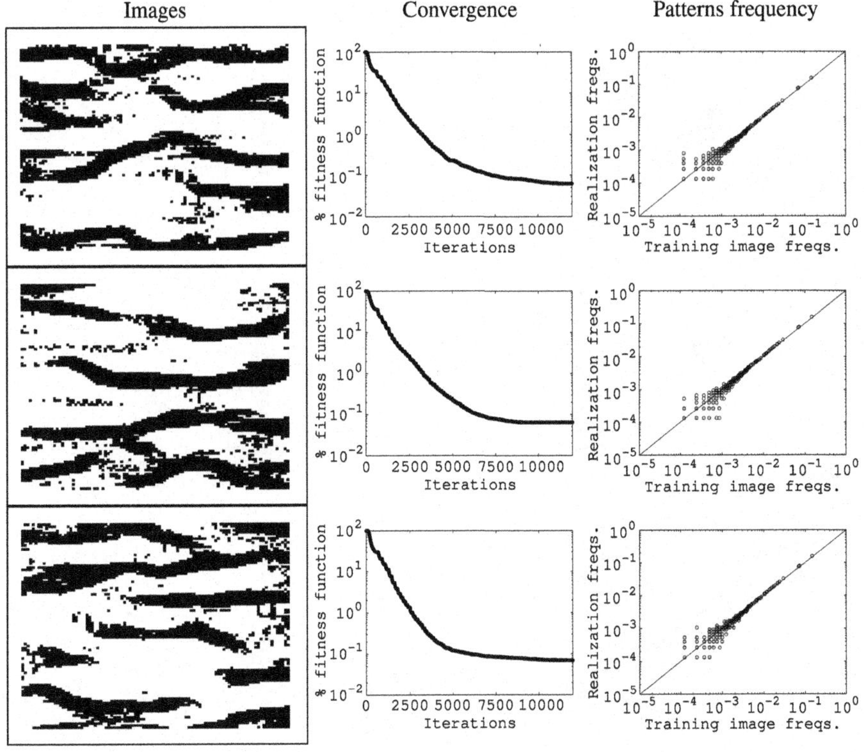

Fig. 4 Simulated images for training image in Fig. 3-center, fitness function $f_{2'}$

Table 1 Avg., max. and min. execution speedup for the training image 3: channels and ellipses 45

Test/procs.	2	4	8	16	32	64
avg channels	1.778	2.950	4.064	4.920	5.230	5.659
avg ellip45	1.871	2.730	4.253	4.898	5.984	6.020
max channels	1.952	3.325	4.913	5.851	7.277	7.311
max ellip45	1.991	3.304	5.064	6.133	7.494	7.250
min channels	1.559	2.478	3.426	4.213	4.332	4.690
min ellip45	1.775	2.404	3.135	3.863	4.357	4.386

6 Conclusions

We have shown an implementation of a genetic algorithm to reproduce multiple-point statistics that uses a shared-memory model to speed up the execution time. The proposed method generates reasonable models, which was one of our main objectives. An interesting feature concerns the usability of the population scheme. The population of individuals can be interpreted as a database of realizations. If we

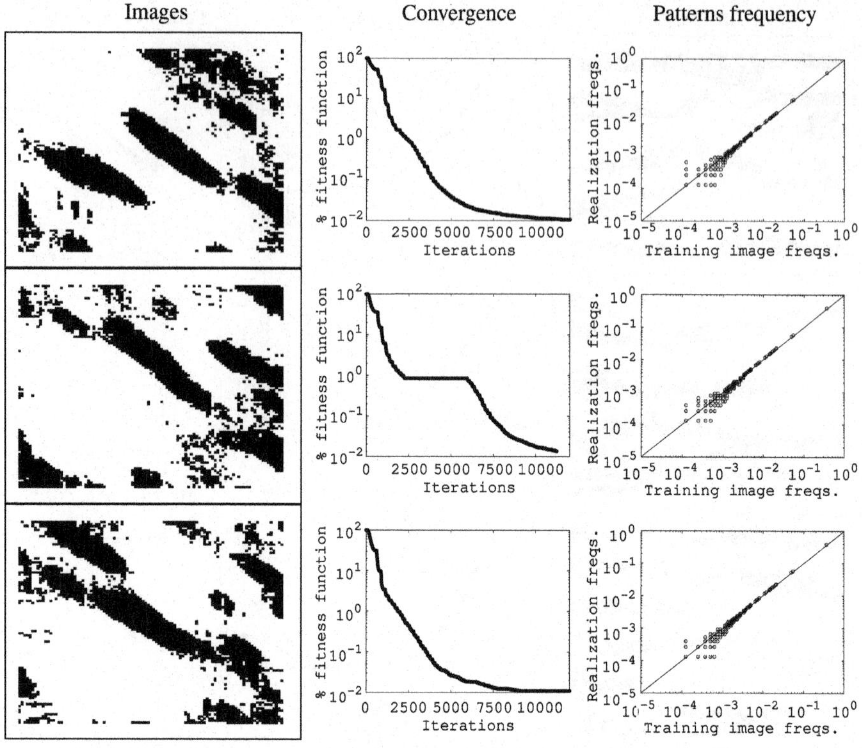

Fig. 5 Simulated images for training image in Fig. 3-right, fitness function $f_{2'}$

use an improved initial database which include previously generated realizations, better results could be obtained.

The simplicity of the method seems to be overshadowed by its excessive execution time, which depends on the number of processes, the clock rate of the CPU (t_{cycle}, seconds per cycle), the cycles per instruction (CPI) and the number of instructions processed (N). As a reference, in the SGI Altix system, the sequential execution time is approximately 32 hours and with 64 processes is approximately 5 hours (1.6 GHz clock rate with compiler `ifort` version 11.0). In a laptop with Intel i7 Core CPU the sequential execution time is 6 hours and with 4 processes is 1.5 hours (2.6 GHz clock rate with compiler `ifort` version 12.0). This dramatic difference can be explained using the following equation (described in [14]) to estimate the execution time with P processes:

$$T_{\text{exec}}^{P} = (1-\alpha) \times N \times \text{CPI} \times t_{\text{cycle}} + \frac{\alpha}{P} \times N \times \text{CPI} \times t_{\text{cycle}},$$

with α the percentage of parallelizable code. With this expression, we can compare execution times in different systems and different number of processes. For example, to explain the difference in the execution times described previously (4 processes with 2.6 GHz versus 64 processes with 1.6 GHz), if approximately 80 %

of our code is parallelizable, the clock cycle times are related as $t_{2.6\text{ GHz}} \approx 0.615 \times t_{1.6\text{ GHz}}$ and based in our results we can assume that $\text{CPI}_{2.6\text{ GHz}} \approx 0.22 \times \text{CPI}_{1.6\text{ GHz}}$, the following relation holds $T^4_{2.6\text{ GHz}}/T^{64}_{1.6\text{ GHz}} = 0.256 \approx 1.5\text{ hours}/5\text{ hours}$. This gives us an idea of the potential performance of our code in systems with advanced chip-set and memory hierarchy, technological factors that can modify the values of CPI and t_{cycle}.

In future versions of the code we will focus our work in decreasing the speed of calculation of the fitness function using the shared-memory framework described in this work plus the addition of GPU accelerators and code optimization.

Acknowledgements The authors thankfully acknowledges the computer resources provided by the Barcelona Supercomputing Center—Centro Nacional de Supercomputación (The Spanish National Supercomputing Center). This research was funded by the National Fund for Science and Technology of Chile (FONDECYT) and is part of the project number 1090056. The authors would also like to acknowledge the funding provided by the Codelco Chair on Ore Reserve Estimation at the Mining Engineering Department, University of Chile.

References

1. Arpat B, Caers J (2007) Stochastic simulation with patterns. Math Geol 39(2):177–203
2. Boisvert JB, Lyster S, Deutsch CV (2007) Constructing training images for veins and using them in multiple-point geostatistical simulation. In: Magri EJ (ed) 33rd international symposium on application of computers and operations research in the mineral industry, APCOM
3. Caers J, Journel AG (1998) Stochastic reservoir simulation using neural networks trained on outcrop data. In: SPE annual technical conference and exhibition, New Orleans, LA, September 1998. Society of Petroleum Engineers. SPE Paper #49026
4. Caers J, Ma X (2002) Modeling conditional distributions of facies from seismic using neural nets. Math Geol 34(2):143–167
5. Chandra R, Dagum L, Kohr D, Maydan D, McDonald J, Menon R (2001) Parallel programming in OpenMP. Morgan Kaufmann, San Francisco, USA
6. Daly C (2005) Higher order models using entropy, Markov random fields and sequential simulation. In: Leuangthong O, Deutsch CV (eds) Geostatistics Banff 2004. Springer, Berlin, pp 215–224
7. Daly C, Knudby C (2007) Multipoint statistics in reservoir modelling and in computer vision. Petroleum Geostatistics A(32)
8. De Jong KA (1980) Adaptive system design: a genetic approach. IEEE Trans Syst Man Cybern 9:566–574
9. Deutsch CV (1992) Annealing techniques applied to reservoir modeling and the integration of geological and engineering (well test) data. Unpublished doctoral dissertation. PhD thesis, Stanford University
10. Eiben AE, Aarts EHL, Van Hee KM (1991) Global convergence of genetic algorithms: a Markov chain analysis. In: Proceedings of the 1st workshop on parallel problem solving from nature. Springer, Berlin, pp 4–12
11. Eskandari K, Srinivasan S (2007) Growthsim—a multiple point framework for pattern simulation. Petroleum Geostatistics A(06)
12. Goldberg DE (1989) Genetic algorithms in search, optimization, and machine learning. Addison-Wesley, Reading
13. Guardiano F, Srivastava M (1993) Multivariate geostatistics: beyond bivariate moments. In: Geostatistics Troia, vol 1, pp. 133–144

14. Hennessy J, Patterson D (2003) Computer architecture—a quantitative approach. Morgan Kaufmann, San Mateo
15. Holland J (1975) Adaptation in natural and artificial systems. University of Michigan Press, Ann Arbor
16. Hong S, Ortiz JM, Deutsch CV (2008) Multivariate density estimation as an alternative to probabilistic combination schemes for data integration. In: Ortiz JM, Emery X (eds) Geostats—proceedings of the eighth international geostatistics congress. Gecamin, Santiago, pp 197–206
17. Lyster S, Deutsch CV (2008) MPS simulation in a Gibbs sampler algorithm. In: Ortiz JM, Emery X (eds) Geostats 2008—proceedings of the eighth international geostatistics congress. Gecamin, Santiago, pp 79–88
18. Mariethoz G (2010) A general parallelization strategy for random path based geostatistical simulation methods. Comput Geosci 36(7):953–958
19. NVIDIA (2008) NVIDIA CUDA programming guide 2.0
20. Ortiz JM (2003) Characterization of high order correlation for enhanced indicator simulation. Unpublished doctoral dissertation. PhD thesis, University of Alberta
21. Ortiz JM, Deutsch CV (2004) Indicator simulation accounting for multiple-point statistics. Math Geol 36(5), 545–565
22. Ortiz JM, Emery X (2005) Integrating multiple point statistics into sequential simulation algorithms. In: Leuangthong O, Deutsch CV (eds) Geostatistics Banff 2004. Springer, Berlin, pp 969–978
23. Parra A, Ortiz JM (2009) Conditional multiple-point simulation with a texture synthesis algorithm. In: IAMG 09 conference. Stanford University
24. Peredo O, Ortiz JM (2011) Parallel implementation of simulated annealing to reproduce multiple-point statistics. Comput Geosci 37(8):1110–1121
25. Rudolph G (1994) Convergence analysis of canonical genetic algorithms. IEEE Trans Neural Netw 5:96–101
26. Snir M, Otto S, Huss-Lederman S, Walker D, Dongarra J (1998) MPI—the complete reference, vol 1: The MPI core, 2nd edn. MIT Press, Cambridge
27. Strebelle S (2002) Conditional simulation of complex geological structures using multiple-point statistics. Math Geol 34(1):1–21
28. Strebelle S, Journel AG (2000) Sequential simulation drawing structures from training images. In: 6th international geostatistics congress, Cape Town, South Africa Geostatistical Association of Southern Africa
29. Zhang T, Switzer P, Journel A (2006) Filter-based classification of training image patterns for spatial simulation. Math Geol 38(1):63–80

The Edge Effect in Geostatistical Simulations

Chaoshui Xu and Peter A. Dowd

Abstract The problem of edge effects in sequential simulation is widely acknowledged but usually overlooked in geostatistical simulations. They can, however, have significant effects on simulations, especially for situations in which the variogram range is relatively large compared to the size of the simulation area (volume). Failure to account properly for the issue can bias simulations by generating values with ranges of correlation and statistical characteristics that differ from those specified. In this paper, we investigate edge effects in detail using sequential Gaussian simulations and derive the critical threshold at which an edge effect becomes significant. Edge correction techniques, such as guard areas, are discussed as a means of mitigating the effects.

1 Introduction

The edge effect in sequential geostatistical simulation is a neglected research area. The effect is generally acknowledged (e.g. [2, 5]) but never properly investigated. In this paper we use sequential Gaussian simulation to demonstrate the various aspects of the edge effect. Some remedies to mitigate the effects are discussed.

The edge effect in general refers to the influences of boundaries of the simulation area (volume) on the characteristics being simulated, whether it is an image (or pattern) or the statistics of variables. The effect is widely reported in the point process literature [1, 3, 9] because of its significant influence on the point pattern generated, but not so in geostatistics as the main focus is on the second order statistics—the variogram—which is less directly connected to the effects.

An edge effect is caused by the unavailability of data outside the area (volume) being simulated and therefore any characteristics estimated close to an edge or boundary are biased. The severity of the bias depends on many factors including

C. Xu (✉) · P.A. Dowd
University of Adelaide, Adelaide, SA 5005, Australia
e-mail: chaoshui.xu@adelaide.edu.au

P.A. Dowd
e-mail: peter.dowd@adelaide.edu.au

P. Abrahamsen et al. (eds.), *Geostatistics Oslo 2012*,
Quantitative Geology and Geostatistics 17,
DOI 10.1007/978-94-007-4153-9_10,

the characteristics (pattern or statistics) being investigated, the size and shape of the area (volume) and the number of available data used for conditioning.

The edge effect is limited to a region around the boundary of a simulation. We have not attempted to quantify this region as the focus is, rightly, on the impact of the edge effect on the statistics of the total simulation area or volume.

2 Numerical Experiment

In Sects. 2.1–2.5 we use two-dimensional unconditional sequential Gaussian simulations to demonstrate the edge effect on three aspects of simulation:

- variogram ($\gamma(h)$) reproduction;
- correlations among supposedly independent simulations; and,
- distribution ($g(x)$) reproduction.

The influence of conditioning is dealt with in Sect. 2.6.

The edge effect is usually negligible if the correlation range, a, of the variable is small compared with the dimensions, L, of the simulation area (volume). The study investigates the variations of the above statistics for different ratios of a/L, representing different degrees of edge effect.

Two-dimensional simulations on a grid size of $L \times L = 100 \times 100$ are used for most simulations and the potential effect for three-dimensional simulations is discussed. The specified range, a, of the variogram model varies from 5 to 80, i.e., a/L varies from 0.05 to 0.8. An isotropic spherical model was used and all output was generated by the SGSIM program from GSLIB [2]. A total of 100 independent realizations were generated for each simulation set-up. The average summary statistics (histogram, variogram or correlation coefficient) ± 96 % confidence intervals are derived from the simulations. Neighborhood search analysis [7, 11] was used to derive a suitable search scheme for the experiment. The most important search neighborhood parameter is the number of samples (or the number of previously simulated node points in this case) used for kriging. We have investigated the average squared difference in the variogram values between the model specified ($\gamma(h)$) and the average experimental variogram ($\gamma_s(h)$) calculated from 100 simulations. The standard deviations of the differences are also calculated from the simulations. For a range of $a = 30$ ($a/L = 0.3$), the results are given in Fig. 1a and for $a = 70$, Fig. 1b. It is clear for $a = 30$, that a suitable number of samples is 20. The number is not obvious for $a = 70$ where the variation is less sensitive to changes in the number. The standard deviations in both cases do not show any significant trend. In the following experiments, the maximum number of previously simulated neighborhood values used for kriging is set to 20. The search radius is set to the range of the variogram model specified.

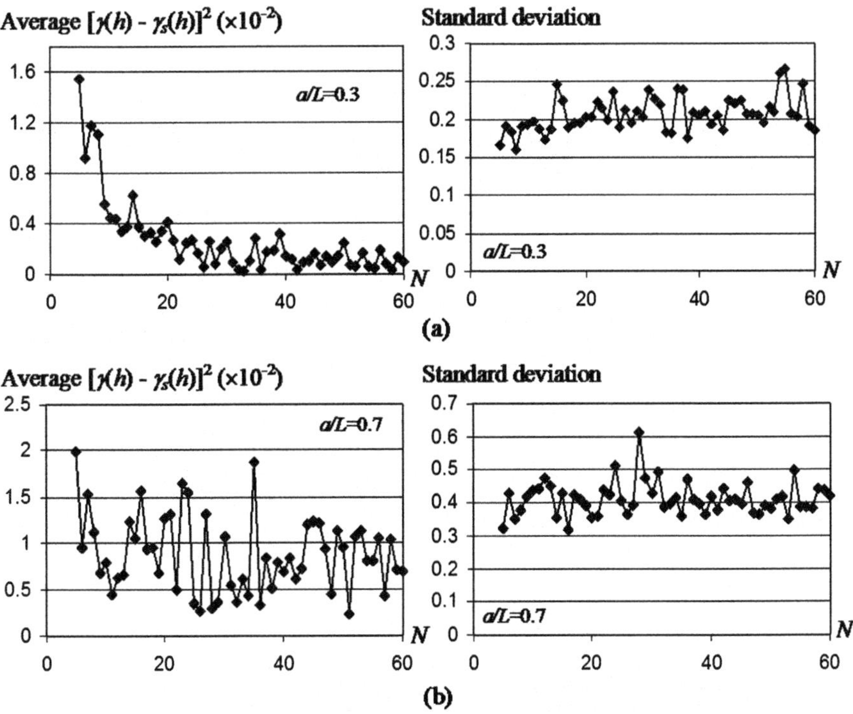

Fig. 1 Differences in variograms $[\gamma(h) - \gamma_s(h)]$ for simulations with different neighborhood search scheme

2.1 Edge Effect on Variograms

Figure 2 shows the difference between the variogram specified, $\gamma(h)$, and that produced by the unconditional simulations, $\gamma_s(h)$, for different variogram ranges. The average squared differences between $\gamma(h)$ and $\gamma_s(h)$, for $h \leq a$ are shown in Fig. 3 and clearly show a significant increase in the squared difference once the variogram range reaches approximately 33 units (or $a/L = \frac{1}{3}$), indicating the considerable increase in the edge effect on the variogram of simulated values. The average difference increases from approximately 3 % when $a/L \leq \frac{1}{3}$ to around 9 % when $a/L \geq 0.6$. This is also evident from Fig. 2 where the averages of the variograms of simulated values tend to have much longer ranges and much lower sills than the model specified once the range of the variogram is greater than 30 units. The effects on variogram range and sill value for different ratios of a/L are shown in Fig. 4. The average values of the simulations agree reasonably well with those specified by the model for variogram ranges up to about $a/L = \frac{1}{3}$. The average sill value reduces from 0.98 (2 % difference to the specified model) when $a/L \leq \frac{1}{3}$ to 0.84 (16 % difference) when $a/L \geq 0.6$. The average difference between the specified variogram range and that of the simulated values increases from 4 when $a/L \leq \frac{1}{3}$ to 27 when $a/L = 0.4$. These observations agree well with those discussed above.

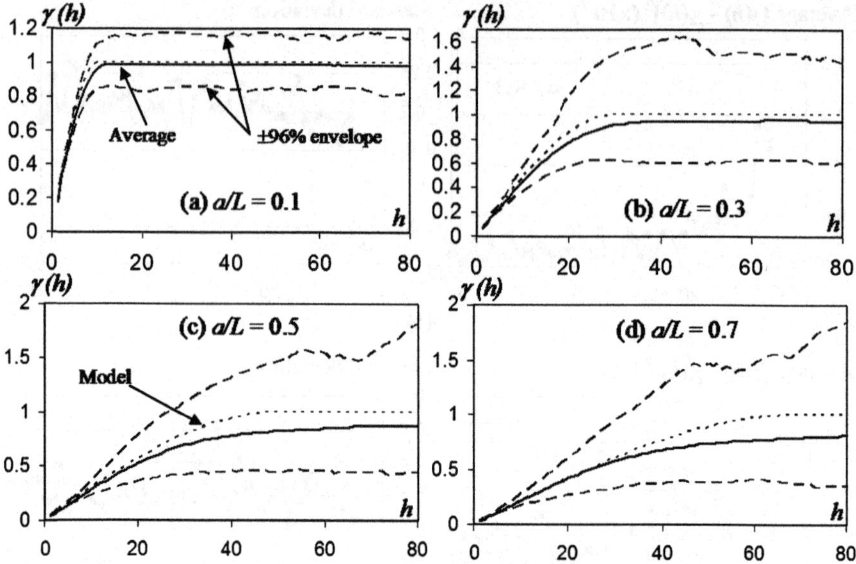

Fig. 2 Difference between variogram specified and produced by simulations

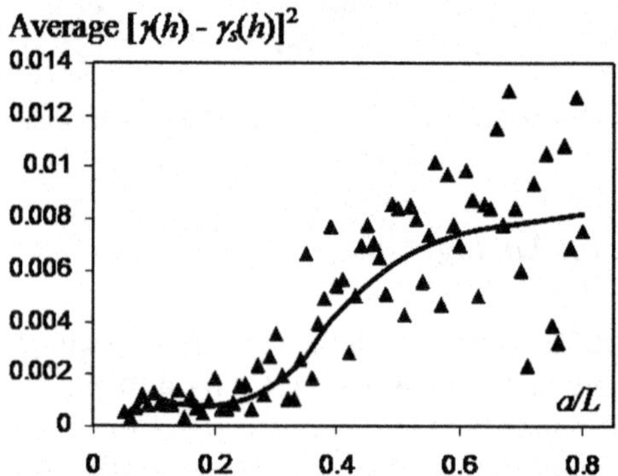

Fig. 3 Average $[\gamma(h) - \gamma_s(h)]$ vs a/L

2.2 Edge Effect on the Nugget Effect of the Variogram of the Simulated Values

The ratio C_0/sill is a critical parameter as it has a significant effect on kriging weights. We have studied the edge effects on the nugget effect for the simulated values for values of $C_0/\text{sill} = 0.1$, 0.3, 0.6 and 0.8. The results are shown in Fig. 5. The edge effect is negligible for small values of the ratio, but it becomes significant once a/L is greater than 0.3. The effect is more obvious if the actual average sill values generated in simulations are used to calculate the ratios, see Fig. 5b. For a specified $C_0/\text{sill} = 0.1$, the generated ratio hardly changes for different a/L. For

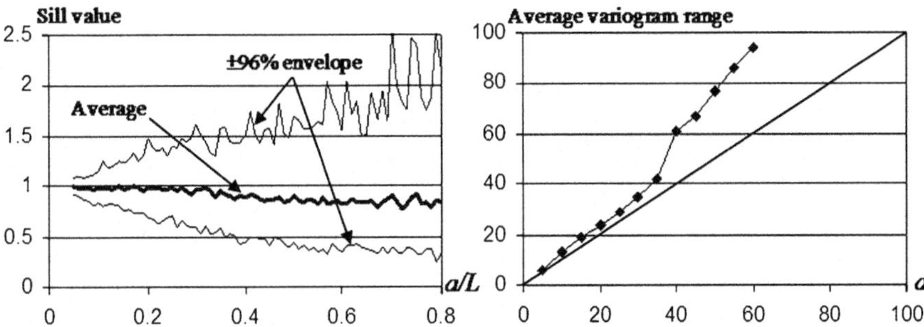

Fig. 4 Edge effects on variogram range and sill values produced by simulation

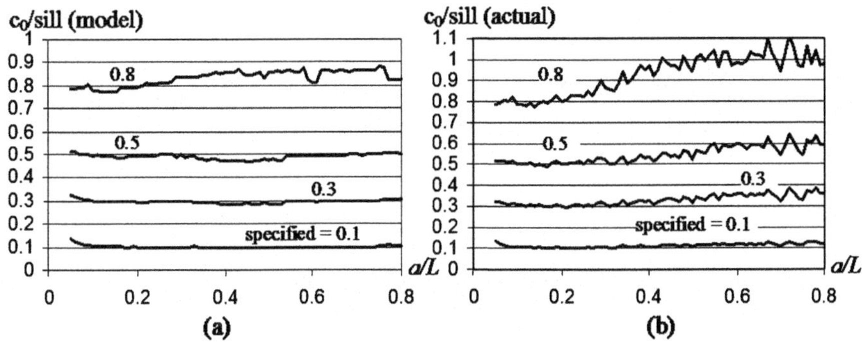

Fig. 5 Edge effect on C_0/sill for different variogram ranges a/L

a specified value of $C_0/\text{sill} = 0.8$, the average generated ratio increases from 0.82 for $a/L \leq 1/3$ to 1.0 (i.e., a pure nugget effect) for $a/L \geq 0.6$, which means the correlation structure is completely eliminated by the edge effect.

2.3 Edge Effect on Distributions

Figure 6 shows the difference between the distributions of standard Gaussian and simulated values produced by the unconditional simulations for different variogram ranges. The average histogram begins to depart from the desired distribution once $a > 30$. The 96 % confidence envelope of the distribution of the simulation also deteriorates considerably. The average squared differences ($[g(x) - g_s(x)]^2$) between the two distributions are shown in Fig. 7, which confirms the threshold of $a/L = 1/3$ beyond which the difference between the model and generated distributions becomes significant. For $a/L \leq 1/3$, the difference between the distributions is negligible. For $a/L \geq 0.6$, the average difference is greater than 0.02, or more than 13 %.

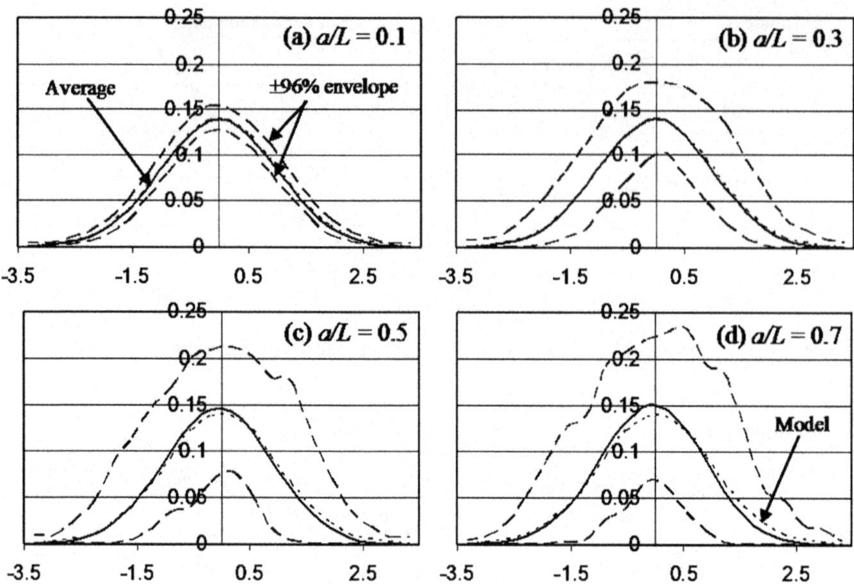

Fig. 6 Difference between distributions specified and produced by simulations

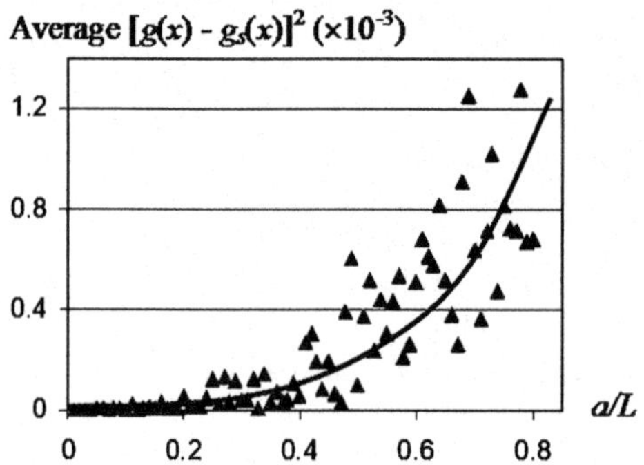

Fig. 7 Average $[g(x) - g_s(x)]^2$ vs a/L

2.4 Edge Effect on Correlations among Simulations

It is common practice to use a set of independent simulations for risk assessment, i.e., different simulations with the same parameter values are generated independently and the outputs are subjected to a specified process (e.g., resource/reserve estimation, open-pit design); the variations in the process outputs are assessed statistically to quantify the uncertainty and the potential risk involved [4, 6]. This procedure is only valid if the different simulations are truly independent.

In practice, however, correlations among different simulations may become significant because of the edge effect. An edge effect is likely to increase as the correla-

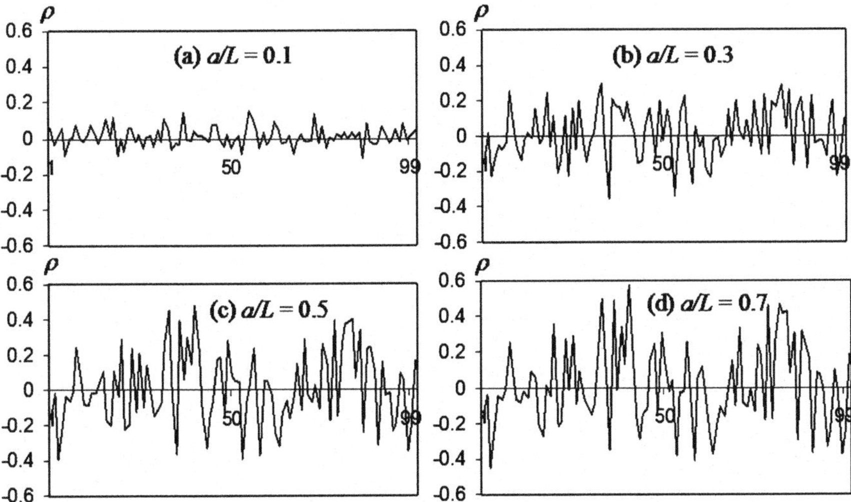

Fig. 8 Correlation coefficients between consecutive simulations

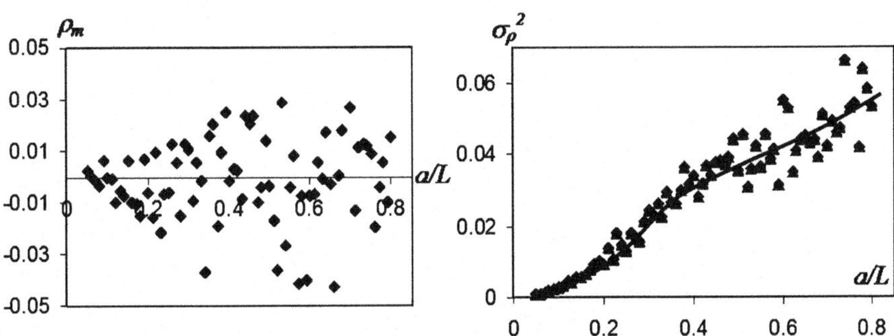

Fig. 9 Mean and standard deviation of correlation coefficients

tion length increases. To demonstrate this effect on the correlations between different independent simulations, we generated 100 pairs of independent realizations for one set of parameters by using different random number seeds for each realizations. The correlation coefficients between each pair of realizations are shown in Fig. 8 for different values of a/L. It is obvious that, on average, the realizations are independent. However, for individual pairs of realizations with large variogram range a, the correlation coefficient, ρ, can exceed ± 0.4, and realizations are no longer truly independent.

To summarize this effect, the mean and variance of the correlation coefficients, ρ_m and σ_ρ^2, calculated from the simulations for different ratios of a/L, are plotted in Fig. 9. Correlations among simulations are unavoidable, though the average behavior still suggests independence. The variance of the correlation coefficients increases from almost zero to greater than 0.04 once $a/L \geq 0.6$, suggesting the

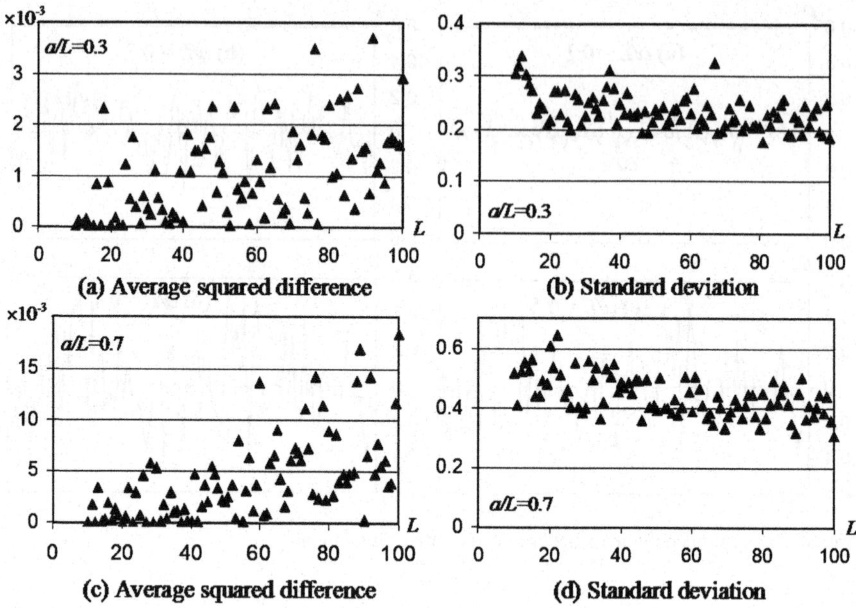

Fig. 10 Differences in variograms $[\gamma(h) - \gamma_s(h)]$ for different areas

generated correlation coefficients will vary within the range of ±0.4 at the 95 % confidence interval. The threshold of $a/L = 1/3$ derived above can be regarded as a naive cut-off above which the correlation between simulations becomes significant.

2.5 Influence of the Absolute Size of the Area of the Simulation

The simulation studies thus far are for a simulation grid $L \times L = 100 \times 100$. To investigate the influence of the absolute size of the simulation area, the simulation grid has been varied from 10×10 to 100×100 while keeping a/L constant. The average squared differences in variograms ($[\gamma(h) - \gamma_s(h)]^2$) for $h \leq a$ and distributions ($[g(x) - g_s(x)]^2$) from 100 independent simulations for each value of L are summarized in Figs. 10 and 11. The figures also show the spread of the variograms and the distributions of the 100 independent simulations around the specified models. Spread is quantified by the standard deviation of the differences for different a/L. The size of the area (volume) has some effect on the reproduction of statistics and variogram, but the influence is limited. The trend of the variation is not obvious except for a very small area ($L < 20$), which in general is too small to be used in practice.

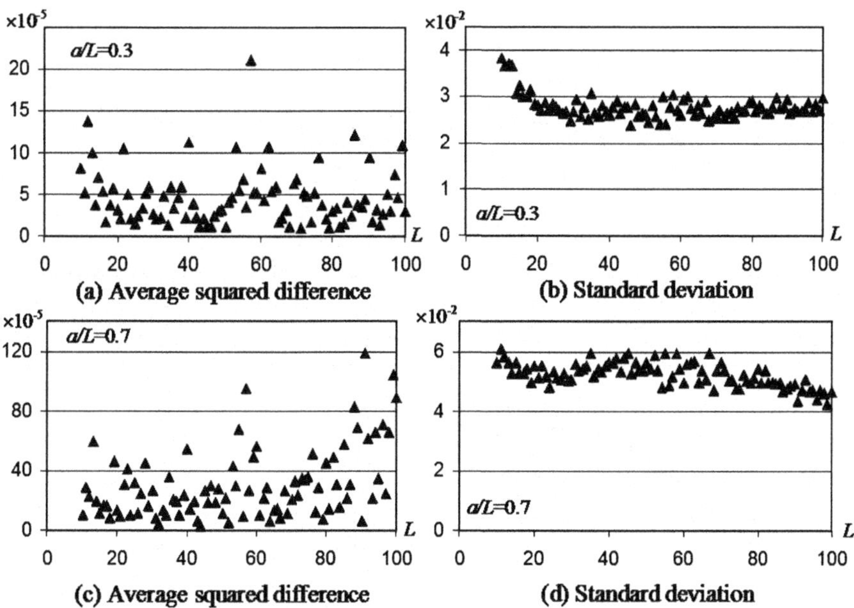

Fig. 11 Differences in distributions $([g(x) - g_s(x)])$ for simulations with different areas

2.6 Conditional Simulations

The output generated by conditional sequential simulation is essentially a compromise between the model specified and the implicit characteristics of the conditioning data. In general, the characteristics described by the model and implied by the data should be identical as the model is based on the experimental variogram calculated directly from the conditioning data. However, as pointed out by [8], the experimental variogram conditional on a particular set of realizations could differ from the underlying random field variogram. The edge effect also contributes to the potential discrepancy, as demonstrated in the previous sections. Techniques are available in marked point process theory to include edge corrections in the experimental mark variogram calculation [1, 12].

As more conditioning data are used, the simulation becomes more data driven and the reproduction of the characteristics implied by the data will become more dominant. For this reason, the edge effect would be expected to decrease, as the number of conditioning data increases.

To investigate the effect of conditioning data on the edge effect, we sampled randomly the simulated images of size $L \times L = 100 \times 100$ generated by variogram ranges of $a = 30$ and 70. The number of samples used ranges from 50 to 2000. Comparisons between the generated variograms and distributions using different numbers of samples are given in Figs. 12 and 13. It is clear from these graphs that once the number of samples is greater than 400, the differences become negligible.

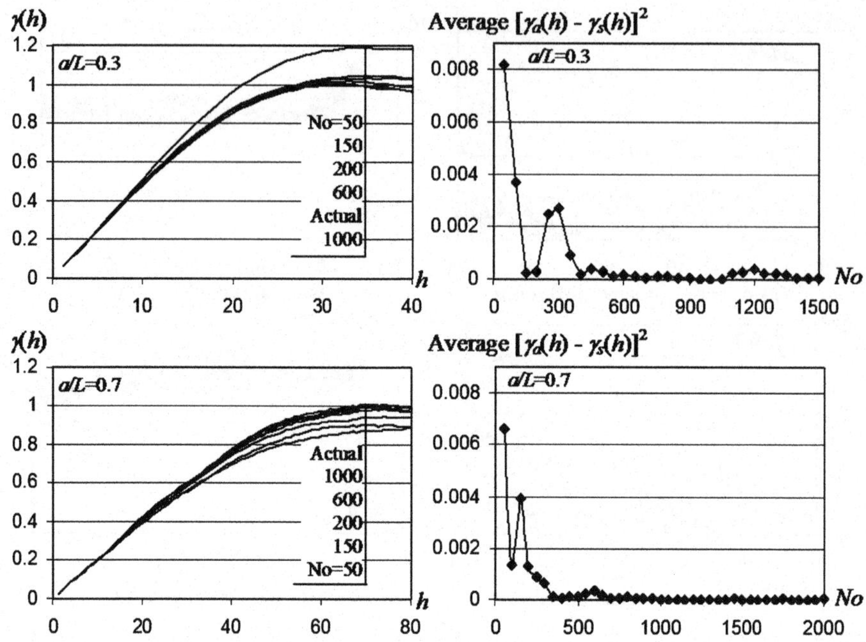

Fig. 12 Variograms reproduced with different number of conditioning data

The behaviors are consistent for both variogram ranges, suggesting that the edge effect is insignificant in conditional simulations because of the conditioning data.

3 Three-Dimensional Applications

It is to be expected that the edge effect for three-dimensional applications will be more significant than for two-dimensional cases, as a result of the additional dimension. To investigate the difference in the edge effect for 2D and 3D applications, 3D simulations using a grid $L \times L \times L = 64 \times 64 \times 64$ were generated for the ratios of $a/L = 0.3$ $(a = 19)$ and 0.7 $(a = 45)$. The variograms and distributions generated are shown in Fig. 14.

The average variogram structure is different to that shown in the 2D examples. It has a longer range and a higher sill value. The average squared differences between the specified variogram and the average experimental variogram are given in Fig. 16a, which should be compared to Fig. 3. The squared differences are significantly higher for the 3D example, supporting the observation of more significant edge effects in 3D applications. Based on the limited number of simulations, the ratio of $a/L = 1/3$ can still be regarded as an approximate range beyond which the edge effect becomes significant. The average squared differences for the 3D distributions are smaller than those of the 2D distributions, which is also evident from Figs. 15 and 16b.

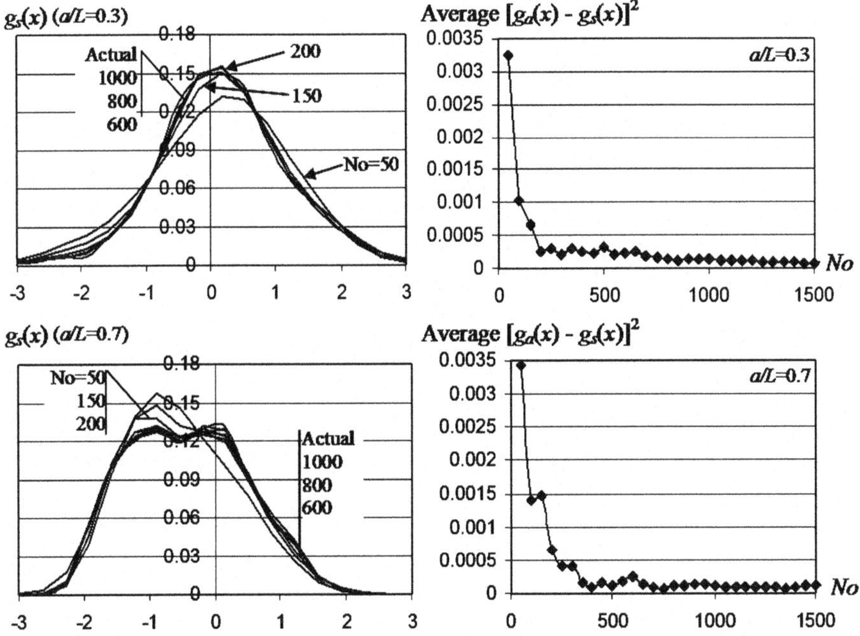

Fig. 13 Distributions reproduced with different number of conditioning data

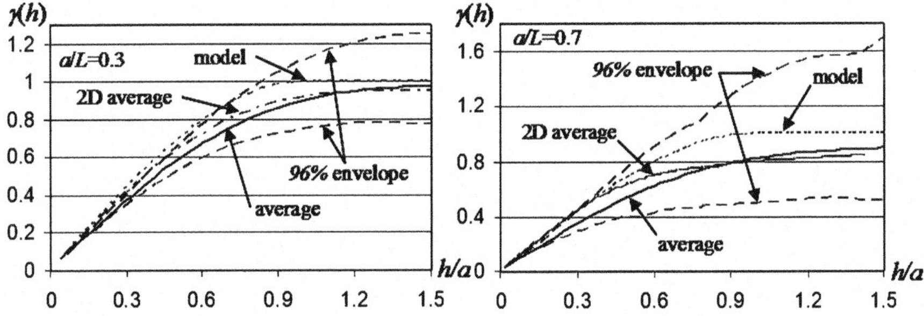

Fig. 14 Difference between variogram specified and produced by simulations (3D case)

4 Mitigations

This investigation indicates that the edge effect is not significant in sequential geostatistical simulations if the variogram range is less than $1/3$ of the size of the region. For longer variogram ranges, some mitigation is required for the simulations satisfactorily to reproduce the correlation structure in a realization or to eliminate the correlations between supposedly independent realizations.

Two commonly used methods to mitigate the edge effect in point process modeling are the use of a guard area and periodical boundaries [1, 10]. The guard area approach involves the construction of an area surrounding the region (volume) being

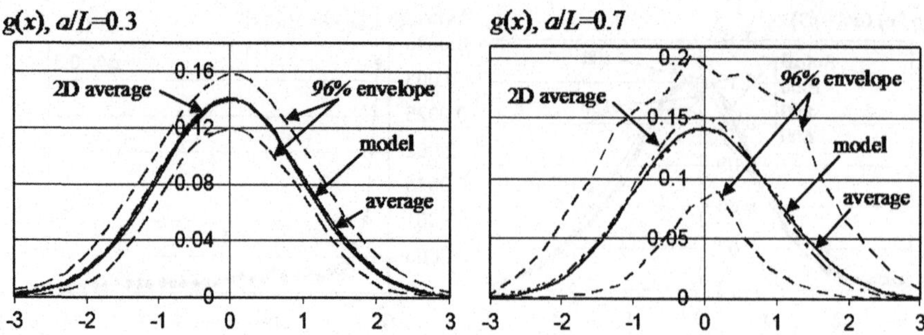

Fig. 15 Difference between distributions specified and produced by simulations

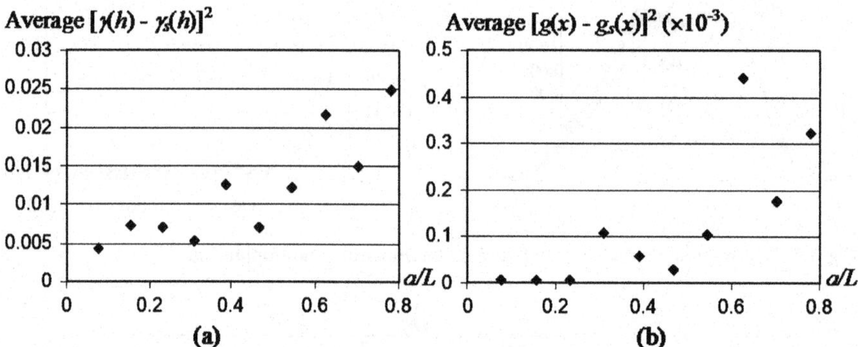

Fig. 16 Average squared differences for variograms and distributions (3D case)

simulated. The actual area (volume) of the simulation is effectively enlarged. This approach can be used for geostatistical simulation. The basic idea is to ensure that, after the enlargement, the ratio of the variogram range to the size of the region is less than $1/3$. The drawback of this approach is the increase in computing cost due to the addition of the guard area. For example, if the ratio of the variogram range to the size of the simulation region is 0.5, the size of the region, L, will have to be increased by 50 % to ensure the required ratio is satisfied.

The use of periodical boundaries effectively turns a rectangular region into a torus so that opposite boundaries are considered joined. The direct consequence of this virtual connection is that distances between points near two opposite boundaries can be much smaller than the distances measured directly within the region. This mitigation method is, therefore, only suitable for reducing the edge effect for the variogram range, a to less than $1/2$ of the size of the region, L.

If the implementation of the guard area becomes impractical (when a/L is too large), other simulation techniques may be required. Simulated annealing can also be used to refine the image generated by sequential simulation to mitigate the edge effect, but this will only improve the correlation structure and not the distribution.

5 Conclusions

The edge effect has a significant impact on sequential geostatistical simulations, especially when the variogram range is greater than $1/3$ of the size of the study region. The impact includes inaccurate reproduction of the variogram model, the statistical distribution and correlations between the supposedly independent simulations. The absolute size of the volume/area has little influence on the edge effect. The edge effect is significantly reduced in conditional simulation. For 3D simulations, the edge effect on variogram reproductions is much more significant than 2D applications. The edge effect can be mitigated by the use of a guard area or periodical boundaries, but both methods have drawbacks and limitations.

References

1. Cressie N (1993) Statistics for spatial data. Wiley, New York
2. Deutsch CV, Journel AG (1998) GSLIB geostatistical software library and user's guide, 2nd edn. Oxford University Press, London
3. Diggle P (2003) Statistical analysis of spatial point patterns, 2nd edn. Arnold, London
4. Dowd PA (1997) Risk in minerals projects: analysis perception and management. Trans Inst Min Metall Ser A, Min Ind, 107:A9–A20
5. Journel AG, Kyriakidis PC (2004) Evaluation of mineral reserves, a simulation approach. Oxford University Press, London
6. Li S, Dimitrakopoulos R, Scott J, Dunn D (2007) Quantification of geological uncertainty and risk using stochastic simulation. In: Proc. orebody modelling and strategic mine planning—uncertainty and risk management models, 2nd edn. AusIMMM, Melbourne
7. Riviorard J (1987) Two key parameters when choosing the kriging neighbourhood. Math Geol 19(8):851–856
8. Stoyan D, Wälder O (2000) On variograms in point process statistics, II: models of markings and ecological interpretation. Biom J 42(2):171–187
9. Stoyan D, Kendall W, Mecke J (1995) Stochastic geometry and its applications, 2nd edn. Wiley, New York
10. Upton G, Fingleton B (1985) Spatial data analysis by example. Wiley, New York
11. Vann J, Jackson S, Bertoli O (2003) Quantitative kriging neighbourhood analysis for the mining geologist—a description of the method with worked case example. In: Proceedings of the 5th international mining conference, Bendigo, Nov. 2003. AusIMMM, Melbourne
12. Xu C, Dowd PA, Mardia KV, Fowell RJ, Taylor CC (2007) Simulating correlated marked point processes. J Appl Stat 34(9):1125–1134

Extensions of the Parametric Inference of Spatial Covariances by Maximum Likelihood

Peter A. Dowd and Eulogio Pardo-Igúzquiza

Abstract The limitations of the maximum likelihood method for estimating spatial covariance parameters are: the assumption that the experimental data follow a multi-dimensional Gaussian distribution, biased estimates, impracticality for large data sets and the common assumption of a polynomial drift. The advantages are easy evaluation of parameter uncertainty, no information loss in binning and the ability to include additional information using a Bayesian framework. We provide extensions to overcome the disadvantages whilst maintaining the advantages. We provide an algorithm for obtaining covariance estimates for non-Gaussian data using Gaussian maximum likelihood. We provide a means of generating unbiased estimates of spatial covariance parameters without increasing the estimation variance. We overcome the impracticality for larger data sets by an approximation to the complete maximum likelihood. Finally, we extend the polynomial drift to other forms.

1 Introduction

Mardia and Marshal [5] were the first to use maximum likelihood estimation (MLE) in geostatistics and [3, 9] promoted its use in earth science applications. Despite the skeptical views of some authors [8, 14], MLE applications have increased considerably over the past 20 years assisted by increasing computing power. MLE is restricted by very large data sets but for smaller data sets it is an appealing alternative for semi-variogram inference for a number of reasons. The semi-variogram (or covariance) parameters are directly estimated without having to calculate an experimental semi-variogram and then fit a model to it. The method provides uncertainty measures of the semi-variogram parameters and thus, in addition to obtaining measures of the reliability of estimates, it provides interval estimates, confidence inter-

P.A. Dowd (✉)
University of Adelaide, Adelaide, SA 5005, Australia
e-mail: peter.dowd@adelaide.edu.au

E. Pardo-Igúzquiza
Instituto Geológico y Minero de España, 28003 Madrid, Spain
e-mail: e.pardo@igme.es

P. Abrahamsen et al. (eds.), *Geostatistics Oslo 2012*,
Quantitative Geology and Geostatistics 17,
DOI 10.1007/978-94-007-4153-9_11,

vals and the ability to conduct statistical tests. The likelihood value can be used for model selection. The Bayesian extension is a straightforward coding into the prior distribution all of parameter information independently of the experimental data. MLE does not use binning or two-point statistics and thus is suitable for complex sampling designs. There are, however, several difficulties with MLE. It is a parametric method, implying the assumption of a multivariate probability density function (mpdf), the most common being Gaussian. However, many spatial variables in geosciences are not Gaussian. Estimates may be biased, especially for small data sets or non-Gaussian data or a spatial trend in the mean of the random function or misspecified models. The drift or trend represents a regional, or low frequency, variation that is usually modeled by a polynomial of low degree. For n data the method requires the inversion of an $n \times n$ matrix multiple times. This is time consuming and may make the method impractical. When $n > 1000$ an approximate method with less computational requirements is needed.

2 Brief Theoretical Review

The geostatistical model considered here is the second order stationary covariance random field. This model corresponds to the spatial linear model defined by

$$Z(\mathbf{u}) = m(\mathbf{u}) + R(\mathbf{u}), \tag{1}$$

where $\mathbf{u} \subset \chi$, χ is a region of R^d or a region of a lattice Z^d; d is the dimension of the region; and $Z(\mathbf{u})$ is a random function with a deterministic component (the drift) accounting for long distance correlation and a stochastic component (the residual) accounting for local correlation; $m(\mathbf{u})$ is the drift, most commonly polynomial:

$$m(\mathbf{u}) = \sum_{i=1}^{p} \beta_i f_i(\mathbf{u}). \tag{2}$$

In (2) there are p unknown drift coefficients β_i and p known monomials that are functions of the coordinates $\mathbf{u}$. For two dimensions, $\mathbf{u} = \{x, y\}$, drifts up to order 2 are sufficient (i.e., quadratic polynomial) as itemized in Table 1. The drift is the mathematical expectation of the random function:

$$E\{Z(\mathbf{u})\} = m(\mathbf{u}). \tag{3}$$

The residual $R(\mathbf{u})$ has zero mean, and second order stationary covariance

$$E\{R(\mathbf{u})R(\mathbf{u}+\mathbf{s})\} = C(\mathbf{s}; \theta^*), \tag{4}$$

where $\mathbf{s}$ is a vector of distances between two arbitrary locations; $C(\mathbf{s}; \theta^*)$ is a covariance model, a function of the distance between two locations defined by parameters θ^*. It is also the covariance of the random field $Z(\mathbf{u})$:

$$C(\mathbf{s}; \theta^*) = E\{[Z(\mathbf{u}) - m(\mathbf{u})][Z(\mathbf{u}+\mathbf{s}) - m(\mathbf{u}+\mathbf{s})]\}. \tag{5}$$

The parameter $\theta^* = (\beta, \sigma^2, \theta)$ is the extended set of parameters comprising: β is a $p \times 1$ vector of drift coefficients; σ^2 is a scalar variance, and σ is the standard

Table 1 Basis functions for a two-dimensional polynomial drift up to order 2

Drift	k	p	Basis function
Constant	0	1	1
Linear	1	3	$1, x, y$
Quadratic	2	6	$1, x, y, x^2, y^2, xy$

deviation; θ is a $t \times 1$ vector of correlogram parameters. The covariance, where $\rho(.)$ is the correlogram, can be parameterized as:

$$C\left(\mathbf{s}; \theta^*\right) = \sigma^2 \rho(\mathbf{s}; \theta). \tag{6}$$

For simplification the dependence on the drift parameters has been omitted in the notation. There is a simple correspondence between covariance and semi-variogram:

$$\gamma\left(\mathbf{s}; \theta^*\right) = \sigma^2 - C\left(\mathbf{s}; \theta^*\right) = \sigma^2\left(1 - \rho(\mathbf{s}; \theta)\right). \tag{7}$$

For the most common correlogram models (spherical, exponential and Gaussian), the minimum number of correlogram parameters is one, the range. The Matérn model [8, 10] is more general and has a minimum of two parameters, the range and the shape parameter. This corresponds to the isotropic case with no nugget variance. If the model is anisotropic with no nugget variance, the number of parameters is three, the longest range, the spatial direction of the longest range and the anisotropy ratio. A nugget variance increases the number of parameters by one.

In geostatistics the purpose of ML is, given a realization of a random field $Z(\mathbf{u})$ known only at n experimental locations, to: choose between the stationary or the intrinsic random field model; choose the most appropriate drift order, $k = 0$ or $k = 1$ or $k = 2$; and estimate the parameters β, σ^2, θ. For optimal estimation by universal kriging there is no need to estimate the drift coefficients β but it may be useful to do so. ML is a parametric procedure that requires an mpdf. The Gaussian mpdf is completely defined by its vector of means $\mathbf{m}$ and covariance matrix $\mathbf{C}$. $\mathbf{m}$ is an $(n \times 1)$ vector of means with general element $[\mathbf{m}]_i = m(\mathbf{u}_i)$ and $\mathbf{C}$ is an $(n \times n)$ matrix of covariances with general element $[\mathbf{C}]_{ij} = C(\mathbf{s}_{ij}; \theta^*)$. This matrix is factorized as in (6) where ρ is an $(n \times n)$ matrix of correlograms with general element $[\rho]_{ij} = \rho(\mathbf{s}_{ij}; \theta)$. The obvious choice is the Gaussian mpdf. Given n data $\mathbf{z}' = \{z(\mathbf{u}_1), z(\mathbf{u}_2), \ldots, z(\mathbf{u}_n)\}$ the Gaussian density is

$$p\left(\mathbf{z}; \beta, \sigma^2, \theta\right) = (2\pi)^{-n/2} \sigma^{-n} |\rho|^{-1/2} \exp\left[-\frac{1}{2\sigma^2}(\mathbf{z} - X\beta)' \rho^{-1} (\mathbf{z} - X\beta)\right], \tag{8}$$

where X is the $n \times p$ matrix of drift basis functions. The ML estimates of the set of parameters $(\beta, \sigma^2, \theta)$ are the values that maximize the log-likelihood function (LLF), defined from (8) as:

$$L\left(\beta, \sigma^2, \theta; \mathbf{z}\right) = -\frac{n}{2} \ln 2\pi - n \ln \sigma - \frac{1}{2} \ln |\rho| - \frac{1}{2\sigma^2}(\mathbf{z} - X\beta)' \rho^{-1} (\mathbf{z} - X\beta). \tag{9}$$

L is minimized by $\hat{\beta} = (X'\rho^{-1}X)^{-1} X'\rho^{-1}\mathbf{z}$ and $\hat{\sigma}^2 = \frac{1}{n}(\mathbf{z} - X\hat{\beta})'\rho^{-1}(\mathbf{z} - X\hat{\beta})$. The LLF can now be written as a function of the parameters θ alone:

$$l(\hat{\beta}, \hat{\sigma}^2, \theta; \mathbf{z}) = -\frac{n}{2}\ln 2\pi - \frac{n}{2}\ln\left[(\mathbf{z} - X\hat{\beta})'\rho^{-1}(\mathbf{z} - X\hat{\beta})\right] + \frac{n}{2}\ln n - \frac{1}{2}\ln|\rho| - \frac{n}{2}. \tag{10}$$

In general, there is no closed form for the ML estimates of θ but they can be obtained by numerical maximization of LLF in (10). In the simplest case of isotropic covariance with no nugget variance there is only one parameter $\theta =$ range and (10) defines a likelihood profile that provides information about the likelihood itself (smoothness), the potential minimization problems (unimodality or multimodality) and the uncertainty of the estimate (curvature of the LLF at the minimum).

3 ML Estimation in the Non-Gaussian Case

For non-Gaussian data there are two possibilities: either transform the data or use the Gaussian MLE anyway. There are two possibilities for transforming data.

- The interest is in the transformed variable. For example, in hydrogeology, hydraulic conductivity is assumed to be lognormal and thus the logarithm of the variable follows a Gaussian distribution. As the transformed variable is used in applications, Gaussian MLE is appropriate for the transformed variable.
- The interest is in the original variable. Gaussian MLE is applied to the normal scores and the estimated parameters are back-transformed to the space of the original variable. This is done by simulating on a regular grid a realization of a Gaussian field with the estimated parameters from the normal scores, applying the back transform to the realization and then estimating from this realization the parameters of the untransformed variable. The latter is easy as there are as many data as desired on a regular grid. Figure 1a summarizes the approach.

The second option is to use Gaussian MLE irrespective of the mpdf of the data; this is known as pseudo-MLE [1]. This implies that Gaussian MLE is a robust procedure for parameter estimation even when data are not Gaussian. Gaussian MLE converges to minimum variance unbiased quadratic iterative estimation whether or not data are Gaussian [4]. The main problem of pseudo-MLE is that Gaussian MLE is a biased estimator. Several factors induce bias in MLE estimates:

- The number of data is small. Gaussian MLE is asymptotically unbiased.
- The data are not Gaussian.
- There is a drift in the spatial mean of the data. There is a bias in the simultaneous estimation of drift and covariance parameters because some of the variability of the residual is modeled by the drift.
- Incorrect model specification. For example, if it is assumed there is no drift when one is present in the data, the range of the covariance will be overestimated.
- The ratio of the dimensions of the study area to the practical range, i.e., the number of times that the range is repeated in the study area.

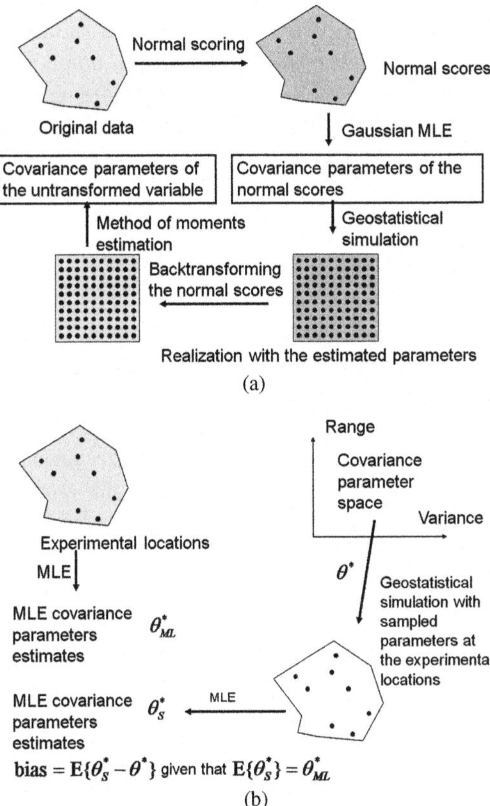

Fig. 1 (**a**) Algorithm for estimating covariances of non-Gaussian data using Gaussian MLE, (**b**) bias estimation by Monte Carlo computationally intensive procedure

Restricted MLE, or REML, works with data contrasts rather than the original data and filters a polynomial drift giving unbiased estimates of the covariance parameters. As this increases the estimation variance MLE may be preferable to REML in terms of mean square error (estimation variance plus squared bias). The ideal is to decrease bias without increasing the estimation variance but there is always a penalty in doing so: an increase in complexity [2] or computational time as in Quenouille's jackknife [7]. Another method is Monte Carlo simulation in which, for a given data configuration, the bias is estimated for each set of parameters until the estimates coincide with those obtained from the sample, as illustrated in Fig. 1b.

4 Extension to Non-polynomial Drifts

The drift in (2) is widely used because of its simplicity. An extension is to use a polynomial drift with basis functions of an external variable similar to kriging with an external drift. For example, when kriging rainfall as a function of altitude the monomials in (2) are not coordinates but altitudes; for linear external drift, $m(\mathbf{u}) = \beta_1 + \beta_2 H(\mathbf{u})$, where $H(\mathbf{u})$ is the external variable at location $\mathbf{u}$. A more flexible approach is to define the trend by a smoothing thin-plate spline where the drift may be complex but is defined by a unique smoothing parameter. The problem is the inference step because, by maximizing the likelihood, the estimated smoothing parameter tends to zero and the smoothing spline converges towards an interpolation spline. A strategy is proposed for solving this problem in Sect. 6.

5 Large Data Sets

In geosciences applications the number (n) of data can range from several tens of observations to very large data sets with more than a million values. We distinguish two cases as a function of the number n of experimental data. For a moderate number of data, $n < N$ (e.g. $N = 1000$), the full likelihood is used. This implies a large number of inversions of $n \times n$ matrices (one for each ML evaluation during optimization). In large data sets, $n > N$, the approximate ML (Vecchia's approximation) is used. This implies a large number of inversions of n $m \times m$ matrices with $m \ll n$. One obvious way to reduce computing time is to use special methods for lattice data. There is also an elegant and extremely simple idea suggested by [12], which allows MLE to be implemented for any sample size. The mpdf for $\mathbf{y} = (y_1, y_2, \ldots, y_n)$ is sequentially decomposed as:

$$p(\mathbf{y}) = p(y_1) \prod_{i=2}^{n} p(y_i \mid y_1, \ldots, y_{i-1}). \tag{11}$$

As some information provided by the data will be redundant, Vecchia [12] suggests the following approximation for the conditional probability

$$p(y_i \mid y_1, \ldots, y_{i-1}) \cong p(y_i \mid y_{i-m}, \ldots, y_{i-1}), \tag{12}$$

with $m < (i-1)$. This can be written more generally as $p(y_i \mid \mathbf{y}_{i-1}) \cong p(y_i \mid \mathbf{y}_m)$, where $\mathbf{y}_{i-1} = (y_1, y_2, \ldots, y_{i-1})$ and $\mathbf{y}_m$ has m elements sampled without replacement from $(y_1, y_2, \ldots, y_{i-1})$. The application of Vecchia's approximation requires a strategy for choosing the m data for the approximation in (12). Pardo-Igúzquiza and Dowd [6] provide software for the approximation and a strategy for monitoring the estimates as m increases. There at least two possibilities which are: place the data in a random sequence and take the m values previous values; or use a unilateral random path, where data are visited in their natural order (e.g. along rows in satellite images) and the m values are taken at random from values already visited.

6 Case Studies

For these studies we use synthetic data sets in which true values of the parameters are known. Such data sets provide more rigorous tests of the validity and performance of the methods than do real data sets in which the true values of the parameters remain unknown. Figure 2a shows a realization of a zero-mean Gaussian

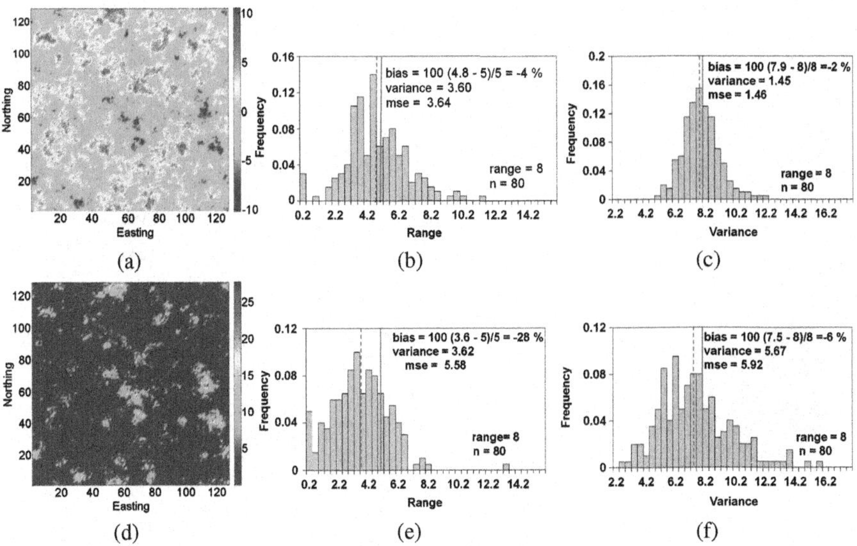

Fig. 2 (**a**) Simulated zero-mean Gaussian field with exponential covariance of range 10 and variance 2. Sampling distribution for (**b**) range and (**c**) variance for Fig. 2**a**. (**d**) Simulated squared Gaussian field with exponential covariance of range 5 and variance 8. Sampling distribution for (**e**) range and (**f**) variance for Fig. 2**d**

field with exponential covariance of range 10 units and variance 2. A non-Gaussian random field can be obtained by squaring a Gaussian random field:

$$Y(\mathbf{u}) = \left[Z(\mathbf{u})\right]^2. \tag{13}$$

This non-Gaussian field is a chi-squared field and, as the transform is not one-to-one, the Gaussian field cannot be recovered by back transformation. This is a difficult case and provides a challenging test for the methods. The covariance is:

$$C_Y(s) = 2\left[C_z(s)\right]^2. \tag{14}$$

Figure 2d is a squared Gaussian realization of Fig. 2a with exponential covariance with range 5 and variance 8. The chi-square field is retained and a new zero-mean Gaussian field with an exponential covariance of range 5 and variance 8 is simulated. We now have realizations of two random fields with the same covariance structure but different probability densities. For illustration we used $n = 80$ and repeatedly (200 times) selected n randomly located values. The covariance parameters are estimated for each of the 200 simulated samples and the bias is calculated as the differences between the means of the 200 estimates and the theoretical values. Figures 2b and 2c show the sampling distributions for range and variance. The sampling distribution means are 4.8 and 7.9 for range and variance respectively. As a percentage of the true values the biases are -4 % and -2 %. Applying the same process to the chi-squared field (Figs. 2e and 2f), increases the bias to -28 % and -6 %.

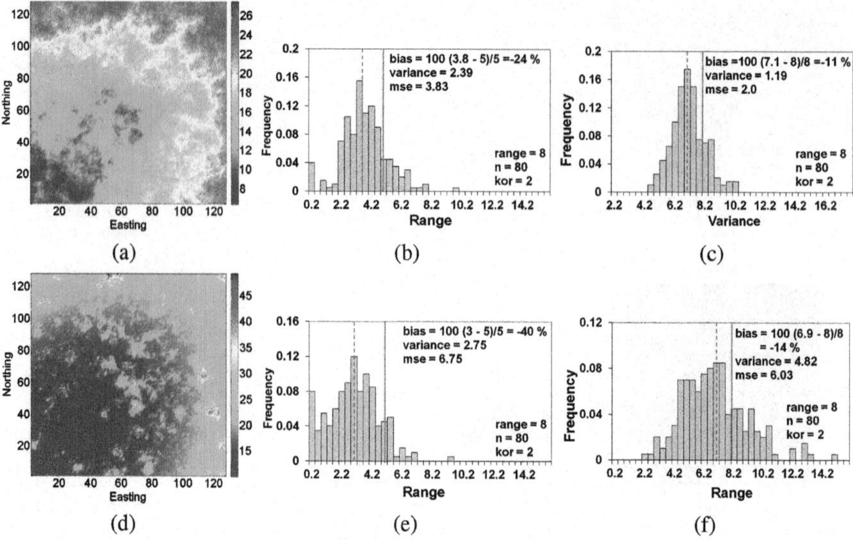

Fig. 3 (**a**) Quadratic drift plus Gaussian field in Fig. 2a. (**b**) Sampling distribution for (**b**) range and (**c**) variance for Fig. 3**a**. (**d**) Quadratic drift plus squared Gaussian field in Fig. 2d. Sampling distribution for (**e**) range and (**f**) variance for Fig. 3**d**

A drift will increase the bias. Figure 3a shows a quadratic drift added to the Gaussian field of Fig. 2a. In Figs. 3b and 3c the bias increases to -24 % for range and -11 % for variance. When drift is added to the chi-squared field (Fig. 3d), the resulting bias is -40 % for range and -14 % for variance. Thus, when there is a drift and the random field is not Gaussian the bias increases almost as the sum of the two factors that influence the bias: the drift and the non-Gaussian character of the random field. In addition, for biased estimates the estimation variance from the Fisher information matrix underestimates the probability coverage of interval estimates using the biased estimate and the estimation variance. For example, for the chi-squared field with no drift (Fig. 2d), the estimate ± 1 standard deviation has a coverage of 49 % and 38 %, which differ significantly from the nominal 68 % Gaussian coverage. It is thus of interest to correct the bias of a realization of a non-Gaussian field. As the normal scores transform cannot recover the original Gaussian realization from the chi-squared field, methods other than that depicted in Fig. 1a are required. For polynomial drift REML [4] can be used to obtain unbiased estimates at the expense of increasing the estimation variance and thus the mean square error. However, REML is of no use when bias is introduced by a non-Gaussian density and for this case we propose the most likely unbiased estimator, illustrated in Fig. 4a. ML provides an estimate and the standard deviation of the estimation error. Assuming a Gaussian distribution for the error, the interval of the estimate ± 0.68 standard deviations, covers a probability of 50 %. As the bias is negative the parameter target value is the upper limit of the interval and this is taken as the most likely value. The results for this estimate are given in Figs. 4b and 4c for range and variance. The order of the bias is similar to the Gaussian case and therefore acceptable.

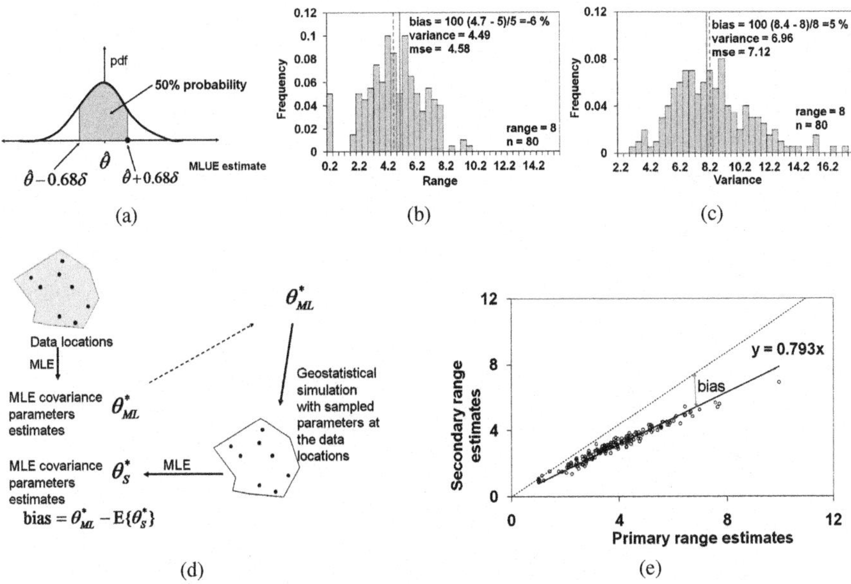

Fig. 4 (**a**) Most likely unbiased estimator. Sampling distribution for most likely unbiased estimator of (**b**) range and (**c**) variance for Fig. 3a. (**d**) Simplified version of the algorithm in Fig. 1b. (**e**) Primary and secondary estimates of range for Fig. 3d

For bias in the drift, the alternative to REML proposed here is illustrated in Fig. 1b. This method is computationally expensive and a simplification is given in Fig. 4d. Assuming MLE estimates of the drift and covariance parameters are correct, we can calculate the experimental bias of MLE by generating a large number of realizations using the parameters and a given sampling configuration and estimating the same parameters by MLE. The bias can then be estimated and the parameters can be corrected. The original (primary) estimates and the (secondary) estimates from realizations with those parameters are shown in Fig. 4e for the range. As expected, the bias increases as the range increases as the longer the range, the higher will be the variability modelled by the quadratic drift and hence the bias will be higher. Thus by estimating bias as the difference between the actual estimate and the mathematical expectation of the realizations generated using those parameters, the parameter estimates can be corrected for the bias introduced by the drift. As shown in Fig. 5 the biases are −8 % for range and 9 % for variance. For the uncorrected estimates shown in Fig. 3b the biases are −24 % for range and −11 % for variance.

The third extension is related to the trend or drift of the random function. A low order polynomial trend as in (2) is sufficient for most applications. The monomials in Table 1 can be replaced by trigonometric functions giving a trend with trigonometric polynomial [11]. However, in some cases a more convoluted model may be

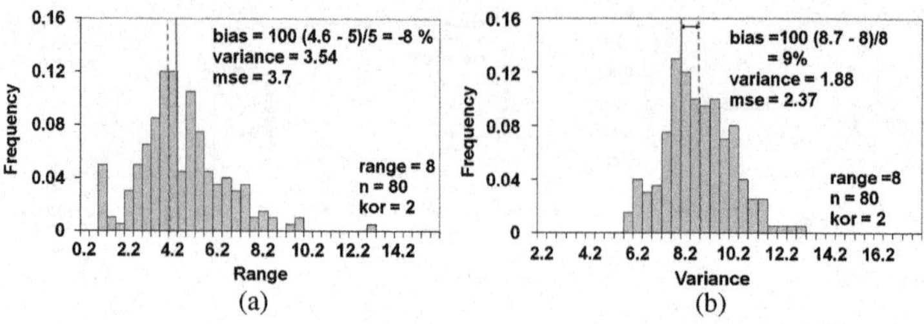

Fig. 5 (**a**) Sampling distributions (**a**) for range and (**b**) for variance for Fig. 3a using method in Fig. 4d and e

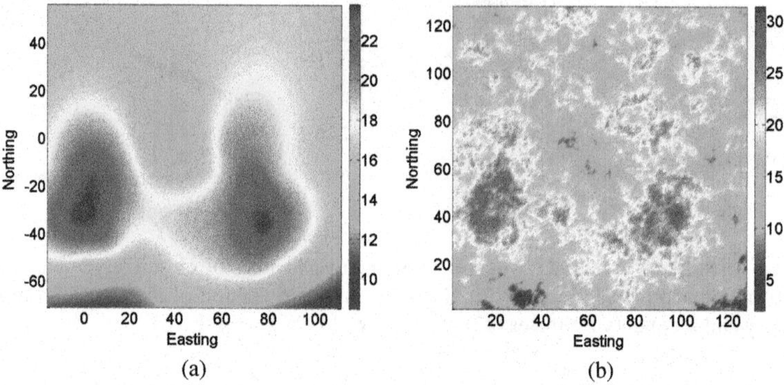

Fig. 6 (**a**) Non-polynomial drift. (**b**) Realization of drift and residual (Figs. 6**a** + 2a)

present and a more flexible model may be desirable. For example, a smoothing thin plate spline [13], $\mathbf{m}$, could be used. The corresponding MLE estimate is

$$l(\hat{\sigma}^2, \theta; z) = -\frac{n}{2}\ln 2\pi - \frac{n}{2} + \frac{n}{2}\ln n - \frac{1}{2}\ln|\rho| - \frac{n}{2}\ln(\mathbf{z}-\mathbf{m})'\rho^{-1}(\mathbf{z}-\mathbf{m}), \quad (15)$$

where the variance is estimated as $\hat{\sigma}^2 = \frac{1}{n}(\mathbf{z}-\mathbf{m})'\rho^{-1}(\mathbf{z}-\mathbf{m})$ and the smoothing thin plane spline $[\mathbf{m}]_i = g_i$ minimizes the function:

$$O(\lambda) = \sum_{i=1}^{n} \|z_i - g_i\|^2 + \lambda \iint \left(\left(\frac{\partial^2 g}{\partial x^2}\right)^2 + 2\left(\frac{\partial^2 g}{\partial x \partial y}\right)^2 + \left(\frac{\partial^2 g}{\partial y^2}\right)^2 \right) \mathrm{d}x\,\mathrm{d}y, \quad (16)$$

where λ is the smoothing parameter or the trade-off between a smooth function and a less smooth function that is a better fit to the data. This parameter cannot be estimated using MLE because the estimate will tend to zero making $\mathbf{m}$ as close to $\mathbf{z}$ as possible. Here we have run MLE with a quadratic drift and then estimated λ as the parameter that produces a smoothing thin plate spline drift that explains the same amount of variability (variance) as the quadratic drift estimated by MLE.

An example non-polynomial drift is shown in Fig. 6a. Figure 6b shows a realization of this drift plus the realization of the residual in Fig. 2a. The quadratic drift estimated from $n = 80$ data selected at random (Fig. 7a) is shown in Fig. 7b.

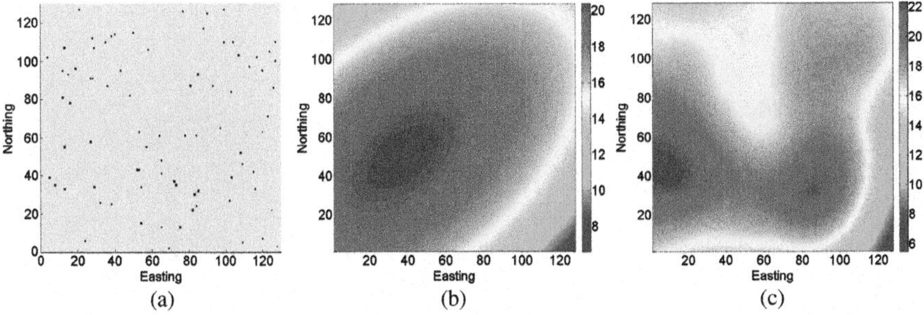

Fig. 7 (**a**) Locations at which the field in Fig. 6b is sampled. (**b**) Quadratic drift estimated from the values in Fig. 6b at the data locations in Fig. 6a. (**c**) Thin plate spline drift estimated from the same data and with $\lambda = 90$

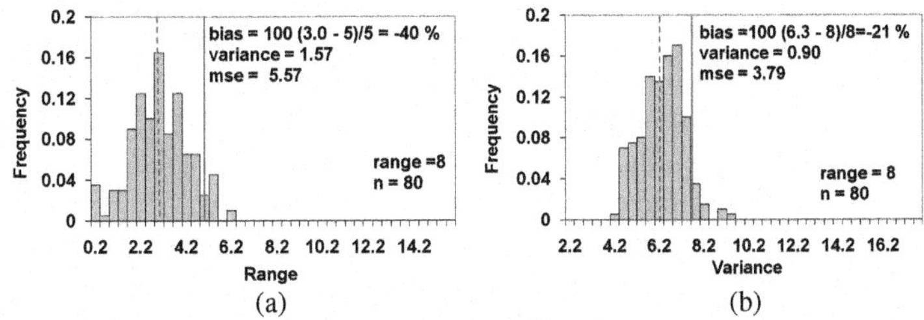

Fig. 8 Sampling distributions for (**a**) range and (**b**) variance for the field in Fig. 6b using a thin plate spline model for the drift

This drift explains a variance of 5.94 which is also explained by the smoothing thin plate spline drift in Fig. 7c for $\lambda = 90$. Other values of λ give trends that explain other amounts of variability; e.g., for $\lambda = 10$ the explained variance is 10.62 and for $\lambda = 500$ it is 3.18. Repeating the estimation using $\lambda = 90$ and for $n = 80$ data generates the sampling distributions in Fig. 8. The penalty for using a non-parametric adaptive drift is an increase in the bias of the covariance parameters. For large data sets an efficient approximation strategy in (12) is important. There are several possibilities for choosing the m neighbors. One is to order the elements of an array or image at random and then in (12) use the m previous elements in the new sequence. The advantage is that all types of distances are included in the approximation. This is not so if the original data ordering is used with the m nearest neighbors. Unilateral random ordering applied to the original order will include all types of distances. A 200×200 image was simulated with an exponential covariance of range 8 and unit variance. Figure 9 shows: (a) the realization, (b) the semi-variogram of the realization and the model and (c) the results of Vecchia approximation. Note, that as m increases, the estimate approaches the target value; convergence is faster for the unilateral random strategy than for random ordering.

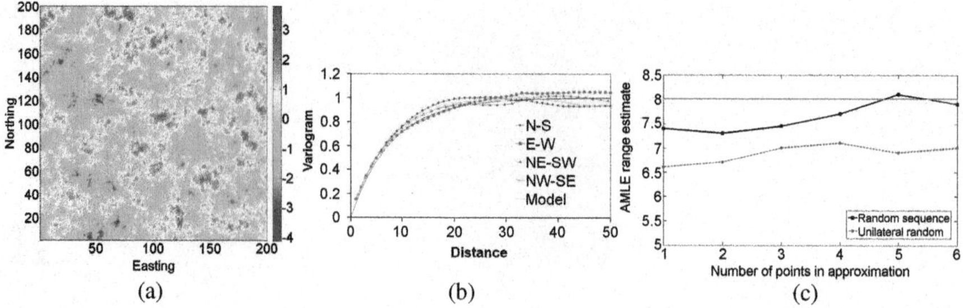

Fig. 9 (**a**) Simulation with exponential semi-variogram with range 8 and unit variance. (**b**) Semi-variogram of simulated values and model semi-variogram specified for simulation. (**c**) Vecchia approximation

7 Conclusions

MLE is an appealing method for the inference of the parameters of a covariance or semi-variogram model from a set of sparse data. The ideal situation for applying MLE is when the data follow a multivariate Gaussian distribution, there is no trend in the data and there are sufficient data. Gaussian MLE is asymptotically biased and thus with small data sets the estimator may be slightly biased. For Gaussian MLE with no drift and a sample size of $n = 80$, the bias for range and variance is of the order of 5 % of the value of the parameter. We have discussed four circumstances that increase the bias of Gaussian MLE and provided methods for bias correction for each. For non-multivariate Gaussian data the estimate plus 0.68 times the standard deviation (as given by MLE from the Fisher information matrix) decreases the bias to values of the order of the Gaussian case (around 5 %). For polynomial drift we have provided a computationally intensive algorithm that yields an acceptable bias (around 10 %). Finally, the Vecchia approximation can be applied with any of the strategies presented here and the value of m can be increased until there are no changes in the estimate (usually reached for m between 10 and 20). Thus, MLE can always be applied to infer the covariance parameters provided some additional computation is done in order to have an unbiased estimate.

Acknowledgements This work was supported by research project CGL2010-15498 from the Ministerio de Ciencia e Innovación Spain. The authors acknowledge the funding provided by Australian Research Council Grant DP110104766.

References

1. Bard Y (1974) Nonlinear parameter estimation. Academic Press, New York
2. Firth D (1993) Bias reduction of maximum likelihood estimates. Biometrika 30(1):27–38
3. Kitanidis PK, Lane RW (1985) Maximum likelihood parameter estimation of hydrologic spatial processes by the Gauss-Newton. J Hydrol 79(1–2):53–71
4. Kitanidis PK (1987) Parametric estimation of covariances of regionalized variables. Water Resour Bull 23(4):671–680

5. Mardia KV, Marshall RJ (1984) Maximum likelihood estimation of models for residual covariance in spatial statistics. Biometrika 71(1):135–146
6. Pardo-Igúzquiza E, Dowd PA (1997) AMLE3D: a computer program for the statistical inference of covariance parameters by approximate MLE. Comput Geosci 23(7):793–805
7. Quenouille MH (1956) Notes on bias estimation. Biometrika 43:353–360
8. Ripley BD (1981) Spatial statistics. Wiley, New York
9. Russo D, Jury WA (1987) A theoretical study of the estimation of the correlation scale in spatially variable fields: 1. Stationary fields. Water Resour Res 23(7):1257–1268
10. Stein ML (1999) Interpolation of spatial data: some theory for kriging. Springer, New York
11. Thorpe WR, Rose CW, Simpson RW (1979) Areal interpolation of rainfall with a double Fourier series. J Hydrol 42(1–2):171–177
12. Vecchia AV (1988) Estimation and model identification for continuous spatial processes. J R Stat Soc B 50(2):297–312
13. Wahba G (1990) Spline models for observational data. SIAM, Philadelphia
14. Warnes JJ, Ripley BD (1987) Problems with the likelihood estimation of covariance functions of spatial Gaussian processes. Biometrika 74(3):640–642

Part II
Petroleum

Constraining a Heavy Oil Reservoir to Temperature and Time Lapse Seismic Data Using the EnKF

Yevgeniy Zagayevskiy, Amir H. Hosseini, and Clayton V. Deutsch

Abstract Proper understanding of the distribution of porosity and vertical permeability is required for optimal extraction of hydrocarbons. While porosity defines the amount of oil resources, permeability determines the fluid flow in the reservoir from drainage area to production wells. Steam assisted gravity drainage (SAGD) is a method for thermal recovery of heavy oil and bitumen that is widely employed in Northern Alberta. Vertical permeability is important in SAGD since it determines communication between the reservoir and paired horizontal injector and producer wells. Additional data will improve reservoir characterization. The Ensemble Kalman Filter (EnKF) is proposed to constrain the spatial distributions of porosity and permeability by integrating core measurements, dynamic temperature observations and time-lapse seismic attributes. The proposed methodology is demonstrated with a synthetic 2D case study. Implementation details are discussed. Integration of temperature is presented for a realistic case study based on the Tucker thermal project. The EnKF is shown to be an effective and promising modeling technique.

1 Introduction

Petroleum reservoir characterization is a continuous and vital part of oil field development. Ultimate oil recovery depends on the quality of the reservoir model and especially on estimates of porosity and permeability. While the spatial distribution of porosity determines the amount of deposited hydrocarbon, the distribution

Y. Zagayevskiy (✉) · C.V. Deutsch
Center for Computational Geostatistics, University of Alberta, 3-133 NREF Building, Edmonton, Alberta, Canada, T6G 2W2
e-mail: zagayevs@ualberta.ca

C.V. Deutsch
e-mail: cdeutsch@ualberta.ca

A.H. Hosseini
Husky Energy Corporation, 707-8 Ave. SW, Station D, Box 6525, Calgary, Alberta, Canada, T2P 3G7
e-mail: amir.hosseini@huskyenergy.com

P. Abrahamsen et al. (eds.), *Geostatistics Oslo 2012*,
Quantitative Geology and Geostatistics 17,
DOI 10.1007/978-94-007-4153-9_12,

of permeability defines flow zones, baffles, and barriers [3]. Proper estimation of vertical permeability is of the highest importance in modeling of oil sands reservoirs in Northern Alberta, which are operated by the thermal steam assisted gravity drainage (SAGD) oil extraction method [2]. Permeability defines the communication level between the producers and injectors and how efficiently heated bitumen drains to the producer [4].

The spatial distributions of porosity and permeability are usually estimated using core data, but dynamic data such as production rates can additionally constrain distributions of these static properties. Continuous temperature observations and time-lapse seismic attributes can help characterize reservoirs even further [18]. Reservoir temperature profiles are continuously measured by permanent thermocouple gauges installed along vertical observation wells in the SAGD deposits of McMurray and Clearwater formations [11]. The spatial distribution of induced temperature is coupled with the distribution of hot injected steam and, thus, temperature profiles can be used to estimate permeability and predict steam chamber growth [2, 15, 22]. Time-lapse seismic surveys are frequently conducted over SAGD fields [11, 16]. High resolution 4D seismic depicts reservoir architecture and provides knowledge about the porosity distribution [24]. The difference between seismic attributes from a baseline survey and any subsequent survey can reveal changes in fluid saturations that helps understand dynamic changes in the reservoir [22].

Geological properties of the reservoir are usually modeled by conventional geostatistical methods like sequential Gaussian simulation (SGS) [12]. However, these methods cannot continuously assimilate valuable soft data, unless the semivariogram model is recalculated and the entire model is rebuilt every time new data becomes available [3, 12]. Thus, the Ensemble Kalman Filter (EnKF) is proposed for simultaneous integration of core data, continuous reservoir temperature observations, and pressure- and temperature-dependent time-lapse seismic attributes in order to estimate the spatial distributions of porosity and permeability [6]. While integration of seismic attributes is an old idea [8, 17, 20], temperature data assimilation is a relatively novel research direction [15]. The integration of exhaustive seismic data is a challenging task, which is resolved by localization techniques [8, 20]. The main advantage of EnKF is that it handles the relationship between static and dynamic variables as a black box and, thus, there is no need to solve analytical equations of flow, mass, and energy conservations like in gradient-based and master point history matching techniques [10, 19].

This paper is organized as follows. First, the theoretical background of EnKF is presented. Second, the implementation details of EnKF are discussed. Third, the proposed methodology for integration of temperature data and seismic attributes is thoroughly explained. Fourth, this methodology is applied to a synthetic 2D SAGD case study and then to a realistic 3D SAGD case study adapted from the Tucker thermal project [11]. Finally, results are discussed and conclusions are made.

2 Theoretical Background of EnKF

Petroleum reservoirs are complex dynamic systems that can be effectively approximated by a numerical model U_t conditioned to available hard and soft data D_t. A dynamic model U_t consists of two distinct types of spatial variables: static model parameters M that do not vary in time, such as porosity and permeability, and dynamic state variables S_t that do vary in time like temperature and time-lapse seismic. Dynamic variables depend on the model parameters, initial S_0 and boundary S_t^b conditions. A mathematical expression of a dynamic system is shown in (1) in vector form. Both hard and soft data information about the model parameters can be used for estimation of porosity and permeability distributions [18]:

$$U_t = \begin{pmatrix} M \\ S_t \end{pmatrix}_{N_v \times 1} = \begin{pmatrix} M \\ F(M, S_0, S_t^b; t) \end{pmatrix}_{N_v \times 1}, \tag{1}$$

$$D_t = G(M; t)_{N_{d_t} \times 1}, \tag{2}$$

where N_v is the sum of N_m model parameters and N_s states, and F and G are the operators that relate state variables and data respectively to model parameters. The model operator F can be represented by governing flow, mass, and energy conservation equations or simply by a flow simulator coupled with a petroelastic model.

The Ensemble Kalman Filter (EnKF) technique is proposed to continuously assimilate both static hard and dynamic soft data into petroleum reservoir models in order to estimate model parameters and history match state variables [6]. The EnKF is a numerical extension of the Kalman Filter (KF) [13]. It utilizes the concept of Monte Carlo simulation (MCS). The set of N_e statistically probable realizations represents different forms of a dynamic system, which are stored in $N_v \times N_e$ matrix $\mathbf{U}_t$. The number of realizations N_e is referred to as the ensemble size. This modeling method derives the best estimate of model parameters, forecasts state variables, and reports associated uncertainty. The covariance function is replaced with the sample covariance matrix derived from the ensemble members.

In modeling, an optimization problem is solved by adjusting the model to the data. The objective function O is expressed as a weighted sum of the squared differences between the estimated and true values of the model [10]:

$$O = \sum_{t=1}^{N_t} \sum_{i=1}^{N_{d_t}} w_{i,t} \times (u_{i,t} - d_{i,t})^2, \tag{3}$$

where $d_{i,t}$ and $u_{i,t}$ are the ith datum and corresponding model estimate at time t; $w_{i,t}$ is the weight that represents the importance of the ith datum; N_t is the number of time steps at which data are sampled; N_{d_t} is the number of data assimilated at t.

The EnKF solves a Bayesian optimization problem where the unconstrained objective function is represented by a product of prior Gaussian probability and data likelihood functions [6]. The solution of this problem leads to the analysis equation of EnKF:

$$\mathbf{U}_t^a = \mathbf{U}_t^f + \mathbf{K}_t\left(\mathbf{D}_t - \mathbf{H}_t\mathbf{U}_t^f\right), \tag{4}$$

where f and a are the superscripts indicating model at forecast and analysis steps. Expression (4) linearly updates the ensemble $\mathbf{U}_t$ based on a covariance matrix and the difference between model estimates and integrated data. The nonlinear forecast equation

$$\mathbf{U}_t^f = F\left(\mathbf{U}_{t-1}^a\right) + \mathbf{E}_t \tag{5}$$

predicts states of the model at future time steps. Since Gaussian prior and likelihood distributions are assumed, the EnKF finds its best application in modeling of Gaussian systems. Hence, nonlinear forecast and linear analysis equations of EnKF form a continuous two-step data assimilation algorithm [6] when we include the two final equations:

$$\mathbf{K}_t = \hat{\mathbf{C}}_t^f \mathbf{H}_t^T \left(\mathbf{H}_t \hat{\mathbf{C}}_t^f \mathbf{H}_t^T + \mathbf{R}_t\right)^{-1}, \tag{6}$$

$$\hat{\mathbf{C}}_t^f = \left(\mathbf{U}_t^f - \bar{\mathbf{U}}_t^f\right)\left(\mathbf{U}_t^f - \bar{\mathbf{U}}_t^f\right)^T / (N_e - 1), \tag{7}$$

where $\mathbf{E}_t$ is the error matrix of the model, which is assumed to be zero; $\mathbf{K}_t$ is the weighting matrix or Kalman gain, it is a function of $N_v \times N_v$ sample covariance matrix $\hat{\mathbf{C}}_t^f$ of the forecasted model, $N_{d_t} \times N_{d_t}$ diagonal covariance matrix $\mathbf{R}_t$ of measurement error and $N_{d_t} \times N_v$ observation matrix $\mathbf{H}_t$, which relates forecasted model values to corresponding data; $\bar{\mathbf{U}}_t^f$ is the mean of forecasted model over realizations. Every realization j of single i^{th} datum $d_{j,i,t}$ consists of a true value $d_{i,t}$, an unknown measurement error $\varepsilon_{i,t}$, and a normal perturbation $\nu_{j,i,t}$ with zero mean and variance of measurement error that defines matrix $\mathbf{R}_t$:

$$d_{j,i,t} = d_{i,t} + \varepsilon_{i,t} + \nu_{j,i,t}. \tag{8}$$

Since the covariance matrix is used only partially at each analysis step, (4) and (6), EnKF can be easily applied to large systems. However, its main drawback lies in the computational cost of the forecast step, which is determined by N_e flow simulation runs. A shortcut is proposed in the next section to reduce the computational time.

The conventional EnKF algorithm can be described as follows:

1. Generate an initial ensemble of model parameters $\mathbf{M}_{t=0}^a$ (e.g., geological properties) that follows a Gaussian distribution using all prior information (data, conceptual model, geological interpretation, etc.), and run a first forecast step (5) to get an initial ensemble of state variables $\mathbf{S}_{t=1}^f$. A full model $\mathbf{U}_{t=1}^f$ is defined.
2. Execute the analysis step (4), where the entire model $\mathbf{U}_{t=1}^a$ is linearly updated with all assimilated data $\mathbf{D}_{t=1}$ sampled at time step t.
3. Go back to step 1 and use updated values of model parameters $\mathbf{M}_{t=1}^a$ as initial ensemble to forecast state variables $\mathbf{S}_{t=2}^f$ at next time step $t = t + 1$. Then again update entire ensemble $\mathbf{U}_{t=2}^a$ with newly acquired data $\mathbf{D}_{t=2}$ according to step 2. Continue the procedure until all data from all time steps are integrated into the model and the objective function is minimized. Last, the forecasted ensemble represents predicted values of states where no data are available.

3 Characteristics and Implementation Details of EnKF

Systems Modeled by EnKF The EnKF is good for linear systems that follow normal or Gaussian distribution. Modifications are required for highly nonlinear systems. An iterative procedure is implemented, in which data are iteratively integrated into the model to obtain a desirable history match [21]. The Ensemble Kalman Smoother (EnKS) may be applied as well, in which data from all time steps are assimilated at the same time [7]. The Randomized Maximum Likelihood (RML) method may be used for highly nonlinear systems, but the method is very sensitive to the initial ensemble and requires a large number of realizations. For highly non-Gaussian models, normal score transformation may be performed [3]. Multimodal categorical variables, such as facies, can be approximately modeled by a Gaussian mixture model (GMM) [5].

Initial Ensemble The initial ensemble is important. All prior information should be used for its generation. An improper ensemble will lead to slow convergence or may even produce estimates that are far from reality. If SGS is chosen as the tool for generation of the initial ensemble, a variogram $\gamma(h)$ and global mean must be chosen.

Computational Cost Computational cost of the analysis step is determined by the number of assimilated data and ensemble size. The ratio of the number of data to the ensemble size should be around 1:10 [8], otherwise there are insufficient degrees of freedom, which lead to long-range spurious covariances and ensemble collapse. This is a crucial issue for integration of exhaustive data like seismic responses. Localization of updating and covariance matrices, (9) and (10), is proposed to resolve this problem and to decrease the number of required realizations [8]. The localization of the covariance matrix produces better results without artifacts in contrast to estimates with localized updating matrix [23]:

$$\mathbf{K}_t^u = \hat{\mathbf{C}}_t^f \mathbf{H}_t^T [\mathbf{L}_t^u]^T (\mathbf{H}_t \hat{\mathbf{C}}_t^f \mathbf{H}_t^T + \mathbf{R}_t)^{-1}, \tag{9}$$

$$\mathbf{K}_t^c = [\mathbf{L}_t^c \circ \hat{\mathbf{C}}_t^f] \mathbf{H}_t^T (\mathbf{H}_t [\mathbf{L}_t^c \circ \hat{\mathbf{C}}_t^f] \mathbf{H}_t^T + \mathbf{R}_t)^{-1}, \tag{10}$$

where $\mathbf{L}_t^u$ is the $N_v \times N_{d_t}$ updating localization matrix of zeros and ones that determines the part of the ensemble to be updated; $\mathbf{L}_t^c$ is the $N_v \times N_e$ covariance localization matrix represented by a continuous function; and $\circ$ is the element-wise product.

The computational cost of the forecast step is mainly determined by the flow simulation time and the number N_e of flow simulation runs. A shortcut based on the ensemble mean and co-simulation of expensive dynamic state variables using SGS is developed to use only one flow simulation run [23]. The ensemble mean of model parameters is used to derive an ensemble mean of state variables by one flow simulation run. The N_e realizations of random values with zero mean and covariance structure of the state variables are generated by SGS conditional to model parameters. The forecasted ensemble mean of the states is added to every realization to obtain an ensemble of state variables. The obtained realizations are geostatistical approximations of the state variables, which should not be used separately to describe spatial distributions of the states. This modification to EnKF gives an approximate solution with reduced computational time.

4 Methodology for Integration of Static and Dynamic Data by Means of EnKF

The methodology for integration of static core measurements of porosity ϕ and permeability K along with dynamic temperature T data and the difference in P-wave acoustic impedances ΔZ_p into a petroleum reservoir model is developed (Fig. 1). While the difference in impedances is assimilated into the model by EnKS, the rest of the data are assimilated by EnKF. Iterative data assimilation may be required to honor data from all time steps. The sampling frequency is the same for all data.

The proposed static and dynamic data assimilation algorithm is as follows. An initial porosity ensemble is generated using SGS and the proper semivariogram model $\gamma(h)$ derived from porosity data at time step $t = 0$. If unconditional porosity realizations are considered, the mean of the initial ensemble should be homogeneous and realizations should follow a normal distribution with mean and variance of the porosity data.

The model operator F is defined by three sub-models: porosity-log permeability transform, thermal flow simulator, and the petroelastic model. Values of horizontal permeability are obtained from probabilistic porosity-log permeability relationship [4]:

$$\log(K_{xx}) = a + b\phi + \varepsilon. \tag{11}$$

Other components of the permeability tensor K are derived using the known relationship with K_{xx}:

$$K_{yy} = \alpha(K_{xx}) \quad \text{and} \quad K_{zz} = \beta(K_{xx}), \tag{12}$$

where K_{xx}, K_{yy}, and K_{zz} are the diagonal components of the permeability tensor K, all other non-diagonal components are assumed to be zero; a and b are the linear regression coefficients; and ε is the probabilistic error in regression model. The relationships are found empirically.

Generated porosity and permeability ensembles along with initial and boundary conditions of reservoir pressure, temperature, water, oil, and gas saturations are used as input to thermal flow simulator, e.g. CMG's Steam, Thermal, and Advanced Processes Reservoir Simulator (STARS), to predict state variables [3]. The pressure- and temperature-dependent Gassmann's fluid substitution model is selected as a petroelastic model for generation of synthetic seismic attributes [9]. The mineralogical content of the rock, its physical and elastic properties, porosity, reservoir pressure, temperature, and fluid saturations at initial and current time steps are required to generate the difference between P-wave acoustic impedances [1, 14].

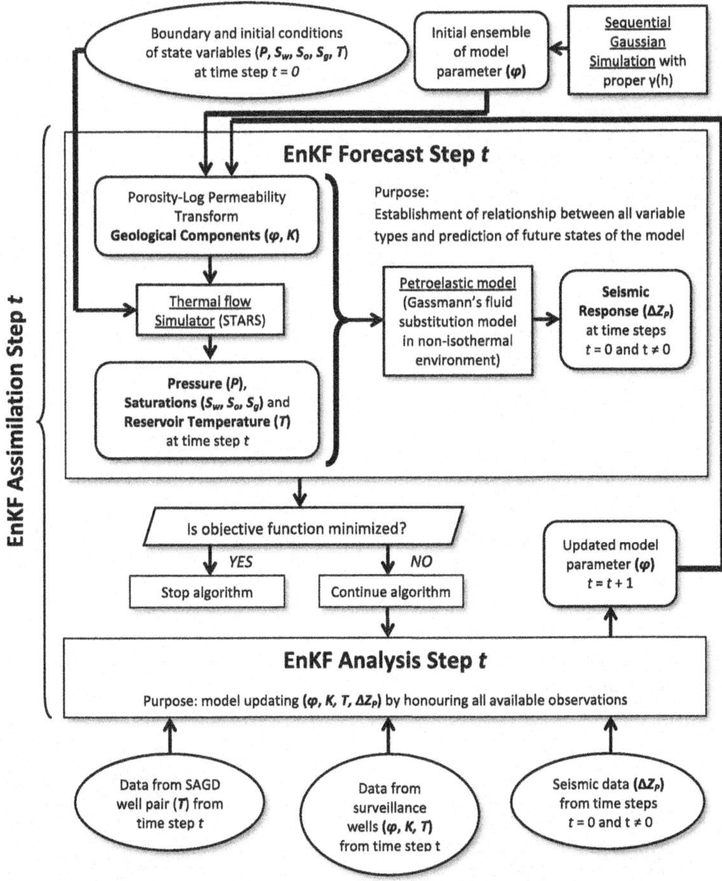

Fig. 1 Methodology for integration of static and dynamic data by means of EnKF

At this forecast step t the entire model $\mathbf{U}_t^f$ is defined. If a good match between estimates and data is achieved, the algorithm is terminated. Otherwise the analysis step is performed, at which data from various sources are integrated into the model (4). Porosity, permeability, temperature and the difference in acoustic impedances are linearly updated. The updated porosity ensemble is used to generate permeability and to forecast states at the next time step $t + 1$. The procedure is repeated until the objective function is optimized.

5 Synthetic 2D SAGD Case Study

The methodology is applied to synthetic 2D SAGD case study with a single well pair and two observation wells. The model grid consists of 1500 (50 $\times$ 30) blocks of size

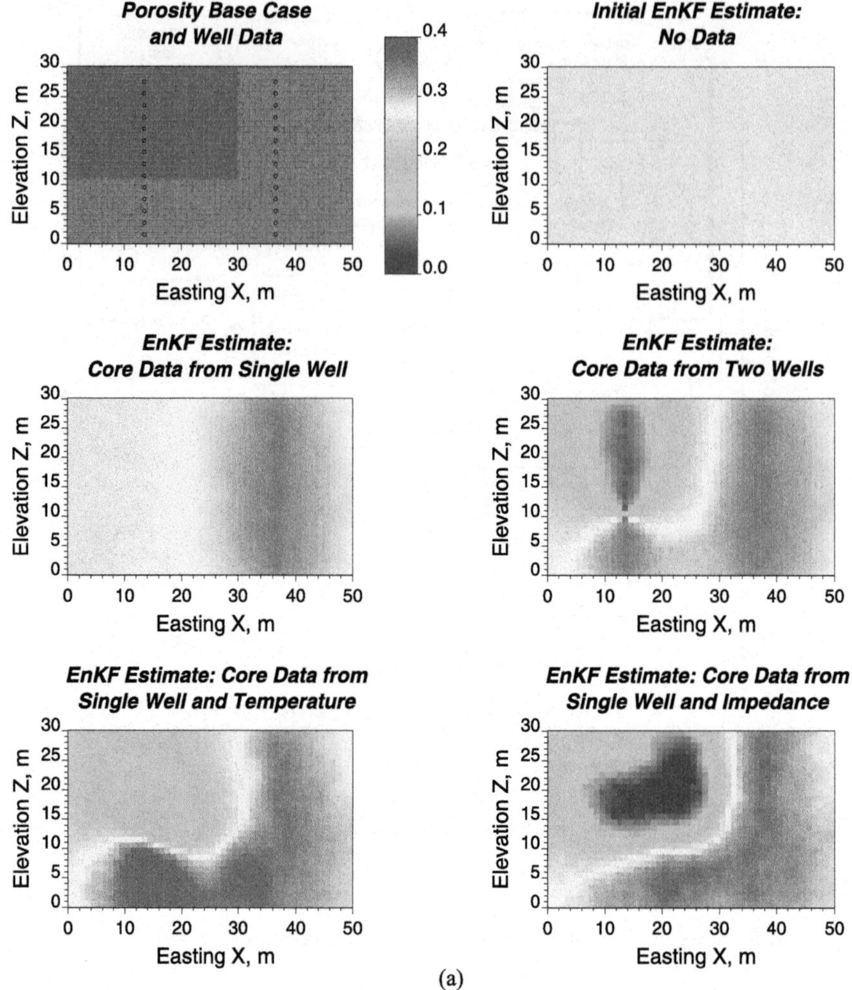

Fig. 2 Base case of porosity and realization means of EnKF porosity estimates (**a**) with associated estimation variances (**b**): no data are used in estimation (initial ensemble), core porosity data from single eastern well, core porosity data from two wells, core data from single well and temperature data, and core porosity from two wells and difference in seismic impedance from both time steps

1.0×1.0 m. It is assumed that porosity and K_{xx} are perfectly correlated, and that $K_{yy} = K_{xx}$ and $K_{zz} = 0.5K_{xx}$. The base case of porosity with observation locations are shown in the top left corner of Fig. 2. The goal of this example is to identify low and high value zones of the porosity field using available static and dynamic data. Twenty eight porosity data are sampled from two vertical observation wells. Same number of temperature data is observed along the same vertical observation wells and SAGD well pair from two time steps: 180 and 270 days after production begins. Seismic attribute is sampled exhaustively: 250 samples for each 180 and 270 days

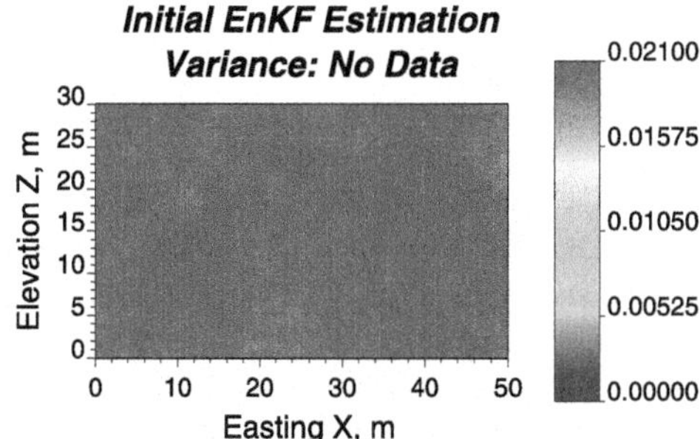

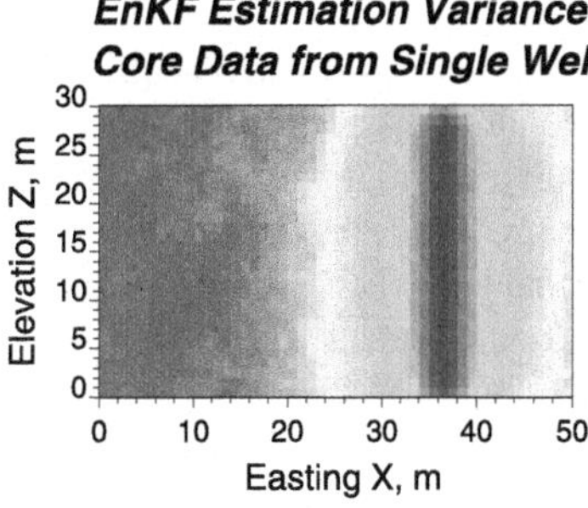

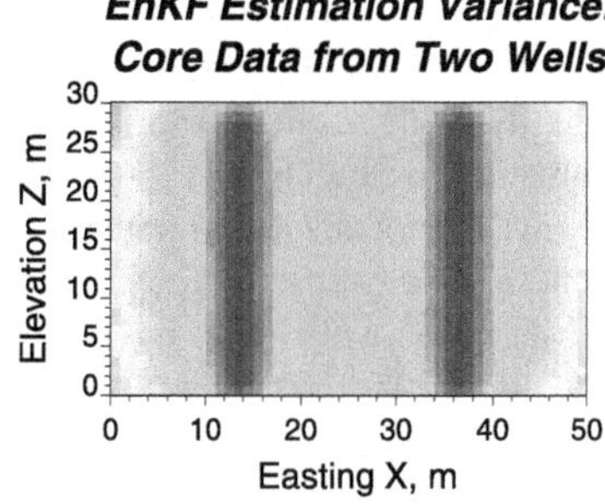

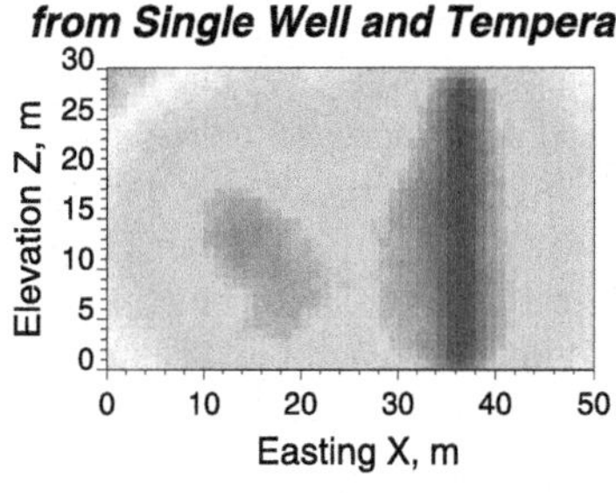

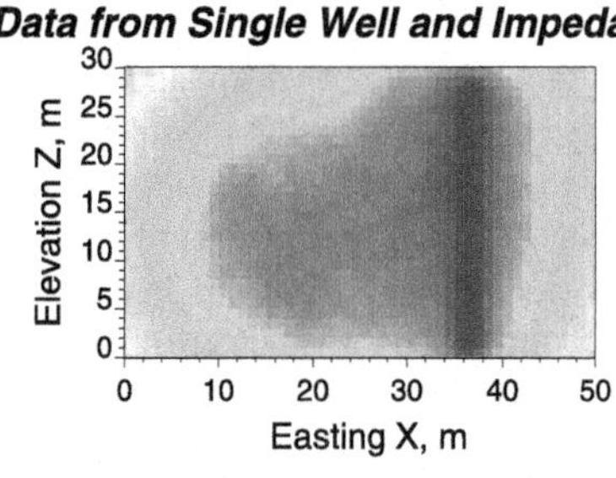

(b)

Fig. 2 (continued)

time step. No permeability data are retained, since they are redundant with porosity data.

Five cases are examined to show the value of additionally assimilated data and the importance of continuous soft data integration. The cases are listed in Table 1 with corresponding validation results for porosity estimates. Root mean squared error (RMSE), mean standard deviation (MSD) (14), and linear correlation coefficient (CC) are used as validation measures. Smaller value of the RMSE and the CC closer to 1.0 indicate a better match between porosity estimate and base case. Smaller value of the MSD represents better convergence of the realizations. A homogeneous initial

Table 1 Comparison of EnKF porosity estimates derived from different datasets

Case	Description	Data	RMSE*	MSD†	CC
1	Homogeneous conceptual model (initial ensemble)	0	0.192	0.142	0.254
2	ϕ data from single eastern observation well	14	0.172	0.115	0.506
3	ϕ data from both observation wells	28	0.117	0.089	0.835
4	ϕ data from single well and T data from time step 1	42	0.143	0.096	0.800
	ϕ data from single well and T data from time steps 1, 2	70	0.115	0.085	0.864
5	ϕ data from single well and ΔZ_p data from time step 1	264	0.110	0.090	0.855
	ϕ data from single well and ΔZ_p data from time steps 1, 2	514	0.091	0.076	0.899

porosity ensemble consisting of 1000 realizations is generated honoring the mean and spatial distribution pattern of core data. The data are assimilated sequentially for cases 4 and 5. The localization of the covariance matrix is applied with an exponential weighting function (13) in order to effectively assimilate exhaustive seismic data. However, localized estimates possess less accuracy and a smoother spatial distribution than conventional EnKF estimates. For instance, when porosity data from both observation wells (case 3) are assimilated without localization, RMSE, MSD, and CC are 0.105, 0.084, and 0.866 respectively, which indicate better estimation results (see Table 1):

$$\mathbf{L}_t^c = \exp\left(-3\frac{h}{\sqrt{B_x^2 + B_y^2 + B_z^2}}\right), \tag{13}$$

where h is the distance between assimilated datum and estimated location; B_x, B_y, and B_z are the length, width, and height of the model grid respectively.

These results show that incorporation of additional data improves the porosity estimate and decreases uncertainty in the estimation (Fig. 2). Core porosity data from the eastern well does not capture the low value zone and, therefore, the barrier is not reproduced. However, if core data from both wells are employed, then better knowledge about this poor quality region is obtained. Temperature and the difference in acoustic impedances observations from two time steps along with core data from single well improves the understanding of the porosity distribution and shows the boundary with higher precision and less uncertainty. Even though exhaustive seismic data produces the best porosity estimate, it might be more feasible to consider only well data due to the sampling cost. Note that integration of continuous data from both time steps leads to better porosity estimates in comparison to estimates derived using data only from a single time step (cases 4 and 5 in Table 1). Thus, EnKF is seen as a simple and effective estimation method for continuous integration of core data, dynamic temperature observations, and 4D seismic attributes.

$$^{*}RMSE = \sqrt{\frac{\sum_{i=1}^{N_b}\sum_{j=1}^{N_e}(\phi_{i,j}^{est} - d_i^{true})^2}{N_b N_e}} \quad \text{and}$$
$$^{\dagger}MSD = \sqrt{\frac{\sum_{i=1}^{N_b}\sum_{j=1}^{N_e}(\phi_{i,j}^{est} - \bar{\phi}_i^{est})^2}{N_b N_e}}, \tag{14}$$

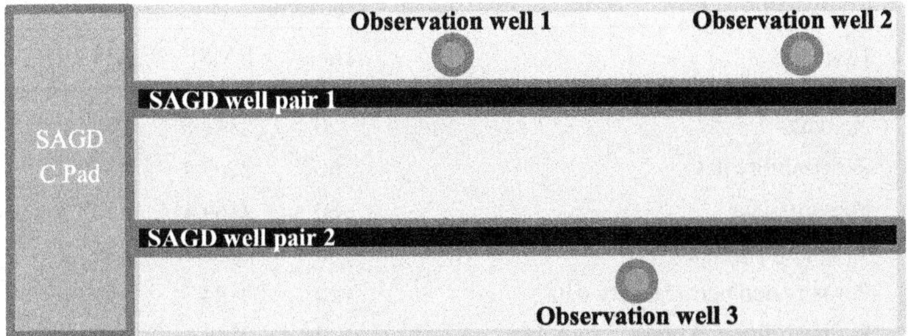

Fig. 3 Schematic plan view of reservoir model around two SAGD well pairs of C Pad of the Tucker thermal project. *While black horizontal bars* represent horizontal well pairs, *grey circles* indicate observation wells

where $\phi_{i,j}^{est}$ and d_i^{true} are the jth realization of porosity estimate and true value at grid block i; $\bar{\phi}_i^{est}$ is the ϕ average at grid block i over realizations; and N_b is the number of blocks in the model.

6 Realistic 3D Case Study

This realistic 3D case study demonstrates integration of static core data and dynamic temperature observations to constrain the spatial distribution of permeability. The case study is adapted from the Tucker SAGD thermal project located in Northern Alberta [11]. Oil is represented by bitumen with high viscosity and density of 9–10° API. The reservoir is deposited in the Clearwater formation and mainly consists of sand-shale sequences. The reservoir is formed by stacked incised valleys created in deltaic and partially marine environments. The average thickness of net pay, porosity, horizontal permeability, and oil saturation are 45.0 m, 0.31, 3000 mD and 0.56, respectively.

A plan view of the reservoir model is shown in Fig. 3. The model grid consists of 12 800 ($16 \times 40 \times 20$) blocks of size $50 \times 5 \times 2$ m^3. Data are only sampled from vertical observation wells. Temperature changes are observed in wells 2 and 3. Thus, there is a possible baffle or barrier zone near well 1. Sixty porosity, permeability, and temperature data are integrated into the model from one time step. The bivariate distribution of porosity and permeability for the base case is generated in a way that honors the spatial distributions of core data and temperature observations. Porosity and horizontal permeability are highly correlated. Homogeneous horizontal permeability is assumed, which is twice the vertical permeability.

Eight different cases are used to show the benefit of each data type and their combined effect on the estimate of horizontal permeability. Vertical permeability can be found from (12). One thousand realizations are generated with no localization honoring spatial distribution of the data. Geostatistical shortcut for integration of temperature data is applied, where ensemble mean of temperature is generated by a single flow simulation and variations are added by SGS co-simulation [23].

Table 2 Comparison of EnKF estimates of horizontal permeability

Case	Description	Data	RMSE	MSD	CC
1	No data	0	2447.8	2136.0	−0.189
2	Permeability data	60	2263.4	1860.4	0.231
3	Porosity data	60	2164.1	1743.5	0.247
4	Temperature data	60	2382.8	1992.2	0.173
5	Porosity and permeability data	120	2194.7	1817.6	0.174
6	Permeability and temperature data	120	2245.8	1771.9	0.243
7	Porosity and temperature data	120	2146.2	1686.3	0.254
8	Porosity, permeability and temperature data	180	2155.7	1723.1	0.185

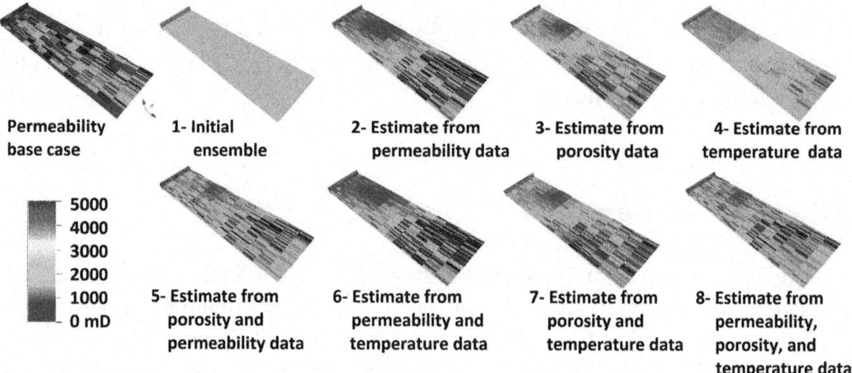

Fig. 4 Base case and various EnKF estimates of horizontal permeability for slice 8

Estimation results are summarized in Table 2 and a base case with means of EnKF estimates of horizontal permeability for slice eight are shown in Fig. 4. The incorporation of any data improves the estimate in comparison to the initial ensemble. Use of redundant core porosity, permeability or temperature data produces similar results. Much of the data is redundant; however, coupled use of temperature data with any core data improves estimation.

7 Conclusions

An algorithm for integration of static core data, dynamic temperature, and acoustic impedance observations into reservoir models is shown. The spatial distribution of porosity and permeability are constrained. The methodology was demonstrated with synthetic 2D and realistic 3D case studies. The incorporation of additional data improves porosity and permeability estimates. Hard data produces good estimation results, which are further improved by continuous soft observations. Exhaustive seismic data leads to the best estimates, but their additional value should be assessed against sampling cost. Implementation details for effective use of the method are discussed. There are approximations that significantly decrease computational time with little harm to estimation quality. The sample locations and frequency, and number of observations are crucial for understanding the geology. Overall, EnKF is shown to be a simple and effective approach for continuous data integration for petroleum reservoir characterization.

References

1. Batzle M, Wang Z (1992) Seismic properties of pore fluids. Geophysics 64:1396–1408
2. Butler R (1991) Thermal recovery of oil and bitumen. Prentice Hall, Englewood Cliffs
3. Deutsch CV (2002) Geostatistical reservoir modeling. Oxford University Press, New York
4. Deutsch CV (2010) Estimation of vertical permeability in the McMurray formation. J Can Pet Technol 49(12), 10–18
5. Dovera L, Rossa D (2007) Ensemble Kalman filter for Gaussian mixture models. In: EAGE pet geostat con. Available via EAGE. http://www.earthdoc.org/detail.php?pubid=7824. Cited 23 Nov 2011
6. Evensen G (2009) Data assimilation: the ensemble Kalman filter, 2nd edn. Springer, Berlin
7. Evensen G, van Leeuwen PJ (2000) An ensemble Kalman smoother for nonlinear dynamics. Mon Weather Rev 128(6):1852–1867
8. Fahimuddin A (2010) 4D seismic history matching using the ensemble Kalman filter (EnKF): possibilities and challenges. PhD Thesis, University of Bergen, Norway
9. Gassmann F (1951) Über die Elastizität poröser Medien. Vierteljahrsschr Nat.forsch Ges Zür 96:1–23. Available via DIALOG. http://sepwww.stanford.edu/sep/berryman/PS/gassmann.pdf. Cited 09 Oct 2011
10. Gomez-Hernandez JJ, Sahuquillo A, Capilla JE (1997) Stochastic simulation of transmissivity fields conditional to both transmissivity and piezometric data. 1. The theory. J Hydrol 203:162–174
11. Husky Oil Operations Limited (2010) Annual performance presentation for Tucker thermal project (Commercial Scheme # 9835). Energy Resources Conservation Board (ERCB). Available via ERCB. http://www.ercb.ca. Cited 09 Oct 2011
12. Isaaks EH, Srivastava RM (1989) Applied geostatistics. Oxford University Press, New York
13. Kalman RE (1960) A new approach to linear filtering and prediction problems. J Basic Eng 82:35–45
14. Kumar D (2006) A tutorial on Gassmann fluid substitution: formulation, algorithm and Matlab code. Geohorizons 11(1):4–12
15. Li Z, Yin J, Zhu D, Datta-Gupta A (2011) Using downhole temperature measurements to assist reservoir characterization and optimization. J Pet Sci Eng 78(2):454–463
16. Lumley D (2001) Time-lapse seismic reservoir monitoring. Geophysics 66(1):50–53
17. Myrseth I (2007) Ensemble Kalman filter adjusted to time-lapse seismic data. In: EAGE pet geostat con. Available via EAGE. http://www.earthdoc.org/detail.php?pubid=7847. Cited 09 Oct 2011
18. Oliver DS, Chen Y (2010) Recent progress on reservoir history matching: a review. Comput Geosci 15(1):185–221
19. Oliver DS, Reynolds AC, Bi Z, Abacioglu Y (2001) Integration of production data into reservoir models. Pet Geosci 7(S):65–73
20. Skjervheim JA, Evensen G, Aanonsen SI, Johansesn TA (2007) Incorporating 4D seismic data in reservoir simulation models using ensemble Kalman filter. SPE J 12(3):282–292
21. Wen X-H, Chen WH (2006) Real-time reservoir model updating using ensemble Kalman filter with confirming option. SPE J 11(4):431–442

22. Zagayevskiy YV, Deutsch CV (2011) Temperature and pressure dependent Gassmann's fluid substitution model for generation of synthetic seismic attributes. Cent Comput Geostat 13:210-1–210-19
23. Zagayevskiy YV, Deutsch CV (2011) Temperature and seismic data integration to a petroleum reservoir model by means of ensemble Kalman filter (EnKF). Cent Comput Geostat 13:209-1–209-24
24. Zhang W, Youn S, Doan Q (2007) Understanding reservoir architectures and steam chamber growth at Christina Lake Alberta, by using 4D seismic and crosswell seismic imaging. SPE Reserv Eval Eng 10(5):446–452

Micro-modeling for Enhanced Small Scale Porosity-Permeability Relationships

Jeff B. Boisvert, John G. Manchuk, Chad Neufeld, Eric B. Niven, and Clayton V. Deutsch

Abstract Accurate modeling of vertical and horizontal permeability in oil sands is difficult due to the lack of representative permeability data. Core plug data could be used to model permeability through the inference of a porosity-permeability relationship. The drawbacks of this approach include: variability and uncertainty in the porosity-permeability scatter plot as a result of sparse sampling, and biased core plug data taken preferentially from sandy or homogeneous intervals. A two-step process can be used where core photographs and core plug data are used to assess small scale permeability followed by upscaling to a representative geomodeling cell size. This paper expands on a methodology that utilizes core photographs to infer porosity-permeability relationships. This methodology is robust because there is abundant core photograph data available compared to core plug permeability samples and the bias due to preferential sampling can be avoided. The proposed methodology entails building micro-scale models with 0.5 mm cells conditional to 5 cm × 5 cm sample images extracted from core photographs. The micro-models are sand/shale indicator models with realistic permeability values ($k_{sand} \approx 7\,000$ mD, $k_{shale} \approx 0.5$ mD). The spatial structure of the micro-model controls the resulting porosity-permeability relationships that are obtained from upscaling. Previously, these models were generated with sequential indicator simulation (`SIS`). However, `SIS` may not capture the spatial structure of the complex facies architecture observed in core photographs. Models based on multiple point statistics and object

J.B. Boisvert (✉) · J.G. Manchuk · E.B. Niven · C.V. Deutsch
Centre for Computational Geostatistics, University of Alberta, Edmonton, Alberta, Canada
e-mail: jbb@ualberta.ca

J.G. Manchuk
e-mail: jmanchuk@ualberta.ca

E.B. Niven
e-mail: eniven@ualberta.ca

C.V. Deutsch
e-mail: cdeutsch@ualberta.ca

C. Neufeld
Statios Software & Services Inc., Fernie, British Columbia, Canada
e-mail: chad@statios.com

P. Abrahamsen et al. (eds.), *Geostatistics Oslo 2012*,
Quantitative Geology and Geostatistics 17,
DOI 10.1007/978-94-007-4153-9_13,

based techniques are proposed to enhance realism. Micro-models are upscaled to the scale of the log data (5 cm in this case) with a steady-state flow simulation to determine the porosity-permeability relationship. The porosity-permeability relationships for geomodeling, or flow simulation, can be determined with subsequent mini-modeling and further upscaling. The resulting porosity-permeability relationship can be used to populate reservoir models and enhance traditional core data. Wells from the Nexen Inc. Long Lake Phase 1 site in the Alberta Athabasca oil sands region are used to demonstrate the methodology.

1 Introduction

Interest in the Alberta oil sands has been growing steadily since the late 1960's. The Alberta oil sands are a vast resource with proven reserves of 169.9 billion barrels [6] mostly consisting of bitumen located in northern Alberta and Saskatchewan. The oil bearing formation in the Athabasca area is the McMurray formation and ranges in thickness from 0 m to 110 m [11]. Although surface mining techniques have been used in the oil sands since the late 1960's and are still used today, the majority of reserves are too deep for mining to be economical. As a result, in-situ recovery methods are required. Since bitumen is more viscous than conventional oil, steam is often injected to raise the temperature of the bitumen, reducing its viscosity and allowing it to be pumped to surface. The most widely used in-situ recovery method in the oil sands is steam-assisted gravity drainage (SAGD). Engineers use flow simulation to make predictions of steam rise and oil and water drainage. A critical input parameter in the flow simulation of SAGD operations is vertical permeability.

There are two main reservoir facies associations of concern in the McMurray. The first is the massive cross-stratified coarse sands with high porosity, permeability and oil saturation. This is the most desirable reservoir facies. The second facies association is inclined heterolithic stratification (IHS). These heterogeneous deposits form as a result of lateral growth of point bars within meandering channels of freshwater rivers and creeks draining inter-tidal mudflats. IHS deposits are generally decimetre to meter-thick repetitive sets of inclined beds of sand and mud and can range in quality from mostly sandy to mostly muddy [16]. A third facies association, Breccia, is also found in small amounts near the base of a channel succession due to the erosion and collapse of previous muddy point bars.

The geology of the McMurray formation has a large influence on its vertical permeability profile. Permeability is controlled by grain size, sorting and sediment type [10, 12]. Vertical permeability is typically inversely related to the horizontal continuity of the sediments. For example, where horizontal continuity is low, there are flow paths around low permeability units. Where horizontal continuity is high, vertical permeability is decreased because there are few flow paths around the low-permeability units.

The scale of core plugs compared to the scale of geomodeling or flow simulation must be accounted for [17]. The permeability of core plugs taken from clean sand is on the order of Darcies while the permeability of core plugs taken from mudstone is on the order of milli-Darcies. The vertical permeability of the whole core interval is what would be most important for SAGD and would be somewhere in between the permeability of the mudstone and clean sand.

Preferential sampling of core is another issue. Core plug samples are often taken preferentially from the clean sand in all facies types; otherwise, samples would have a high tendency to break or deteriorate prior to lab testing. The preferential sampling usually results in bias and an incomplete permeability to porosity relationship that is difficult to infer.

Normally, the core plug porosity and permeability data could be calibrated with well-test data. This is impossible in oil sands since the bitumen is immobile under in-situ temperature and pressure conditions. There are a number of common approaches to estimate permeability in oil sands:

1. A constant horizontal and vertical permeability could be assigned within each facies. If the permeability variation within a single facies is small compared to the permeability variation between facies, this simplification could be acceptable. Furthermore, the flow simulation grid cells are quite large and short-scale variations will be averaged out to some degree.
2. Regression models of the log of permeability versus porosity can be constructed for each facies. Then the log derived porosity data can be transformed to permeability within each facies. The main limitation of regression approaches is that they do not account for the uncertainty in permeability for a given porosity.
3. A cloud transformation [7] technique could be used to attempt to account for the statistical variation in permeability. Under a cloud transform permeability values are assigned to the grid by visiting a location, finding the collocated porosity and facies, determining the distribution of permeability for a given porosity and drawing a permeability value with Monte Carlo simulation. The permeability model will have more spatial variation than the porosity model because of the uncorrelated random drawing, but that is considered reasonable.
4. A p-field simulation technique [4] could be used where the values from a correlated random field are used in the Monte Carlo simulation to draw permeability values. The realizations of permeability will have the correct spatial variation.

Although the aforementioned techniques are valid and useful, the overall estimation of vertical permeability can be improved by considering all available information. It should be noted that well test data is unavailable to calibrate permeability modeling because of the high in-situ viscosity. Core photographs or full-bore formation micro-images (FMIs) are one source of fine scale information. Facies and permeability can be assigned to the core photograph on a pixel-by-pixel basis. Individual pixels can be classified as either sand or shale. There is no mixing of facies to consider at this resolution. The micro-scale model can be upscaled to an effective porosity and permeability at the scale corresponding to the image size. With enough core photograph based micro-models, a reasonable porosity-permeability relationship can be defined. The advantage of this methodology is that it considers all available data, permitting improved predictions of vertical permeability, while resolving some of the aforementioned scaling and non-representative sampling issues.

This paper presents a methodology to use the 2D core photographs to calculate 3D geostatistical models of porosity and permeability for each facies at the scale of the image pixels. The micro-models are upscaled using flow simulation to calculate an effective vertical and horizontal permeability at a 5 cm scale, corresponding with the scale of conventional wireline geophysical log data. Subsequent mini-modeling addresses the change from the well log scale to the scale of geological modeling or flow simulation. The methodology is applied to wells located in the Nexen, Inc. Long Lake Phase 1 Site.

2 Methodology

The goal of this work is to determine upscaled porosity (ϕ)-horizontal permeability (k_h) and k_h-vertical permeability (k_v) relationships that can be used in a cloud transformation for inference of reservoir properties for flow simulation. The upscaled ϕ–k_h and k_h–k_v relationships for a particular facies are inferred by analyzing core photographs. The main idea is to generate a 3D model of permeability at a scale where the model cell size is such that each cell can be assumed entirely sand or entirely shale which simplifies permeability assignment. In this work, the models are built at the approximate resolution of the core photographs (0.5 mm × 0.5 mm × 0.5 mm blocks). This resolution is used to capture the small scale variations in sand/shale; clearly, the size of the sand grains would make the actual flow properties of such a small volume difficult to infer.

The core photograph provides the necessary data for assignment of sand and shale in a 3D indicator micro-model. A distribution of permeability is assumed within the sand and shale categories. Flow simulation is used to determine an upscaled porosity-permeability relationship for the sand/shale mixture. Core photographs and FMI data provide the only sources of information for the small scale spatial structure of sand and shale within a given facies. The methodology for generating upscaled porosity-permeability relationships from core photographs or FMI data and the methodology is summarized in this section [5]. This work considers core photographs but these techniques could be extended to consider FMI data. The methodology:

1. Digitize the core photograph and select a cutoff value
2. Infer a 3D model of permeability with `SIS` followed by `SGS`
3. Flow simulation to determine the upscaled permeability
4. Repeat for multiple core photographs

The main driver of flow is the sand/shale spatial arrangement. The goal of this work is to improve upon the generation of the sand/shale indicator models to better capture the flow behavior of different facies. In past work, a variogram was automatically inferred from the 2D core photograph and `SIS` was used to generate a 3D indicator model. The appropriateness of using `SIS` to generate the indicator models depends on the facies considered; the focus here is specific to the McMurray Formation and the facies present can be broadly classified as sand, IHS and breccia.

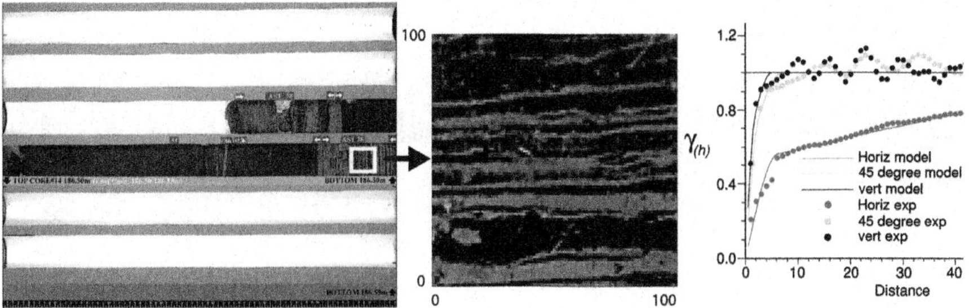

Fig. 1 Gray scale image and indicator variogram using the appropriate cutoff

Specifically, the following techniques are considered for micro-model generation. Technique 1, *Sand*: SIS accurately captures the spatial structure of the sand facies because the small proportion of shale present is typically in isolated, discontinuous regions. Technique 3, *Breccia*: The spatial arrangement of sand and shale within breccia is more complex and is not well modeled by SIS. Breccia micro-models can be improved upon by using multiple point simulation techniques with an appropriate training image (TI). Technique 3, *IHS*: The spatial arrangement of sand and shale in IHS is a sequence of layers of sand and shale dipping between 8° and 15°. SIS can capture this linear relationship but it is not possible to determine the true dip of the layers. Typically, the apparent dip seen in the core photograph is assumed to be the true dip. The impact of this assumption is addressed in a number of numerical experiments where the unknown true dip is assumed to be different than the apparent dip observed in the core photograph. Details of the micro-modeling methodology are expanded upon below.

Step 1: Digitize the Core Photograph The locations of the extracted models are manually selected from the core photographs (Fig. 1). A representative range of core photograph samples should be selected such that the range of porosity values in the log data are represented in the micro-models. This is accomplished by selecting models with various proportions of shale. The images are 5 cm by 5 cm with 100 cells in each direction. It is more convenient to work in pixels and models are shown to be 100×100 with 0.5 mm blocks throughout. Gray scale cutoff values are selected to assign sand and shale categories. The appropriate cutoff value for a model varies for each photograph because of local lighting conditions, water saturations, etc. For each model this cutoff value is selected by examining a range of cutoffs and determining the most visually appropriate value. The cutoff value is a key parameter as it controls the spatial distribution of categories; past experience with automatic cutoff selection has shown that lighting conditions and core quality vary sufficiently to require a visual cutoff selection. This is typically a simple process in the Alberta oil sands region as the visual contrast between sand and shale is clear (Fig. 2). In situations where this visual distinction in not apparent it is not recommended to apply the proposed methodology as an incorrect cutoff value would introduce error and potentially a bias in the upscaled permeability results.

Fig. 2 Core photograph exemplars for the facies modeled. *Above*: Sand. *Middle*: IHS. *Below*: Breccia

Step 2: Infer a 3D Model of Permeability The result of Step 1 is a 2D indicator model of sand and shale. While this model could be used in a 2D flow simulation, the three dimensional characteristics of flow would not be captured. `SIS` is typically used to generate the 3D categorical model and requires a variogram model. The conditioning data, in the form of a 2D model of sand/shale, provides the necessary data for inference of the vertical and horizontal variograms (Fig. 1). The continuity in the second horizontal direction (into the page in Fig. 1) is assumed to be the same as the horizontal direction in the core photograph. The validity of this assumption is the focus of the first part of this work and is discussed further below.

`SIS` is implemented with the variogram model and the conditioning data provided by the core photograph. Ten realizations of sand/shale are generated to assess the uncertainty in the upscaled results. The number of realizations could be increased; however, multiple core photographs are considered for each facies and the large number of flow simulations (~6 000 realizations in the case study below) quickly becomes CPU demanding.

Flow simulation requires a model of permeability. `SGS` is used to populate the categorical models with realistic permeability values. Due to the small scale of the individual cells in this model and the assumption that each cell is either entirely sand or entirely shale, a k_v/k_h ratio of ~1.0 is reasonable. The permeability distribution for sand is assumed to approximately match the core samples taken in sand $N(7\,000\text{ mD}, 2\,500\text{ mD})$. There are no core samples taken in shale; a realistic distribution of $N(0.5\text{ mD}, 0.1\text{ mD})$ is assumed. Adjusting these distributions has an effect on the upscaled porosity-permeability relationships and is roughly calibrated to existing core samples and previous experience with similar deposits.

Step 3: Flow Simulation to Determine Upscaled Permeability A steady state flow simulation using `FLOWSIM` is used to calculate the upscaled k_v and k_h values for each 3D micro-model [3]. However, there is no porosity value for each model and it must be inferred. It would be inappropriate to use log porosity data as the volume of influence is much larger than the 5 cm × 5 cm models considered. The porosity for shale is assumed to be 1 % and sand is assumed to be 40 %; the proportion weighted average (1) provides the porosity for each realization. Expression (1) is a good approximation for porosity but assumes perfectly clean sand without in-

Table 1 Available data

Facies	Core samples	Average k_h (mD)	Average k_v (mD)	Average ϕ	Number of core photograph models
Sand	66	6 622	5 669	0.35	100
IHS	34	5 479	5 115	0.33	99
Breccia	31	7 696	7 123	0.36	99
Total	131	6 405	5 668	0.34	297

terstitial clays and could be modified given site specific considerations if whole core scale tests are available:

$$\phi_{\text{realization}} = 0.4 p_{\text{sand}} + 0.01 p_{\text{shale}}. \tag{1}$$

3 Case Study

The methodology as presented above relies on `SIS` for the generation of the sand/shale categories within each facies modeled. In this section, modeling considerations specific to the sand/shale geometry of each facies are incorporated into micro-modeling. Core photographs, core samples and log data from 12 wells in the Long Lake Phase 1 project in the Athabasca oil sands region of Alberta, Canada are used to demonstrate the methodology. There are three identified facies of interest: sand, IHS and breccia. Examples of each facies are shown in Fig. 2. Based on log data, the proportions of facies in the reservoir are 60 % sand, 29 % IHS and 11 % Breccia.

The core data is not used explicitly for calibration nor in selecting micro-model locations as it is not typically representative of in-situ reservoir properties (i.e. porosity) due to the difficulty of sampling high shale proportions. This issue is highlighted in Table 1 where the average porosity (and even permeability) of IHS and breccia are similar to the samples in sand. Samples are preferentially located in sand regardless of facies shale content. Core data alone cannot provide sufficient information to fully infer the relationship between porosity and permeability.

Modeling IHS The layered nature of IHS results in thin sand beds that are the main conduit of flow at the scale considered. The geometric orientation of these layers can affect upscaled k_v and k_h. Literature suggests that the typical dip of IHS sets in the McMurray formation can range from 8° to 15° with minimums and maximums observed between 3° and 30° [2, 9, 13]; however, the distribution of dips as measured by the automatic variogram fitting is much smaller because the apparent dip, not the true dip, is measured. As the true dip increases, k_h decreases and k_v increases. It is important to fully understand what effect a larger true dip has on the permeability relationships inferred. In this section, numerical experiments are conducted to assess the sensitivity of flow properties to the unknown true dip. It should

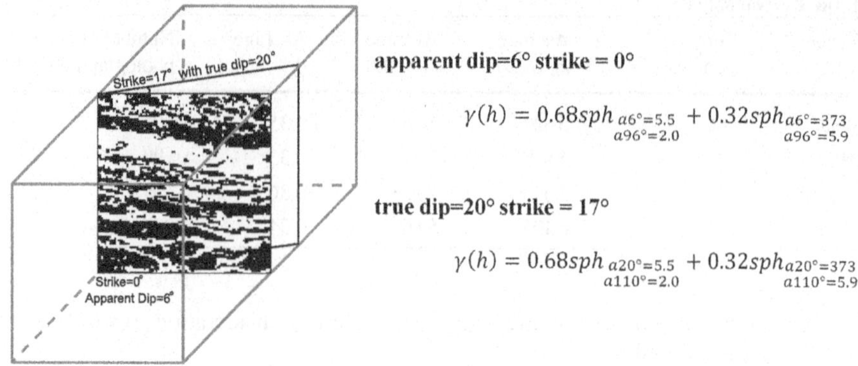

Fig. 3 *Left*: Conditioning data showing strike of 17° assuming a true dip of 20°. *Right*: automatic variogram fitting dip = 6°. Variogram is isotropic in both horizontal directions

be noted that in the presence of FMI data this is less of a concern as FMI provides 360° coverage of the borehole wall that can be used to infer the true dip of the IHS set [15]; however, it is not common practice to obtain FMI data for all wells. The use of core photographs to assess flow characteristics is applicable to a wider range of reservoirs.

The impact of dip on the ϕ–k_h–k_v relationships is assessed by assuming a range of true dips. First, consider the apparent dip to be the true dip. The variogram from the 2D core photograph is fit automatically and the dip determined (Fig. 3 right). This variogram is used with `SIS` to generate sand/shale models and results in the relationships shown in Fig. 4. Of interest to modeling is the actual ϕ–k_h and ϕ–k_v relationships and how they are affected when the true dip is not measured.

The true dip is assumed to be 5°, 10°, 15° and 20° and the relationships (Fig. 4) are recalculated. A corrected variogram (Fig. 3 right) is calculated by determining the strike relative to the core photograph orientation (2) where the apparent and true dips are known. The corrected variogram is used in `SIS` and the methodology repeated for each core photograph. Rather than repeat all plots (Fig. 4) for each true dip considered, the relationships are fit and compared (Fig. 5). For high porosity values, the difference when considering the true dip is small. However for porosity values that are common in the McMurray formation there are significant differences in the modeled horizontal and vertical permeability values when considering the true dip. Higher porosity models are less affected because there is less connectivity of the small proportions of shale and the orientations of these disconnected features are not as relevant to the flow response of the model. FMI data, when available, can be used to infer the true dip in the IHS beds.

$$\sin(strike) = \frac{\tan(apparent\,dip)}{\tan(true\,dip)}. \tag{2}$$

Modeling Breccia `SIS` effectively models the linear spatial orientation of sand and shale in the IHS and sand facies. Typically breccia is observed to have large

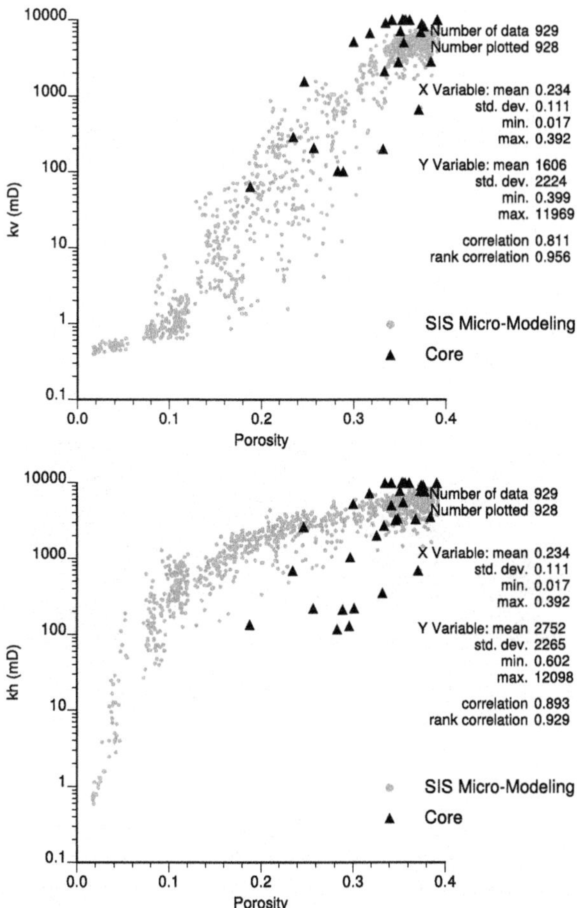

Fig. 4 Relationships of interest for IHS using the apparent dip

clasts with some variation in size (Fig. 6). If `SIS` is used to model breccia, the resulting sand/shale realizations are not consistent with the known breccia geometry (Fig. 6) resulting in an inaccurate flow response assessment.

Multiple point statistics are used to generate realizations that better honor breccia geometry. First a library of `TI`'s that represent different geometries and shale proportions is generated. `TI`'s are created using a randomized object based modeling algorithm. Objects are breccia clasts that are randomly seeded throughout the micro-model. The geometry of the clasts is randomly grown until the model has the correct fraction of shale. Clasts are grown according to anisotropy and volume statistics that are derived from core photographs of breccia facies so that the models reflect the observed geometry. Generated clasts can be convex and nearly elliptic shaped to highly non-linear non-convex shaped by allowing growth to occur from the original seed point or by allowing the seed point to move using a random walk process.

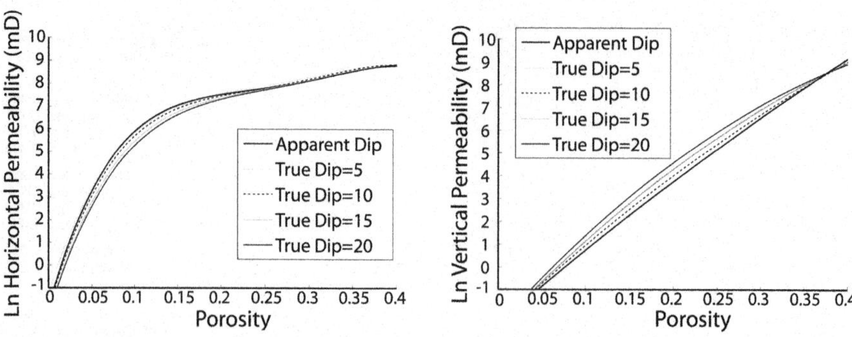

Fig. 5 ϕ–k_h and ϕ–k_v relationships for different true dip in IHS facies

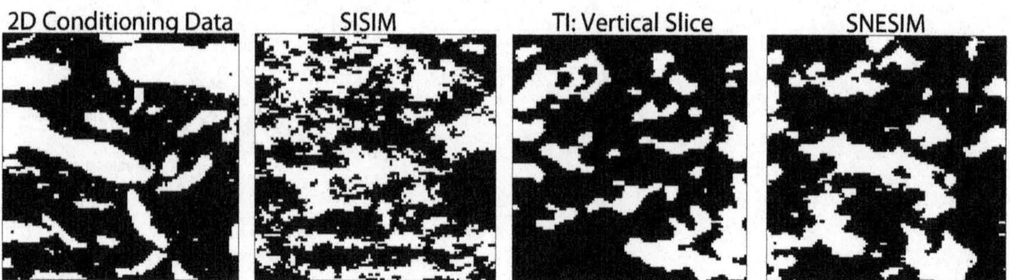

Fig. 6 From *left*: 2D Breccia model from the core photograph. One slice of an `SIS` realization of breccia. Slice from the `TI` used in `SNESIM`. `SNESIM` realization slice

Consider the 2D indicator model shown in Fig. 6. To select the most appropriate `TI` for this set of conditioning data, the distribution of runs [8] for the conditioning data is compared to the distribution of runs of each `TI` in the library with similar proportions. The `TI` most similar to the conditioning data in a minimum squared error sense, is used in `SNESIM` [14] to generate the categorical models for Breccia [1].

4 Results

The bivariate relationships provided by micro-modeling are largely consistent with the core samples available (Fig. 7). In general, the horizontal continuity due to the layered shale significantly reduces vertical permeability as indicated by the micro-modeling results. A significant advantage of micro-modeling results over deterministic functions is that an understanding of the variability in permeability for a given measure of porosity is quantified.

A new contribution to the micro-modeling workflow made in this work is accounting for the lack of orientation data from the core photographs. This source of uncertainty is quantified by assuming the dip observed in the photographs of IHS facies is an apparent dip. Introducing variability in the true dip of the micro-models

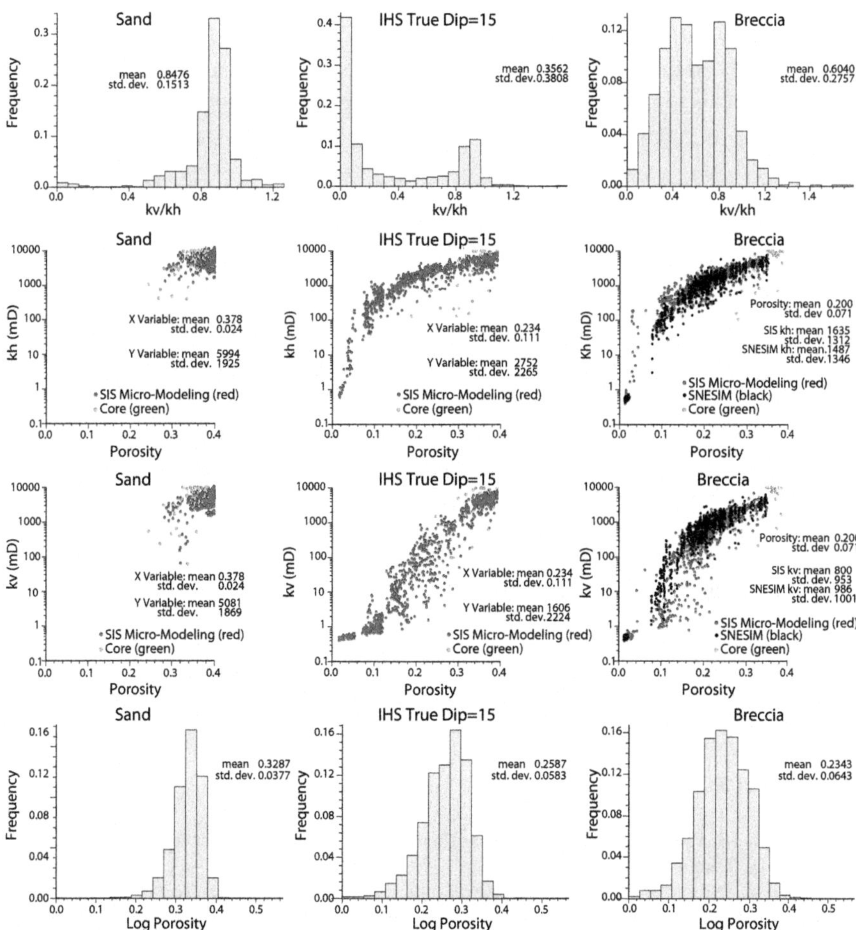

Fig. 7 Summary relationships

provides a more complete understanding of the variability in ϕ, k_h, and k_v. The effects can be significant. Consider Fig. 5: the vertical permeability for a porosity of 20 % is roughly an order of magnitude higher for a true dip of 20 degrees than for the apparent dip. This would lead to large differences in flow performance prediction.

The focus of this work is establishing the micro-model scale ϕ–k_h and ϕ–k_v relationships. Normally, these relationships are then used in mini-modeling to obtain the porosity-permeability relationships at the scale of geomodeling or flow simulation. Mini-modeling is not considered in this work due to space constraints. It is recommended that unconditional realizations of porosity at the geomodeling scale of interest (usually $\sim 1 \times 1 \times 1$ m^3 with $\sim$10 cm blocks) be generated. Using a cloud transformation and the micro-modeling relationships developed, 3D realizations of permeability can be generated and upscaled. This provides the ϕ–k_h–k_v relationship at a geomodeling scale. Deutsch [5] provides additional details on mini-modeling.

The micro-modeling results have a direct impact on the upscaled geomodel scale ϕ–k_h–k_v relationships. Obtaining reasonable micro-scale results allows the larger scale geomodels to better reflect the small scale behavior of complex facies such as IHS.

5 Conclusions

Integrating the micro-modeling methodology into a typical reservoir characterization workflow takes advantage of the large number of core photographs available. Beyond core photographs and FMI data there is little information available for the ϕ–k_h–k_v inference; biased core samples cannot be relied upon to obtain a reasonable understanding of the geocellular modeling scale relationship for all porosity ranges. The small scale spatial arrangement of sand and shale dominates the overall behavior of each facies and can be incorporated in micro-modeling. The techniques presented generate and assess the reasonableness of micro-models for better inference of porosity-permeability relationships, which is a critical and data-poor aspect of any reservoir modeling work flow.

Acknowledgements The authors would like to thank Nexen Inc. in general, and Don Dodds specifically, for providing the data used in this study and for helpful feedback on the work.

References

1. Boisvert JB, Pyrcz MJ, Deutsch CV (2007) Multiplepoint statistics for training image selection. Nat Resour Res 16(4):313–321
2. Crerar E, Arnott R (2007) Facies distribution and stratigraphic architecture of the Lower Cretaceous McMurray Formation, Lewis Property, Northeastern Alberta. Bull Can Pet Geol 55(2):99–124
3. Deutsch CV (1987) A probabilistic approach to estimate effective absolute permeability. PhD thesis, Stanford University, Stanford, CA, 165 pp
4. Deutsch CV (2002) Geostatistical reservoir modeling. Oxford University Press, New York
5. Deutsch CV (2010) Estimation of vertical permeability in the McMurray formation. J Can Pet Technol 49(12):10–18. doi:10.2118/142894-PA
6. Government of Alberta (2011) Alberta Energy: the oil sands facts and statistics page. Available via: http://www.energy.alberta.ca/OilSands/791.asp. Cited 20 Feb 2011
7. Kolbjornsen O, Abrahamson P (2004) Theory of the cloud transform for applications. In: Proceedings of the seventh international geostatistics congress, Banff, Alberta, pp 45–54
8. Mood AM (1940) The distribution theory of runs. Ann Math Stat 11(4):367–392
9. Mossop GD, Flach PD (1983) Deep channel sedimentation in the Lower Cretaceous McMurray Formation, Athabasca oil sands, Alberta. Sedimentology 30(4):493–509
10. Olson JE, Yaich E, Holder J (2009) Permeability changes due to shear dilatancy in uncemented sands. In: Proceedings of the 43rd US rock mechanics symposium, Asheville, NC
11. Ranger MJ, Gingras MK (2003) Geology of the Athabasca oil sands—field guide and overview. Canadian Society of Petroleum Geologists, Calgary
12. Shang R, Wang WH (2011) Sedimentary texture characterization and permeability anisotropy quantification of McMurray oil sands. Abstract. In: Gussow geoscience conference, Banff, Alberta

13. Smith DG (1987) Meandering river point bar lithofacies models: modern and ancient examples compared. In: Ethridge FG, Flores RM, Harvey MD (eds) Recent developments in fluvial sedimentology. Society of Economic Paleontologists and Mineralogists. Special Publication, vol 39, pp 83–91
14. Strebelle S (2000) Sequential simulation drawing structures from training images. PhD thesis, Stanford University, Stanford, CA, 187 pp
15. Strobl R, Ray S, Shang R, Shields D (2009) The value of dipmeters and borehole images in oil sands deposits—a Canadian study. CSPG CSEG CWLS convention. 7
16. Thomas RG, Smith DG, Wood JM, Visser J, Calgerley-Range EA, Koster EH (1987) Inclined heterolithic stratification-terminology, description, interpretation and significance. Sediment Geol 53(1–2):123–179
17. Tran TT (1996) The missing scale and direct simulation of block effective properties. J Hydrol 182:37–56

Applications of Data Coherency for Data Analysis and Geological Zonation

John G. Manchuk and Clayton V. Deutsch

Abstract A measure of coherency in facies between nearby wells is developed to aid in the processes of quality control of well data and geological zonation. Coherency measures the agreement between a well and its immediate neighbors based on structural markers and facies interpretations. The calculation can be done incrementally on sets of wells to identify incoherencies caused by errors in interpretation, measurement differences, data acquisition problems or actual changes caused by geological differences. Attributes may include the year a well was interpreted or included in a database, the interpreter, or what logging tool was used. Coherency is also used as a similarity metric in a hierarchical clustering algorithm for geological zonation. Results are applied in several examples that demonstrate the use of coherency and clustering for quality control and grouping data into geologically similar objects.

1 Introduction

Quality control of well data can be a time-consuming process in reservoir characterization and geomodeling studies [2, 8]. Data may be collected over many years prior to production. During this time, technology for data acquisition changes and multiple geologists and well log analysts handle the data, undoubtedly with some variation in the subjective process of interpretation. Two particular types of data that are prone to inconsistencies are structural markers and facies due to the potentially subjective nature of these data types [3]. When variations or incoherencies can be detected during a quality control study, the database may require adjustment to improve the resulting geological models and engineering decisions.

Variations in a database may be subtle and go unrecognized, especially when the database contains hundreds of wells and spans more than a decade of data acquisition. Such a scenario is typical of large oilsands mining projects, mature fields

J.G. Manchuk (✉) · C.V. Deutsch
University of Alberta, Edmonton, Alberta, Canada
e-mail: jmanchuk@ualberta.ca

C.V. Deutsch
e-mail: cdeutsch@ualberta.ca

P. Abrahamsen et al. (eds.), *Geostatistics Oslo 2012*,
Quantitative Geology and Geostatistics 17,
DOI 10.1007/978-94-007-4153-9_14,

heading into enhanced recovery stages of production and of in-situ production of heavy oil. Subtle incoherences may be perceived as inconsequential; however, they can have a significant impact on parameters for geomodeling, such as the variogram and local accuracy of prediction. In this work, a measure of coherency is defined to help detect wells that are inconsistent within a database. Coherency is calculated between a well and its immediate neighbors based on several parameters including the spatial position of wells, the facies interpretations along the wells, and the similarity between facies. The calculation is fast and automatic making it possible to detect incoherency in very large databases.

Resulting coherency measures for each well depend on the space considered. There are often two spaces in geomodeling including physical space where the wells exist in present time and modeling space where the effects of time on the depositional environment have been accounted for. Modeling space may also be referred to as stratigraphic space or geo-chronologic space [2, 4]. The spatial correlation of properties is typically higher in modeling space; therefore, so is the coherency. Another use of the coherency measure is to assess the quality of the transformation from physical space to modeling space. No improvement in coherency should raise some concern. To help assess how good an improvement is made, the coherency can be maximized by adjusting the vertical position of wells in a database. In cases where maximization does not change the coherency, the modeling space is optimal in reference to the coherency measure. In cases where large changes in coherency are observed, it may be necessary to re-evaluate the transformation process from physical to modeling space. A large change may also indicate a well bust.

Incoherencies in a database are not necessarily due to variations in interpretation or differences in technology. Rather, they may be a product of the depositional environment. For example, a middle estuarine environment with sinuous channels, inclined heterolithic stratification (IHS), breccias, and other complexities may appear highly incoherent due to the heterogeneity in facies [5, 6]. In this case, the measure of coherency has a secondary use, that is, to aid in the identification of geological zones based on the facies designations. Facies intervals along wells are clustered with nearby wells into geological objects, where the coherency is used as a similarity metric. Such geological zonation is important for delineating regions that are stationary for geostatistical modeling. The clustering process can also be utilized for quality control, detecting boundaries between different geological successions, and trend modeling.

2 Methodology

The measure of coherency between two vertical wells was derived from the covariance function:

$$C(\mathbf{h}) = \frac{1}{N(\mathbf{h})} \sum_{i=1}^{N(\mathbf{h})} z(\mathbf{u}_i) z(\mathbf{u}_i + \mathbf{h}) - \mu_0 \mu_{\mathbf{h}}, \tag{1}$$

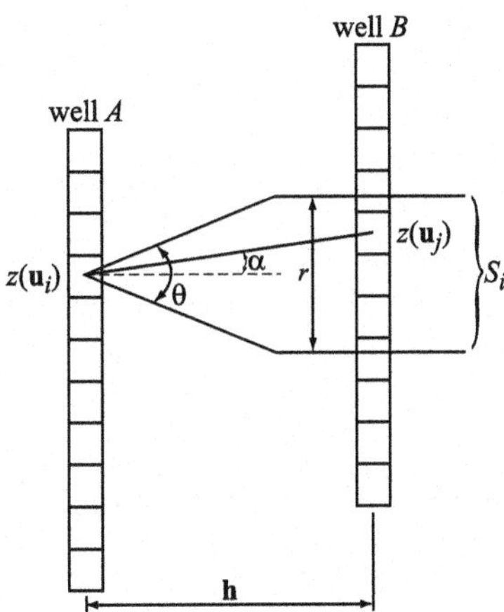

Fig. 1 Parameters involved in the coherency calculation

where z is a random function, $\mathbf{h}$ is a vector separating two points, $\mathbf{u}_i$ and $\mathbf{u}_i + \mathbf{h}$, $N(\mathbf{h})$ is the number of pairs separated by $\mathbf{h}$, and μ_0 and $\mu_{\mathbf{h}}$ are the expected value of z at $\mathbf{u}$ and $\mathbf{u} + \mathbf{h}$.

The covariance function measures how two sets of data separated by $\mathbf{h}$ relate, or how coherent they are. Considering z as a Gaussian random function with unit variance, a covariance of zero indicates no coherency whereas a covariance of 1 indicates full coherency. The covariance between two vertical wells, A and B, separated by $\mathbf{h}$ in an aerial plane is evaluated with (1) by considering all $\mathbf{u}_i$ along well A and all $\mathbf{u}_i + \mathbf{h}$ along well B. If the two wells are at the same depth, sample the same depth interval, and $z(\mathbf{u}_i) = z(\mathbf{u}_i + \mathbf{h})\ \forall i$, the covariance or coherency is 1 when z is standardized. However, not all pairs between the wells have the same $\mathbf{h}$ when the depth coordinate is considered because only those pairs that exist in the same aerial plane are separated by $\mathbf{h}$.

The covariance measure in (1) does not allow for any flexibility due to differences in depth between two wells and $\mathbf{h}$ must be permitted to vary within some limits. A search window is used to allow some variation in $\mathbf{h}$ to account for stratigraphic variations between pairs of wells (Fig. 1). Search parameters include a search angle, θ, and search radius, r. For each point $\mathbf{u}_i$ along well A, a set $\mathbf{u}_j \in S_i$ along well B are found within the search constraints such that $|\mathbf{u}_j - \mathbf{u}_i| \geq |\mathbf{h}|,\ \forall \mathbf{u}_j \in S_i$. Even though some variation in depth is accounted for this way, pairs that exceed $\mathbf{h}$ are penalized. For wells A and B that have identical z values, the coherency should be zero if their depths differ by an amount greater than r so that problems related to well busts or space transformations can be identified. A weight function is introduced

$$\lambda(\mathbf{u}_i, \mathbf{u}_j) = \cos(\pi \alpha_{ij} / \theta), \tag{2}$$

where α_{ij} is the dip angle between $\mathbf{u}_i$ and $\mathbf{u}_j$ and π/θ is a scaling factor so that $\lambda = 0$ when $\alpha = \theta$.

Because the search parameters yield a set of points along well B for each point along well A, only the pair with maximum weighted coherency is retained. Otherwise, the weight function could lead to a reduced average coherency measure when actual coherent intervals between the wells are smaller than the search radius. The final equation for coherency between wells A and B is

$$C(A,B) = \frac{1}{N}\sum_{i=1}^{N_A} \max_{\forall j \in S_i}\left[\lambda_{ij} M(z_i, z_j)\right] + \frac{1}{N}\sum_{j=1}^{N_B} \max_{\forall i \in S_j}\left[\lambda_{ji} M(z_j, z_i)\right], \quad (3)$$

where N is the number of samples along the wells and $\mathbf{u}$'s have been dropped for clarity.

Expression (3) involves the computation from well A to well B and B to A so that the coherency function is symmetric. Unlike the covariance equation where $z(\mathbf{u}_i)z(\mathbf{u}_j = \mathbf{u}_i + \mathbf{h}) = z(\mathbf{u}_j)z(\mathbf{u}_j - \mathbf{h})$, the max function is non-symmetric, that is, $\max(z_i z_j)\ \forall j \in S_i = \max(z_j z_i)\ \forall i \in S_j$ does not necessarily hold. The parameter, $M(z_i, z_j)$, is a similarity metric that replaces the product, $z(\mathbf{u}_i)z(\mathbf{u}_i + \mathbf{h})$ in (1) and measures the similarity between two values. For continuous variables, M is defined by

$$M(z_i, z_j) = 1 - (z_i - z_j)^2, \quad z \in [0,1], \quad (4)$$

where z should be transformed to the [0, 1] interval so that the coherency measure ranges from 0 to 1. Alternatively, the covariance function of the random function z could be used:

For categorical variables, M is a symmetric user-defined matrix with entries that define the similarity between different facies and ranges from 0 to 1. For example, if M is an identity matrix, categories are completely dissimilar. Assigning similarity > 0 may be the case for multiple facies that are found in the same geological feature such as a point bar, or if there is considerable overlap between the underlying rock property distributions.

Computing the coherency of a well, A, with n of its immediate neighbors, B_k, $k = 1, \ldots, n$, is the average coherency between A and each B_k defined by

$$C(A) = \frac{1}{\omega_k}\sum_{k=1}^{n} \omega_k C(A, B_k), \quad (5)$$

where ω_k are weights that can be computed in a variety of ways such as equal or inverse distance. Weights should be computed so that wells further away have less impact on the coherency calculation.

Expression (5) provides a quantitative assessment of the agreement between a well and its local neighborhood. A high coherency indicates wells have a similar arrangement of facies or other reservoir properties. Low coherency may indicate a few different issues or characteristics of the data including: wells that have poor agreement between reservoir properties; wells that have a significant depth offset such as well busts; and reservoir properties that have a correlation length less than the well spacing. The last result is a characteristic of the data and would likely result in low coherency measures for all wells in a database assuming the property was stationary.

2.1 Maximizing Coherency

Determining the maximum coherency is an optimization problem. A gradient descent algorithm is developed that changes the vertical position of wells until the average coherency of all wells is maximized. This problem is not significantly different than methods used in automatic well correlation [1]. The gradient is approximated for each well pair by evaluating the coherency at two points: after shifting the well by Δz in the negative z direction and again in the positive z direction:

$$\frac{\mathrm{d}C(A, B_k)}{\mathrm{d}z} = \frac{1}{2\Delta z}\left[C_{+\Delta z}(A, B_k) - C_{-\Delta z}(A, B_k)\right], \quad k = 1, \ldots, n, \tag{6}$$

where z is the vertical coordinate in stratigraphic space. Since shifting well A down is equivalent to shifting well B_k up, we get

$$\frac{\mathrm{d}C(A, B_k)}{\mathrm{d}z} = -\frac{\mathrm{d}C(B_k, A)}{\mathrm{d}z}. \tag{7}$$

The total gradient for well A is then computed using

$$\frac{\mathrm{d}C(A)}{\mathrm{d}z} = \sum_{k=1}^{n} \frac{\mathrm{d}C(A, B_k)}{\mathrm{d}z} - \sum_{l} \frac{\mathrm{d}C(B_l, A)}{\mathrm{d}z} \tag{8}$$

for all wells B_l that have A as an immediate neighbor.

Vertical positions of wells are updated by taking a step in the gradient direction defined by

$$z_i^m = z_i^{m-1} + \alpha \frac{\mathrm{d}C_i^m}{\mathrm{d}z}, \quad i = 1, \ldots, N, \tag{9}$$

where m is the iteration, z_i is the top of a well, α is the step size, and N is the number of wells.

A line search algorithm is used to determine the step size that maximizes the coherency for the given gradient. As wells are adjusted, the gradient changes since different facies intervals align at different angles. The algorithm is as follows:

1. Compute initial coherency.
2. While $\alpha > 0$ and $\|\mathrm{d}C_i^m/\mathrm{d}z\| > \varepsilon$.
   a. Approximate the gradient using (6) and (8).
   b. Find α to maximize the coherency using a line search.
   c. Update z using (9).

Because the step size is computed to maximize the average coherency of all wells together, it does not guarantee that the coherency of each well will increase. This implementation of gradient descent is also not a global optimization algorithm. Assuming the data and transformation from physical to modeling space is primarily of good quality, a local maximum is likely close to the global maximum and optimized well positions will provide an equal amount of information about data quality and space as a global optimal would.

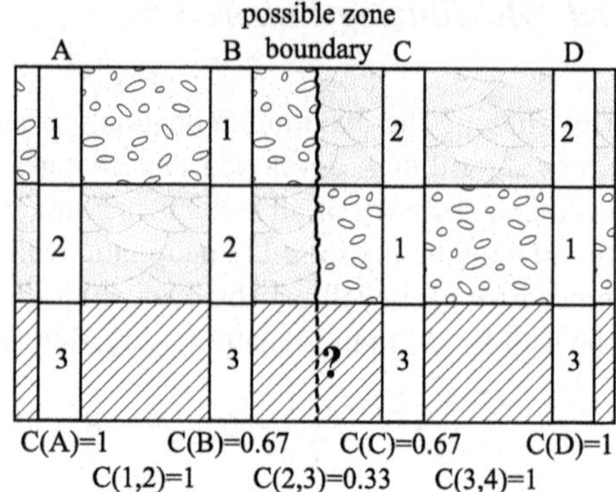

Fig. 2 Basic concept of geological zonation using coherency. Wells are labeled *A* through *D* and facies *1* to *3*. The difference between coherency of each well using the left and right neighbors and coherency between well pairs is shown. Determining if a zone boundary exists in facies *3* based on well pair coherency is problematic

The optimization procedure is intended to provide information about the data including quality of facies and of accuracy or correctness of well positions in the depth coordinate. It may be used to check the goodness of a transformation from physical space to modeling space. For example, a large increase in coherency for a well in modeling space may warrant updating the transformation as long as such a change is justified. A large increase in coherency may also indicate a well bust or problematic formation marker elevation. Optimized well positions should not be used blindly to define the modeling space since results are not related to any particular geological environment or process and there is some danger in over-fitting.

2.2 Coherency-Based Clustering

The coherency measure defined by (3) and (5) can also be used as a similarity metric for clustering data into geological objects or zones with similar properties. Developments made towards geological zonation involve categorical data. For clustering, (3) is used since it provides the similarity between two wells as opposed to a well with its surroundings as in (5). The latter does not provide useful information to determine if wells belong to the same geological zone. A simple example with four wells and three facies demonstrates the concept of zonation and the different information provided from (3) and (5) (Fig. 2). Coherency between well pairs is used to detect where changes in zone occur; however, this poses problems when the coherency is greater than zero (Fig. 2).

Coherency must be computed between smaller intervals along the wells to more accurately determine where geological zone boundaries exist. The smallest possible intervals to consider are individual sample points and the coherency is defined by (10), where z_i and z_j are from different wells, *A* and *B*, that are immediate neighbors,

$$C(z_i, z_j) = \max_{\forall j \in S_i} \left[\lambda_{ij} M(z_i, z_j) \right]. \quad (10)$$

The number of pairs depends on the number of samples, N, and number of nearest neighbors, n, to consider and is equal to $R = N \times n$. Evaluating (10) for all pairs yields a sparse symmetric $R \times R$ matrix, $\mathbf{C}$. The matrix can become substantial for databases with many wells having a small sampling interval. Developing the clustering algorithm in this fashion is an area of future study. To maintain a smaller matrix and accurately detect zone boundaries, wells are broken into intervals that have the same facies. In Fig. 2, each well would be separated into three intervals and the number of coherency values increases from 3 to 9. Boundaries between well B and C in the upper and middle interval are detected since the coherency is zero and no boundary is assigned in the bottom interval since the coherency is 1.

In actual well databases, the zonation problem is three dimensional with much more variation in well elevations, geometry of nearest neighbor sets, facies, and geological architecture. A hierarchical clustering algorithm [7] is used to process the sparse coherency matrix into a set of possible geological zones or objects and proceeds as follows:

1. Initialize set of clusters, $G_i = 1, \ldots, R$, that refer to facies intervals.
2. While $\max[\mathbf{C}] > u$.
   a. Find i and j such that $C_{ij} = \max[\mathbf{C}]$.
   b. Merge cluster G_j with cluster G_i, saving the result in G_i.
   c. Update all entries in $C_{kl}, k = 1, \ldots, R, k \neq i$, where $l = i$ or $l = j$.
   d. Zero all entries $C_{jl} > 0$ in $\mathbf{C}$.

In this algorithm, u is a coherency cutoff that defines when clustering is stopped. Because $\mathbf{C}$ is sparse, setting $u = 0$ will not necessarily lead to one cluster as in typical hierarchical algorithms. The $\max[\mathbf{C}]$ operation finds the entry in $\mathbf{C}$ with maximum coherency. The merging operation in step 2 stores all items (facies intervals) contained in G_j within G_i. The update operation in step 3 finds all entries in $\mathbf{C}$ that have a positive coherency with the items in the new cluster G_i. If for a particular cluster, k, both entries C_{ki} and C_{kj} exist, the coherency with G_i is computed using one of the linkage types for hierarchical clustering: average linkage that takes $(C_{ki} + C_{kj})/2$; single linkage that takes $\max[C_{ki}, C_{kj}]$; and complete linkage that takes $\min[C_{ki}, C_{kj}]$.

Several pieces of information are collected as clustering progresses including: the final cluster identifier; initial cluster identifier; and the coherency involved in merging an item with a cluster. Final cluster identifiers are assigned to every sample present in the database for visualization and further processing. Initial cluster identifiers are assigned to items in the order they are first merged together. As items are merged with existing clusters, they are assigned the initial cluster identifier of that cluster; however, when two clusters are merged, there is no change in the initial cluster identifier. This provides information about the growth patterns of the clustering process. Lastly, the coherency involved in tying an item to a cluster is recorded and can be used as a measure of the probability that an item belongs to a cluster.

3 Examples

The coherency measure, maximization, and clustering are demonstrated using a set of vertical wells that sample three different types of media: 1—uniform random facies; 2—an estuarine training image; 3—a fluvial training image. The first data set is used as a control to observe coherency values that indicate poor data. Data sets 2 and 3 are used to assess the expected coherency of fluvial and estuarine deposition under ideal conditions. For maximization, the estuarine data set is used with a few wells shifted to simulate well busts or post-depositional deformation. Coherency clustering is applied to the fluvial data set.

3.1 Coherency

Data sets for demonstrating the coherency calculation are shown in Fig. 3, which also shows the well positions shaded by their coherency. Coherency was computed using the three nearest neighbors to each well. An estuarine set with cross stratified sand (CSS), pointbar shale (PBSH), pointbar sand (PBS), breccia (BR) and channel fill (CF) and a fluvial set with flood plain (FP), channel sand (CH), levee (LV) and crevasse splay (CS) facies are considered. Data sets each consist of 256 wells that sample different grids with dimensions provided in Table 1. Differences in the scale of geological variation between the estuarine and fluvial data are observed through the local variation in coherency, with more local variation for the estuarine data. Because there is no stratigraphic deviation among the wells, the search parameters were set to $\theta = 0.01$ degrees and $r = 1$ meters. All facies were dissimilar for the random and fluvial data. The estuarine data involved a similarity of one between the point bar facies because they exist in the same geological object. Histograms of the resulting coherency are shown in Fig. 4. Coherency from the estuarine data had a variance substantially higher than the fluvial data, likely due to differences in channel wavelength, width, and sinuosity.

3.2 Maximization

Coherency maximization is demonstrated using the estuarine data set with some of the wells shifted to simulate well busts or stratigraphic variation in well markers. In the case of stratigraphic variation, the purpose of the maximization procedure is to identify wells that may require adjustment to optimize the space for geostatistical modeling, that is, to improve the transformation from physical space to stratigraphic space. Twenty wells were selected with a variety of coherency values from the estuarine data. The z coordinates of the wells were shifted in the range $(-4, 4)$ using uniform random numbers.

Parameters for coherency included an angle of $\theta = 0.4$ degrees, a search radius $r = 0.4$ meters, and the number of nearest wells equal to five. Results are shown in

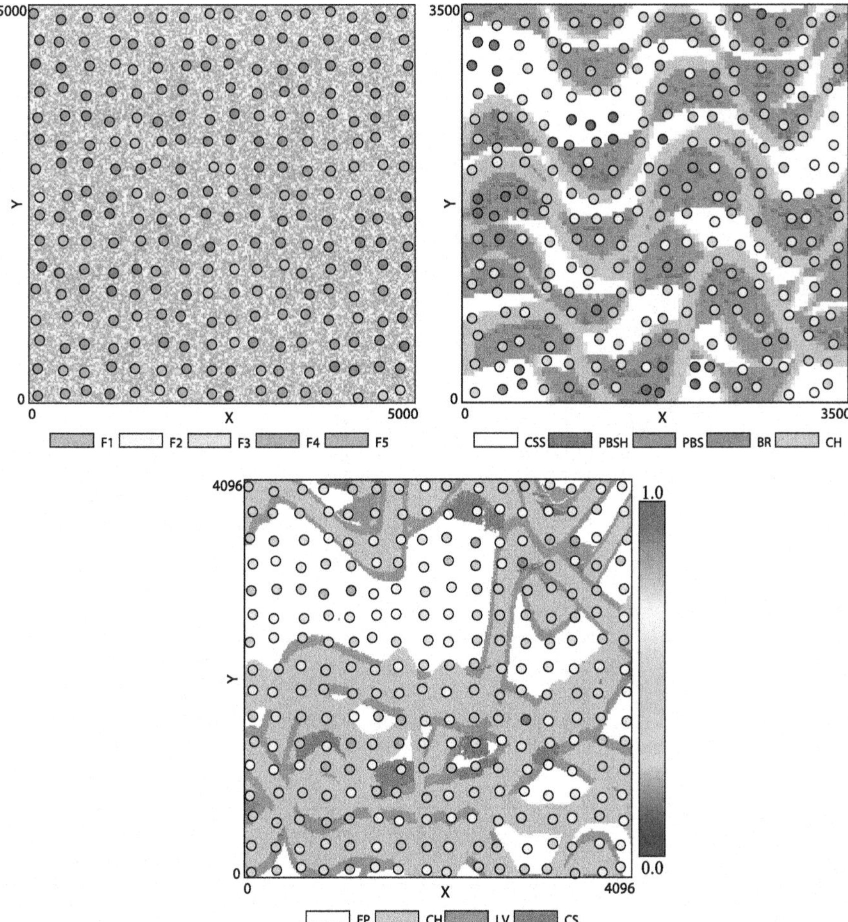

Fig. 3 Coherency results for random (*top left*), estuarine (*top right*) and fluvial (*bottom*) data sets also showing geology

Table 1 Grid sizes for example data sets

Data set	Nx, Ny	Nz	dx, dy	dz
Random	200	100	25	0.1
Estuarine	140	80	25	0.25
Fluvial	256	128	16	0.16

Fig. 5 and summary statistics are provided in Table 2. Wells were shifted close to their original positions. A consequence of the optimization procedure is that many wells were shifted to reach the maximum coherency, not only the 20 wells that were artificially shifted for the test. This information would normally not be available.

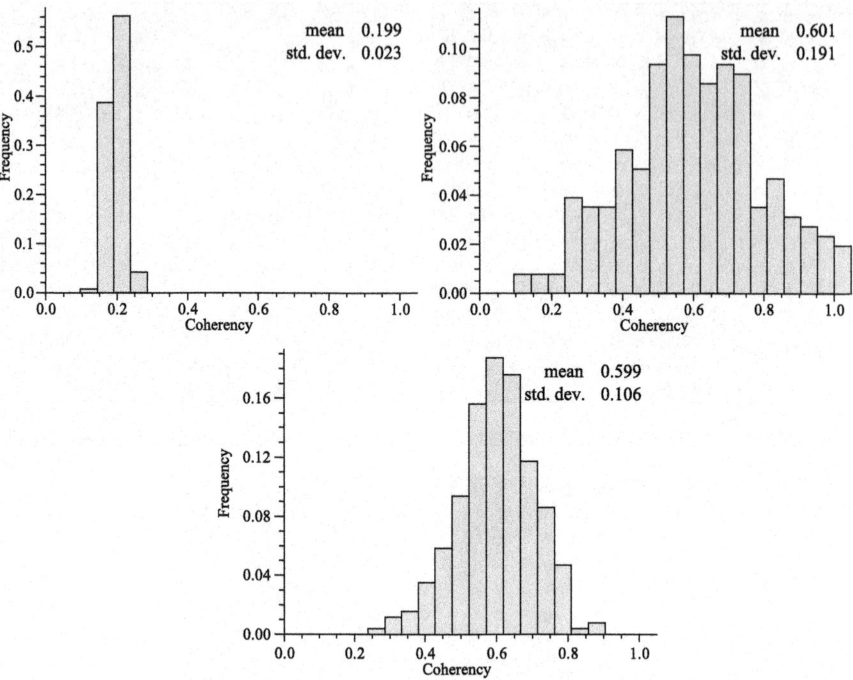

Fig. 4 Histograms of coherency for random (*top left*), estuarine (*top right*) and fluvial (*bottom*) data sets

Table 2 Shifted and optimized well position statistics

Case	Mean	Standard deviation	Minimum	Maximum
Shifted	0.05	0.74	−3.96	3.94
Optimized	0.05	0.24	−0.60	1.45

In cases where the wells are known to be accurately positioned in physical and stratigraphic space, the maximization procedure can be used to guide the gridding process. In this case, the grid conforms to the shift indicated by the maximization procedure and wells are not actually shifted.

3.3 Clustering

Coherency based clustering is demonstrated using the fluvial data set. Results are displayed in the same plan view as Fig. 2 for comparison with lines drawn around data that were clustered together (Fig. 6). Note the lines do not represent the actual position of boundaries between facies, but are used strictly for visualization

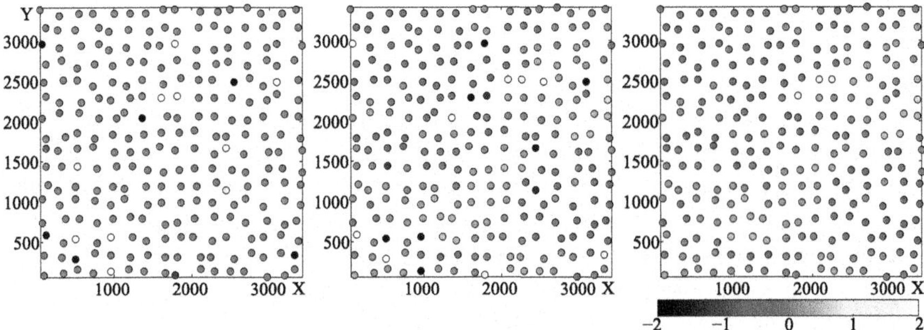

Fig. 5 Example of coherency maximization showing initial shifted wells (*left*), the shift that leads to maximum coherency relative to the initial shift (*middle*) and relative to the original well positions (*right*)

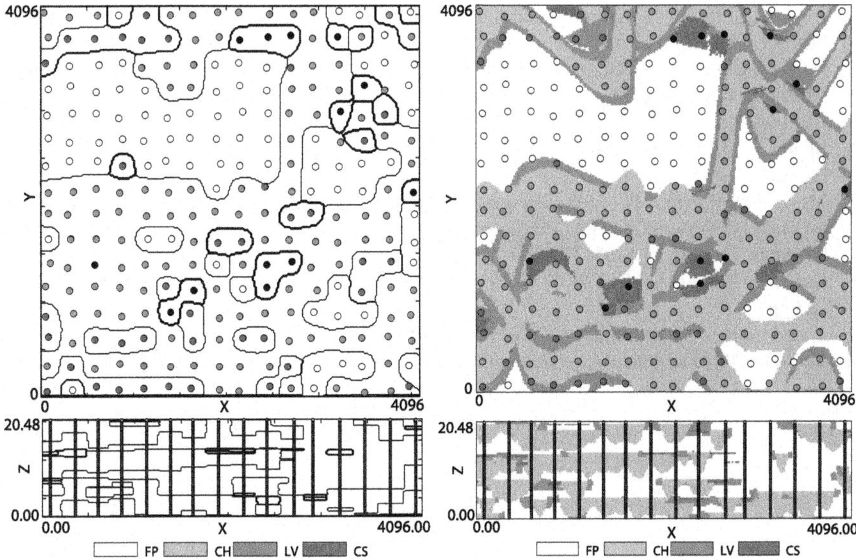

Fig. 6 Clustering results summarized using polygons (*left*) compared to the actual facies distributions (*right*)

of clusters. Data that belong to particular geological objects tend to be clustered together due to the high coherency between the data. This does not provide any information as to the location of the boundaries between neighboring geological objects or their exact shape, which is an area of future work.

Clustering resulted in 363 clusters, 200 of which only contain one facies interval. For quality control purposes, these single item groups could be checked with surrounding data to make sure the facies are correct. In cases where there is no explanation for the single item clusters, then such results are indicative of data with scales of heterogeneity smaller than the well spacing.

4 Conclusions

This work introduced a measure of coherency that is useful for quality control and to aid in geological modeling, specifically for categorical variables. Quality control is an important stage of geomodeling, but can be quite demanding for large databases. The coherency measure can be used to quickly identify potential problems with facies designations or well markers. Another use of coherency is clustering data into groups having similar facies. The approach is non-linear and resulting groups of facies can form objects with shapes that would otherwise be unattainable using traditional geostatistical approaches for categorical data.

References

1. Agterberg FP (1990) Automated stratigraphic correlation. Elsevier, Amsterdam
2. Deutsch CV (2002) Geostatistical reservoir modeling. Oxford University Press, London
3. Hein FJ, Cotterill DK, Berhane H (2000) An atlas of lithofacies of the McMurray formation, Athabasca oil sands deposit, Northeastern Alberta: surface and subsurface. Alberta Energy and Utilities Board, Edmonton, 217 pp
4. Mallet J-L (2004) Space-time mathematical framework for sedimentary geology. Math Geol 36(1):32
5. McPhee D, Ranger MJ (1998) The geological challenge for development of heavy crude and oil sands of Western Canada. In: Proceedings 14th international conference on heavy crude and tar sands, Beijing, pp 1765–1777
6. Ranger MJ, Gingras MK (2003) Geology of the Athabasca oil sands: field guide & overview, 123 pp
7. Theodoridis S, Koutroumbas K (2009) Pattern recognition. Elsevier, Amsterdam
8. Theys P (1999) Log data acquisition and quality control. Editions Technip, Paris

Multiscale Modeling of Fracture Network in a Carbonate Reservoir

Sanjay Srinivasan and Ankesh Anupam

Abstract Discrete Fracture Networks (DFN) models have long been used to represent heterogeneity associated with fracture networks. Realistic representation of DFN in field scale models have been impossible due to two reasons: First because the representation of extremely large number of fractures requires significant computational capability and second, because of the inability to represent fractures on a simulation grid, due to extreme aspect ratio between fracture length and aperture. This paper presents a hierarchal approach for fracture modeling and a novel particle tracking algorithm to upscale the fracture permeability. The modeling approach entails developing effective flow characteristics of discrete fractures at micro- and macrofracture scales without explicitly representing the fractures on a grid. A particle tracking algorithm is used that moves particles along implicit fractures honoring the intersection characteristics of the fracture network. The permeabilities of the microfractures are then used as base matrix permeabilities for particle tracking simulation of macrofractures.

1 Introduction

Characterizing fracture network is crucial for understanding the flow behavior of carbonate reservoirs. This task has proven extremely difficult over the past years. The difficulty comes from the complexity of the network topology and the wide disparity in the length scales of fractures and matrix blocks. The sparse data available to model the fracture network also implies that it is not possible to come up with a single reliable model. The attempt is rather to bracket the uncertainty in permeability due to fractures.

Usually very little information is available about the sub-surface fractures during field development. Direct evidence of fractures comes from FMI (Formation Micro Imager) logs and cores. However this information is very limited, as only those fractures that intersect the wellbores can be observed through logs and cores. Other

S. Srinivasan (✉) · A. Anupam
Petroleum and Geosystems Engineering Dept., The University of Texas at Austin, Austin, USA
e-mail: sanjay.srinivasan@engr.utexas.edu

P. Abrahamsen et al. (eds.), *Geostatistics Oslo 2012*, 185
Quantitative Geology and Geostatistics 17,
DOI 10.1007/978-94-007-4153-9_15,

indirect information about fractures can be obtained from seismic data, production data, well testing etc. Based on the available data, the best which can be done is to generate a suite of models each of which honors the available data. Little has been done so far in an attempt to rigorously model fractures at the field scale using multiple realizations. Fracture networks generally exhibit organizations over a hierarchy of scales. In this particular study, microfractures is the term assigned to fractures exhibiting apertures of 0.3 mm or less and that are completely contained within a grid cell used for flow simulation. Macrofractures on the other hand have apertures ranging from 0.3 mm to 1.1 mm and typically span several grid blocks.

Finding effective permeability of fracture network entails characterization of intersections between the fractures and interactions between the fractures and the matrix. Flow based upscaling of fractured network using conventional simulators require discretization of fractures on a gridded system. The extreme aspect ratio between the fracture aperture and fracture length makes it impossible to represent fractures with realistic apertures in a gridded domain. Few other methods [2, 5] have also been proposed for upscaling fracture networks, but none of them is rigorous enough to allow upscaling without any simplifying assumptions. In this work, we chose a particle tracking approach since that does not require explicit flow simulations and also can be used to take into account the influence of micro-scale fracture apertures.

2 Particle Tracking Simulation

Our model of microfractures consists of grids which are either fractures or matrix. We randomly place random particles on one of the face of the 3D gridded model. The coordinates of the injection point are picked at random. Then we assume a potential or pressure difference between the two opposite faces. The particles will tend to follow a path of least resistance. If the particle is moving within a fracture it will move much faster compared to the particle inside the matrix. We can track the particles at the other face in two ways. Either we can count the total number of particles reaching the other side at different time or we can track the time each particle takes to reach the other face. Qualitatively speaking if a particle is initially placed in fracture which connects to the other end, the particle should reach the outlet end very fast, however if it does not finds connecting fractures the time taken to reach the other end will be much larger.

However in these traditional implementation of particle tracking algorithm, field scale reservoir models with realistic distribution of fractures are a necessity. Given the extremely thin apertures of the micro-fractures explicit representation of such fractures may not be possible. Consequently, a new approach was developed where fractures are represented using planes (basically ellipses) with the thickness equal to the actual fracture aperture. The filling criterion for the fractures is the surface area of the fractures per unit bulk volume. The dip and azimuth of the fractures is assumed to follow a normal distribution with mean and variance taken for each fracture set as used before. The length of the fracture was assumed to follow a power law distribution. The aperture distribution was assumed to follow a uniform distribution.

3 Implicit Representation of Fractures

Instead of dealing explicitly with fractures represented on a grid, we will directly use equations of the ellipse within the random walk simulation to track the position of the particles. We can derive the equation of the ellipse for each fracture based on the dip and azimuth of each fracture. The centers of the fracture can be anywhere in the domain and do not have to conform to any grid. The assumption is made that if a fracture cuts the boundary of the domain it gets replicated on the other side of the domain. Particles are moved continuously in space and the extent of their movement is established by comparing their position to that of fracture ellipses whose center coordinates and lengths are known (just as in object simulation). Furthermore, a realistic aperture is also associated with each ellipse, although that ellipse is not explicitly mapped on to a grid. The probability of transition of a particle's position is calculated based on a transition within the same fracture, a jump to a nearby fracture or a transition to the matrix. A random number is drawn and compared against the computed transition probabilities. A transition path is selected and the next position of the particle is ascertained.

4 Calculation Details

The movement of a random particle through the medium is controlled by the permeability distribution along the path of the particle. In order to assign the permeability of fractures in the model, the fractures themselves are considered as planar objects with finite apertures. For each move of the particle, it is first ascertained if the particle will be within a fracture or outside it. In order to ascertain this, some simple trigonometric calculations are performed that are described next.

The equation of an ellipse with a given azimuth and dip can be written as

$$\frac{(x\cos\theta - y\sin\theta)^2}{a^2\cos^2\varphi} + \frac{(x\sin\theta - y\cos\theta)^2}{b^2} = 1, \tag{1}$$

$$z = (x\cos\theta - y\sin\theta)\tan\varphi. \tag{2}$$

These two equations satisfactorily define the locus of points along an ellipse that has been rotated by an azimuth angle θ and a dip angle φ. The ellipse has major axis a and minor axis b.

4.1 Simulation of Fracture Network

As shown above we can represent each fracture using the equations of an ellipse in 3D. Thus for the fracture simulation we need to simulate the following properties of the fracture:

1. Center of the Fracture: (X_c, Y_c, Z_c).
2. Length of the Fracture: The length of the fracture is the length of the major axis of the ellipse. The length distribution can follow for example a power law.
3. Aperture: The aperture can be assumed from a uniform distribution with a linear dependence on the length of the fracture:

$$\text{Aperture} = A_{\text{min}} + (A_{\text{max}} - A_{\text{min}})(\text{Uniform}_{(0-1)} \times L_{\text{frac}})/(1 + L_{\text{frac}}), \quad (3)$$

   where A_{min} = Minimum Aperture, A_{max} = Maximum Aperture, and L_{frac} = Length of Fracture.
4. Dip and Azimuth: Both of these are assumed to follow a normal distribution with mean and variance specific for a particular fracture set.

All the properties of the fracture listed above can be generated stochastically to fill up the 3D space [3]. When the desired fracture density is attained we can stop generating the ellipses. The number of fractures generated will depend upon the fracture density specified as an input.

It is to be emphasized that for tracking the movement through the micro-fractured media, it is not necessary to explicitly simulate the fracture network. Instead, the equations above can be used to determine the position of the particle and the probability of the particle moving to that position can be computed by taking suitable averages of fracture and matrix permeability. In order to make our domain more representative of the actual scenario and also to adhere to the specified filling fraction, we use a periodic boundary condition.

4.2 Periodic Boundary Condition

In order to make a small domain representative of the large system extending infinitely, we use periodic boundary condition. In reality any sub-volume of a fractured system always has fractures cutting through its edges and surfaces at all angles. Thus we can never have a standalone volume/domain in which all the fractures lie completely inside the system. Thus for our system whenever a fracture cuts any of the boundaries/surface it is replicated exactly on the opposite side as continuation of the truncated fracture. This simulates both the fractures going out of the domain and also the fractures cutting in to the domain from outside.

In order to implement the periodic boundary condition we need to do two things:

1. Check whether the fracture cuts any of the boundaries.

2. If it cuts then generate the replicated ellipse.
   If it cuts only one boundary, we will have one replicated ellipse.
   If it cuts 2 boundaries, we will have 3 replicated ellipses.
   If it cuts all the 3 boundaries, we will have 7 replicated ellipses.

The detailed procedure to check whether an ellipse intersects a boundary is presented in [1].

4.3 Tracking a Particle's Position

The intensity of fracture in the model is decided by the filling fraction which is specified as an input. The next step is to find the percolation properties of the network and if they percolate, then find the effective permeability of the system. Consider an ellipse with an arbitrary dip φ and azimuth $(90-\theta)$. The equation of the plane containing this ellipse in terms of its dip and azimuth is given by

$$(X - X_c)\sin\varphi\cos\theta + (Y - Y_c)\sin\varphi\sin\theta + (Z - Z_c)\cos\varphi = 0. \tag{4}$$

4.4 Distance of a Point from the Fracture Plane

For any point to lie inside the fracture its distance from the plane (containing the fracture) must be less than the aperture of the fracture. If the distance is greater than the aperture, the point lies in the matrix. Let $(X1, Y1, Z1)$ be the co-ordinates of the point. The distance from plane of the fracture is given by

$$\begin{aligned}\text{Distance} &= \frac{(X1 - X_c)\sin\varphi\cos\theta + (Y1 - Y_c)\sin\varphi\sin\theta + (Z1 - Z_c)\cos\varphi}{\sqrt{(\sin\varphi\cos\theta)^2 + (\sin\varphi\sin\theta)^2 + (\cos\varphi)^2}} \\ &= (X1 - X_c)\sin\varphi\cos\theta + (Y1 - Y_c)\sin\varphi\sin\theta + (Z1 - Z_c)\cos\varphi.\end{aligned}$$

In the above equation the distance can be both negative and positive depending on which side of the plane the point lies. For comparison to fracture aperture, absolute value of the distance is taken.

4.5 Test if a Point Lies in a Fracture

Even if the distance of the point from the fracture plane is less than the aperture, the point may not lie in the fracture. It depends on the location of the fracture on the plane. To check if the point lies inside the fracture, the equation of the ellipse is used. The following two steps are used to check if the point lies inside fracture.

1. Find a point on fracture plane such that the line joining the test point and the point on the plane is normal to the plane. Let its co-ordinates, $(X1', Y1', Z1')$, be

$$X1' = X1 - \text{Distance} \times \sin\varphi\cos\theta,$$
$$Y1' = Y1 - \text{Distance} \times \sin\varphi\sin\theta,$$
$$Z1' = Z1 - \text{Distance} \times \cos\varphi.$$

2. For the point $(X1', Y1', Z1')$ on the plane, find if it lies in the fracture or not by using the equation of ellipse. If

$$\frac{((X1' - X_c)\cos\theta + (Y1' - Y_c)\sin\theta)^2}{a^2\cos^2\varphi} + \frac{((X1' - X_c)\sin\theta + (Y1' - Y_c)\cos\theta)^2}{b^2} - 1 < 0,$$

then the point lies inside the fracture. It must be noted that the distance used in the equation should be used with its proper sign.

4.6 Intersection of Two Ellipses in 3D

Possible fracture intersections that might lie in the path of the particle are another complication that may arise when modeling the movement of particles through a continuous system of coordinates. The intersection of two ellipses in 3D will be a line with both of its ends lying on anyone of the ellipses. It is difficult to solve the two quadratic equations simultaneously, so we implement a step by step process details of which are presented in [1]. The intersection between two fractures has to satisfy the equations of the ellipses as well as the equation of the plane jointly. In order to accomplish this, the line of intersection between two fracture planes is solved first using the equation of the plane. Subsequently, the intersection of that line with the two fracture ellipses is calculated. The result of this check can be either no common point when none of the points of intersection lies inside the other ellipse or the two common points defining the line of intersection of two fractures.

4.7 Percolation Analysis

The next task is to upscale the micro fracture models and calculate the upscaled permeability value for each flow simulation grid block. For each value of fracture intensity, multiple realizations of micro fracture model are considered in order to obtain a distribution of upscaled permeability and quantifying the uncertainty in the effective permeability value for that grid block. In order to calculate this effective permeability, percolation analysis [4] is performed on each micro fracture model to assess if the microfracture network in the block forms a connected pathway across the block. The effective permeability of the block is calculated proportional to the arrival time characteristic of the particles for those blocks that exhibit percolation.

A fracture network is said to be percolating if there exists at least one connected path through the fractures from one side of the domain to the other. The detailed algorithm used for moving the particles through the network is outlined below. The assumption for this analysis is that the matrix is non-conducting and fluid flow is only through the fractures. This assumption applies very well for a carbonate reservoir where fluid flow is only through the fractures and matrix has a very low permeability.

1. The fracture parameters are specified. The number of fractures depends on the fracture intensity.
2. All intersection points among the fractures are calculated including all the replicated ellipses. If the particle has to percolate then it needs to starts from one of the fractures that intersect the face.
3. At the start the random particles are placed in one of the fractures on the inlet face of the domain. The starting fracture is selected randomly among all the fractures intersecting that face.
4. All possible paths for the next movement of the particle are analyzed. The only constraint is that the movement should be in the direction of the pressure gradient across the block. The number of paths depends on the number of fractures intersecting the fracture currently occupied by the particle.
5. Among the available paths one path is selected at random.
6. The same process as in steps 4 & 5 are followed again to find the possible paths for further movement of particle to the next fracture. At each step it is tested if the fracture on which the particle resides intersects the other side of the domain or not.
7. If the particle reaches a fracture that intersects the other face of the domain, the particle percolates. On the other hand if the particle is somewhere in the middle and has no possible path for moving further, it cannot percolate. This is true also if the particle reaches a face that is not the outlet face.
8. The above steps (1–7) are repeated for all subsequent particles. The number of particles should be large enough to explore all possible connected paths through the fracture network to get a statistically stable result.
9. At the end of the simulation, the number of percolating particles is compiled. For non-percolating systems this number is zero and the effective permeability due to microfractures is assigned to be zero md. For percolating system the number is non-zero and higher the number, the greater is the number of connected paths in the system and consequently, higher the effective permeability.

4.8 Results of Particle Tracking Simulation

The aim of percolation analysis is to find the upscaled permeability of microfractures for each grid block based on its fracture intensity. It is assumed that there

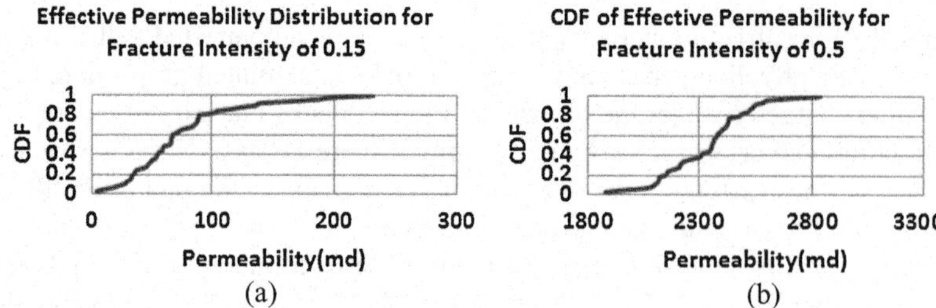

Fig. 1 Cumulative Distribution Function of effective permeability for fracture intensity of 0.15. The plot shows uncertainty corresponding to (**a**) fracture intensity of 0.15, and (**b**) fracture intensity of 0.50

is a linear correspondence between the upscaled permeability and the fraction of particles percolating, as both of them are the indicators of connectivity of the system. This linear correspondence is used to express the effective permeability as a weighted average of matrix and fracture permeability, where the weights are the fraction of particles percolating:

$$\begin{aligned}\text{Upscaled Permeability} = {} & (\text{Fraction of percolating walkers}) \times K_{\text{frac}} \\ & + (\text{Fraction of non-percolating walkers}) \times K_{\text{mat}},\end{aligned}$$

where K_{mat} is average matrix permeability and K_{frac} is average fracture permeability. For the present case we use the microfracture fracture permeability as 10 000 mD and matrix permeability as 0.1 mD, which are typical for carbonate fracture-matrix systems. Here the fracture permeability is considered to be constant, but this condition can be relaxed and the permeability can be related to the aperture. The permeability of individual microfractures is a function of fracture aperture and roughness which may also vary along the length of the fracture.

There can be many different distributions of fracture network having the same fracture intensity. This is represented by multiple realizations of fracture network for the same fracture intensity. When particle simulation is performed on all these realizations, the output is not a single permeability but a range of permeability values, each one corresponding to a particular realization. Figure 1(a) and (b) show the CDF of effective permeability distribution for fracture intensities of 0.15 and 0.5. The same process is repeated for each value of fracture intensity to give an uncertainty envelope for all fracture intensities.

The percolation study can be performed in all three directions on the same fracture network. The same network can be percolating in one direction but not in other directions. This may lead to permeability anisotropy in the system. Therefore to completely characterize the network it is important to study percolation in all the three directions. Percolation study has been performed on microfractures in all the three directions to quantify the permeability anisotropy of the system.

Figure 2 shows the variation in maximum, minimum and average effective permeability in the X direction with change in fracture intensity. Similar plots can be prepared for average permeability in the Y and Z directions.

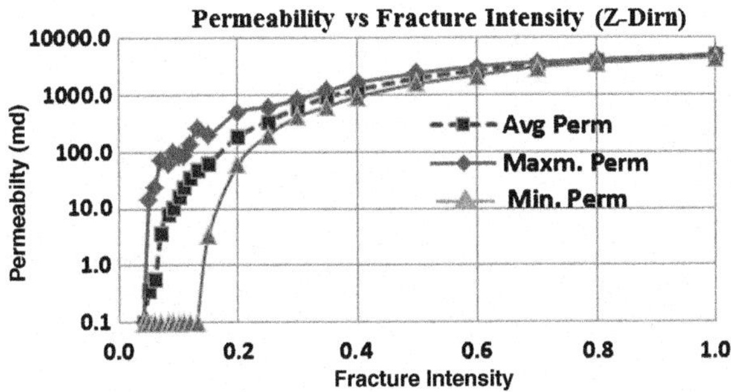

Fig. 2 Plot shows maximum, minimum and average upscaled permeability as a function of fracture intensity for flow in the z-direction

Table 1 Anisotropic permeability values as a function of fracture intensity

Fracture intensity	Kx (md)	Ky (md)	Kz (md)
0.05	0.10	0.10	0.37
0.1	0.15	0.17	17.14
0.15	0.69	1.94	62.24
0.2	6.08	12.64	185.24
0.3	54.07	94.18	587.74
0.4	218.54	302.00	1 228.00
0.5	528.37	692.35	1 968.85
0.6	938.00	1 209.00	2 677.26

The minimum, maximum and average permeability values corresponding to a particular value of fracture intensity are computed over several realizations constrained to that value of fracture intensity. The maximum and minimum values come from the realization that has the maximum and minimum numbers of percolating particles. While the average permeability, is the permeability corresponding to the average number of particles percolating computed over all the realizations.

Another important point to note is that the variation in permeabilities is higher at lower fracture intensities and all the three permeabilities merge to a single point at higher intensities and remain constant with further increase in fracture intensity. As the number of fractures increase beyond a limit in a percolating medium, the effective permeability is limited by the maximum permeability of individual fracture and remains constant.

The values obtained above can be used to populate microfracture permeability in all the grids. Based on the fracture intensity in each grid block, the permeability tensor due to microfractures can be assigned. Based on this, the anisotropy in the permeability values is presented in Table 1.

The z-direction permeability is much higher compared to that in the x and y directions. We expect the x and y direction permeabilities to be equivalent because the azimuth distribution of the fracture sets span 360°. However in the z-direction, the mean dip direction for all the six fracture sets is around 60°. Thus, there is a directional preference in the dip distribution of the fractures and consequently, the permeability obtained in the z-direction is quite different from that in the x and y directions. This results in the permeability in z direction to be much higher than that in the x and y directions.

5 Particle Tracking Simulation Through Macrofractures

The ultimate objective of the particle tracking simulation is to find the upscaled permeability of fracture network integrating both the microfracture and the macrofracture models. Explicit representation of fractures at both scales is not possible given the wide range of length and aperture distributions. Therefore to incorporate the effect of microfractures, the upscaled microfracture permeability is used as matrix permeability when simulating the movement of particles through the macrofracture. With the introduction of finite permeability for matrix, the random particles can now move through both fractures and matrix. This is unlike the particle tracking simulation used for upscaling microfractures where particles could move only through the fractures.

The time taken by the random particle to reach the other end of the domain is calculated. The path of each particle emulates the actual flow streamline and the time taken is a measure of the tortuosity of that streamline. Therefore it is expected that the average of the time taken by the particles should directly correlate with the upscaled permeability of the system.

5.1 Algorithm for Moving Particles Through Macrofractures

The random particles are moved from one side of the domain to the other side and the time taken by each particle is calculated. The following step-by-step procedure is implemented:

1. The macro-simulation parameters (center of ellipsoids, aperture, length, orientation etc.) are input to the particle tracking simulation.
2. Using these input intersections between the fractures are calculated. These calculations are done using the analytical equation of ellipse. They give the coordinates of all the points of interaction between the fractures.
3. Initially the particle is placed on the starting face of the domain. An arbitrary point on the starting face is selected randomly to start the movement of the particle. A uniform pressure field is assumed between the starting and the ending face between which the particle has to travel.
4. There are two options for the starting point. The particle can either lie in a fracture or in the matrix. If it lies in a fracture, all the fracture intersections at the location of the starting point are found. It is extremely rare to find particle lying in multiple fractures simultaneously at the starting point. However if such a case arises, one of the fractures in which the particle lies is selected randomly as the starting fracture.

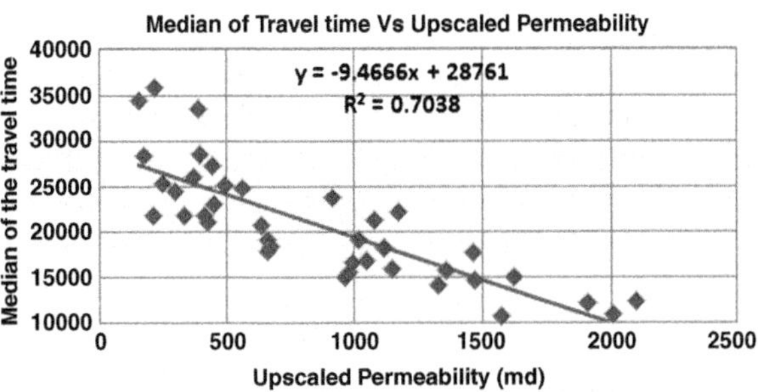

Fig. 3 Calibration of the median travel time of random particles to flow based upscaled permeability (mD)

5. If the particle lies in the matrix, distances to the nearest fractures are calculated. The number of nearest fractures for which the distance is stored is an input parameter and can be varied depending on the fracture intensity. The distances calculated is the distance joining the present position of the particle to the nearest point on the fracture. The co-ordinates of that closest point on the fractures are also calculated. For this calculation only those fractures are considered which allow the particle to move in the direction of positive pressure/potential gradient.

5.2 Results of Macrofracture Simulation

Figure 3 shows a scatter plot between the median travel time and the upscaled permeability obtained by fine scale flow simulation. The median of the travel time shows a negative correlation with effective permeability of the medium. Increase in median travel time indicates an increase in average resistance offered by the medium and hence a decrease in effective permeability of the medium. The straight line fit is quite satisfactory except for very low values of permeabilities. A perfect correlation cannot be expected, for the simple reason that the fractures on the gridded system used for flow based upscaling is only an approximate representation of the actual fracture network.

6 Conclusions

A hierarchical modeling approach on a continuous space domain was adopted for computing the effective permeability of fractures at two different length scales: microfractures and macrofractures. Using a continuous space domain, fractures are represented as elliptical discs with realistic apertures. Analytical equations of ellipse are used to represent the fractures and to determine their intersections. Then on this implicit representation of fractures, the movement of random particles is modeled and the percolation characteristic of the microfracture model is ascertained. The presence of fractures affects the effective permeability of the system only when the intensity of fractures is above a critical threshold such that fluid flow can occur across the block. The critical threshold is a function of the number of fracture sets and their orientation and is also direction dependent.

Next, the effective permeability of macrofractures is integrated together with the macrofractures to arrive at the effective flow characteristics of the multiscale network. It was found that the median of the travel time distribution is a perfect measure for the upscaled permeability of the system.

References

1. Anupam A (2010) Hierarchical modeling of naturally fractured reservoirs. M.S. Thesis, University of Texas at Austin
2. Arbogast T (1993) Gravitational forces in dual-porosity systems: vols I & II. Computational validation of the homogenized model. Transp Porous Media, 13(2):205–220
3. Billaux D, Chiles JP, Hestir K, Long J (1989) Three-dimensional statistical modelling of a fractured rock mass—an example from the Fanay Augeres mine. Int J Rock Mech Min Sci Geomech Abstr 26(34):281–299
4. Drory A, Balberg I, Berkowitz B (1994) Random-adding determination of percolation threshold in interacting systems. Phys Rev E, 49(2):949–952
5. Oda M (1985) Permeability tensor for discontinuous rock masses. Geotechnique 35(4):483–495

Uncertainty Quantification and Feedback Control Using a Model Selection Approach Applied to a Polymer Flooding Process

Sanjay Srinivasan and Cesar Mantilla

Abstract Polymer flooding is economically successful in reservoirs where the water flood mobility ratio is high, and/or the reservoir heterogeneity is adverse, because of the improved sweep resulting from mobility-controlled oil displacement. The performance of a polymer flood can be further improved if the process is dynamically controlled using updated reservoir models and by implementing a closed-loop production optimization scheme. This paper presents a feedback control framework that: assesses uncertainty in reservoir modeling and production forecasts, updates the prior uncertainty in reservoir models by integrating continuously monitored production data, and formulates optimal injection/production rates for the updated reservoir models. The value of the feedback control framework is demonstrated with a synthetic example of polymer flooding where the economic performance was maximized.

1 Feedback Control System for Polymer Flooding

Polymer flooding is a mobility-controlled enhanced oil recovery process often used when the water-oil mobility ratio is too high for an efficient water flood and/or the reservoir geology is highly heterogeneous, i.e. extensive stratification or high permeability contrast between channels and the rest of the formation [11]. Reservoir heterogeneity is one of the major contributors to uncertainty in reservoir modeling, and it is highly dependent on the geological depositional environment of the reservoir. Feedback control approaches for reservoir management have demonstrated improved recovery and economics of water flooding projects [3, 10]. This is accomplished by continuous updating of the prior reservoir models and optimization of well controls on the updated model [7]. The objective of model updating is to modify prior reservoir model(s) such that the posterior model(s) better reproduce the observed production data.

Despite the benefits of closed-loop approaches for managing water flooding processes, such schemes had not been implemented in polymer flooding processes yet.

S. Srinivasan (✉) · C. Mantilla
Petroleum and Geosystems Engineering Dept., The University of Texas at Austin, Austin, USA
e-mail: sanjay.srinivasan@engr.utexas.edu

P. Abrahamsen et al. (eds.), *Geostatistics Oslo 2012*,
Quantitative Geology and Geostatistics 17,
DOI 10.1007/978-94-007-4153-9_16,

Special characteristics of polymer flow in porous media requires specialized simulation models like UTCHEM—a three-dimensional chemical flood simulator developed at the University of Texas at Austin—that prevents straightforward implementation of model updating and optimization schemes applied for water flooding. The main goal of this paper is to develop an approach to maximize an economic index such as net present value by controlling the injection and production rates in a polymer flood.

The process starts with an initial large ensemble of reservoir models that reflects the prior uncertainty in reservoir description, including all plausible geologic scenarios conditioned to the available static data. In the absence of production data all prior reservoir models are equally probable, thus the model selection algorithm is applied to select a few reservoir models that are more consistent with the observed production characteristics. The optimal control settings are computed over the retained subset of reservoir models by an optimization algorithm and then implemented in the field through flow control devices that regulate injection and production rates. The well responses are monitored with pressure sensors at injectors and producers, and fluid samples are analyzed to determine the water cut and polymer concentration at the producers. At the same time, a reservoir simulation model is constrained to the current well controls for simulating the production data during that control period. Then, the field production data is compared to the simulated production data. If the mismatch between simulated and actual production data is within a user defined tolerance, the operation continues with the already formulated control settings. On the contrary, if the mismatch is more than a tolerance threshold, the observed production data is utilized to further refine the selected reservoir models, and the control settings are revised with the optimization algorithm. In this way the control system is a closed-loop operation that continuously updates the reservoir representation and revises the production strategy. The development of this protocol is illustrated with the example of polymer injection in a two-well reservoir as described in the next sections.

2 Case Description

The feedback control framework is illustrated using a synthetic reservoir example that has one polymer injector and one producer. The geology of the reference reservoir consists of high permeability channels embedded in a low permeability background as shown in Fig. 1. The permeability contrast between channel and non-channel facies is 1000:10 approximately. An injector and producer are completed in the high permeability regions and communicate through a high permeability streak that causes flow channeling. The reservoir is flat at a depth of 3500 ft, with a width 800 ft, length 800 ft long and thickness 8 ft. Porosity is uniform within the reservoir and equal to 0.1. The initial water saturation is equal to the irreducible water saturation (0.2). The reservoir was discretized into 160×160 grid-blocks each of dimension $5 \times 5 \times 8$ ft^3. The initial reservoir pressure is 2500 psi.

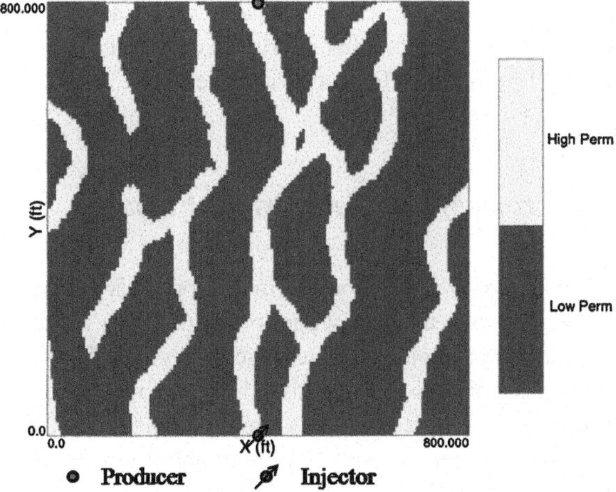

Fig. 1 Facies map of reference reservoir with two wells

The water-oil mobility ratio at the end points (= 10.9) and channel heterogeneity render water flood unfavorable for oil recovery, making this reservoir a good candidate for polymer flooding. Partially hydrolyzed polyacrylamide (HPAM) polymer is injected at a concentration of 750 ppm. The polymer viscosity at the injected concentration is 20 cP at zero shear rate. The polymer solution injected is a shear thinning fluid whose viscosity is reduced to 10 cP at a shear rate of 280 s^{-1}. The polymer retention is 60 μgr/gr of pore volume at the injected concentration. The salinity of the aqueous phase is 0.006096 meq/ml. The modeling parameters of this polymer solution were taken from [5].

The injection and production rates are to be optimized during 2000 days of operation such that the final net present value is maximized. The total operation time is divided into 8 control periods of 250 days. The control vector consists only of injection rates because the production rates are assumed equal to the injection rates at any time. The production data used to guide the model selection process are the water cut, the injection pressure and the producer pressure observed during polymer flooding. These data are monitored every 10 days.

3 Assessing Prior Geologic Uncertainty

The uncertainty in reservoir geology and heterogeneity is represented by an initial ensemble of reservoir models. All reservoir models in the initial ensemble honor the conditional data available, i.e. reservoir properties at sampled locations (well logs) and indirect information (seismic). Geostatistical algorithms like *sisim* or *snesim* [6] are used to generate multiple reservoir models. The initial ensemble should be large enough to include variations in well connectivity because the model selection process is based on differences in well connectivity. The objective is to organize the initial ensemble by groups of similar reservoir models in terms of well connectivity.

To do that it is required to have a measure of heterogeneity related to well connectivity and production characteristics of polymer flooding. For this purpose, a proxy model for polymer flooding was developed.

3.1 Proxy Model for Polymer Flooding

A robust proxy model for polymer flooding is crucial in the feedback control framework. For this, the proxy has to capture flow characteristics due to heterogeneity, and provide a rapid estimate of recovery functions given a geologic model and operating conditions at the wells. The proxy model employed in this work is a particle tracking algorithm to quickly propagate particles from injectors to producers mimicking the displacement of polymer solution from the injector to the producer. The paths of random particles placed in the reservoir are traced from their initial state (injection points) to their final state (production points). The probability rules for particles to transition from the current position to the next position are governed by the velocity field obtained as a solution of the pressure equation for a single phase, steady-state displacement, considering the heterogeneity of a given reservoir model. In the case of polymer flooding, the particle tracking algorithm needs to take into consideration the difference in velocity of the water front and the polymer front, because of the retention of polymer in porous media [11].

The word "particles" employed in this paper refers to the random walkers. Its meaning should not be confused with a physical particle of water or polymer. According to the fractional flow theory explained by [9], the velocity of the water front v_w is proportional to the slope of the tangent line to the fractional flow f_w curve of a water-oil system drawn from the initial water saturation S_w i.e. $(\mathrm{d}f/\mathrm{d}S)_{ow}$. For a single-phase system, the fractional flow curve is simply a 45 degrees straight line with slope equal to 1. Given the single phase fluid velocity v_{sp}, the velocity of water particles for the water-oil system is updated using

$$v_w = v_{sp} \times \left(\frac{\mathrm{d}f_w}{\mathrm{d}S_w}\right)_{ow}, \tag{1}$$

where it is assumed that the ratio of velocities, v_w/v_{sp}, is equal to the ratio of the slopes of the tangent lines $(\mathrm{d}f/\mathrm{d}S)_{ow/1}$ in each system. Similarly, the velocity of polymer particles is computed using

$$v_p = v_{sp} \times \left(\frac{\mathrm{d}f_w}{\mathrm{d}S_w}\right)_{op}, \tag{2}$$

where it is assumed that the ratio of the velocity of the polymer front to the velocity in a unit-mobility ratio system is equal to the ratio of slope of the tangent lines. The subscript *op* in the following denotes the oil-polymer system.

As a result, random walkers tagged to be the water phase propagate in the medium faster than polymer particles and breakthrough at the producers earlier. The position of water and polymer particles is related to the position of the water and polymer fronts respectively. Estimated recovery factor and breakthrough time are outputs of the proxy model to characterize a reservoir model. The recovery factor is calculated by performing a material balance considering the particles injected, the particles initially in the medium and the particles recovered. Additional measures of reservoir heterogeneity can be considered according to their relation to the production data available.

In summary, the proxy model yields a few characteristic measures of well connectivity for each reservoir model from the initial ensemble. The computational cost of processing a large number for reservoir models with this proxy model is relatively low because the pressure equation is solved only once, and the particles are moved using fractional flow theory. Another alternative for measuring effective reservoir heterogeneity is streamline simulation adapted to the process of interest, in this case polymer flooding.

3.2 Metric Space for Assessment of Uncertainty in Reservoir Geology

Once we have an ensemble of reservoir models representative of the prior geologic uncertainty, the next step is to group them according to their similarity. This implies that we need a protocol for putting multiple reservoir models side by side and a measure for similarity or dissimilarity. An ample set of reservoir models that adequately sample the prior uncertainty are the bases for constructing the metric space. This metric space is sometimes referred to as the uncertainty space in this paper.

The term distance between reservoir models was introduced by [1], referring to a measure of dissimilarity between any two different reservoir models. Multidimensional scaling (MDS) was used by [4] to project reservoir models in a two-dimensional space. Another option for constructing a metric space is to use principal component analysis (PCA), which is a mathematical procedure that transforms a number of correlated variables into a smaller number of linearly uncorrelated variables or components. The reservoir models are represented on a 2-D plot of the first principal component versus the second.

The procedure is illustrated by processing 600 reservoir models with different geologic characteristics. All reservoir models are defined on a 40×40 grid, and show high or low permeability values. The well arrangement is simple: polymer is injected at a location at the bottom of the model (see Fig. 1) and fluids are produced at a location at the top of the model.

Reservoir models can be associated within clusters or groups using the k-means clustering algorithm [2] that partitions an entire set of points into clusters. Eight clusters of reservoir models were formed using k-means as indicated in Fig. 2 in order to analyze the correspondence between the position in the metric space and well connectivity. These eight clusters were selected only to show the orderly transition of models across the metric space and in the subsequent section a criterion for picking the optimum number of clusters is presented. The weighted average properties

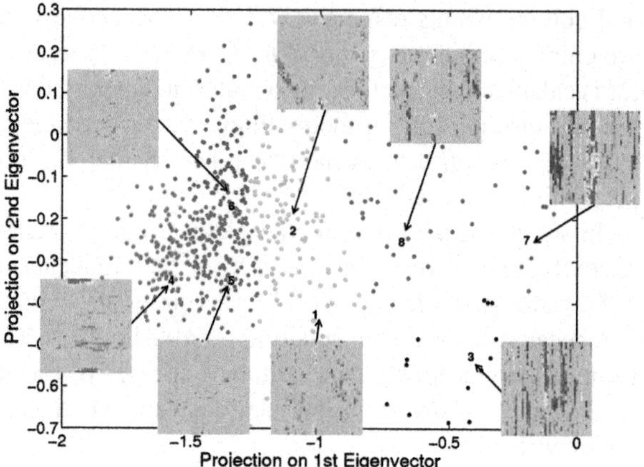

Fig. 2 Clustering of reservoir models in the space of the top 2 eigenvectors. Representative permeability models for each cluster are also shown

of models in a cluster such as the average permeability of each cluster are calculated where the weights for each realization (*nr*) within a cluster (*nc*) is inversely proportional to the distance to the center of the cluster. The ensemble of reservoir models organized into eight clusters show an orderly transition between extremes in that space. Reservoir models smoothly transition from high to low connectivity in the N-S direction through this map.

3.3 Updating Prior Probability Using Production Data

All reservoir models were processed with the proxy model in order to obtain responses correlated to well pressures and water cut and then projected on the basis of the top 3 eigen-vectors of the covariance matrix, and partitioned into three clusters. This is the optimal number of clusters in the sense that the distance of each model within a cluster from the cluster mean is minimized and simultaneously the distance between clusters is maximized. Based on the density of points in each cluster, the prior probability of cluster 1 is 0.3583, 0.5033 for cluster 2 and 0.1383 for cluster.

A weighted average model for each cluster was calculated (with weights calculated as previously) and subject to flow simulation using UTCHEM in order to simulate injection and production pressure and water cut during 250 days. The objective of this step is to assess the conditional probability for a particular cluster given the observed production history. The posterior probability is obtained by application of Bayes' rule and comparing representative production data of each cluster to the reference production data:

$$\text{Prob}(\mathbf{x} \in \text{cluster}\, i\,|\mathbf{D}_{\text{ref}}) = \frac{\text{Prob}(\mathbf{D}_{\text{ref}}|\mathbf{x} \in \text{cluster}\, i)\,\text{Prob}(\mathbf{x} \in \text{cluster}\, i)}{\sum_i \text{Prob}(\mathbf{D}_{\text{ref}}|\mathbf{x} \in \text{cluster}\, i)}, \tag{3}$$

where $\mathbf{D}_{\text{ref}}$ refers to the production data of the reference reservoir. Now, the question is how to obtain $\text{Prob}(\mathbf{D}_{\text{ref}}|\mathbf{x} \in \text{cluster}\, i)$? Suppose, the well pressure has been

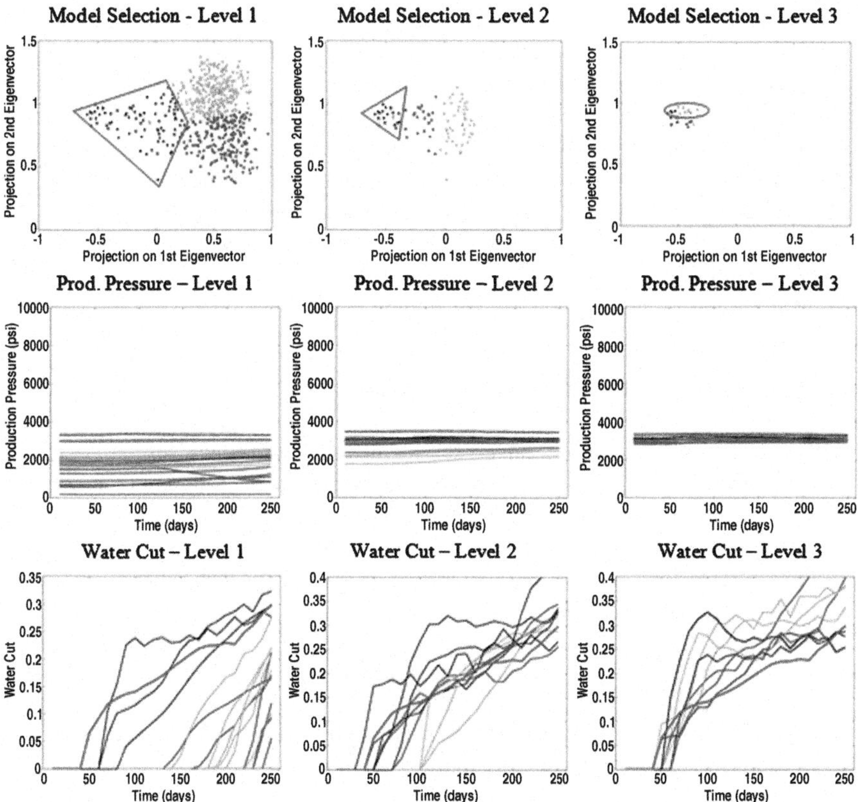

Fig. 3 Model selection process in three levels. *Top row* shows the clustering and selection process at each level. *Subsequent rows* show the production data corresponding to each cluster used to evaluate the updated cluster probability. *Red lines* represent the production data of the reference reservoir model. *Other lines* are *color coded* according to the clusters

acquired for a time period, an uncertainty envelope can be delineated assuming a Gaussian distribution with the measured pressure history as mean. The assumption of Gaussianity is reasonable since a Gaussian measurement error model is commonly assumed. The choice of the variance for that distribution is based on the statistics of the measurement error. The likelihood of observing the measured data given a particular model is given by the probability envelope that brackets the response simulated corresponding to that reservoir model.

The posterior probabilities of the clusters after incorporating the reference production data are 0.13, 0.20 and 0.67. The cluster with the highest probability is selected for a second level of selection as illustrated in Fig. 3. Production data starts converging towards the reference at the second level of selection. At the third level, few reservoir models remain. There are two stopping criteria: either the minimum number of reservoir models remaining is reached, or the posterior probability distribution does not show preference for any cluster. In this case the process is stopped when the minimum number of reservoir models in the final cluster is fifteen.

Finally, each member of the final set was subject to flow simulation using UTCHEM. The injection and producer pressure from the final set is close to the reference as seen in Fig. 3. The selection method narrowed the uncertainty in production data significantly using a minimum number of simulation runs. In this case 26 flow simulations with UTCHEM were carried out to calculate the posterior probability and for calibrating the random walk procedure. The permeability maps of the final set have in common a high permeability streak connecting the injector to the producer (similar to the ones in Fig. 5). This is also a prominent characteristic of the reference model (Fig. 1).

4 Production Optimization

Once a set of reservoir models have been identified that produce similar responses as observed in history, the next step is to find the optimal production strategy for maximizing the economics of the project. The production optimization problem is formalized as follows: find the optimal control vector **u**, composed of injection and production rates for NT time periods, such that the net present value at final time ($J = \mathrm{NPV}$) is maximized:

$$\arg\max_{\mathbf{u}} J = \int_0^{t_f} f(\mathbf{x}, \mathbf{u})\, \mathrm{d}t, \tag{4}$$

where **x** is a vector of state variables: porosity, permeability, pressure, saturation etc. Expression (4) is maximized subject to the following constraints:

$$\mathbf{x}^i = g\left(\mathbf{x}^{i-1}, \mathbf{u}^i\right), \tag{5}$$

$$\mathbf{LB} \leq \mathbf{u} \leq \mathbf{UB}, \tag{6}$$

$$\sum_{\text{inj}} \mathbf{u}_{\text{inj}}(t) = \sum_{\text{prod}} \mathbf{u}_{\text{prod}}(t), \tag{7}$$

where (5) are flow equations for polymer flooding, (6) are operational constraints, and (7) is material balance between cumulative injection and cumulative production.

Here, $\mathbf{x}^i$ is the updated state vector obtained by executing the transfer function $g(\cdot)$. The number of control periods refers to the frequency of adjustments of well constraints during the operation of the wells. The number of control variables (nu) for this problem is expressed as

$$nu = (NW - 1) \times NT, \tag{8}$$

where, NW is the number of wells and NT is the number of control periods. Note the loss of one degree of freedom indicated in (8) because of the overall material balance condition imposed by (7).

The objective function is the net present value using the economic parameters shown in Table 1. These economic parameters were chosen only for demonstration purposes—the aim was to obtain a net present value that makes the trivial solution (maximum production rates) sub-optimal. For that reason, the assumed cost of water

Table 1 List of economic parameters used in optimization process

Economic parameter	Value
Oil Price	\$35
Discount Rate	\$10 %/year
Production Facility Cost	\$10000/month
Water Injection Cost	\$1/bbl
Water Production Cost	\$10/bbl
Oil Production Cost	\$4/bbl
Polymer Injection Cost	\$1/lb
Royalty Tax	38.5 %
Ad Valorem Tax	\$0.046/bbl

treatment penalizes excessive water production. The physical constraints are: the injection/production rates are bounded between 0 and 200 ft^3/day, and the injection rate is equal to the production rate at all times. Yearly cash flows are calculated knowing the fluid production rates, the volume of water and polymer injection and the facility operating cost. Royalty and *ad valorem* taxes are added to make the economics realistic. Cash flows are discounted to present time using the discount rate and the net present value (which is the objective function) is calculated as an aggregate of the present value of all the cash flows.

4.1 Response Surface for Production Optimization

Experimental design and response surfaces have been commonly used in optimization and uncertainty assessment when the relationship between control variables and objective function is difficult to express mathematically, particularly when the transfer function is numerical. A response surface is an empirical model that relates the model parameters (such as geologic variables and/or well controls) to corresponding responses. The generated empirical model is often represented as a linear regression model:

$$J^* = \beta_o + \beta_1 u_1 + \cdots + \beta_n u_n + \sum_{i<j}\sum_{j=2}^{n} \beta_{ij} u_i u_j + \sum_{i=1}^{n} \beta_{ii} u_i^2 + \epsilon, \tag{9}$$

where J^* is the response, β^i's are a set of regression coefficients, $\mathbf{u}_i$'s are the independent variables (control parameters), n is the degree of the polynomial fit and ε is an error term [8].

Denoting *nruns* as the minimum number of function evaluations required to solve for the regression coefficients in (9), *nruns* is equal to $(n^2 + 3n)/2 + 1$. Although adding more terms to the equation of the response surface yields a better fit of the true responses, over-fitting problems may arise. To avoid that, a test of significance is performed to identify the statistically significant regression parameters, i.e. regression coefficients statistically correlated to the response. Fewer terms in (9) may

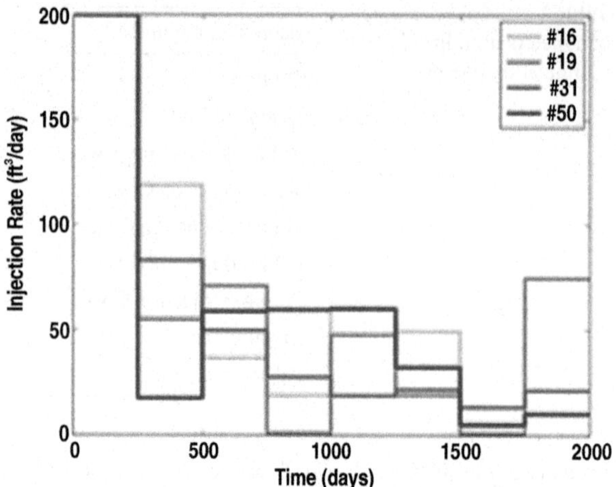

Fig. 4 Optimal control settings for the reduced set of reservoir models after using production data for 250 days. The optimization starts from 250 to 2000 days

remain after the test of significance and the effect of the variables eliminated is absorbed in the error term.

Figure 4 shows the optimal control settings obtained using four of the reservoir models in the final selection. In general, the optimal injection rate reduces gradually as the reservoir models predict high water production due to the high connectivity between wells. The optimization results are consistent with the objective of delaying water production as much as possible while producing oil to generate revenue. The question now is: what control settings to implement on the field? The optimal settings obtained for any of the four reservoir models or the expected value of the optimal control settings? A series of test cases consistently indicated that $J * (E\{\mathbf{u}_{\text{opt}}\})$ is greater or equal than $E\{J * (\mathbf{u}_{\text{opt}})\}$ i.e. the NPV profile obtained corresponding to the effective optimum control expressed as the average of the control strategy for several realizations is more than the expected NPV obtained by implementing the optimum control for each realization.

5 Updating Reservoir Model Selection for the Second Period

The model selection process is repeated using production data up to 500 days. The four reservoir models with the best match in production data (Fig. 5) among the final set were selected to reevaluate the control scheme for the remaining 1500 days. The connection between wells is through channels with a sinusoidal and branched shape. The selected realizations resemble better the underlying geology of the reference reservoir (Fig. 1). Figure 6 shows that the production data from the four selected reservoir models continue being consistent with the observations from the reference. Although the residual uncertainty in water production remains, those predictions do not deviate significantly from the reference. As a result, no additional revisions to the control strategy are required after the two updates.

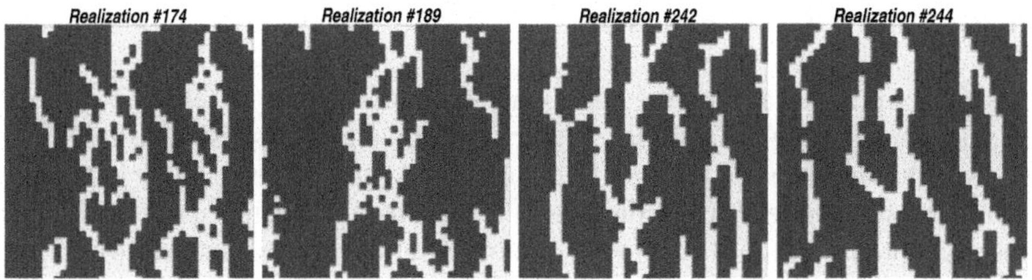

Fig. 5 Four models from the final set after incorporating production data for 500 days

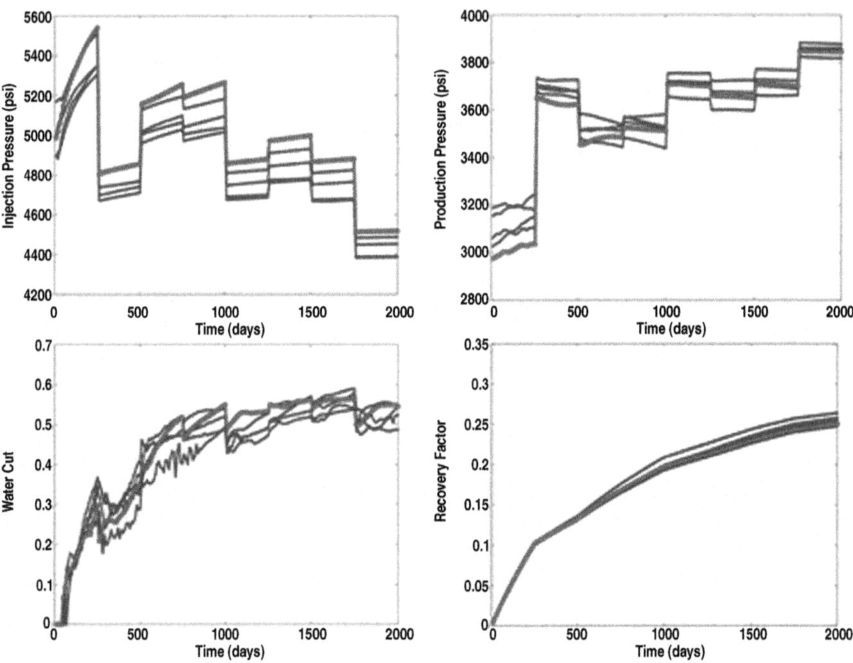

Fig. 6 Data after 8 control periods (2000 days) for the selected reservoir models (*blue*) and the reference (*red*)

The NPV obtained after feedback control and the NPV obtained if we would know the "true" reservoir (called optimal NPV) are close. In order to quantify the value of the information, the NPV for the feedback control case is compared to the NPV obtained with the optimal control settings of a reservoir model chosen without updating the prior uncertainty (initial ensemble). Using the optimal NPV as a reference, the feedback control system results in a loss of 0.78 %, whereas the optimal control settings for a random reservoir model results in a loss of 7.9 %.

6 Summary and Conclusion

This paper presents a novel method for assessing uncertainty in reservoir geology and updating it using dynamic data. Groups of reservoir models distinguished on the basis of distance to each other can be employed to assess the uncertainty in production forecasts. Furthermore, a comparison of simulated production data to the actual reservoir production history allows selecting a set of reservoir models with more accurate production forecasts. As a result, the posterior uncertainty in production forecasting is quantified. The entire methodology leads to a shift in the paradigm of history matching from perturbing a single reservoir model to the selection of a set of reservoir models that represent the posterior uncertainty in geology and future production.

Using the selected set of reservoir models, optimal well control settings are derived. The complex optimization problem for accomplishing this can be simplified using a response surface, which can be constructed by evaluating the responses of the reservoir to different control inputs. The added value of this feedback control framework has been demonstrated on a synthetic reservoir example.

References

1. Arpat BG, Caers J (2005) A multiple-scale, pattern-based approach to sequential simulation. In: Geostatistics Banff 2004, vol 1. Quantitative geology and geostatistics. Springer, Dordrecht, pp 255–264
2. Bishop C (2006) Pattern recognition and machine learning. Springer, New York
3. Brouwer DR, Nvdal R, Jansen JD, Vefring EH, van Kruijsdijk J (2004) Improved reservoir management through optimal control and continuous model updating. SPE 90149. SPE annual technical conference and exhibition, Houston, Oct 2–5
4. Caers J, Park K, Scheidt, C (2009) Modeling uncertainty in metric space. In: International association of mathematical geology meeting, Stanford University
5. Dakhlia H (1995) A simulation study of polymer flooding and surfactant flooding using horizontal wells. Ph.D. Dissertation, The University of Texas at Austin
6. Deutsch CV (1997) Geostatistical software library and user's guide. Oxford University Press, New York
7. Jansen JD, Brouwer DR, Naevdal G, van Kruijsdiik CP (2005) Closed-loop reservoir management. First Break 23:43–48
8. Myers RH, Montgomery DC (2002) Response surface methodology: process and product in optimization using designed experiments. Wiley, New York
9. Pope GA (1980) Application of fractional flow theory to enhanced oil recovery. In: SPEJ, June 1980, pp 191–205
10. Sarma P, Aziz K, Durlofsky LJ (2005) Implementation of adjoint solution for optimal control of smart wells. Presented at the SPE reservoir simulation symposium, Houston
11. Sorbie KS (1991) Polymer-improved oil recovery. CRC Press, Boca Raton

Efficient Conditional Simulation of Spatial Patterns Using a Pattern-Growth Algorithm

Yu-Chun Huang and Sanjay Srinivasan

Abstract Reproduction of complex 3D patterns is not possible using algorithms that are constrained to two-point (covariance or variogram) statistics. A unique pattern-growth algorithm (GrowthSim) is presented in this paper that performs multiple point spatial simulation of patterns conditioned to multiple point data. Starting from conditioning data locations, patterns are grown constrained to the pattern statistics inferred from a training image. This is in contrast to traditional multiple-point statistics based-algorithms where the simulation progresses one node at a time. In order to render this pattern growth algorithm computationally efficient, two strategies are employed—(i) computation of an optimal spatial template for pattern retrieval, and (ii) pattern classification using filters. To accurately represent the spatial continuity of large-scale features, a multi-level simulation scheme is implemented. In addition, a scheme for applying affine transformation to spatial patterns is presented to account for local variation in spatial patterns in a target reservoir. The GrowthSim algorithm is demonstrated for developing the reservoir model for a deepwater turbidite system. Lobes and channels that exhibit spatial variations in orientation, density and meandering characteristics characterize the reservoir. The capability of GrowthSim to represent such non-stationary features is demonstrated.

1 Introduction

Pattern recognition and accurate reproduction is important in many scientific applications such as remote sensing, reflection seismology, and subsurface reservoir modeling. Traditionally, pattern reproduction algorithms are constrained to simple two-point covariance or variogram statistics [3]. Variogram-based modeling of spatial patterns is mathematically consistent and convenient (mathematical representa-

Y.-C. Huang (✉) · S. Srinivasan
University of Texas at Austin, 1 University Station CO300, Austin, TX 78712, USA
e-mail: jack.huang@utexas.edu

S. Srinivasan
e-mail: sanjay.srinivasan@engr.utexas.edu

P. Abrahamsen et al. (eds.), *Geostatistics Oslo 2012*,
Quantitative Geology and Geostatistics 17,
DOI 10.1007/978-94-007-4153-9_17,

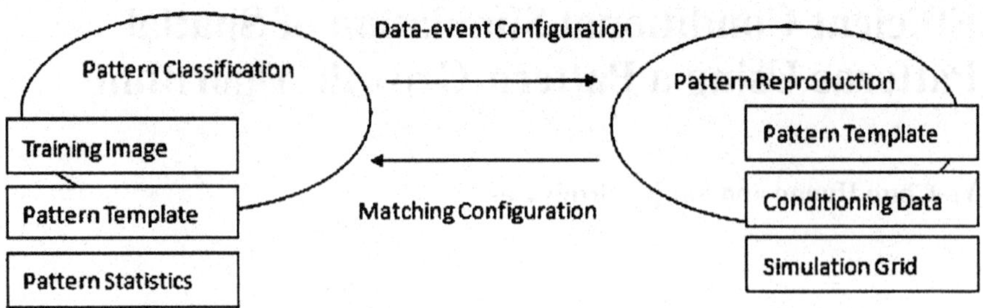

Fig. 1 GrowthSim components and interactions

tion of physical concepts), and its application is appropriate considering the lack of conditioning information in typical modeling scenarios [4].

However, the traditional covariance or variogram-based modeling approaches fail to appropriately retain the required conditional information to reproduce complex curvilinear features, generating amorphous realizations that exhibit maximum entropy instead of systematically organized structures and patterns, as would be expected from prior knowledge about the phenomena. Modeling such features calls for a different approach utilizing multiple-point statistics instead of variogram models (two point statistics) to capture the required conditional information, thereby generating fields that exhibit more structural organization and preserve reservoir heterogeneity [5]. This approach requires a training image, a numerical representation of the spatial law, and an explicit non-conditional conceptual description of the patterns to be reproduced. Training images can be obtained from analogs (such as geological outcrops) and photographs. Patterns to be reproduced can be explicitly depicted in the training model, as opposed to the hidden higher-order statistics implicit within traditional variogram-based models.

The general objective of this paper is to implement a new simulation approach that is based on feature identification and reproduction using a unique growth-based simulation algorithm. Features are grown starting from conditional data locations based on multiple-point statistics, which is inferred using optimized spatial templates. It must be emphasized that though there are numerous algorithms available in the literature for reproducing complex patterns, what sets the technique presented apart is the treatment of the simulation event as a multiple-point event sampled conditioned to the multiple-point data event.

We will begin with an introduction to the algorithm for GrowthSim on a conceptual level, focusing on pattern reproduction. Then we will discuss various modifications to the basic algorithm in order to represent long range connectivity and handle non-stationary features.

2 GrowthSim

The GrowthSim algorithm contains two main parts: pattern classification and pattern reproduction, as illustrated in Fig. 1. In pattern classification, we scan a training

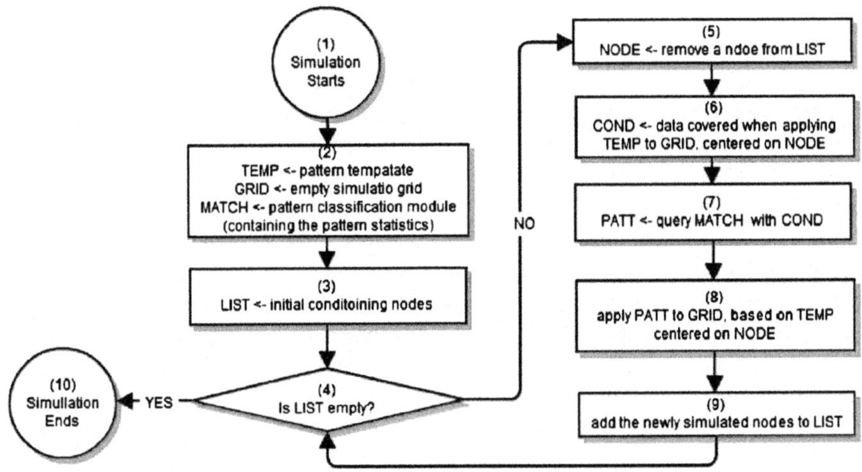

Fig. 2 GrowthSim pattern reproduction algorithm

image with a template, compiling the observations into a pattern statistics. In pattern reproduction, we retrieve patterns from the database and paste them onto the simulation grid, which originally contains the conditioning data only. We will first discuss the pattern reproduction process, since it is where novelty of GrowthSim lies. The pattern classification process borrows heavily from previously proposed algorithms such as *filtersim*. We will present our choices for implementing pattern classification in later sections.

2.1 Growth Algorithm

GrowthSim differs from other simulation algorithms by adding a "growth" aspect to the simulation. Instead of randomly selecting a simulation node to reproduce a pattern, the pattern reproduction begins from points around the conditioning data. We assume for now that we have already finished the pattern classification process and have the pattern statistics at our disposal. We begin by placing the template on the grid, centered on the point with conditioning data. Then we query the pattern statistics for a configuration that matches the conditioning data covered by the template (how a matching pattern is retrieved will be discussed in Sect. 3) and paste that pattern onto the simulation grid. This effectively creates new conditioning data. We then center the template on the newly filled points and repeat this process. This simulation halts when either the whole grid is filled, or when there is no more new place to center the template on.

Figure 2 illustrates the algorithm for pattern reproduction in GrowthSim. In Step 2 we initialize various modules needed, such as the simulation grid, pattern template, and pattern statistics database. In Step 3 we initialized a list and add to it the initial conditioning nodes (i.e. points containing the conditioning data). In Step 5 we remove a node randomly from the list. In Step 6 we center the pattern template on the node and gather the surrounding conditioning data. Suppose the template is consisted of 9 nodes, and out of the 9 nodes it covers on the simulation grid, 3 have already been defined. In Step 7 we query the pattern statistics for a pattern configuration that matches the three conditioning data (in their respective template position). In Step 8 we paste the matching pattern on to the simulation grid. This action will fill in some values for the previously 6 undefined nodes. In Step 9 we added the 6 newly simulated nodes into the list. Because only the newly simulated nodes are added, the list will eventually be empty when the simulation grid is completely filled, at which time Step 4 evaluates to be true and the simulation halts.

At this point, it is useful to contrast GrowthSim with other popular algorithms for multiple point simulation such as *simpat* [1] and *filtersim* [7]. In *simpat*, based on the configuration of data at any step during the simulation, a pattern is retrieved from the database and pasted at the simulation location. However, since the simulation grid is traversed along a random path, such pasting of patterns can result in artifact discontinuities at the border between patches. *Filtersim* attempts similar patching of patterns, but in order to make the process faster, the pattern database is organized using filter scores in a manner similar to that implemented in GrowthSim (discussed in Sect. 3). However, because the emphasis is on patching of patterns from the database in both these algorithms, discontinuities in the simulated image are possible and require cumbersome servo-correction schemes. In contrast, because the emphasis in GrowthSim is on pattern growth, such discontinuities are avoided. The process of creating and maintaining the node lists, where the next simulation step would occur, is the essential feature of this pattern growth scheme that renders this algorithm unique.

2.2 GrowthSim Example

Figure 3 captures the process of GrowthSim over successive stages of the simulation. From a single conditioning data location (1), the simulation queries the pattern statistics for a matching pattern and applies it to the grid (2). The process repeats itself as the algorithm continues to query and apply patterns to nodes in the simulation grid, until the grid is fully filled (7). The final simulated image exhibits the salient patterns observed in the training image (TI) and yet is not identical due to the local conditioning data (1) used for the simulation.

3 Pattern Classification

There are two steps to pattern classification. The first is to compile a pattern statistics using a template and a training image. The second is to provide a scheme for finding matching patterns from the statistics given some conditioning data. Here we will present some implementation schemes that are borrowed heavily from existing algorithms.

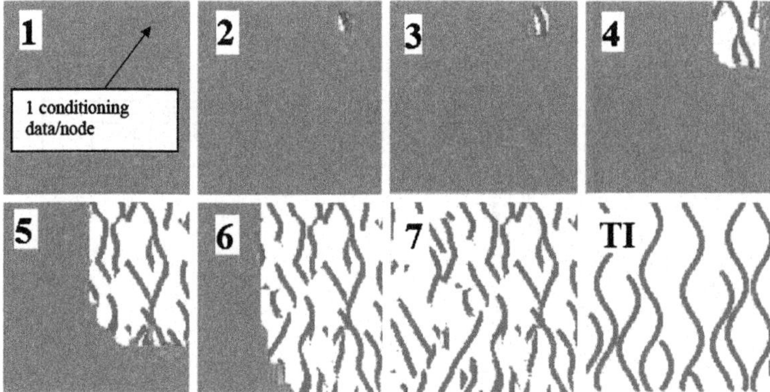

Fig. 3 Snapshots of GrowthSim in process

3.1 Pattern Statistics

Pattern statistics is the compiled database of the observed pattern configurations in the training image. It is obtained by simply moving a pattern template inside the training image, observing the pattern configuration at each valid location. For example, using a 100×100 TI and 9×9 template, there will be $(100-9+1)^2 = 8\,281$ pattern observations. Out of these configurations, some must be identical, allowing us to draw a histogram on the configurations occurrences. Figure 4 shows one such configuration occurrences. (This TI, called TI_SINE, will be used several times in this paper as the example.)

Even for a clean training image the number of distinct configurations is large, which will translate into more time required during the pattern matching process. For more complex training image, the size the statistics grows tremendously, demanding more CPU time for matching process.

We believe that the default rectangular/cubic pattern template captures too much noise in the pattern statistics. One idea to reduce the statistics size while retaining the salient features is to optimize the pattern template for a given TI. The optimization involves calculating the covariance of each node (u') in the default rectangular pattern template grid to the center node (u_c). The covariance between the two template node locations u_c and u' describes how likely a pattern is to span between those two nodes. By scanning the TI with the reduced two-point template (u_c and u'), we can obtain the covariance between every node on the template and the central node. Then, we simply select a percentage of the nodes with the highest covariance values. Figure 5 shows the covariance values and the optimized templates for the TI_SINE.

The result of using the optimized template on TI_SINE is a 30 % reduction in statistics size (from 1 396 to 960). On the more complex training image we used (3-category, $200 \times 200 \times 20$, using $9 \times 9 \times 9$ template), the statistics reduction is 19 %.

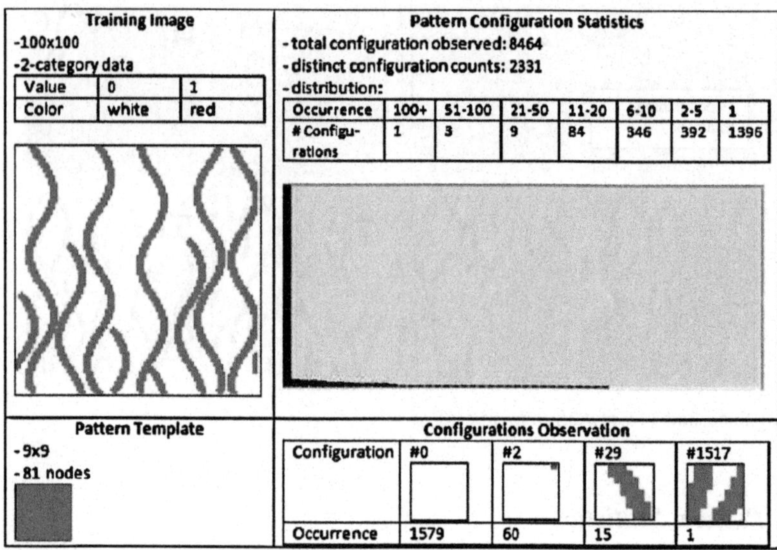

Fig. 4 Pattern statistics of 2-category, 100 × 100 TI_SINE and 9 × 9 templates

Original Pattern Template	Covariance Values	Optimized Template
- 9x9 - 81 nodes	- Value at grayscale (white as lowest and black as highest)	- top 50% correlated nodes (41 nodes) - Black indicates unused nodes

Fig. 5 Covariance values and optimized template for TI_SINE

3.2 *Pattern Matching*

During pattern reproduction, the template is centered on a node in the simulation grid to retrieve the surrounding value as the conditioning data. Based on those conditioning data, we need a way to retrieve a matching pattern from the previously compiled statistics. This is the most time-consuming portion of the whole simulation process, due to the large size of statistics database (even after applying the optimal template scheme).

Since most of the time there will be more than one configuration matching the conditioning data, the probability that one particular configuration will be chosen should be proportional to its predominance in the TI. More precisely, we will make all the matching configurations into a *cumulative-density function* (*cdf*), weighted on their occurrences in the TI, and then randomly draw one. If there is no matching pattern, we can either reduce the matching criteria until we find one, or simply return no pattern, at which point the simulation will skip over to the next node in the list.

However, the task of finding all matching configurations is not trivial. If we were to go through the entire statistics one by one, the simulation process becomes im-

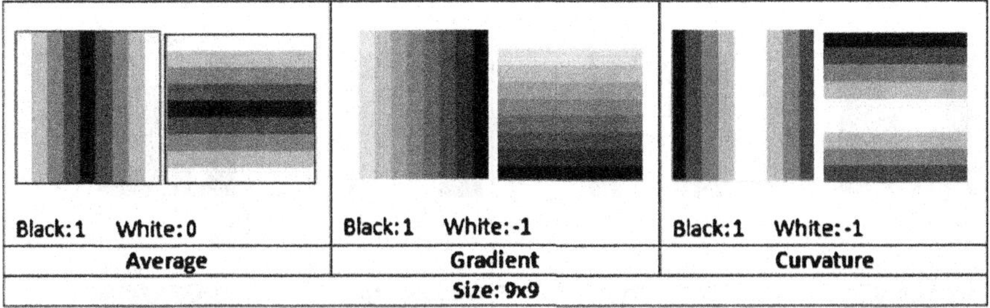

Fig. 6 Standard filters for 9 × 9 pattern template

practically long. A tree-like structure can be used to store the statistics such that the search time can be theoretically reduced to $\log(N)$; nevertheless, we were unsuccessful in such implementation due to memory limit.

We resolved to a pattern classification approach. Instead of having one big set of all the possible configurations, we group similar configurations into clusters, each represented by a point-wise averaged configuration called the prototype [7]. During the matching process, we first compare the conditioning data to these prototypes; the one cluster (prototype) that is "closest" to the conditioning is used to draw the matching pattern using the *cdf* scheme discussed earlier. Assuming that all clusters have approximately the same size, this approach would reduce the matching time required by a factor equal to the number of clusters created.

There are a number of clustering algorithms, but they all require some distance function to assign the clusters. Therefore, we use the mechanism in *filtersim* to compute the filter scores (as a vector) for each pattern configuration on the template [7]. Figure 6 shows the standard filters for a 9 × 9 pattern template.

Each filter captures a specific feature. The filter score of a pattern configuration is obtained by applying the filter weights on the pattern, node by node and then computing the weighted sum. For example, a configuration exhibiting a gradient trend in y-direction (i.e. it has higher values in the nodes at the bottom than in the nodes at the top) would have a higher filter score on the gradient-x filter. Similar configurations would be expected to have closer filter scores (i.e. closer distance between the respective vectors). Then, we cluster the filter scores using *k-mean clustering* [2]. Figure 7 shows the prototypes after the 960 configurations observed in SINE_TI have been grouped into 8 clusters. It is possible that two clusters may not be distinct from each other enough that they have the same prototype (plus the fact that we use integer, not real number, for categorical data). In such case, the configurations in each cluster will all be considered. Figure 8 displays the effects of this pattern classification scheme on the simulation process. The time required for the simulation decreases with more clusters. However, there is some deterioration in the connectivity of the feature reproduced.

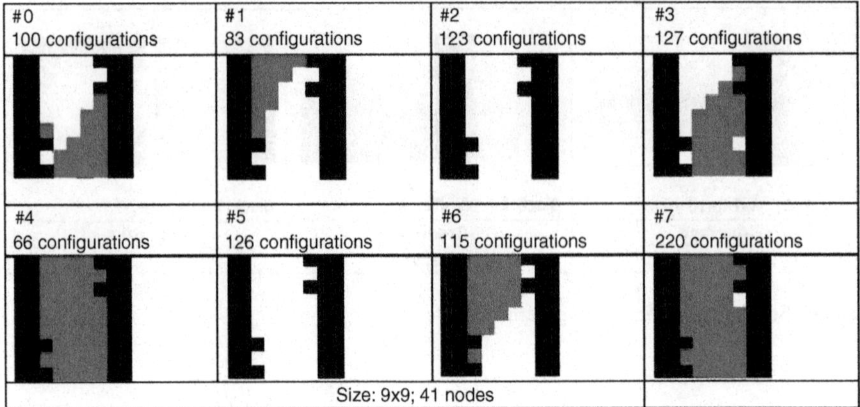

Fig. 7 Cluster prototypes for TI_SINE

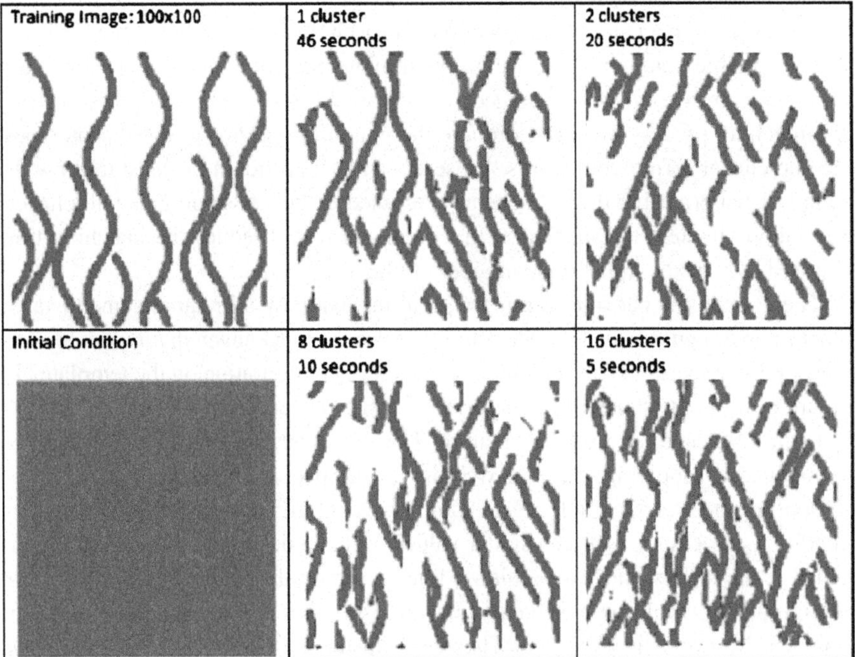

Fig. 8 Implementation of pattern classification/clustering on the simulation process

4 Multigrids of Pattern Statistics

The pattern template must remain small in order to keep the pattern statistics database manageable. However, a small template is unable to capture long-range features. To cope with this problem, we use a multigrid approach to performing the

Level	Size	Original Template	Covariance	Optimized Template (50%)
1	9x9			
2	17x17	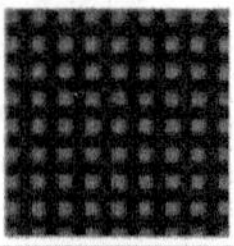	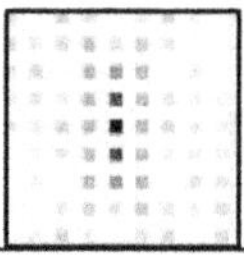	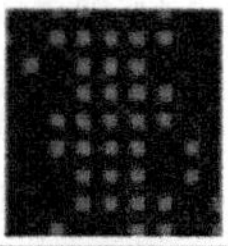
3	25x25	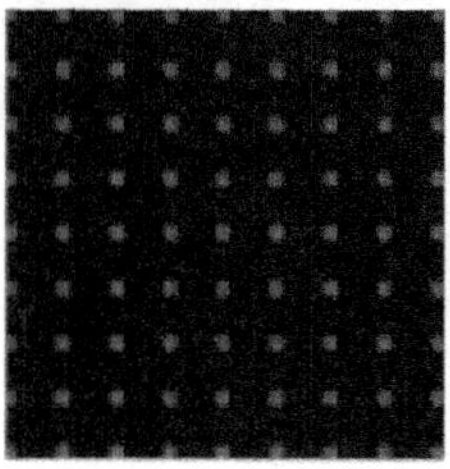	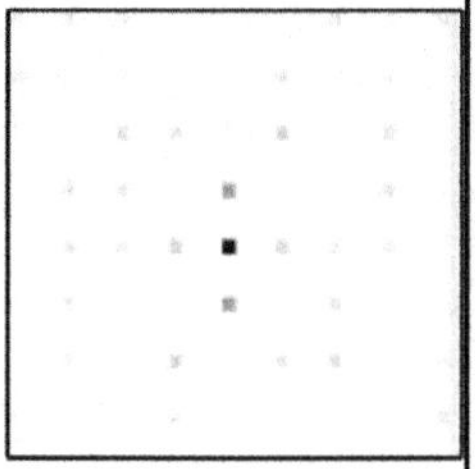	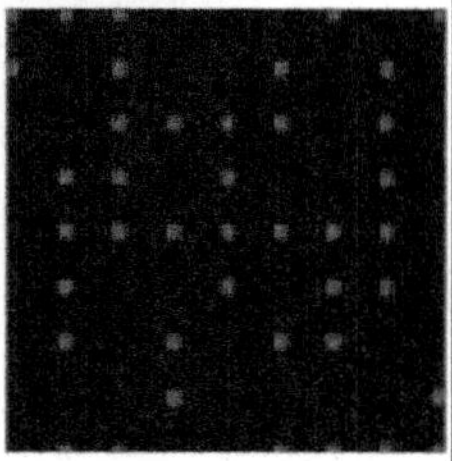

Fig. 9 Multigrid template for TI_SINE

simulation [3]. The simulation algorithm is implemented multiple times, from the coarsest (largest) template to the finest (smallest) one. The coarser template captures large scale features, while the finer template fills in the smaller scale features.

To define a coarser template, we simply take the existing optimized template and apply an affine expansion. This effectively puts "gaps" in between the template nodes. The number of template nodes remains the same regardless of the template size, preventing the pattern statistics from increasing out of control. Instead of applying affine transformation to a single optimized template, it is possible to use a separate optimized template at each simulation level and compile the corresponding pattern database. Figure 9 shows the expanded templates and their optimized versions for TI_SINE (refer to Fig. 4). The optimized templates correctly reflect the salient feature of TI_SINE at their respective levels. However, as the template grows larger, the correlations of the nodes become weaker, as indicated by the decrease or "whitening" of covariance values.

The simulation result obtained by running multiple template levels is shown in Fig. 10. The long-distance features of TI_ SINE are preserved better in (b), which consists of two levels of templates. The features in (a) obtained using only one level of template reproduce short-scale features well, but at the longer scale deviate from TI_SINE's sinuous features. The simulation obtained using three levels of template shown in (c) seems to be more noisy that the models in (a) and (b). This suggests that there is a practical limit on the number of multigrids to use for simulation and with an increase in the number of coarser grids, long range continuity is preserved at the expense of poor reproduction of short scale features.

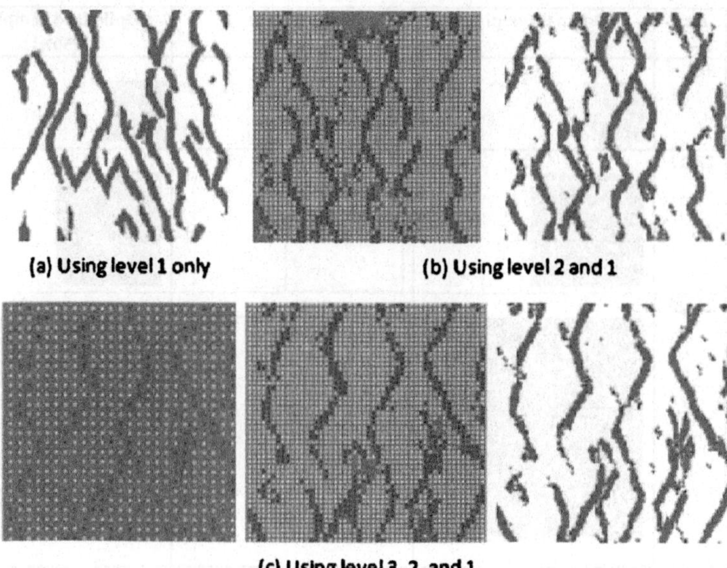

Fig. 10 Pattern reproduction using multiple levels of template

5 Pattern Transformation

In most practical cases, the features observed in a reservoir will exhibit non-stationarity in the form of azimuthal or dip variations and/or variations in the dimensions of the reservoir features. Thus the pattern statistics inferred from a stationary training image would have to be modified in order to account for the non-stationarity. We use the concept of affine transformation to accomplish this modification of target statistics [6]. The template is first transformed by applying a suitable rotation or linear rescaling and the then is used with the original pattern statistics to perform pattern simulation. In order to achieve this, we generate a node-wise mapping whereby each node in the original template maps to exactly one node in the transformed template (but not necessary the other way around). Because of the data abstraction between the modules, the whole pattern statistics and pattern classification process can be left unchanged. Revisiting the algorithm presented in Sect. 2, we can add the pattern transformation with the modifications to Steps 5 and 9, as shown in Fig. 11.

Figure 12 shows the simulation results of applying scaling (b), rotation (a), and both (c). Note that in theory, this one-way mapping works only for scaling with factor less than 1 or rotation with a multiple of 90 degree. Other transformation results in templates possibly with "holes" in them (not every point can match perfectly onto another point in integer coordinate system), thus requiring additional interpolation for values at these locations.

To address this problem, we need an algorithm that finds all the holes in the expanded template and assigns them to nodes from the original template. For our

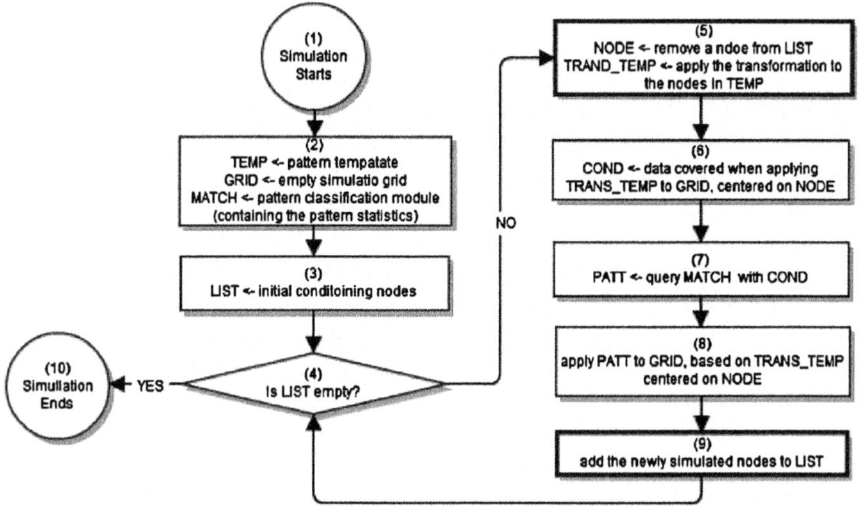

Fig. 11 Algorithm for pattern reproduction with pattern transformation

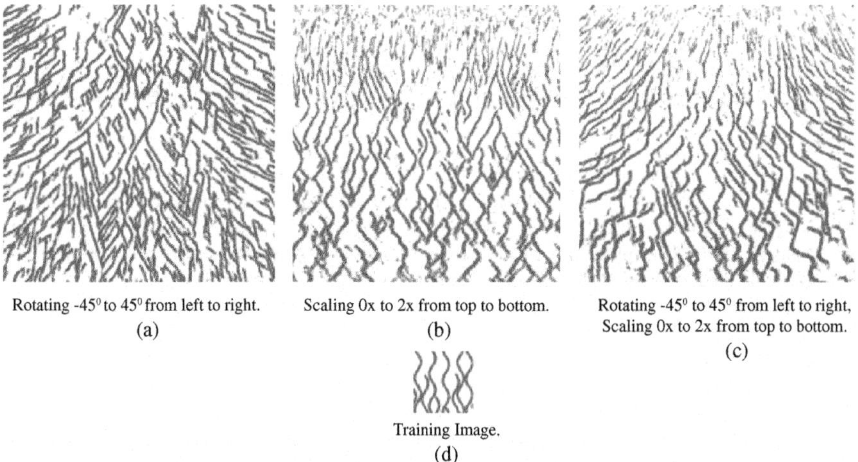

Rotating -45° to 45° from left to right.
(a)
Scaling 0x to 2x from top to bottom.
(b)
Rotating -45° to 45° from left to right,
Scaling 0x to 2x from top to bottom.
(c)
Training Image.
(d)

Fig. 12 Simulation result with linear transformation on patterns

experiment purpose, it suffices to approximate the effect of pattern transformation using the simplistic one-way mapping.

6 Conclusion

GrowthSim is a unique pattern classification and simulation algorithm. It reproduces the salient features observed in the training image and conditions the simulated image to available local conditioning information. If the pattern template is not big enough to capture the large-scale feature, we implement a multi-level simulation approach using coarser templates. Linear transformation of the patterns can also be implemented (with ease) with a node-wise one-way mapping from the original to the translated template, although a complete way accounting all nodes in the transformed template is necessary for a more robust implementation.

The computational issue with GrowthSim is its efficiency. The perfect implementation of GrowthSim would be the combination of unmodified pattern statistics (pattern template as it is) and direct pattern matching (searching through the whole pattern statistics). However, this requires so much computing power that we have to make some compromises. To render the simulation more efficient, we use the optimized template to reduce statistics size and use pattern classifications with clusters to reduce matching time. While these measures make the simulation algorithm deviate from the perfect implementation, they are necessary for making GrowthSim practical. The resultant simulation models exhibit the connectivity characteristics observed in the training image.

References

1. Arpat G, Caers j (2004) A multiple-scale, pattern-based approach to sequential simulation. In: Proceedings of the 7th international geostatistics congress, GEOSTAT 2004, Banff, Canada, October 2004
2. Coleman DA, Woodruff DL (2000) Cluster analysis for large datasets: an effective algorithm for maximizing the mixture likelihood. J Comput Graph Stat 9(4):672–688
3. Deutsch C, Journel A (1998) GSLIB: geostatistical software library and user's guide. Oxford University Press, London
4. Isaaks E (1990) The application of Monte Carlo methods to the analysis of spatially correlated data. PhD thesis, Stanford University
5. Strebelle S (2002) Conditional simulation of complex geological structures using multiplepoint statistics. Math Geol 34(1):1–21
6. Strebelle S, Zhang T (2005) Non-stationary multiple-point geostatistical models. In: Geostatistics Banff 2004, vol 1. Quantitative Geology and Geostatistics, vol 14, pp 235–244. doi:10.1007/978-1-4020-3610-124
7. Zhang T (2006) Filter-based training pattern classification for spatial pattern simulation. PhD thesis, Stanford University, Stanford, CA

Multiple-Point Statistics in a Non-gridded Domain: Application to Karst/Fracture Network Modeling

Selin Erzeybek, Sanjay Srinivasan, and Xavier Janson

Abstract Paleokarst reservoirs exhibit complex geologic features comprised of collapsed caves. Traditionally, cave structures are modeled using variogram-based methods and this description does not precisely represent the reservoir geology. Therefore, a quantitative algorithm to characterize the reservoir connectivity of karst features is needed. Multiple-point statistics (MPS) algorithms have been used for modeling complex geologic structures such as channels and fracture networks. Statistics required for these algorithms are inferred from "gridded" training images but information about structures like modern cave networks and paleokarst facies are available only in the form of cave surveys or "point" data sets. Thus, it is not practical to apply "rigid and gridded" templates and training images for the simulation of such features. In this study, a new MPS analysis technique and a pattern simulation algorithm are presented to characterize and simulate connected geologic features such as cave networks and paleokarst reservoirs. The algorithm uses non-gridded flexible templates with various distance and angle tolerances on non-gridded training images for inferring the pattern statistics. The calibrated statistics are subsequently used to constrain pattern simulation of connected geologic features. In order to validate the proposed technique, it is first implemented on 3D synthetic data sets. Subsequently, the algorithm is also applied on the Wind Cave data set. The cave data is used as non-gridded training images for performing MPS analysis. It is demonstrated that the proposed non-gridded MPS analysis and pattern simulation algorithms are successful at characterizing and modeling connected features for which information is only available in the form of point sets.

S. Erzeybek (✉) · S. Srinivasan
Petroleum and Geosystems Engineering Dept., The University of Texas at Austin, Austin, USA
e-mail: selin@utexas.edu

S. Srinivasan
e-mail: sanjay.srinivasan@engr.utexas.edu

X. Janson
The Bureau of Economic Geology, The University of Texas at Austin, Austin, USA
e-mail: xavier.janson@beg.utexas.edu

P. Abrahamsen et al. (eds.), *Geostatistics Oslo 2012*,
Quantitative Geology and Geostatistics 17,
DOI 10.1007/978-94-007-4153-9_18,

1 Introduction

In paleokarst reservoirs, extensive dissolution and cave collapse due to continuous burial result in complex structures such as collapsed cave facies and sinkholes. Also, there exists significant faulting around the collapsed caves and associated heterogeneities. In order to understand and characterize the paleokarst and paleocave features, stratigraphical and sedimentological structures should be examined thoroughly. Although some of the important hydrocarbon resources are in paleokarst reservoirs (e.g. the Ellenburger Group carbonate reservoirs, Elk Basin and Yates fields), detailed reservoir characterization and methods for realistically modeling cave features are limited [4, 18]. On the other hand, most of the studies in paleokarst reservoirs are restricted to structural, stratigraphical and sedimentological analysis and these studies yield valuable information on paleokarst formation mechanisms and associated facies [15, 16]. Since several mechanisms control the formation of these complex geological features, it is difficult to describe the spatial distribution of cave structures and the corresponding effects on fluid flow by only using the geological interpretations. In some cases, the distribution of karstic structures is controlled by the presence of fractures. In such cases, one approach could be to model fractures as disks defined by their centroid, shape, size and orientation. In "Random Disk" models [2], fractures are represented as two-dimensional convex circular disks located randomly in space. The locations of the disks, radii and orientation are assumed uncorrelated from one disk to the next. A spatial density function can be utilized to represent clustering of fractures [3, 8].

Traditional Multiple-Point Statistics (MPS) algorithms have been widely used for modeling geologic structures such as channels and fracture networks. Recently developed MPS algorithms are capable of representing complex geologic features as well as integrating data from different sources [5–7, 10, 20]. In these MPS algorithms, statistics of patterns are inferred using customized spatial templates and realistic analog/training images. Complex geological features such as fractures can be represented accurately using MPS [14].

While MPS algorithms successfully represent complex geologic structures, in traditional implementations, a gridded training image is required for inferring the statistics. However, modern caves that are analogous to collapsed cave features are represented by surveys reporting XYZ coordinates of cave central line. It is not practical to perform gridding and apply a rigid gridded template to calibrate statistics using such "point" data surveys. There are some recent studies focused on the determination of spatial distribution of cave size and stochastic modeling of cave formation [12, 17], however, these studies do not address the problem of inference of pattern statistics from sparse data. There is thus the necessity of a methodology for characterizing and modeling paleokarst reservoirs so as to provide better description of such complex facies in reservoir simulation studies for understanding flow mechanisms. In order to address these issues, we have developed an MPS algorithm that works on a non-gridded basis to characterize cave and paleokarst networks using "point-set" information. The algorithm basically consists of MPS analysis using flexible spatial templates and subsequent pattern simulation using the

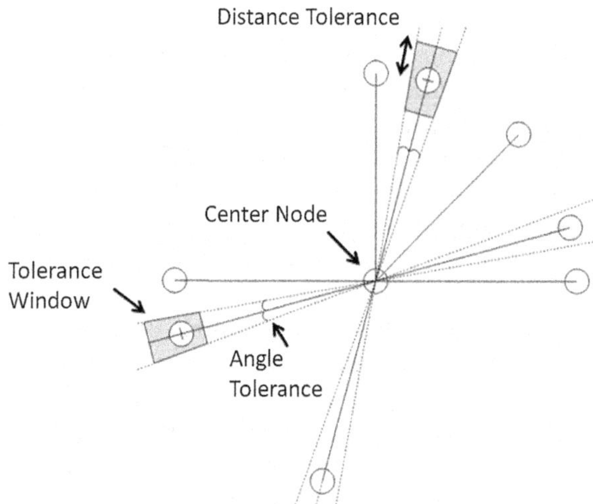

Fig. 1 8-node template with tolerance window

calibrated statistics. Application of the proposed algorithm on 2D synthetic data sets has recently presented by [11]. This paper provides further details on the simulation methods and application on the Wind Cave analog data.

2 Fracture Modeling Approach

As mentioned previously, the MPS algorithm presented in this paper adopts non-gridded training images and flexible non-gridded templates that consist of a center node and neighboring nodes with predefined distance and angle tolerances (Fig. 1). The point-set training image data is scanned with templates having various angle and distance tolerances. The spatial template is centered on a data point and all surrounding data-nodes that fall within the tolerance window of any template nodes are determined. If a data point falls into a tolerance window (Fig. 1), the corresponding template node is activated and the observed pattern configuration is stored. By translating the template over the entire training data set, frequencies of the observed configurations are continuously updated and pattern histogram is constructed. In some cases, pattern statistics may be inferred using multiple templates (in lieu of one large, complex spatial template).

Once the pattern histograms are constructed, pattern simulation is performed using the flexible spatial template. Sparse data for conditioning the simulation models are assumed available. The mp statistics inferred from the training image are used to constrain the pattern simulation and for representing connected geologic features. The pattern simulation steps are the following:

1. Randomly select a "simulatable" node such that there is at least one conditioning data location in the vicinity of that node.

2. Apply the pattern with the highest probability corresponding to the configuration of surrounding conditioning data points. When matching the data configuration on the simulation grid with the pattern statistics for the template, the lag and angle tolerances defined for the template are employed.
3. Calculate new list of "simulatable" nodes. These are nodes such that there is at least one conditioning data somewhere in the vicinity of that node.
4. Randomly select a simulation node from the "simulatable" nodes list.
5. Scan the data configuration around the simulation node using either a unique spatial template or in some cases using the entire set of templates.
6. Gather the patterns that are possible on the remaining nodes of the template(s) from the corresponding multiple point histograms.
7. Monte Carlo sample the pattern to apply on the remaining nodes of the spatial template. In the case of multiple spatial templates compare the probabilities of different simulation patterns obtained from multiple histograms and apply the one with the highest probability.
8. Update "simulatable" nodes list and go to Step 4.
9. Continue Steps 4 to 8 until the target statistics (fracture density etc.) are attained.

At the end of the pattern simulation, a set of points that represents a realization of the connected network or features is obtained. Connection of each node to the neighboring ones is transferred by the template used in pattern simulation. Thus, the implemented templates mainly affect connectivity of the simulated structures. The proposed technique enables us to infer MPS from point-set training data using spatially flexible templates with tolerances and perform mesh-less pattern simulation using a pattern-growth concept.

In order to improve the simulation results, pattern filtering is also implemented in the algorithm. In pattern filtering, at each simulation step, the simulation patterns with a frequency below a user defined threshold are eliminated. This serves to reduce the noise in the simulation result. Stopping criterion for the simulation is selected based on the user's preference and could be based on a certain amount of fracture/feature density, a maximum error deviation from the pattern histogram or a maximum number of simulation steps. In the results presented in this paper, the simulation is stopped when the specified maximum number of simulation steps is exceeded.

3 Description of Data Sets

The non-gridded MPS analysis and pattern simulation algorithms are applied to two different data sets; (1) 3D synthetic network which consists of connected features along certain orientations, (2) network survey of Wind Cave located in South Dakota, USA. Detailed descriptions of the data sets are provided in the following sections.

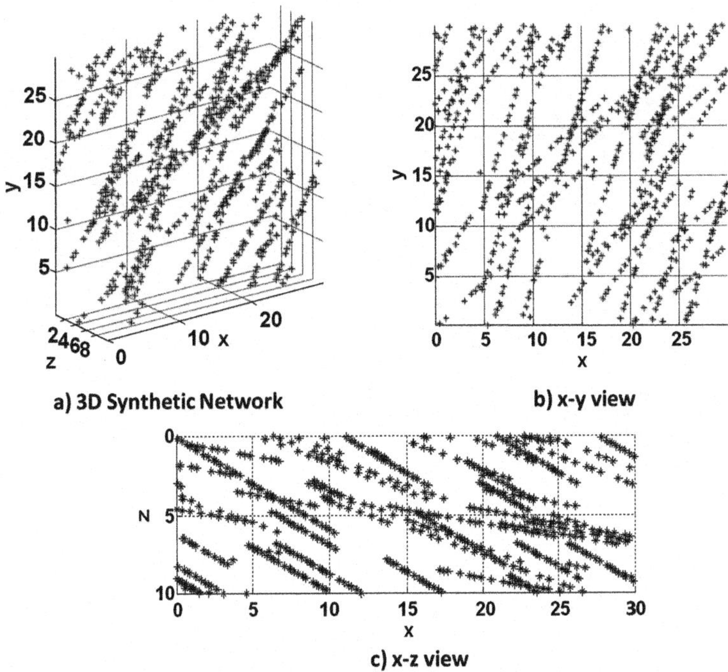

Fig. 2 3D synthetic network

3.1 3D Synthetic Network

In order to test and verify the proposed algorithm, 3D synthetic networks were initially developed (Fig. 2). The synthetic data sets were constructed by modifying the program *ellipsim* in GSLib [9]. First, object center locations are extracted and then, points along the corresponding ellipsoid orientation are calculated. This is to mimic the point-survey data that might be available in the case of a real cave.

Initially, an object simulation of three different types of ellipsoids is performed using the modified ellipsim program and the object center locations are extracted. Ellipsoid center locations with corresponding feature type are stored. The features are in 20°, 35° and 50° azimuth angle direction with 30°, 10° and 20° dip angles respectively. Once the center locations are obtained, points along the corresponding feature orientation are calculated. Points are sampled with an irregular spacing and a user-defined minimum distance separation of 0.5 units. The constructed synthetic network is thus analogous to modern cave network surveys which might have information recorded along irregularly spaced survey points.

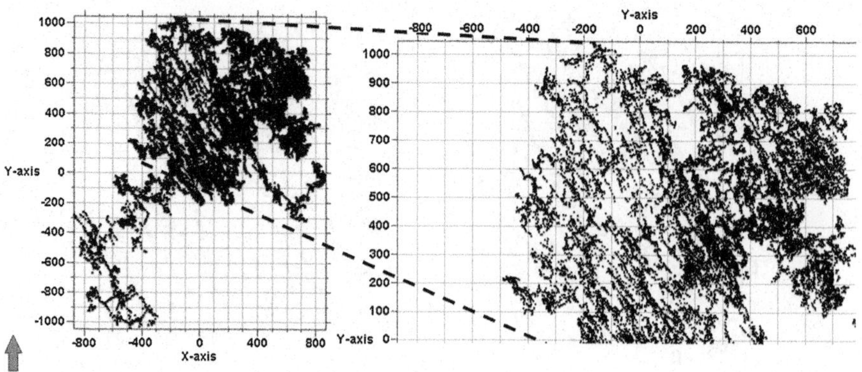

Fig. 3 Line plot of Wind Cave central line

3.2 Wind Cave

Wind Cave located in South Dakota was discovered in 1881 and the initial surveys for exploration of the cave were conducted in 1902 [13]. It is one of the largest cave systems in the world [19]. Wind Cave is formed in Madison limestone with major trend in N52°E slightly plunging to the northeast and with a 4°–5.5° slope to the southeast [13].

The 3D cave central data is available for Wind Cave and a line plot of the data is presented in Fig. 3. The line plot consists of data point coordinates and includes point connection information reporting the connectivity of each node to the neighboring ones.

4 Results

Results of the non-gridded MPS analysis and pattern simulation of the data sets are provided in Sects. 4.1 and 4.2 below.

4.1 3D Synthetic Network

MPS analyses of 3D synthetic network are initially performed using 19-node generic templates (Fig. 4). This generic template is only to determine the azimuth and dip orientation of the main connectivity features. The lag tolerance is specified to be 1/3 of lag separation and, 6° and 10° angle tolerance values are used when computing the statistics.

First, dip angle of the generic template is systematically varied from 0° to 60° and the total numbers of connections captured by each template are compared (Fig. 5).

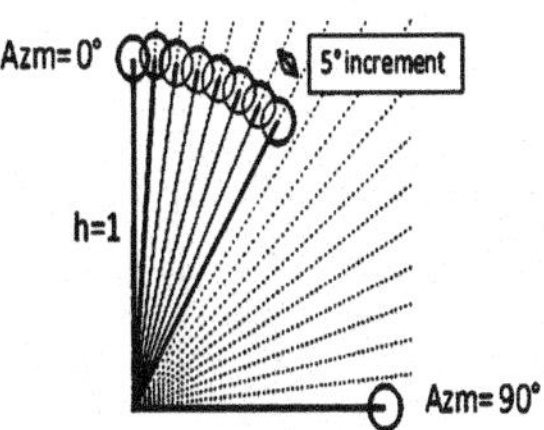

Fig. 4 19-node generic template

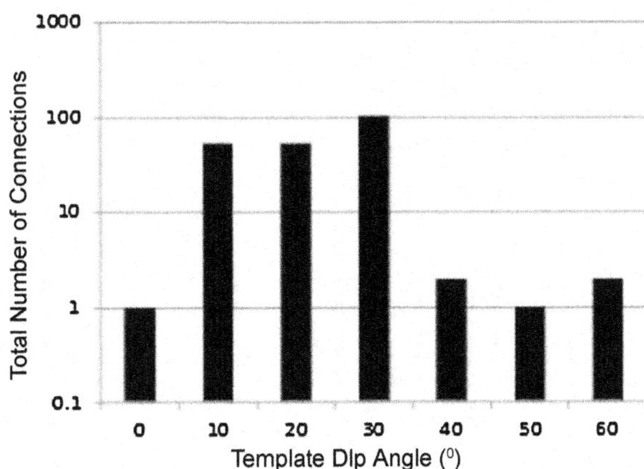

Fig. 5 Comparison of number of connections captured by 19-node generic templates

Significant occurrence of connected features are obtained for templates with 10°, 20° and 30° dip angles and this observation is consistent with the original 3D synthetic network data (Fig. 2). For each dip angle value, the azimuth angles along which significant amount of connections are captured, are determined. Thus, an initial and quick assessment of the feature orientations is performed.

Once the overall dip and azimuth orientations of the connectivity features have been established, now the pattern inference procedure is performed using 4-node templates along the major feature orientations. Templates in 20°, 35° and 50° azimuth directions with 30°, 10° and 20° dip angles respectively, captured significant amount of connections. The corresponding pattern histograms (Fig. 6b–d) are constructed and subsequently used for pattern simulation.

Pattern simulation of the 3D synthetic network data is performed using the inferred statistics. 3D geometry of the three templates are shown in Fig. 7a–c. These are the same templates for which the histograms in Fig. 6b–d were computed. The realizations are conditioned to sparse data randomly selected from the synthetic data set. In order to avoid data clustering, a minimum distance (5 units) between the conditioning data locations is assigned. Several simulation options were implemented such as simulation with and without pattern filtering. The results using these options are discussed next.

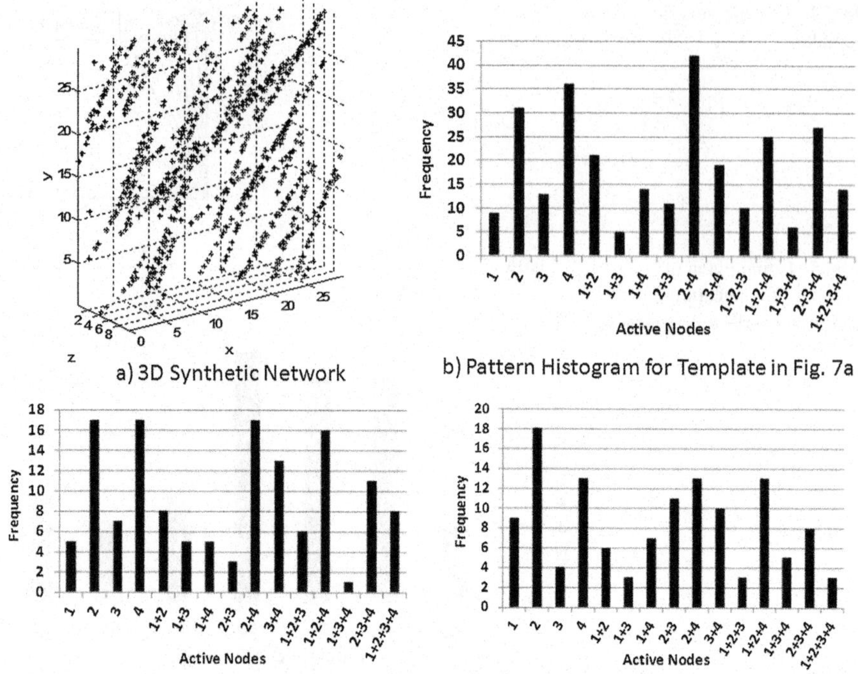

Fig. 6 Pattern histograms of 3D synthetic network

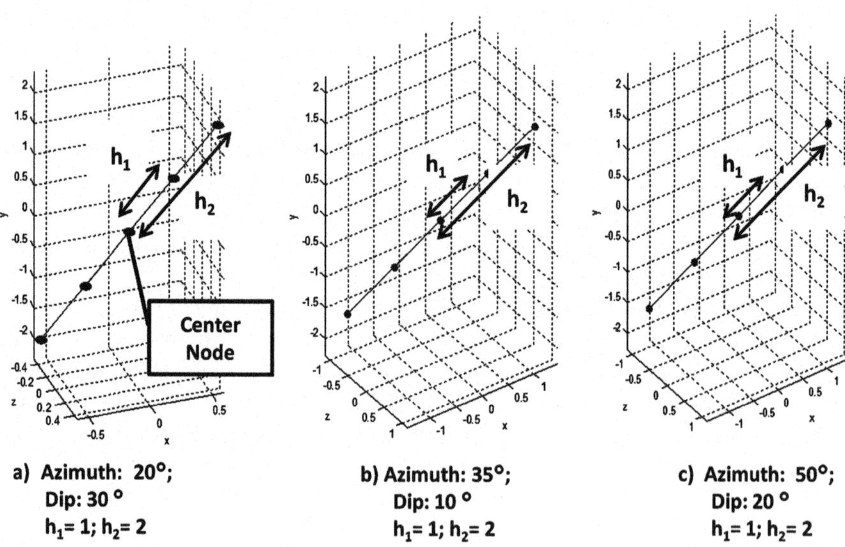

Fig. 7 Templates used in pattern simulation

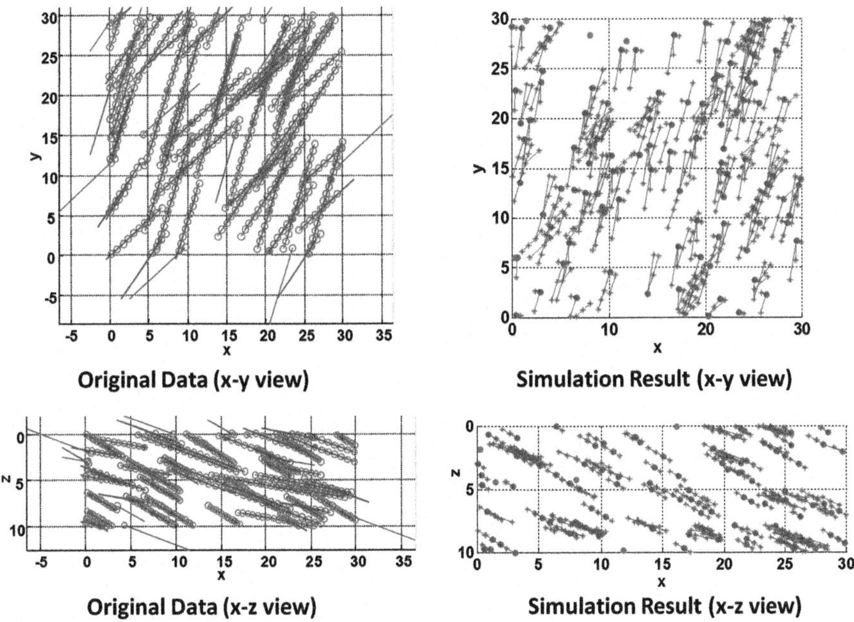

Fig. 8 Pattern simulation results—without pattern filtering (*circles*: conditioning data; $*$: simulated point)

4.1.1 Simulation Without Pattern Filtering

Pattern simulation of the 3D synthetic network data is performed using 100 conditioning data. The simulation is performed using the statistics inferred by the three spatial templates and terminated after 200 simulation steps. At each step of the simulation process, the patterns possible using each of the three templates are retrieved and pattern with the highest probability is selected. The raw histograms shown in Fig. 6b–d were used during the simulation. Pattern simulation results are illustrated in Fig. 8. It is observed that the simulation results are consistent with the original training image; the patterns of connectivity observed in the training set are satisfactorily reproduced.

Since pattern reproduction in terms of the mp histogram is crucial for accurate results, the simulated point set was scanned by a 12-node template which is a combination of the templates implemented in pattern simulation (Fig. 7a–c) and MPS results were compared to the original ones (Fig. 9).

Although the pattern histogram was reproduced satisfactorily, there were significant deviations from the target mp histogram. Some of the patterns were not reproduced and clustering of certain configurations was observed. Thus, the pattern simulation algorithm is modified to overcome these problems by filtering spurious features corresponding to low pattern frequency. Simulation with pattern filtering method and corresponding results are presented in the next section.

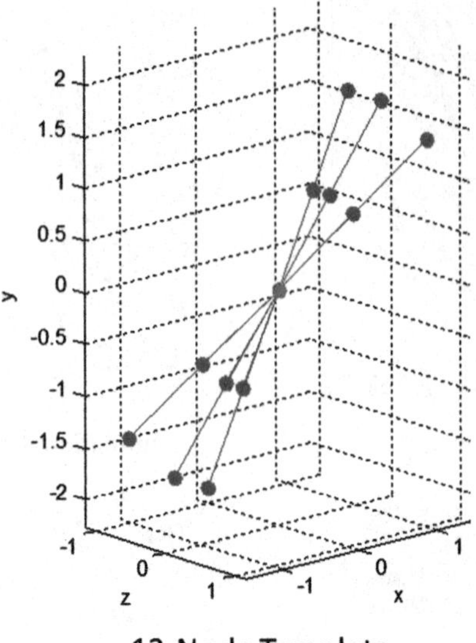

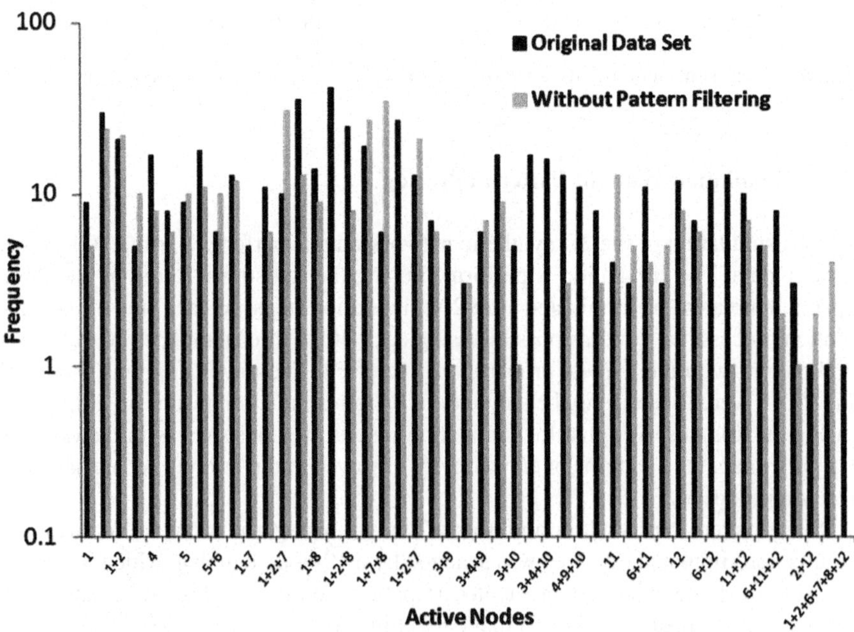

Fig. 9 Pattern histogram comparison (simulation without pattern filtering)

4.1.2 Simulation with Pattern Filtering

As seen from the comparison of the simulated and target mp histograms, a number of spurious features corresponding to low pattern frequency occurrences appear in the

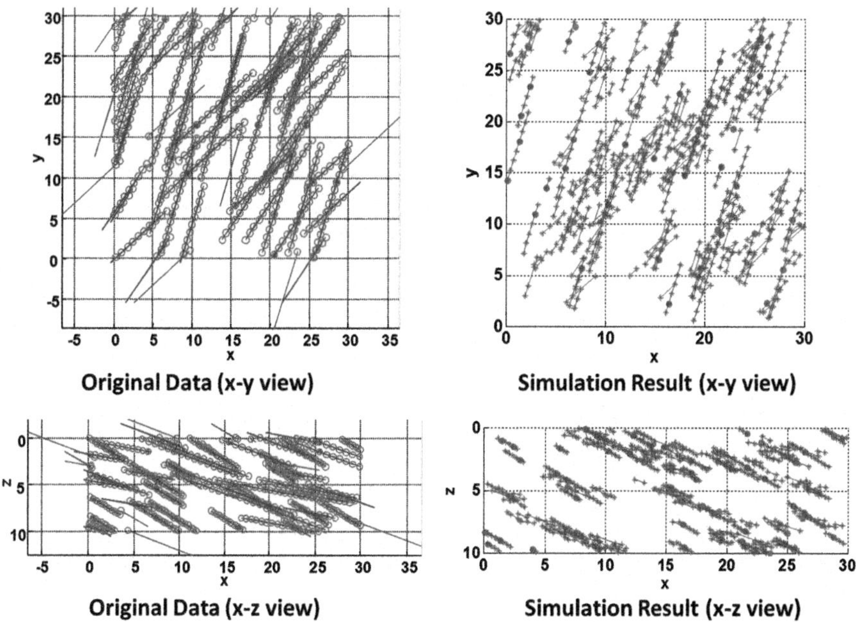

Fig. 10 Pattern simulation results—with pattern filtering (*circles*: conditioning data; ∗: simulated point)

simulated model. A servo-mechanism was implemented using which the multiple point histogram of the simulated image is checked periodically and patterns with very low frequency are eliminated. This pattern filtering is applied after every 100 simulation steps in order to obtain the results in Fig. 10.

The realization shown in Fig. 10 is obtained by using 50 conditioning data locations, 400 simulation steps and a low-frequency threshold value of 5. The number of conditioning points was reduced to demonstrate the capability to reproduce patterns with less conditioning data. Moreover, the simulation was terminated at 400 steps (unlike the previous results obtained by 200 simulation steps) for better application of pattern filtering. Figure 11 shows the reproduction of pattern frequencies using the 12-node template. Results confirm that better representation of spatial patterns is possible using the pattern filtering algorithm.

4.2 Wind Cave Analyses

Because the cave has a large surveyed areal extent, it is practical to perform MPS analysis on a subset of the complete data set and later using the inferred statistics to simulate the entire cave system. After initial examination of the Wind Cave galleries, some short and circular cave segments were eliminated (Fig. 12a). Then, the

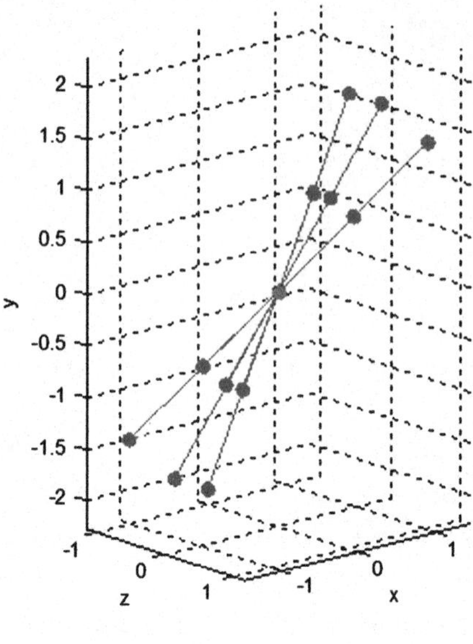

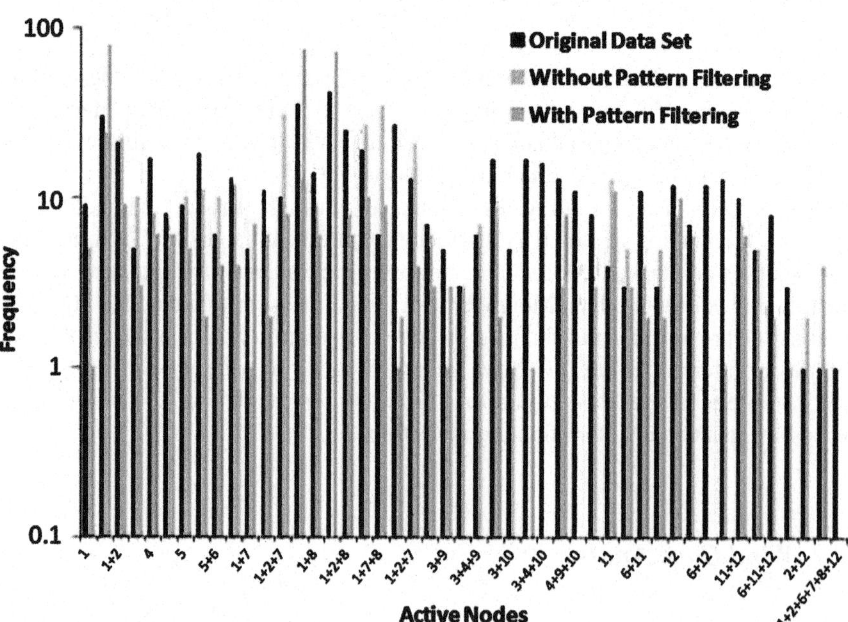

Fig. 11 Pattern histogram comparison

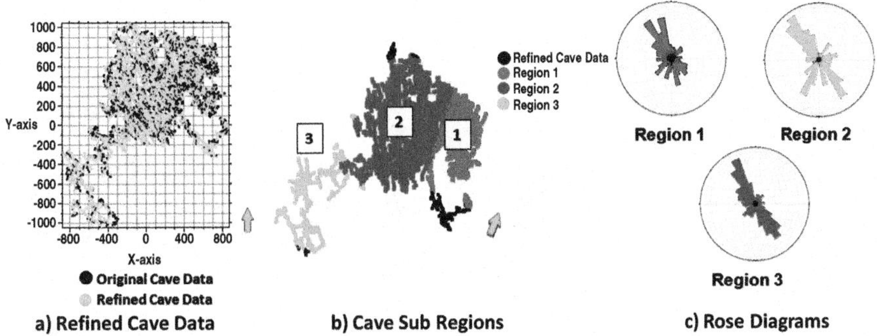

Fig. 12 Cave data and region division

remaining cave passages are divided into 3 main regions based on variations in main azimuth and dip orientations (Fig. 12b).

Since stationary domains are a must for statistical inference, the angle distributions of the cave passages are examined. For each sub region, stereonet plots and rose diagrams of the passages are constructed [1] (Fig. 12c). Rose diagrams are constructed using 10° bins. It is observed that caves in Region 2 have a preferential main orientation and have relatively homogeneous cave branching. Thus, initial MPS analysis of the Region 2 area shown in Fig. 12b was performed.

First, a 19-node generic template by assigning the node azimuth angles in NW-SE direction with 5° increments is constructed. Then, the template dip angle is varied between 0° and 60° along the main dip direction of the Madison Formation that hosts the caves. A significant number of connections are captured by the generic templates in 5° dip direction with node lag distances ranging from 5–40 units and at −20°, −40°, −50° and −70° azimuth angles. One of these templates is shown in Fig. 13a. The template size (16 nodes) is assigned to capture long cave passages as accurately as possible. The azimuth angle tolerance is increased to 10°. Templates in −40° and −50° azimuth direction capture higher amount of connections compared to the ones with −20° and −70° azimuth angle as shown in Fig. 13b. Based on 16-node template analysis, a refined template of 5-node along −40° and 230° azimuth directions with a dip angle of 5° is constructed (Fig. 13c) for detailed capturing of the patterns in Region 2.

The pattern histograms corresponding to 16-node and 5-node the templates are shown in Fig. 14a–b. The pattern histogram of the generic template (Fig. 14a) is mainly spread out and reports a variety of configurations. On the other hand, 5-node template pattern histogram is more representative of the training image since it has certain configurations reflecting the cave pattern with high frequency values (e.g. configurations with active nodes 2, 4, 2 + 4, 2 + 4 + 5). The MPS analysis suggests that optimum sized templates have to be used for the pattern simulation. Since at large lag separation, spurious features and connections are captured, template size should be at an optimum value for accurate representation. Furthermore, the optimum template should inherit the main feature orientation and template nodes will

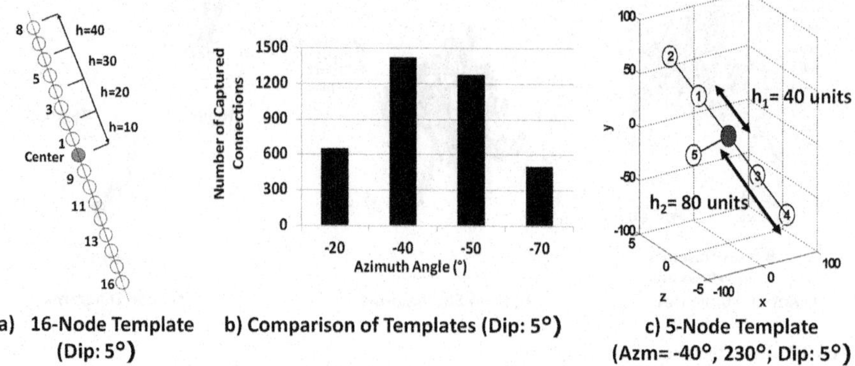

Fig. 13 Templates and template comparison

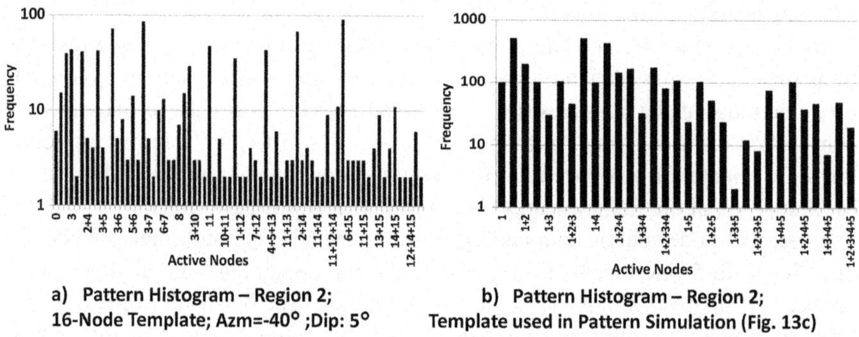

Frequency
Active Nodes
a) Pattern Histogram – Region 2; 16-Node Template; Azm=-40° ;Dip: 5°
b) Pattern Histogram – Region 2; Template used in Pattern Simulation (Fig. 13c)

Fig. 14 Region 2 pattern histograms

be oriented along −40° and 230° azimuth directions with a 5° dip angle. By using a 5-node template (Fig. 13c), pattern simulation of Region 2 is performed.

The realization shown in Fig. 15 is obtained by using 200 conditioning data with 1300 simulation steps and without pattern filtering. The overall cave pattern and node connectivity of Region 2 are successfully reproduced and the accuracy of the pattern simulation is assessed by performing an MPS analysis of the simulated image. The original pattern histogram (Fig. 14b) is satisfactorily reproduced (Fig. 15). Uncertainty of the simulation results can be assessed by obtaining several realizations and implementing different templates.

5 Discussions and Conclusion

The MPS analysis and pattern simulation of 3D synthetic network data are performed by using various templates and different approaches. Besides the visual comparison, the results are scanned by using a 12-node template for pattern his-

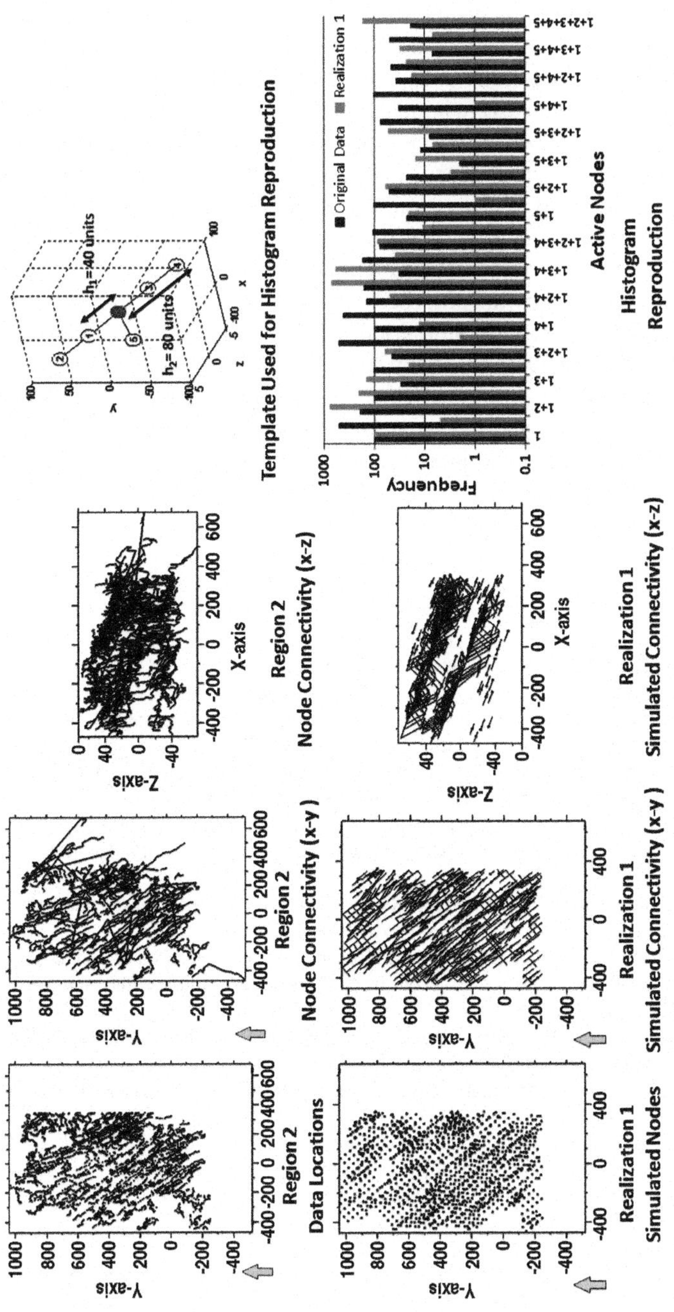

Fig. 15 Pattern simulation of region 2

togram comparison. The MPS analyses of the simulation results show that the proposed algorithm is successful in reproduction of patterns. Moreover, pattern filtering increases the accuracy of the results and results in better reproduction of the target pattern histogram.

The MPS analysis algorithm is also applied on the Wind Cave—Region 2 data set. Several generic templates are initially implemented for a quick assessment of the main connectivity exhibited by the cave network. The analyses show that Region 2 has main connected passages along −40° and −50° azimuth directions with a 5° dip angle. Template used in pattern simulation of Region 2 is selected based on the MPS analysis with 5-nodes along −40° and 230° azimuth directions with a 5° dip angle. It is observed that the overall pattern and node connectivity of Region 2 are successfully reproduced by the proposed methodology.

In conclusion, the proposed non-gridded MPS analysis and pattern simulation algorithms are successful at characterizing and modeling connected features that can only be described by point sets. The methodology is practical; it eliminates the gridding procedure and can be directly applied on network data. In the future, the algorithm will be applied to different modern cave network surveys and extended to model fracture networks.

References

1. Allmendinger RW (2011) Stereonet 7 for Windows Beta
2. Baecher GB, Einstein HH, Lanney NA (1997) Statistical descriptions of rock properties and sampling. In: Proc of the 18th US symposium on rock mechanics, pp 5C1.1–5C1.8
3. Billaux D, Chiles JP, Hestir K, Long J (1989) Three dimensional statistical modeling of a fractured rock mass: an example from the anay-agueres mine. In: International workshop on forced fluid flow through fractured rock masses, vol 26(3–4), pp 281–299
4. Botton-Dumay R, Manivit T, Massonnat G, Gay V (2002) Karstic high permeability layers: characterization and preservation while modeling carbonate reservoirs. In: 10th Abu Dhabi IPEC meeting. SPE 78534
5. Caers J (2002) Geostatistical history matching under training-image based geological modeling constraints. In: 2002 SPE ATCE, San Antonio, Texas. SPE 77429
6. Caers JS, Srinivasan S, Journel AG (2000) Geostatistical quantification of geological information for a fluvial type North Sea reservoir. In: SPE reservoir evaluation and engineering, vol 3, pp 457–467
7. Caers J, Zhang T (2004) Multiple-point geostatisctics: a quantitative vehicle for integrating geologic analogs into multiple reservoir models. In: Grammer GM, Harris PM, Eberli GP (eds) Integration of outcrop and modern analogs in reservoir modeling. AAPG Memoir, vol 80, pp 383–394
8. Chiles JP (1987) Three-dimensional geometric modeling of a fracture network. In: Proceedings of the conference on geostatistical, sensitivity, and uncertainty methods for ground-water flow and radionuclide transport modeling, San Francisco, CA, pp 361–385
9. Deutsch CV, Journel A (1998) GSLIB—geostatistical software library and user's guide, 2nd edn. Oxford University Press, New York, 369 pp
10. Eskandaridalvand K (2008) Growthsim: a complete framework for integrating static and dynamic data into reservoir models. PhD Dissertation, The University of Texas at Austin, 197 pp
11. Erzeybek S, Srinivasan S (2011) Modeling connected geologic features using multiple-point statistics from non-gridded data. In: 14th IAMG conference, Salzburg, Austria

12. Henrion V, Pellerin J, Caumon G (2008) A stochastic methodology for 3D cave system modeling. In: 2008 international geostatistics congress, Santiago, Chile
13. Horrocks RD, Szukalski BW (2002) Using geographic information systems to develop a cave potential map for Wind Cave, South Dakota. J Caves Karst Stud 64(1):63–70
14. Liu X, Srinivasan S (2005) Field scale stochastic modeling of fracture networks. In: Leuangthong O, Deutsch CV (eds) Geostatistics Banff 2004. Springer, Berlin, pp 75–84
15. Loucks RG (1999) Paleocave carbonate reservoirs: origins, burial-depth modifications, spatial complexity and reservoir implications. Am Assoc Pet Geol Bull 83(11):1795–1834
16. Loucks RG, Mescher PK (2001) Paleocave facies classification and associated pore types. In: 2001 AAPG southwest section, annual meeting, Dallas, Texas
17. Madriz DD (2009) Stochastic characterization of carbonate buildup architectures, using two- and multiple-point statistics, and statistical evaluation of these methods. MSc Thesis, The University of Texas at Austin, 212 pp
18. Massonat G, Pernarcic E (2002) Assessment and modeling of high permeability areas in carbonate reservoirs. In: 2002 SPE ATCE, San Antonio, Texas. SPE 77591
19. Miller TE (1989) Evidence of quaternary tectonic activity, and for regional aquifer flow at wind cave, South Dakota. NSS Bull 52:111–119
20. Strebelle S (2002) Conditional simulation of complex geological structures using multiple-point statistics. Math Geol 34:1

Sequential Simulations of Mixed Discrete-Continuous Properties: Sequential Gaussian Mixture Simulation

Dario Grana, Tapan Mukerji, Laura Dovera, and Ernesto Della Rossa

Abstract We present here a method for generating realizations of the posterior probability density function of a Gaussian Mixture linear inverse problem in the combined discrete-continuous case. This task is achieved by extending the sequential simulations method to the mixed discrete-continuous problem. The sequential approach allows us to generate a Gaussian Mixture random field that honors the covariance functions of the continuous property and the available observed data. The traditional inverse theory results, well known for the Gaussian case, are first summarized for Gaussian Mixture models: in particular the analytical expression for means, covariance matrices, and weights of the conditional probability density function are derived. However, the computation of the weights of the conditional distribution requires the evaluation of the probability density function values of a multivariate Gaussian distribution, at each conditioning point. As an alternative solution of the Bayesian inverse Gaussian Mixture problem, we then introduce the sequential approach to inverse problems and extend it to the Gaussian Mixture case. The Sequential Gaussian Mixture Simulation (SGMixSim) approach is presented as a particular case of the linear inverse Gaussian Mixture problem, where the linear operator is the identity. Similar to the Gaussian case, in Sequential Gaussian Mixture Simulation the means and the covariance matrices of the conditional distribution at a given point correspond to the kriging estimate, component by component, of the mixture. Furthermore, Sequential Gaussian Mixture Simulation can be conditioned by secondary information to account for non-stationarity. Examples of applications

D. Grana (✉) · T. Mukerji
Stanford University, 397 Panama Mall, Stanford, CA 94305, USA
e-mail: dgrana@stanford.edu

T. Mukerji
e-mail: mukerji@stanford.edu

L. Dovera · E. Della Rossa
Eni E&P, Via Emilia 1, Milan 20097, Italy

L. Dovera
e-mail: laura.dovera@eni.com

E. Della Rossa
e-mail: ernesto.dellarossa@eni.com

P. Abrahamsen et al. (eds.), *Geostatistics Oslo 2012*,
Quantitative Geology and Geostatistics 17,
DOI 10.1007/978-94-007-4153-9_19,

with synthetic and real data, are presented in the reservoir modeling domain where realizations of facies distribution and reservoir properties, such as porosity or net-to-gross, are obtained using Sequential Gaussian Mixture Simulation approach. In these examples, reservoir properties are assumed to be distributed as a Gaussian Mixture model. In particular, reservoir properties are Gaussian within each facies, and the weights of the mixture are identified with the point-wise probability of the facies.

1 Introduction

Inverse problems are common in many different domains such as physics, engineering, and earth sciences. In general, solving an inverse problem consists of estimating the model parameters given a set of observed data. The operator that links the model and the data can be linear or nonlinear.

In the linear case, estimation techniques generally provide smoothed solutions. Kriging, for example, provides the best estimate of the model in the least-squares sense. Simple kriging is in fact identical to a linear Gaussian inverse problem where the linear operator is the identity, with the estimation of posterior mean and covariance matrices with direct observations of the model space. Monte Carlo methods can be applied as well to solve inverse problems [12] in a Bayesian framework to sample from the posterior; but standard sampling methodologies can be inefficient in practical applications. Sequential simulations have been introduced in geostatistics to generate high resolution models and provide a number of realizations of the posterior probability function honoring both prior information and the observed values. References [3] and [6] give detailed descriptions of kriging and sequential simulation methods. Reference [8] proposes a methodology that applies sequential simulations to linear Gaussian inverse problems to incorporate the prior information on the model and honor the observed data.

We propose here to extend the approach of [8] to the Gaussian Mixture case. Gaussian Mixture models are convex combinations of Gaussian components that can be used to describe the multi-modal behavior of the model and the data. Reference [14], for instance, introduces Gaussian Mixture distributions in multivariate nonlinear regression modeling; while [10] proposes a mixture discriminant analysis as an extension of linear discriminant analysis by using Gaussian Mixtures and Expectation-Maximization algorithm [11]. Gaussian Mixture models are common in statistics (see, for example, [9] and [2]) and they have been used in different domains: digital signal processing [13] and [5], engineering [1], geophysics [7], and reservoir history matching [4].

In this paper we first present the extension of the traditional results valid in the Gaussian case to the Gaussian Mixture case; we then propose the sequential approach to linear inverse problems under the assumption of Gaussian Mixture distribution; and we finally show some examples of applications in reservoir modeling. If the linear operator is the identity, then the methodology provides an extension of the traditional Sequential Gaussian Simulation (SGSim, see [3], and [6]) to a new methodology that we call Sequential Gaussian Mixture Simulation (SGMixSim). The applications we propose refer to mixed discrete-continuous problems of reservoir modeling and they provide, as main result, sets of models of reservoir facies and porosity. The key point of the application is that we identify the weights of the Gaussian Mixture describing the continuous random variable (porosity) with the probability of the reservoir facies (discrete variable).

2 Theory: Linearized Gaussian Mixture Inversion

In this section we provide the main propositions of linear inverse problems with Gaussian Mixtures (GMs). We first recap the well-known analytical result for posterior distributions of linear inverse problems with Gaussian prior; then we extend the result to the Gaussian Mixtures case.

In the Gaussian case, the solution of the linear inverse problem is well-known [15]. If $\mathbf{m}$ is a random vector Gaussian distributed, $\mathbf{m} \sim N(\boldsymbol{\mu}_{\mathbf{m}}, \mathbf{C}_{\mathbf{m}})$, with mean $\boldsymbol{\mu}_{\mathbf{m}}$ and covariance $\mathbf{C}_{\mathbf{m}}$; and $\mathbf{G}$ is a linear operator that transforms the model $\mathbf{m}$ into the observable data $\mathbf{d}$

$$\mathbf{d} = \mathbf{G}\mathbf{m} + \boldsymbol{\varepsilon}, \tag{1}$$

where $\boldsymbol{\varepsilon}$ is a random vector that represents an error with Gaussian distribution $N(\mathbf{0}, \mathbf{C}_{\boldsymbol{\varepsilon}})$ independent of the model $\mathbf{m}$; then the posterior conditional distribution of $\mathbf{m}|\mathbf{d}$ is Gaussian with mean and covariance given by

$$\boldsymbol{\mu}_{\mathbf{m}|\mathbf{d}} = \boldsymbol{\mu}_{\mathbf{m}} + \mathbf{C}_{\mathbf{m}}\mathbf{G}^T\left(\mathbf{G}\mathbf{C}_{\mathbf{m}}\mathbf{G}^T + \mathbf{C}_{\boldsymbol{\varepsilon}}\right)^{-1}(\mathbf{d} - \mathbf{G}\boldsymbol{\mu}_{\mathbf{m}}) \tag{2}$$

$$\mathbf{C}_{\mathbf{m}|\mathbf{d}} = \mathbf{C}_{\mathbf{m}} - \mathbf{C}_{\mathbf{m}}\mathbf{G}^T\left(\mathbf{G}\mathbf{C}_{\mathbf{m}}\mathbf{G}^T + \mathbf{C}_{\boldsymbol{\varepsilon}}\right)^{-1}\mathbf{G}\mathbf{C}_{\mathbf{m}}. \tag{3}$$

This result is based on two well known properties of the Gaussian distributions: (A) the linear transform of a Gaussian distribution is again Gaussian; (B) if the joint distribution $(\mathbf{m}, \mathbf{d})$ is Gaussian, then the conditional distribution $\mathbf{m}|\mathbf{d}$ is again Gaussian.

These two properties can be extended to the Gaussian Mixtures case. We assume that $\mathbf{x}$ is a random vector distributed according to a Gaussian Mixture with N_c components, $f(\mathbf{x}) = \sum_{k=1}^{N_c} \pi_k N(\mathbf{x}; \boldsymbol{\mu}_{\mathbf{x}}^k, \mathbf{C}_{\mathbf{x}}^k)$, where π_k are the weights and the distributions $N(\mathbf{x}; \boldsymbol{\mu}_{\mathbf{x}}^k, \mathbf{C}_{\mathbf{x}}^k)$ represent the Gaussian components with means $\boldsymbol{\mu}_{\mathbf{x}}^k$ and covariances $\mathbf{C}_{\mathbf{x}}^k$ evaluated in $\mathbf{x}$. By applying property (A) to the Gaussian components of the mixture, we can conclude that, if $\mathbf{L}$ is a linear operator, then $\mathbf{y} = \mathbf{L}\mathbf{x}$ is distributed according to a Gaussian Mixture. Moreover, the pdf of $\mathbf{y}$ is given by $f(\mathbf{y}) = \sum_{k=1}^{N_c} \pi_k N(\mathbf{y}; \mathbf{L}\boldsymbol{\mu}_{\mathbf{x}}^k, \mathbf{L}\mathbf{C}_{\mathbf{x}}^k\mathbf{L}^T)$.

Similarly we can extend property (B) to conditional Gaussian Mixture distributions. The well-known result of the conditional multivariate Gaussian distribution has already been extended to multivariate Gaussian Mixture models (see, for example, [1]). In particular, if $(\mathbf{x}_1, \mathbf{x}_2)$ is a random vector whose joint distribution is a Gaussian Mixture

$$f(\mathbf{x}_1, \mathbf{x}_2) = \sum_{k=1}^{N_c} \pi_k f_k(\mathbf{x}_1, \mathbf{x}_2), \tag{4}$$

where f_k are the Gaussian densities, then the conditional distribution of $\mathbf{x}_2|\mathbf{x}_1$ is again a Gaussian Mixture

$$f(\mathbf{x}_2|\mathbf{x}_1) = \sum_{k=1}^{N_c} \lambda_k f_k(\mathbf{x}_2|\mathbf{x}_1), \tag{5}$$

and its parameters (weights, means, and covariance matrices) can be analytically derived. The coefficients λ_k are given by

$$\lambda_k = \frac{\pi_k f_k(\mathbf{x}_1)}{\sum_{\ell=1}^{N_c} \pi_\ell f_\ell(\mathbf{x}_1)}, \tag{6}$$

where $f_k(\mathbf{x}_1) = N(\mathbf{x}_1; \boldsymbol{\mu}^k_{\mathbf{x}_1}, \mathbf{C}^k_{\mathbf{x}_1})$; and the means and the covariance matrices are

$$\boldsymbol{\mu}^k_{\mathbf{x}_2|\mathbf{x}_1} = \boldsymbol{\mu}^k_{\mathbf{x}_2} + \mathbf{C}^k_{\mathbf{x}_2,\mathbf{x}_1}\left(\mathbf{C}^k_{\mathbf{x}_1}\right)^{-1}\left(\mathbf{x}_1 - \boldsymbol{\mu}^k_{\mathbf{x}_1}\right) \tag{7}$$

$$\mathbf{C}^k_{\mathbf{x}_2|\mathbf{x}_1} = \mathbf{C}^k_{\mathbf{x}_2} - \mathbf{C}^k_{\mathbf{x}_2,\mathbf{x}_1}\left(\mathbf{C}^k_{\mathbf{x}_1}\right)^{-1}\left(\mathbf{C}^k_{\mathbf{x}_2,\mathbf{x}_1}\right)^T, \tag{8}$$

where $\mathbf{C}^k_{\mathbf{x}_2,\mathbf{x}_1}$ is the cross-covariance matrix. By combining these propositions, the main result of linear inverse problems with Gaussian Mixture can be derived.

Theorem 1 *Let* $\mathbf{m}$ *be a random vector distributed according to a Gaussian Mixture* $\mathbf{m} \sim \sum_{k=1}^{N_c} \pi_k N(\boldsymbol{\mu}^k_{\mathbf{m}}, \mathbf{C}^k_{\mathbf{m}})$, *with* N_c *components and with means* $\boldsymbol{\mu}^k_{\mathbf{m}}$, *covariances* $\mathbf{C}^k_{\mathbf{m}}$, *and weights* π_k, *for* $k = 1, \ldots, N_c$. *Let* $\mathbf{G} : \Re^M \to \Re^N$ *be a linear operator, and* $\boldsymbol{\varepsilon}$ *a Gaussian random vector independent of* $\mathbf{m}$ *with* $\mathbf{0}$ *mean and covariance* $\mathbf{C}_{\boldsymbol{\varepsilon}}$, *such that* $\mathbf{d} = \mathbf{G}\mathbf{m} + \boldsymbol{\varepsilon}$, *with* $\mathbf{d} \in \Re^N$, $\mathbf{m} \in \Re^M$, $\boldsymbol{\varepsilon} \in \Re^N$, *then the posterior conditional distribution* $\mathbf{m}|\mathbf{d}$ *is a Gaussian Mixture.*

Moreover, the posterior means and covariances of the components are given by

$$\boldsymbol{\mu}^k_{\mathbf{m}|\mathbf{d}} = \boldsymbol{\mu}^k_{\mathbf{m}} + \mathbf{C}^k_{\mathbf{m}}\mathbf{G}^T\left(\mathbf{G}\mathbf{C}^k_{\mathbf{m}}\mathbf{G}^T + \mathbf{C}_{\boldsymbol{\varepsilon}}\right)^{-1}\left(\mathbf{d} - \mathbf{G}\boldsymbol{\mu}^k_{\mathbf{m}}\right) \tag{9}$$

$$\mathbf{C}^k_{\mathbf{m}|\mathbf{d}} = \mathbf{C}^k_{\mathbf{m}} - \mathbf{C}^k_{\mathbf{m}}\mathbf{G}^T\left(\mathbf{G}\mathbf{C}^k_{\mathbf{m}}\mathbf{G}^T + \mathbf{C}_{\boldsymbol{\varepsilon}}\right)^{-1}\mathbf{G}\mathbf{C}^k_{\mathbf{m}}, \tag{10}$$

where $\boldsymbol{\mu}^k_{\mathbf{m}}$ *and* $\mathbf{C}^k_{\mathbf{m}}$, *are respectively the prior mean and covariance of the* k^{th} *Gaussian component of* $\mathbf{m}$. *The posterior coefficients* λ_k *of the mixture are given by*

$$\lambda_k = \frac{\pi_k f_k(\mathbf{d})}{\sum_{\ell=1}^{N_c} \pi_\ell f_\ell(\mathbf{d})}, \tag{11}$$

where the Gaussian densities $f_k(\mathbf{d})$ *have means* $\boldsymbol{\mu}^k_{\mathbf{d}} = \mathbf{G}\boldsymbol{\mu}^k_{\mathbf{m}}$ *and covariances* $\mathbf{C}^k_{\mathbf{d}} = \mathbf{G}\mathbf{C}^k_{\mathbf{m}}\mathbf{G}^T + \mathbf{C}_{\boldsymbol{\varepsilon}}$.

3 Theory: Sequential Approach

Based on the results presented in the previous section, we introduce here the sequential approach to linearized inversion in the Gaussian Mixture case. We first recap the main result for the Gaussian case [8].

The solution of the linear inverse problem with the sequential approach requires some additional notation. Let m_i represent the ith element of the random vector $\mathbf{m}$, and let $\mathbf{m_s}$ represent a known sub-vector of $\mathbf{m}$. This notation will generally be used to describe the neighborhood of m_i in the context of sequential simulations. Finally we assume that the measured data $\mathbf{d}$ are known having been obtained as a linear transformation of $\mathbf{m}$ according to some linear operator $\mathbf{G}$.

Theorem 2 *Let* $\mathbf{m}$ *be a random vector, Gaussian distributed,* $\mathbf{m} \sim N(\boldsymbol{\mu}_{\mathbf{m}}, \mathbf{C_m})$ *with mean* $\boldsymbol{\mu}_{\mathbf{m}}$ *and covariance* $\mathbf{C_m}$. *Let* $\mathbf{G}$ *be a linear operator between the model* $\mathbf{m}$ *and the random data vector* $\mathbf{d}$ *such that* $\mathbf{d} = \mathbf{Gm} + \boldsymbol{\varepsilon}$, *with* $\boldsymbol{\varepsilon}$ *a random error vector independent of* $\mathbf{m}$ *with* $\mathbf{0}$ *mean and covariance* $\mathbf{C}_{\boldsymbol{\varepsilon}}$. *Let* $\mathbf{m_s}$ *be the subvector with direct observations of the model* $\mathbf{m}$, *and* m_i *the* i^{th} *element of* $\mathbf{m}$. *Then the conditional distribution of* $m_i|(\mathbf{m_s}, \mathbf{d})$ *is again Gaussian.*

Moreover, if the subvector $\mathbf{m_s}$ *is extracted from the full random vector* $\mathbf{m}$ *with the linear operator* $\mathbf{A}$ *such that* $\mathbf{m_s} = \mathbf{Am}$, *where the* i^{th} *element is* $m_i = \mathbf{A}_i\mathbf{m}$, *with* $\mathbf{A}_i$ *again linear, then the mean and variance of the posterior conditional distribution are:*

$$\mu_{m_i|(\mathbf{m_s},\mathbf{d})} = \mu_{m_i} + \begin{bmatrix} \mathbf{A}_i\mathbf{C_m}\mathbf{A}^T & \mathbf{A}_i\mathbf{C_m}\mathbf{G}^T \end{bmatrix} (\mathbf{C}_{(\mathbf{m_s},\mathbf{d})})^{-1} \begin{bmatrix} \mathbf{m_s} - \mathbf{A}\boldsymbol{\mu}_{\mathbf{m}} \\ \mathbf{d} - \mathbf{G}\boldsymbol{\mu}_{\mathbf{m}} \end{bmatrix} \tag{12}$$

$$\sigma^2_{m_i|(\mathbf{m_s},\mathbf{d})} = \sigma^2_{m_i} - \begin{bmatrix} \mathbf{A}_i\mathbf{C_m}\mathbf{A}^T & \mathbf{A}_i\mathbf{C_m}\mathbf{G}^T \end{bmatrix} (\mathbf{C}_{(\mathbf{m_s},\mathbf{d})})^{-1} \begin{bmatrix} \mathbf{A}\mathbf{C_m}\mathbf{A}_i^T \\ \mathbf{G}\mathbf{C_m}\mathbf{A}_i^T \end{bmatrix}, \tag{13}$$

where $\mu_{m_i} = \mathbf{A}_i\boldsymbol{\mu}_{\mathbf{m}}$, $\sigma^2_{m_i} = \mathbf{A}_i\mathbf{C_m}\mathbf{A}_i^T$, *and*

$$\mathbf{C}_{(\mathbf{m_s},\mathbf{d})} = \begin{bmatrix} \mathbf{A}\mathbf{C_m}\mathbf{A}^T & \mathbf{A}\mathbf{C_m}\mathbf{G}^T \\ \mathbf{G}\mathbf{C_m}\mathbf{A}^T & \mathbf{G}\mathbf{C_m}\mathbf{G}^T + \mathbf{C}_{\boldsymbol{\varepsilon}} \end{bmatrix}. \tag{14}$$

To clarify the statement we give the explicit form of the operators $\mathbf{A}_i$ and $\mathbf{A}$. In particular, $\mathbf{A}_i$ is written as

$$\mathbf{A}_i = [0 \quad 0 \quad \ldots \quad 1 \quad \ldots \quad 0], \tag{15}$$

with the one in the ith column. If the sub-vector $\mathbf{m_s}$ has size n, $\mathbf{m_s} = \{m_{i_1}, m_{i_2}, \ldots, m_{i_n}\}^T$, and $\mathbf{m}$ has size M; then the operator $\mathbf{A}$ is given by

$$\mathbf{A} = \begin{bmatrix} 0 & 0 & \ldots & 1 & \ldots & 0 \\ 0 & \ldots & 1 & 0 & \ldots & 0 \\ \vdots & \vdots & \vdots & \vdots & \vdots & \vdots \\ 0 & 1 & \ldots & 0 & 0 & 0 \end{bmatrix}, \tag{16}$$

where $\mathbf{A}$ has dimensions $n \times M$ and the ones are in the $i_1, i_2, \ldots, i_n$ columns. Theorem 2 can be proved using the properties (A) and (B) described in Sect. 2 (see [8]). Then, by using Theorem 1, we extend the result to the Gaussian Mixture case.

Theorem 3 *Let* $\mathbf{m}$ *be a random vector distributed according to a Gaussian Mixture,* $\mathbf{m} \sim \sum_{k=1}^{N_c} \pi_k N(\boldsymbol{\mu}_{\mathbf{m}}^k, \mathbf{C}_{\mathbf{m}}^k)$, *with* N_c *components and with means* $\boldsymbol{\mu}_{\mathbf{m}}^k$, *covariances* $\mathbf{C}_{\mathbf{m}}^k$, *and weights* π_k, *for* $k = 1, \ldots, N_c$. *Let* $\mathbf{G}$ *a linear operator such that* $\mathbf{d} = \mathbf{Gm} + \boldsymbol{\varepsilon}$, *with* $\boldsymbol{\varepsilon}$ *a random error vector independent of* $\mathbf{m}$ *with* $\mathbf{0}$ *mean and covariance* $\mathbf{C}_{\boldsymbol{\varepsilon}}$. *Let* $\mathbf{m}_s$ *be the sub-vector with direct observations of the model* $\mathbf{m}$, *and* m_i *the* i^{th} *element of* $\mathbf{m}$. *Then the conditional distribution of* $m_i|(\mathbf{m}_s, \mathbf{d})$ *is again a Gaussian Mixture.*

Moreover, the means and variances of the components of the posterior conditional distribution are:

$$\mu_{m_i|(\mathbf{m}_s,\mathbf{d})}^k = \mu_{m_i}^k + \left[\mathbf{A}_i \mathbf{C}_{\mathbf{m}}^k \mathbf{A}^T \quad \mathbf{A}_i \mathbf{C}_{\mathbf{m}}^k \mathbf{G}^T\right] \left(\mathbf{C}_{(\mathbf{m}_s,\mathbf{d})}^k\right)^{-1} \begin{bmatrix} \mathbf{m}_s - \mathbf{A}\boldsymbol{\mu}_{\mathbf{m}}^k \\ \mathbf{d} - \mathbf{G}\boldsymbol{\mu}_{\mathbf{m}}^k \end{bmatrix} \tag{17}$$

$$\sigma_{m_i|(\mathbf{m}_s,\mathbf{d})}^{2(k)} = \sigma_{m_i}^{2(k)} - \left[\mathbf{A}_i \mathbf{C}_{\mathbf{m}}^k \mathbf{A}^T \quad \mathbf{A}_i \mathbf{C}_{\mathbf{m}}^k \mathbf{G}^T\right] \left(\mathbf{C}_{(\mathbf{m}_s,\mathbf{d})}^k\right)^{-1} \begin{bmatrix} \mathbf{A}\mathbf{C}_{\mathbf{m}}^k \mathbf{A}_i^T \\ \mathbf{G}\mathbf{C}_{\mathbf{m}}^k \mathbf{A}_i^T \end{bmatrix}, \tag{18}$$

where $\mu_{m_i}^k = \mathbf{A}_i \boldsymbol{\mu}_{\mathbf{m}}^k$, $\sigma_{m_i}^{2(k)} = \mathbf{A}_i \mathbf{C}_{\mathbf{m}}^k \mathbf{A}_i^T$, *and*

$$\mathbf{C}_{(\mathbf{m}_s,\mathbf{d})}^k = \begin{bmatrix} \mathbf{A}\mathbf{C}_{\mathbf{m}}^k \mathbf{A}^T & \mathbf{A}\mathbf{C}_{\mathbf{m}}^k \mathbf{G}^T \\ \mathbf{G}\mathbf{C}_{\mathbf{m}}^k \mathbf{A}^T & \mathbf{G}\mathbf{C}_{\mathbf{m}}^k \mathbf{G}^T + \mathbf{C}_{\boldsymbol{\varepsilon}} \end{bmatrix}. \tag{19}$$

The posterior coefficients of the mixture are given by

$$\lambda_k = \frac{\pi_k f_k(\mathbf{m}_s, \mathbf{d})}{\sum_{\ell=1}^{N_c} \pi_\ell f_\ell(\mathbf{m}_s, \mathbf{d})}, \tag{20}$$

where the Gaussian components $f_k(\mathbf{m}_s, \mathbf{d})$ *have means*

$$\boldsymbol{\mu}_{(\mathbf{m}_s,\mathbf{d})}^k = \begin{bmatrix} \mathbf{A}\boldsymbol{\mu}_{\mathbf{m}}^k \\ \mathbf{G}\boldsymbol{\mu}_{\mathbf{m}}^k \end{bmatrix}, \tag{21}$$

and covariances $\mathbf{C}_{(\mathbf{m}_s,\mathbf{d})}^k$.

In the case where the linear operator is the identity, the associated inverse problem reduces to the estimation of a Gaussian Mixture model with direct observations of the model space at given locations. In other words, if the linear operator is the identity, the theorem provides an extension of the traditional Sequential Gaussian Simulation (SGSim) to the Gaussian Mixture case. We call this methodology Sequential Gaussian Mixture Simulation (SGMixSim), and we show some applications in the next section.

4 Application

We describe here some examples of applications with synthetic and real data, in the context of reservoir modeling. First, we present the results of the estimation of a Gaussian Mixture model with direct observations of the model space as a special case of Theorem 3 (SGMixSim). In our example, the continuous property is the porosity of a reservoir, and the discrete variable represents the corresponding reservoir facies, namely shale and sand. This means that we identify the weights of the mixture components with the facies probabilities. The input parameters are then the prior distribution of porosity and a variogram model for each component of the mixture. The prior is a Gaussian Mixture model with two components and its parameters are the weights, the means, and the covariance matrices of the Gaussian components. We assume facies prior probabilities equal to 0.4 and 0.6 respectively, and for simplicity we assume the same variogram model (spherical and isotropic) with the same parameters for both. We then simulate a 2D map of facies and porosity according to the proposed methodology (Fig. 1). The simulation grid is 70×70 and the variogram range of porosity is 4 grid blocks in both directions. The simulation can be performed with or without conditioning hard data; in the example of Fig. 1, we introduced four porosity values at four locations that are used to condition the simulations, and we generated a set of 100 conditional realizations (Fig. 1). When hard data are assigned, the weights of the mixture components are determined by evaluating the prior Gaussian components at the hard data location and discrete property values are determined by selecting the most likely component.

As we previously mentioned, the methodology is similar to [8], but the use of Gaussian Mixture models allows us to describe the multi-modality of the data and to simulate at the same time both the continuous and the discrete variable. SGMixSim requires a spatial model of the continuous variable, but not a spatial model of the underlying discrete variable: the spatial distribution of the discrete variable only depends on the conditional weights of the mixture (20). However, if the mixture components have very different probabilities and very different variances (i.e. when there are relatively low probable components with relatively high variances), the simulations may not accurately reproduce the global statistics. If we assume, for instance, two components with prior probabilities equal to 0.2 and 0.8, and we assume at the same time that the variance of the first component is much bigger than the variance of the second one, then the prior proportions may not be honored. This problem is intrinsic to the sequential simulation approach, but it is emphasized in case of multi-modal data. For large datasets or for reasons of stationarity, we often use a moving searching neighborhood to take into account only the points closest to the location being simulated [6]. If we use a global searching neighborhood (i.e. the whole grid) the computational time, for large datasets, could significantly increase. In the localized sequential algorithm, the neighborhood is selected according to a fixed geometry (for example, ellipsoids centered on the location to be estimated) and the conditioning data are extracted by the linear operator (Theorem 3) within the neighborhood. When no hard data are present in the searching neighborhood and the sample value is drawn from the prior distribution, the algorithm could generate

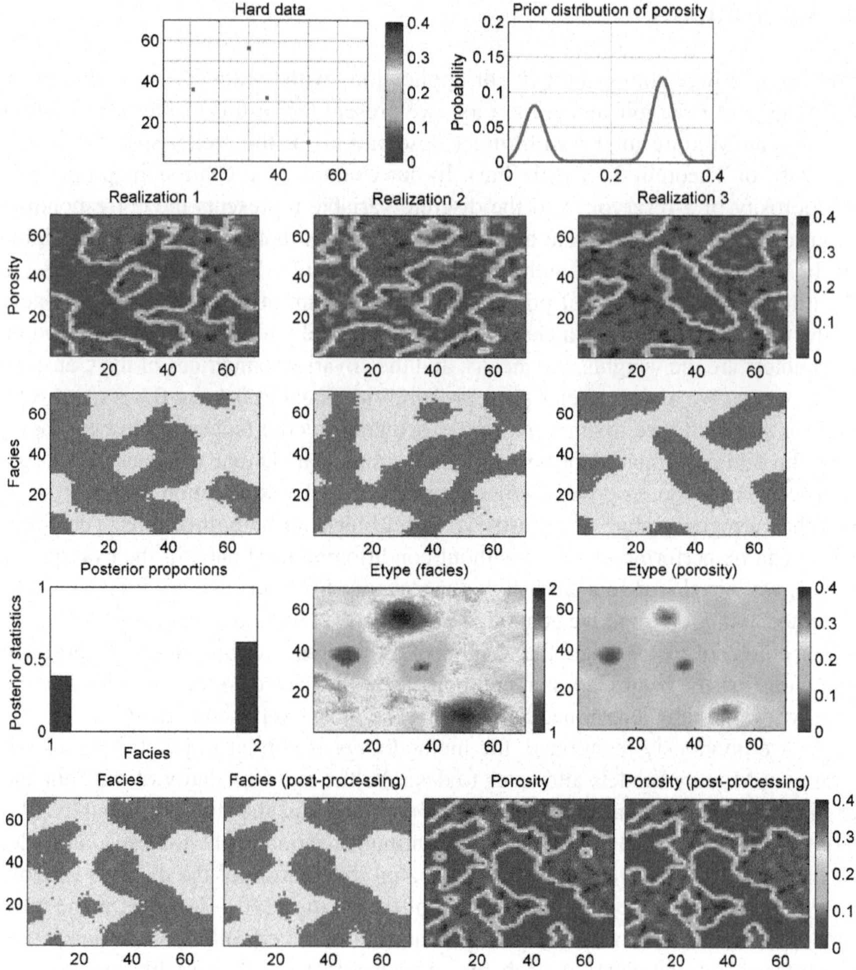

Fig. 1 Conditional realizations of porosity and reservoir facies obtained by SGMixSim. The prior distribution of porosity and the hard data values are shown on *top*. The *second* and *third rows* show three realizations of porosity and facies (*gray* is shale, *yellow* is sand). The *fourth row* shows the posterior distribution of facies and the ensemble average of 100 realizations of facies and porosity. The *last row* shows the comparison of SGMixSim results with and without post-processing

isolated points within the simulation grid. For example, a point drawn from the first component could be surrounded by data, subsequently simulated, belonging to the second component, or vice versa. This problem is particularly relevant in the case of multi-modal data especially in the initial steps of the sequential simulation (in other words when only few values have been previously simulated) and when the searching neighborhood is small.

To avoid isolated points in the simulated grid, a post-processing step has been included (Fig. 1). The simulation path is first revisited, and the local conditional

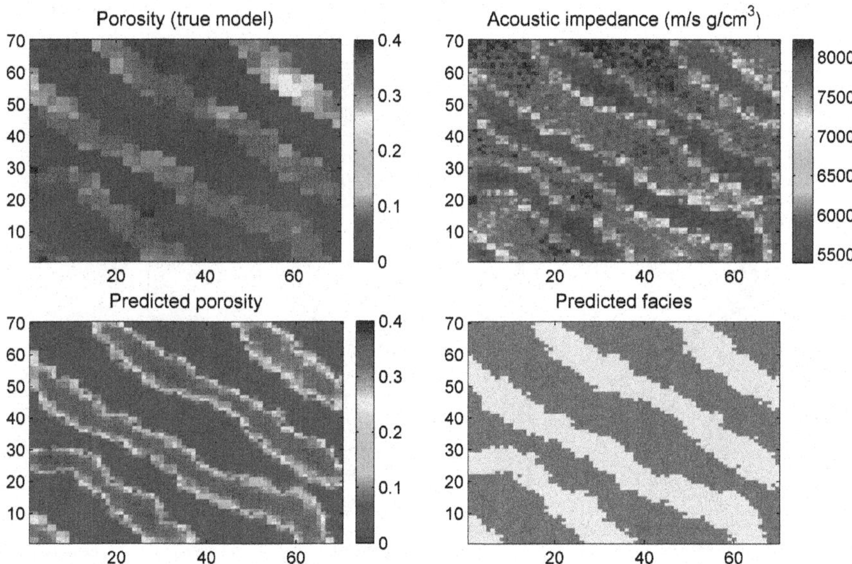

Fig. 2 Linearized sequential inversion with Gaussian Mixture models for the estimation of porosity map from acoustic impedance values. On *top* we show the true porosity map and the acoustic impedance map; on the *bottom* we show the inverted porosity and the estimated facies map

probabilities are re-evaluated at all the grid cells where the sample value was drawn from the prior distribution. Then we draw again the component from the weights of the re-evaluated conditional probability. Finally, we introduce a kriging correction of the continuous property values that had low probabilities in the neighborhood.

Next, we show two applications of linearized sequential inversion with Gaussian Mixture models obtained by applying Theorem 3. The first example is a rock physics inverse problem dealing with the inversion of acoustic impedance in terms of porosity. The methodology application is illustrated by using a 2D grid representing a synthetic system of reservoir channels (Fig. 2). In this example we made the same assumptions about the prior distribution as in the previous example. As in traditional sequential simulation approaches, the spatial continuity of the inverted data depends on the range of the variogram and the size of the searching neighborhood; however, Fig. 2 clearly shows the multi-modality of the inverted data. Gaussian Mixture models can describe not only the multi-modality of the data, but they can better honor the data correlation within each facies.

The second example is the acoustic inversion of seismic amplitudes in terms of acoustic impedance. In this case, in addition to the usual input parameters (prior distribution and variogram models), we have to specify a low frequency model of impedance, since seismic amplitudes only provide relative information about elastic contrasts and the absolute value of impedance must be computed by combining the estimated relative changes with the low frequency model (often called prior model in seismic modeling). Once again, the discrete variable is identified with the reservoir facies classification. In this case shales are characterized by high impedance values,

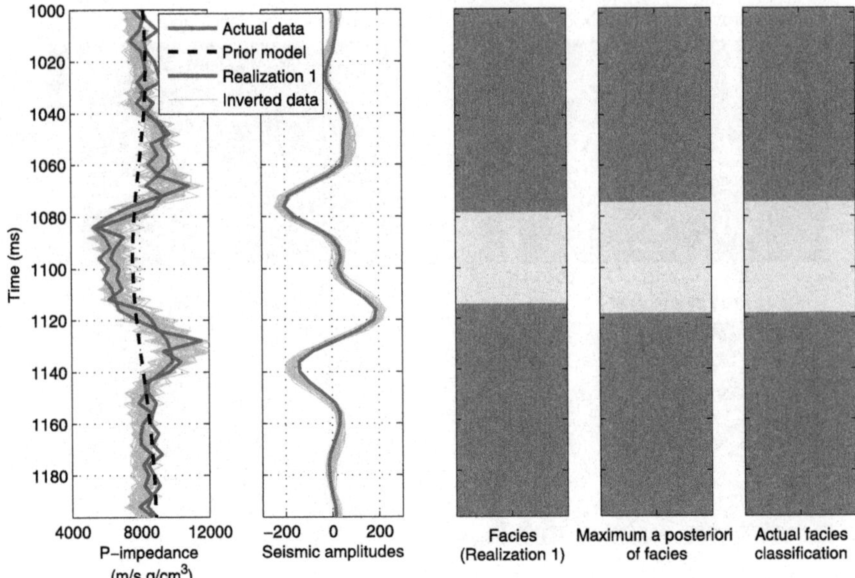

Fig. 3 Sequential Gaussian Mixture inversion of seismic data (ensemble of 50 realizations). From *left* to *right*: acoustic impedance logs and seismograms (actual model in *red*, realization 1 in *blue*, inverted realizations in *gray*, *dashed line* represents low frequency model), inverted facies profile corresponding to realization 1, maximum a posteriori of 50 inverted facies profiles and actual facies classification (sand in *yellow*, shale in *gray*)

and sand by low impedances. The results are shown in Fig. 3. We observe that even though we used a very smoothed low frequency model, the inverted impedance log has a good match with the actual data (Fig. 3), and the prediction of the discrete variable is satisfactory compared to the actual facies classification performed at the well. In particular, if we perform 50 realizations and we compute the maximum a posteriori of the ensemble of inverted facies profiles, we perfectly match the actual classification (Fig. 3). However, the quality of the results depends on the separability of the Gaussian components in the continuous property domain.

Finally we applied the Gaussian Mixture linearized sequential inversion to a layer map extracted from a 3D geophysical model of a clastic reservoir located in the North Sea (Fig. 4). The application has been performed on a map of P-wave velocity corresponding to the top horizon of the reservoir. The parameters of the variogram models have been assumed from existing reservoir studies in the same area. In Fig. 4 we show the map of the conditioning velocity and the corresponding histogram, two realizations of porosity and facies, and the histogram of the posterior distribution of porosity derived from the second realization. The two realizations have been performed using different prior proportions: 30 % of sand in the first realization and 40 % in the second one. Both realizations honor the expected proportions, the multi-modality of the data, and the correlations with the conditioning data within each facies.

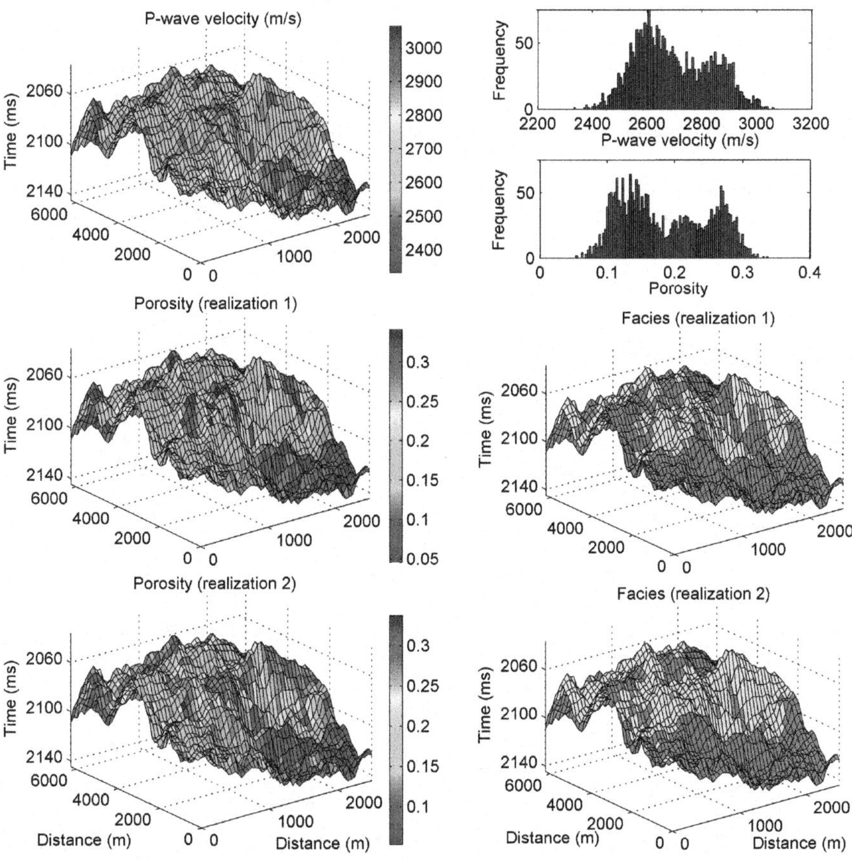

Fig. 4 Application of linearized sequential inversion with Gaussian Mixture models to a reservoir layer. The conditioning data is P-wave velocity (*top left*). Two realizations of porosity and facies are shown: realization 1 corresponds to a prior proportion of 30 % of sand, realization 2 corresponds to 40 % of sand. The histograms of the conditioning data and the posterior distribution of porosity (realization 2) are shown for comparison

5 Conclusion

In this paper, we proposed a methodology to simultaneously simulate both continuous and discrete properties by using Gaussian Mixture models. The method is based on the sequential approach to Gaussian Mixture linear inverse problem, and it can be seen as an extension of sequential simulations to multi-modal data. Thanks to the sequential approach used for the inversion, the method is generally quite efficient from the computational point of view to solve multi-modal linear inverse problems and it is applied here to reservoir modeling and seismic reservoir characterization. We presented four different applications: conditional simulations of porosity and facies, porosity-impedance inversion, acoustic inversion of seismic data, and inversion of seismic velocities in terms of porosity. The proposed examples show that we can generate actual samples from the posterior distribution, consistent with the prior information and the assigned data observations. Using the sequential approach, we can generate a large number of samples from the posterior distribution, which in fact are all solutions to the Gaussian Mixture linear problem.

Acknowledgements We acknowledge Stanford Rock Physics and Borehole Geophysics Project and Stanford Center for Reservoir Forecasting for the support, and Eni E&P for the permission to publish this paper.

References

1. Alspach DL, Sorenson HW (1972) Nonlinear Bayesian estimation using Gaussian sum approximation. IEEE Trans Autom Control 17:439–448. doi:10.1109/TAC.1972.1100034
2. Dempster AP, Laird NM, Rubin DB (1977) Maximum likelihood from incomplete data via the EM algorithm. J R Stat Soc, Ser B, Methodol 39(1):1–38. doi:10.2307/2984875
3. Deutsch C, Journel AG (1992) GSLIB: geostatistical software library and user's guide. Oxford University Press, London
4. Dovera L, Della Rossa E (2011) Multimodal ensemble Kalman filtering using Gaussian mixture models. Comput Geosci 15(2):307–323. doi:10.1007/s10596-010-9205-3
5. Gilardi N, Bengio S, Kanevski M (2002) Conditional Gaussian mixture models for environmental risk mapping. In: Proc. of IEEE workshop on neural networks for signal processing, pp 777–786. doi:10.1109/NNSP.2002.1030100
6. Goovaerts P (1997) Geostatistics for natural resources evaluation. Oxford University Press, London
7. Grana D, Della Rossa E (2010) Probabilistic petrophysical-properties estimation integrating statistical rock physics with seismic inversion. Geophysics 75(3):O21–O37. doi:10.1190/1.3386676
8. Hansen TM, Journel AG, Tarantola A, Mosegaard K (2006) Linear inverse Gaussian theory and geostatistics. Geophysics 71:R101–R111. doi:10.1190/1.2345195
9. Hasselblad V (1966) Estimation of parameters for a mixture of normal distributions. Technometrics 8(3):431–444. doi:10.2307/1266689
10. Hastie T, Tibshirani R (1996) Discriminant analysis by gaussian mixtures. J R Stat Soc B 58(1):155–176. doi:10.2307/2346171
11. Hastie T, Tibshirani R, Friedmann J (2009) The elements of statistical learning. Springer, Berlin
12. Mosegaard K, Tarantola A (1995) Monte Carlo sampling of solutions to inverse problems. J Geophys Res 100:12431–12447. doi:10.1029/94JB03097
13. Reynolds DA, Quatieri TF, Dunn RB (2000) Speaker verification using adapted Gaussian mixture models. Digit Signal Process 10(1–3):19–41. doi:10.1006/dspr.1999.0361
14. Sung HG (2004) Gaussian mixture regression and classification. PhD thesis, Rice University
15. Tarantola A (2005) Inverse problem theory. SIAM, Philadelphia

Accounting for Seismic Trends in Stochastic Well Correlation

Charline Julio, Florent Lallier, and Guillaume Caumon

Abstract Stratigraphic well correlation is a critical step of basin and reservoir analysis and modeling workflows. In this paper, we propose an automatic stratigraphic well correlation method which is based on both borehole data and interwell information extracted from poststack seismic data to constrain stratigraphic well correlation. The presented stratigraphic well correlation method uses the Dynamic Time Warping algorithm. Global correlations are built by combining elementary correlation costs between stratigraphic units or markers identified along studying wells. Whereas various rules can be used to compute the correlation likelihood between well sections, a significant challenge is to compute the cost for an unconformity to occur. Therefore, we use first-order trends extracted from seismic data: a rule based on a 3D scalar field whose gradient is orthogonal to horizons; and a rule based on a seismic attribute which highlights the convergence of seismic reflectors.

1 Introduction

Well data provide a great deal of information on physical and geological properties of underlying rocks. This relatively precise information, sparsely distributed over the study area, does not allow to build stratigraphic correlations of units identified along wells without making strong subjective assumptions [5]. This results in correlation uncertainties that may significantly affect the geometry of the model used for geostatistical studies.

Numerous reproducible automated correlation methods which require a mathematical formalization of the correlation rules have been developed [3, 6, 10, 16]. To

C. Julio (✉) · F. Lallier · G. Caumon
Nancy School of Geology—CRPG, Université Lorraine, ENSG rue du Doyen M. Roubault, BP 40, 54501 Vandoeuvre-les-Nancy, France
e-mail: julio@gocad.org

F. Lallier
e-mail: lallier@gocad.org

G. Caumon
e-mail: caumon@gocad.org

P. Abrahamsen et al. (eds.), *Geostatistics Oslo 2012*,
Quantitative Geology and Geostatistics 17,
DOI 10.1007/978-94-007-4153-9_20,

take into account uncertainties, some algorithms present a stochastic approach aiming at generating numerous correlation models [11]. However, both stochastic and deterministic methods consider only information along the wellbore to establish the stratigraphic correlation rules. Because well data are not sufficient to significantly reduce correlation ambiguities due to the incompleteness of the observations along the borehole and long distances between wells, we introduce new correlation rules based on seismic data which provide interwell information on the continuity of the sedimentary bodies. Due to the difference of resolution between well and seismic data, these rules exploit first-order trends inferred from seismic and should not be substituted to higher order rules between well sections.

In this paper, we introduce a new workflow for stratigraphic well correlation based on the Dynamic Time Warping algorithm which allows to compute likely correlations according to correlation rules built from both borehole and seismic data (Fig. 1, Sect. 2). A first correlation rule (noted $\mathscr{R}_1$) considers gamma ray trends recorded on stratigraphic units in order to evaluate the likelihood of the stratigraphic correlation (Sect. 3). Aiming at integrating seismic information, two additional rules are developed: (1) a rule $\mathscr{R}_2$ built from a 3D draft geometric model, (2) a rule $\mathscr{R}_3$ based on a seismic attribute enhancing stratigraphic sequences (Sect. 4). The proposed stratigraphic correlation method is applied to the sequence stratigraphic correlation of Miocene, Pliocene, and Pleistocene fluvio-deltaic deposits of the Dutch sector of the North Sea (Fig. 8, Sect. 5).

2 One Algorithm for Stratigraphic Correlation

Dynamic Time Warping (DTW) is a simple and rapid algorithm to compute the optimal correlation between two sequences according to elementary correlation rules often used for stratigraphic well correlations [3, 5, 8, 16].

$\mathbf{a} = \{a_1, \ldots, a_n\}$ and $\mathbf{b} = \{b_1, \ldots, b_m\}$ denote two sequences of stratigraphic units, with respectively n and m strata. The dynamic programming framework uses the two sequences $\mathbf{a}$ and $\mathbf{b}$ as axes of a cost matrix D containing all the possibilities of correlation between stratigraphic units (Fig. 2). Indeed, between two stratigraphic units a_i and b_j, there are three possibilities of correlation: a_i and b_j match, a_i is gapped or b_j is gapped. The gap of the unit a_i (respectively b_j) corresponds to the non-association of this unit with any unit of the sequence $\mathbf{b}$ (respectively $\mathbf{a}$) (e.g., units a_2, a_4 and b_3 in Fig. 2). Gaps are associated to the occurrence of stratigraphic unconformities. A cost is associated to each possibility of correlation (match or gap) reflecting its likelihood (i.e. the higher the cost, the lower the correlation likelihood) and is computed according to specific correlation rules (lithostratigraphy, sequence stratigraphy, chemiostratigraphy for instance). Then, the matrix D whose axes correspond to the possible well sections is filled with these costs (Fig. 2).

The algorithm computes the least-cost pathway through the matrix corresponding to the most probable correlation. For this, the search of the minimal cost is applied recursively at each cell location. Waterman and Raymond [16] define three functions:

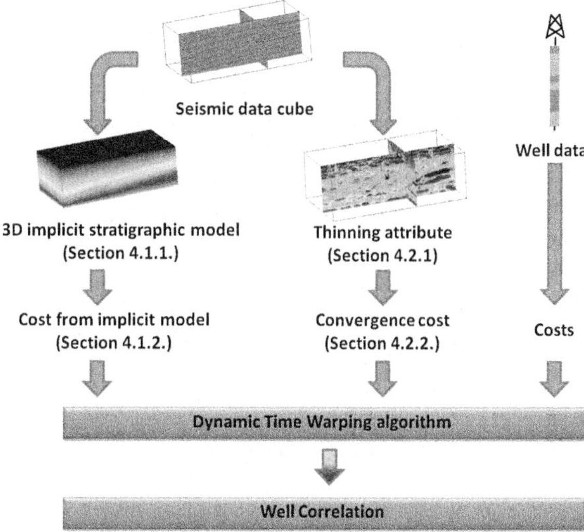

Fig. 1 The successive steps of a proposed workflow which takes interwell information from seismic data into account

- d: distance function which associates a cost to the matching between two units. For instance, if lithostratigraphic correlation rule is considered, d may be equal to the difference of the average grain size between the studied units.
- g: gap function which associates the cost of a gap to each stratigraphic unit. This cost may be stationary or inversely proportional to the unit thickness [16].
- D_{ij}: minimal distance function which associates the minimal sum of costs to the correlation between the strata $a_1, a_2, \dots, a_i$ and $b_1, b_2, \dots, b_j$. This cost is defined in [16] as

$$D_{ij} = \min\{D_{i-1,j} + g(a_i), D_{i,j-1} + g(b_j), D_{i-1,j-1} + d(a_i, b_j)\}, \tag{1}$$

with $D_{00} = 0$, $D_{i0} = \sum_{l=1}^{i} g(a_i)$, and $D_{0j} = \sum_{k=1}^{j} g(b_j)$.

To handle uncertainties on stratigraphic well correlation, a stochastic variation of the Dynamic Time Warping algorithm can be considered [11]. The expression of D_{ij} becomes

$$D_{ij} = S\begin{pmatrix} \alpha = D_{i-1,j} + g(a_i), \\ \beta = D_{i,j-1} + g(b_j), \\ \gamma = D_{i-1,j-1} + d(a_i, b_j) \end{pmatrix}, \tag{2}$$

where $S(\alpha, \beta, \gamma)$ is a value equal to cost α (respectively β or γ) with a probability inversely proportional to the relative cost $\frac{\alpha}{\alpha+\beta+\gamma}$ (respectively $\frac{\beta}{\alpha+\beta+\gamma}$ or $\frac{\gamma}{\alpha+\beta+\gamma}$).

Thus, the DTW algorithm allows either to find the optimal correlation, or to stochastically generate correlation models. In Sect. 3, a cost function based on well log trends is proposed and in Sect. 4 we present correlation rules computed from seismic reflection data for stratigraphic interpretation. The objective is threefold: (1) try to exploit the interwell information provided by seismic data to guide correlation; (2) use higher order trends extracted from the seismic data to define non-

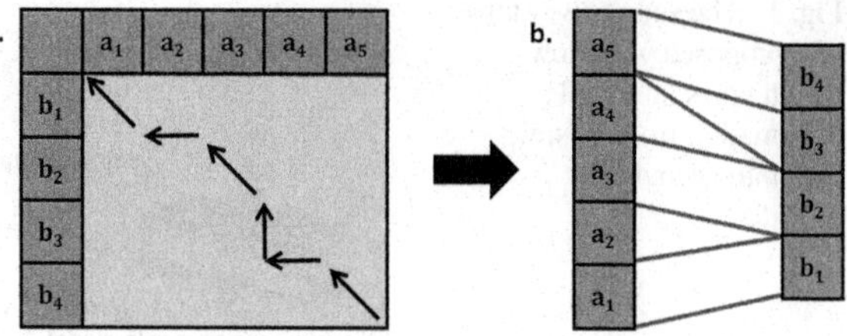

Fig. 2 The DTW algorithm. **a** Correlation pathway computed within the matrix D whose axes correspond to the stratigraphic sequences **a** and **b**. **b** Optimum correlation between the two stratigraphic sequences

stationary gap cost functions; (3) combine both borehole and seismic data to build stratigraphic well correlations.

3 Correlation Rule $\mathscr{R}_1$ Based on Well Log

The well logging tools are very useful to characterize underlying rocks. Several methods based on the comparison of well logs have been proposed to automatically correlate stratigraphic units [5, 8].

In the context of fluvio-deltaic sediments, sequence stratigraphic units could be correlated on the basis of vertical trends of shale contents and thus on gamma ray variation. Therefore, we propose a correlation rule $\mathscr{R}_1$ with a matching cost defined as: $d(a_i, b_j) = |s_i^a - s_j^b|$, where s_i^a (respectively s_j^b) is the linear regression slope of the gamma ray values in function of depth for the unit a_i (respectively b_j). The gap cost is defined constant and equal to the mean of all matching costs computed between considered units.

4 Correlation Rules from Seismic Data

Seismic reflection methods provide a low-resolution and noisy representation of subsurface heterogeneities. The seismic reflectors, that are often modeled as the convolution of sediment acoustic impedance contrast with a wavelet could, in many cases, be interpreted as bounding horizons of stratigraphic sequences [14]. Over the last decade, the development of seismic attributes and automated horizon extraction methods has considerably facilitated the seismic interpretation task. Using these tools, we propose two correlation rules to evaluate the likelihood of the correlation between stratigraphic units and the possibility of unconformities.

4.1 Correlation Rule $\mathscr{R}_2$ Based on a Draft Stratigraphic Model

Seismic reflector tracking is a common way to perform well to well stratigraphic correlation. This requires preliminary seismic-to-well tie (e.g. using checkshot

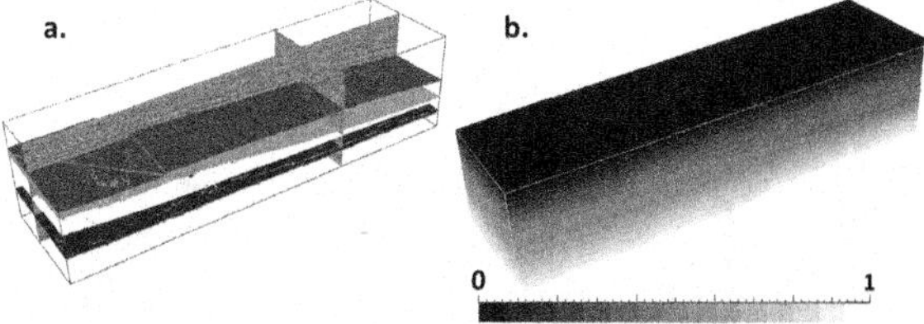

Fig. 3 Building an implicit 3D stratigraphic model highlighting the reflector geometry. **a** Extraction of horizons segments from seismic data. **b** Computation of the scalar field constrained by the extracted segments

data). However, the integration of horizon auto-tracking into automatic well correlation methods may not be simply done. Indeed automatic reflector tracking methods are sensitive to seismic quality and thus do not ensure building continuous horizon between considered wells. We propose to use the geometric information provided by the reflectors extraction to build a draft stratigraphic model to compute correlation cost.

4.1.1 Implicit Draft Stratigraphic Modeling from Seismic Data

The objective is the creation of continuous data between wells characterizing the geometry of the sedimentary deposits. To this end, we model the stratigraphy of the subsurface with an implicit approach, which considers geological interfaces as iso-surfaces of a 3D scalar field [9, 13]. Creating a 3D implicit stratigraphic model from seismic amplitude calls for two main steps (Fig. 3):

- The extraction of horizon segments from seismic cubes [1, 12] (Fig. 3.a).
- Implicit model building. The top and the bottom of the seismic cube are assigned constant scalar value, respectively 0 and 1. Then, a scalar field is built in such way that it respects two types of constraints: (1) a constant gradient, (2) iso-value constraints at the top and the bottom of the seismic cube, and at the extracted horizons segments (Fig. 3.b).

The implicit stratigraphic model reflects the 3D large-scale variations of the thicknesses of the stratigraphic units. In this paper, we suppose that no large-scale erosion occurred in the studied area, since the computation of the scalar field assumes continuous variation of thickness. However, erosions clearly visible at the seismic scale can be handled by using several scalar fields for each conformable sequence [2, 7].

This modeling method presents three main advantages: an implicit stratigraphic model can be rapidly created and only partial seismic interpretation is required. Furthermore, the implicit modeling makes easy to refine or to update the model with additional data.

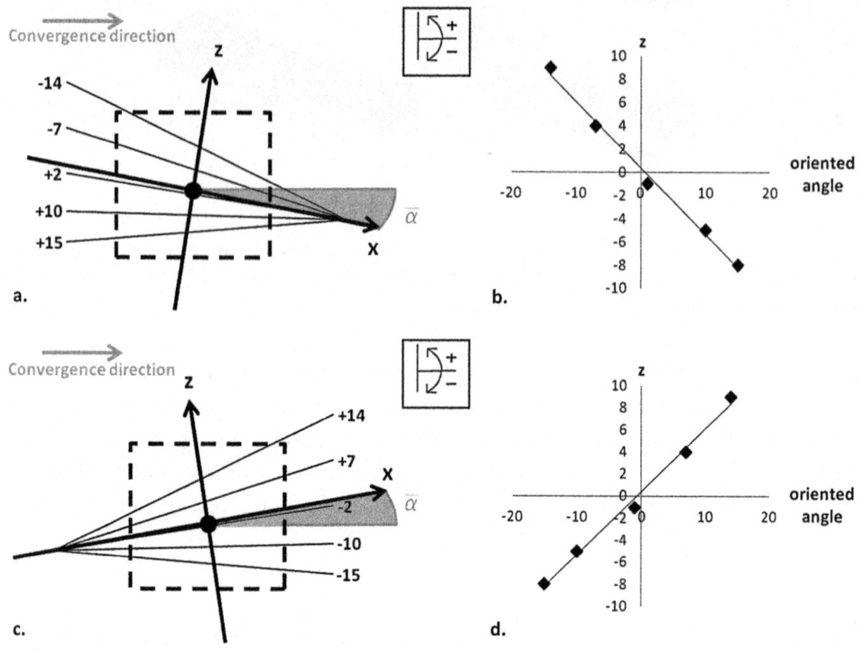

Fig. 4 Computation of the thinning attribute. **a** and **c**. The mean dip value $\bar{\alpha}$ is computed in the neighborhood (*dotted square*) of the studied point (*black dot*). The axis x and z are obtained rotating respectively the horizontal and vertical axes by $\bar{\alpha}$. The *arrows* show the direction considered as convergent. **b** and **d**. The oriented variations of the dip, relative to the dip mean, along the axis z are plotted. For the convergent case (**b**), the slope of the linear regression line is negative. Conversely, in divergent situations (**d**), the slope of the linear regression line is positive

4.1.2 Correlation Cost from Draft Stratigraphic Model

The 3D implicit stratigraphic model reflects the 3D geometry of the stratigraphic horizons, so two units with close values in the scalar field are probably nearer in the stratigraphic sequence than units with very different values.

We use this property to establish a new correlation rule allowing to estimate the likelihood of match, $d(a_i, b_j)$ (1) and (2): the closer the scalar field values of the units a_i and b_j, the lower the correlation cost between a_i and b_j is. The correlation cost between two stratigraphic units is defined as the absolute value of the difference of the mean values of the normalized scalar field in the two units. This rule provides a coarse geometric framework to guide well correlations.

4.2 Correlation Rule $\mathscr{R}_3$ Based on a Stratigraphic Seismic Attribute

The identification of the seismic reflection terminations (onlap, downlap, toplap, erosional truncation) allows to define reflection packages that delimit seismic se-

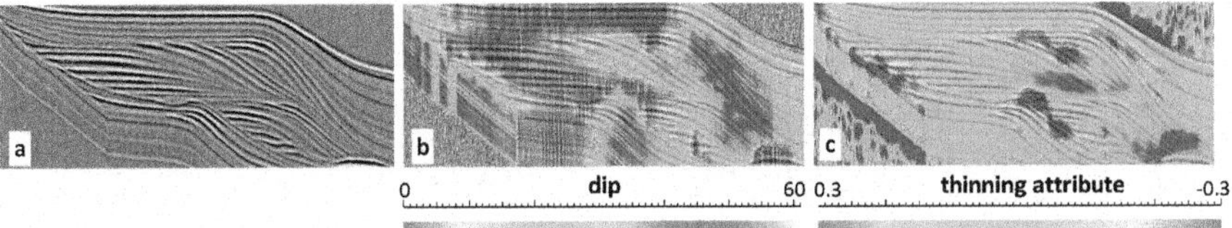

Fig. 5 **a** Synthetic version of Vail's [14] systems tract, from [15]. **b** The dip field. **c** The thinning attribute highlighting reflection terminations (i.e., onlap, downlap)

quences and systems tracts [14]. However, the exhaustive identification of reflection terminations is a tedious manual work. To facilitate the stratigraphic seismic interpretation, [15] introduce the thinning attribute which quantifies the degree of seismic reflection convergence (or divergence) of a seismic package over a moderate-to-large scale hence relates to unconformities. We propose using the convergence of seismic reflectors to compute non-stationary gap likelihood.

4.2.1 The Thinning Attribute

As pointed out by [15], the dip variations reflect the degree of convergence (respectively divergence) of the seismic reflectors. The reflection terminations are specifically highlighted as a result of the important variations of reflector inclinations.

At each point of a seismic section, we compute the value of the thinning attribute in three steps:

1. Computation of the dip mean $\bar{\alpha}$ in the neighborhood of the point.
2. Computation of the oriented variation of dip (relative to $\bar{\alpha}$). A new basis B is created. Its origin is the study point. The x and z axes of B are respectively equivalent to the horizontal and vertical axes, both rotated by $\bar{\alpha}$ (Fig. 4.a–c).
3. Computation of the degree of convergence (or divergence). An arbitrary direction is chosen to differentiate the convergence case from the divergence case (e.g., *the arrows* in Fig. 4 show the direction considered as convergent). From a dip vs. depth plot, the sign of the slope of the regression line allows to differentiate convergent strata from divergent strata. Moreover, the degree of convergence (divergence) is proportional to the slope of the regression line (Fig. 4.b–d).

The thinning attribute computed on a synthetic version of Vail's systems tract is presented in Fig. 5.

4.2.2 Correlation Cost from Thinning Attribute

In opposition to the cost of match which is naturally constrained by the data along the wellbore, the cost of a gap is often difficult to estimate. For instance, [16] define that a gap cost for a unit a_i (or b_j) is proportional to its thickness. However, many field examples show thin units spread over a wide area and thick units with a relatively small lateral extension.

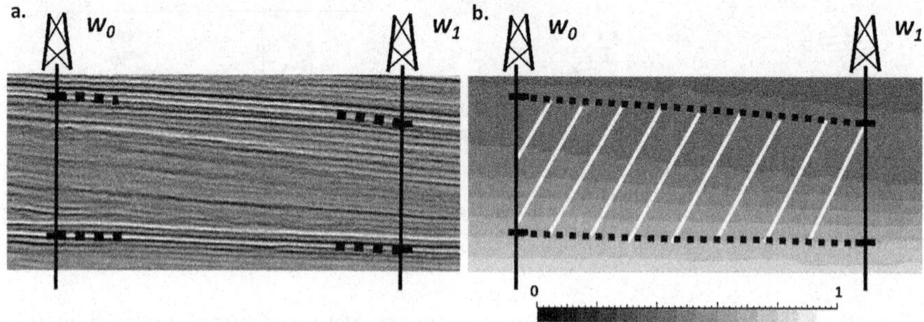

Fig. 6 Building study zone to compute the cost between two units. **a** The reflectors located at the top and the bottom of the two units are extracted locally (*black dotted lines*). The top reflectors are assigned the value v_1. The bottom reflectors are assigned the value v_2. **b** The gradient of the built scalar field is orthogonal to the extracted horizons. Vertically the study zone is delimited by the iso-values v_1 and v_2 (*white hatched area*)

We propose to establish a cost rule relating the likelihood of a gap between two units to the convergence degree in a zone of interest. The definition of this zone is performed for each calculation of cost between units a_i and b_j.

Consider two wells w_0 and w_1 where the stratigraphic units $\{a_1, \ldots, a_i, \ldots, a_n\}$ are identified along w_0 and $\{b_1, \ldots, b_j, \ldots, b_m\}$ along w_1. Currently, we use the following methodology to compute the costs for correlating the units a_i and b_j:

Laterally, the study zone is bounded by the two wells w_0 and w_1. To bound it vertically, the implicit modeling approach is used. For this, the seismic reflectors located at the top and the bottom of the two units are extracted locally (Fig. 6.a). The top horizons of units a_i and b_j are assigned the value v_1 and the bottom horizons are assigned the value v_2. Then, we build a scalar field whose gradient is orthogonal to the horizons previously extracted. Vertically, the study zone is delimited by the iso-values v_1 and v_2 of this scalar field (Fig. 6.b).

The direction of the convergence is assumed from w_0 towards w_1. The ratio between the thickness of a_i and the thickness of b_j is noted r_{ij} and a cut-off value α for r_{ij} is chosen in the interval $]0; 1]$. This value α allows to distinguish three cases:

1. $r_{ij} > \frac{1}{\alpha}$: the unit b_j is significantly thinner than the unit a_i. This case is considered as a convergent case (Fig. 7.a).
2. $r_{ij} < \alpha$: the unit a_i is significantly thinner than the unit b_j. This case is considered as a divergent case (Fig. 7.b).
3. $r_{ij} \leq \frac{1}{\alpha}$ and $r_{ij} \geq \alpha$: the units a_i and b_j have approximately the same thickness. This case is considered as a parallel case (Fig. 7.c).

The total area of the study zone defined between the units a_i and b_j is noted $\mathscr{A}_{ij}^t$. Within this zone, the areas of the convergent, divergent and parallel zones, respectively noted $\mathscr{A}_{ij}^c$, $\mathscr{A}_{ij}^d$ and $\mathscr{A}_{ij}^p$, are defined by two thinning attribute cut-off values β and γ. According to the values of r_{ij} and α, the costs of gap and matching between a_i and b_j are computed as follows:

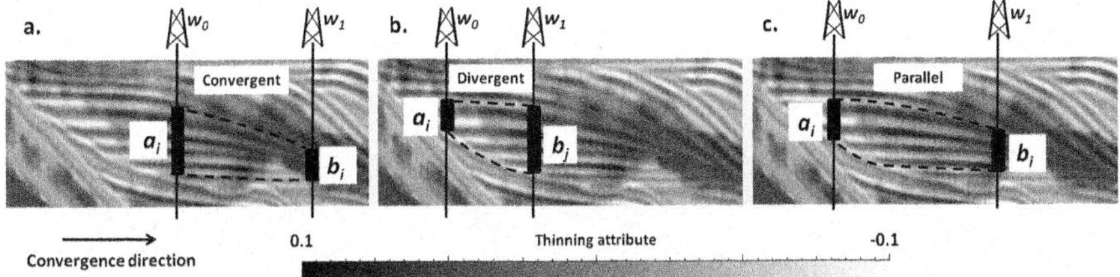

Fig. 7 The three cases for the calculation of a cost. The *arrow* shows the direction considered as convergent. **a** In the convergent case, the unit b_j is significantly thinner than the unit a_i. **b** In the divergent case, the unit a_i is significantly thinner than the unit b_j. **c** In the parallel case, the units a_i and b_j have approximately the same thickness

$$\begin{cases} d(a_i, b_j) = 1 - \dfrac{\mathscr{A}_{ij}^c}{\mathscr{A}_{ij}^t} \\ g(a_i) = 1 - \dfrac{\mathscr{A}_{ij}^c}{\mathscr{A}_{ij}^t} \\ g(b_j) = 1 - \dfrac{\mathscr{A}_{ij}^d}{\mathscr{A}_{ij}^t} \end{cases} \tag{3}$$

$$\begin{cases} d(a_i, b_j) = 1 - \dfrac{\mathscr{A}_{ij}^d}{\mathscr{A}_{ij}^t} \\ g(a_i) = 1 - \dfrac{\mathscr{A}_{ij}^c}{\mathscr{A}_{ij}^t} \\ g(b_j) = 1 - \dfrac{\mathscr{A}_{ij}^d}{\mathscr{A}_{ij}^t} \end{cases} \tag{4}$$

$$\begin{cases} d(a_i, b_j) = 1 - \dfrac{\mathscr{A}_{ij}^p}{\mathscr{A}_{ij}^t} \\ g(a_i) = 1 - \dfrac{\mathscr{A}_{ij}^c}{\mathscr{A}_{ij}^t} \\ g(b_j) = 1 - \dfrac{\mathscr{A}_{ij}^d}{\mathscr{A}_{ij}^t} \end{cases} \tag{5}$$

where (3), (4) and (5) correspond respectively to the convergent, divergent and parallel cases. These costs estimate the likelihood of the matching or gap of two units for each case (convergent, divergent or parallel). For instance, in the convergent case, the likelihood of matching between a_i and b_j is function of the proportion of the convergent area within the study zone. The same reasoning is applied to the other cases.

The computation of costs assumes a precise seismic-to-well calibration to limit uncertainties at the step where seismic horizons are extracted at the top and the

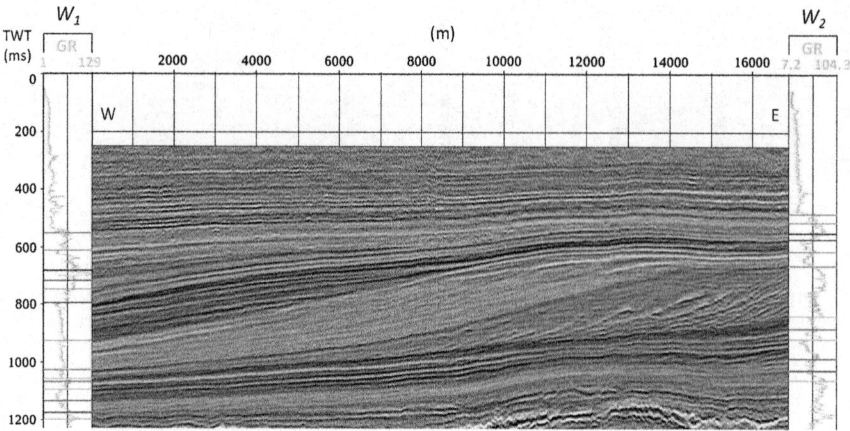

Fig. 8 Reference correlation proposed after the interpretation of well and seismic data. Along the wells w_1 and w_2, respectively 14 and 12 markers are identified

bottom of units. The seismic is also assumed to have a reasonable quality so that the area of interest is determined with confidence.

However, we can question the relevance of the computation of a correlation cost between two units far apart in the sedimentary sequences, and so with a likelihood of correlation very low. Indeed, the defined study zone can intersect the seismic reflectors, and in this case, the defined zone is not representative of the lateral variation of the units. Therefore, we only apply the rule $\mathscr{R}_3$ when the likelihood given by the rule $\mathscr{R}_2$ is deemed acceptable.

5 Application

The study is based on the Dutch North Sea data set provided by the seismic open repository (http://www.opendtect.org/osr/Main/NetherlandsOffshoreF3Block Complete4GB). Two wells w_1 and w_2 are considered, where respectively 13 and 11 sequence stratigraphic units are identified [4]. Along the two boreholes, the gamma ray has been recorded. The stratigraphic model proposed by [4] serves as a base case (Fig. 8).

First, the rule $\mathscr{R}_1$ is applied to the wells w_1 and w_2 in a deterministic way (1). The obtained stratigraphic correlation differs substantially from the base case (Fig. 9.a–b). The gaps are particularly improperly positioned. The well-based correlation does not respect the clinoform structures, clearly identifiable on the seismic data (Fig. 8).

Second, the seismic rules $\mathscr{R}_2$ and $\mathscr{R}_3$, and the well rule $\mathscr{R}_1$, are simultaneously used to compute deterministic, and stochastic, correlations (1) and (2) between the two wells w_1 and w_2. Each stratigraphic correlation is equally weighted by the three rules. The deterministic correlation is close to the base case (Fig. 9.a–c). The clinoform structures are respected. However, between 500 ms and 700 ms (TWT), the

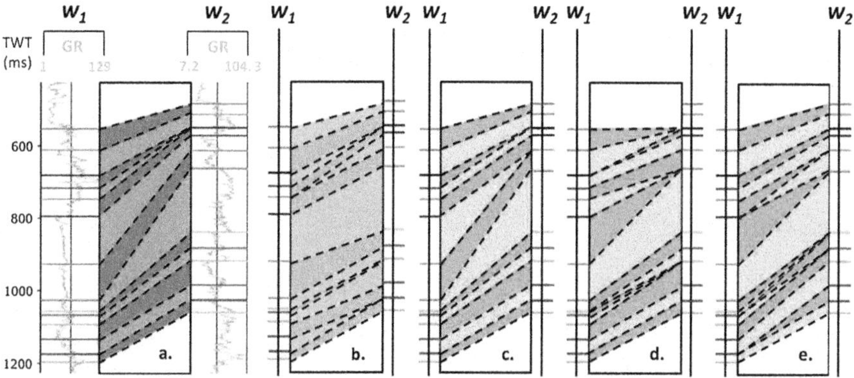

Fig. 9 Stratigraphic well correlations between the wells w_1 and w_2. **a** The reference stratigraphic correlation model [4]. **b** The most likely correlation built from the wellbore-based rule $\mathscr{R}_1$. **c** The most likely correlation built from the wellbore-based rule $\mathscr{R}_1$, and the seismic-based rules $\mathscr{R}_2$ and $\mathscr{R}_3$. **d–e** Two possible correlations stochastically built from the well-based rule $\mathscr{R}_1$, and the seismic-based rules $\mathscr{R}_2$ and $\mathscr{R}_3$

correlation is shifted by gaps improperly positioned. This slight shift highlights the difficulty to precisely correlate close markers from seismic information due to the difference of resolution between well and seismic data. Then different possibilities of stratigraphic correlations are computed with the stochastic approach (Fig. 9.d–e). The various realizations preserve the main sedimentary trends, although a high variability is observed for relatively thin stratigraphic units.

6 Conclusion

The main objective of our correlation method is to use interwell data to constrain the well stratigraphic correlation. The development of this method is made possible thanks to a new generation of geometric seismic attributes. Taking into account seismic data allows to discriminate possible stratigraphic correlation models. Then the use of a stochastic approach provides a sampling of these possibilities, and thus allows to better estimate the parameters characterizing both basin and reservoir.

A number of possible improvements for the seismic-based correlation method can be considered. In terms of uncertainty management, an obvious topic would be to account for well-to-seismic calibration uncertainty and more generally for velocity uncertainty. From a structural standpoint, the proposed method has currently no specific management of faults. A major way of improvement would then be to have a tighter coupling of this method to full-featured seismic interpretation tools so that fault could be handled. Hence, with slight modifications of the cost computation methodology, the cost based on the thinning attribute and the cost from the draft stratigraphic model would become applicable to the case of faulted data set. The poor quality of the seismic data, for instance in the context of highly fractured underlying rocks, remains a limiting factor.

Acknowledgements This work was performed in the frame of the GOCAD research project. The companies and universities members of the GOCAD consortium are acknowledged for their support. Furthermore, we acknowledge dGB Earth Sciences and TNO for the Netherlands F3 data set.

References

1. Bouchet P, Jacquemin P, Mallet JL (2002) Goscope project: extracting geological structures from seismic data. In: 22nd gocad meeting proceedings
2. Calcagno P, Chilès J, Courrioux G, Guillen A (2008) Geological modelling from field data and geological knowledge: part I. Modelling method coupling 3d potential-field interpolation and geological rules. Phys Earth Planet Inter 171(1–4):147–157
3. Collins DR, Doveton JH (1993) Automated correlation based on Markov of vertical successions and Walther's law. In: Computers in geology—25 years of progress, pp 121–132
4. De Bruin G, Bouanga E (2007) Time attributes of stratigraphic surfaces, analyzed in the structural and Wheeler transformed domain. In: 69th EAGE conference & exhibition.
5. Doveton JH (1994) Lateral correlation and interpolation of logs. In: Geologic log analysis using computer methods, pp 127–150
6. Doveton JH (1994) Theory and applications of time series analysis to wireline logs. In: Geologic log analysis using computer methods, pp 97–125
7. Durand-Riard P, Caumon G, Salles L, Viard T (2010) Balanced restoration of geological volumes with relaxed meshing constraints. Comput Geosci 36(4):441–452
8. Fang JH, Chen HC, Shultz AW, Mahmoud W (1992) Computer-aided well log correlation (1). Am. Assoc. Pet. Geol. Bull. 76(3):307–317
9. Frank T, Tertois A, Mallet J (2007) 3d-reconstruction of complex geological interfaces from irregularly distributed and noisy point data. Comput Geosci 33(7):932–943
10. Lallier F, Caumon G, Borgomano J, Viseur S (2009) Dynamic time warping: a flexible efficient framework for stochastic stratigraphic correlation. In: 29th gocad meeting proceedings
11. Lallier F, Viseur S, Borgamano J, Caumon G (2009) 3d stochastic stratigraphic well correlation of carbonate ramp systems. In: Society of petroleum engineers—international petroleum technology conference
12. Mallet JL, Jacquemin P, Labrunye E (2002) On the use of trigonometric polynomials in seismic interpretation. In: 22nd gocad meeting proceedings
13. Moyen R, Mallet JL, Frank T, Leflon B, Jean-Jacques R (2004) 3D-parameterization of the 3D geological space—the GeoChron model. In: Proc. European conference on the mathematics of oil recovery (ECMOR IX)
14. Vail PR (1987) Seismic stratigraphy interpretation using sequence stratigraphy. Part 1: Seismic stratigraphy interpretation procedure. In: Atlas of seismic stratigraphy. Studies in geology, vol 27. AAPG, Tulsa, pp 1–10
15. Van Hoek T, Gesbert S, Pikens J (2010) Geometric attributes for seismic stratigraphic interpretation. Lead Edge 29:1056–1065
16. Waterman H, Raymond R (1987) The match game: new stratigraphic correlation algorithms. Math Geol 19(2):109–127

Some Newer Algorithms in Joint Categorical and Continuous Inversion Problems Around Seismic Data

James Gunning and Michel Kemper

Abstract Conventional geophysical inversion tools often use purely continuous optimization techniques that model rock properties as if they come from some common population, even though geological formations usually have a strong mixture character. Such pooling imposes a strong prior-model footprint on inversion results. Newer hierarchical Bayesian approaches that embed a categorical/facies aspect via discrete Markov random fields, coupled with conditional prior distributions that embed rock-physics relationships, are a tractable way to represent the categorical aspects of geology. We show that maximum a posteriori model inference in joint lithology-fluid/rock-properties problems using seismic is possible using some newer algorithms from computer vision. The optimization is cast as an EM algorithm, using Bayesian Belief Propagation as the "E" step, and conventional large-scale least squares as the "M" step. Very fast approximate alternatives to the "E" step are available using graph-cutting algorithms.

1 Introduction

It is common in the oil and gas business to subject seismic data to a sequence of cascaded inversion calculations, the most ambitious of which seek to make inferences about rock and fluid properties which are only indirectly related to elastic properties. This cascading of inferences is always somewhat distasteful to statisticians, who naturally prefer to work with 'raw data', but this partitioning of processing and inversion workflows is likely to remain a permanent feature of the working environment in oil and gas companies. As such, it is common for reservoir characterization teams to be expected to work with seismic "data" which are inverted images of "true-amplitude bandpass reflectivity", with all 3D effects removed.

J. Gunning (✉)
CSIRO ESRE, Melbourne, Australia
e-mail: James.Gunning@csiro.au

M. Kemper
Ikon Science, Teddington, UK
e-mail: mkemper@ikonscience.com

P. Abrahamsen et al. (eds.), *Geostatistics Oslo 2012*,
Quantitative Geology and Geostatistics 17,
DOI 10.1007/978-94-007-4153-9_21,

Imaging to reflectivity has the advantage of delivering a result that is somewhat more robust to the missing information content in the seismic experiment. The bandwidth of the sources and spatial resolving power of the acquisition are well known to lose both high frequency content (say > 100 Hz) and a low frequency gap typically from 2 to 10 Hz [14]. Further, though great progress has been made in the imaging community in the last 15 years, particularly in the area of full waveform inversion [14], the computational demands of these methods are so great that it is unlikely that very meaningful statements about the image uncertainties will be available. Hence, reservoir characterization groups regard these images as "pseudo-data", and usually proceed with them as the basis for a subsequent "AVO" inversion problem for rock properties using a simplified vertical model of wave propagation, e.g. 1D convolution.

These AVO inversion routines tend to use some kind of least squares minimization based on linearizion of the reflectivity response. In this limit, a linearized forward model for seismic data $\mathbf{y}$ from the finer-scale elastic properties model $\mathbf{m}$ (gridded on a time-lattice) is going to look something like (per seismic stack)

$$\mathbf{y} = WAD\mathbf{m} + \boldsymbol{\varepsilon},$$

where the noise is $\boldsymbol{\varepsilon}$, D represents discrete time differentiation, A contains Aki-Richards-like reflectivity coefficients, and W is the wavelet convolution. The effective design matrix $X = WAD$ has an unstable pseudo-inverse, $(X^T X)^{-1} X^T$, which reflects the frequency content loss mentioned above, but also weak or poor sensitivity to some components of the model vector $\mathbf{m}$, e.g. shear properties.

The normal cure for the instability mentioned above is to frame the inversion problem in a Bayesian setting, writing down approximate beliefs for the rock properties in depth as a prior distribution. The noise distribution is usually modeled as Gaussian $N(\mathbf{0}, C_d)$, and if we are willing to accept a Gaussian model for the prior distribution $N(\bar{\mathbf{m}}, C_p)$, a lot of analytical tricks become available. This second assumption is usually acceptable for a particular rock type and fluid, in a depth window that is not too large: it is often reasonable to embed the dependencies between effective stress (or a depth surrogate), v_p, v_s and density or porosity in a correlated Gaussian distribution. The problem is that most seismic energy comes from the jumps between facies, so to use these analytical tricks, we have to resort to writing the prior distribution as a "umbrella" Gaussian that covers all the rock types: see Fig. 1. The Bayesian least-squares inversion then looks like $\hat{\mathbf{m}} = (X^T C_d^{-1} X + C_p^{-1})^{-1} (X^T C_d^{-1} \mathbf{y} + C_p^{-1} \bar{\mathbf{m}})$, this relation has the property that when we are not close to a seismic event ($\mathbf{y} \approx \mathbf{0}$), the model gets forced back to the mean $\bar{\mathbf{m}}$ which is a compromise or "mongrel" rock chosen only for convenience, and possibly not consistent with any of the rocks in the geology.

Thus, although the use of linear likelihoods and fully multi-Gaussian priors enables impressively fast inversion frameworks like that of [3], the maximum a posteriori inferences from these carry a strong footprint from the prior model, and samples drawn from the posterior will certainly not capture the mixture character typical of sedimentary formations.

It is for this reason that much recent research has been devoted to hierarchical discrete/continuous models for seismic inversion [5, 9, 13]. In these approaches, the

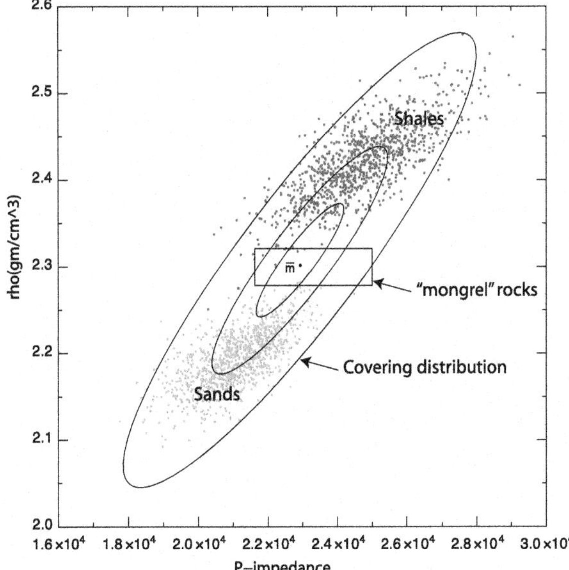

Fig. 1 Typical impedance clusters for sand and shale facies, and representative "umbrella" distribution that is used as a Bayesian prior in purely continuous inversion approaches

elastic properties of particular facies or fluid-substituted rock types are conditionally dependent on a co-located spatial facies label F, so the overall Bayesian hierarchical model looks like $P(\mathbf{m}, F|\mathbf{y}) \sim P(\mathbf{y}|\mathbf{m})P(\mathbf{m}|F)P(F)$. The main difficulty lies in choosing a tractable form for the high dimensional discrete distribution $P(F)$. Of the various popular methods for categorical modeling, very few offer tractable forms for making optimization or sampling efficient in this context. We think marked-point models or discrete Markov Random Fields (MRFs) [16] are among the few choices that are likely to be workable, and concentrate on MRFs in this paper.

The papers [9, 13] use MRFs for the discrete prior model, and describe a block-wise Gibbs sampler for sampling from the posterior. One of their intriguing findings is that the posterior uncertainty in facies from dense seismic information does not appear to be very great. This means that a MAP inference for facies and rock properties is likely to be very useful, furnishing a single "best guess" model which is often required by users. Hence, in this paper, we concentrate on the problem of efficiently locating MAP inferences for these large-scale joint discrete/continuous problems. Parameter inference is a topic we leave aside at present, while acknowledging that it is known to be a difficult problem [16]. Sampling is also not discussed, though we remark that efficient optimization techniques are usually the key to approximate sampling of very high dimensional problems via bootstrapping techniques, and these may be the only practical approximation when the dimensionality-curse associated with MCMC methods becomes unacceptable with large models.

2 Theory

The inversion problem we solve here is for a suite of continuous rock properties $\mathbf{m} = \{\mathbf{v}_p, \mathbf{v}_s, \boldsymbol{\rho}\}$ together with facies labels F, $F_i \in \mathcal{L} \equiv [0, 1, \ldots, N_f - 1]$ at all the sample locations in the volume (usually the xyt lattice defined by the seismic). The prior distribution of rock properties is a joint distribution assembled from the rock physics regressions developed from regional log data. This distribution is facies dependent, so we write this as $P(\mathbf{m}|F)$. The facies labels are equipped with a discrete prior MRF distribution $P(F)$ over the volume. The underlying graph on which the MRF is defined is aligned with the natural stratigraphy in the inversion domain. Thus, given the data $\mathbf{y}$, the forward model $\mathbf{f}(\mathbf{m})$, and a suitable likelihood function yielding the likelihood of the data given the model $L(\mathbf{y}|\mathbf{m}) = L(\mathbf{y}|\mathbf{f}(\mathbf{m}))$, the joint posterior distribution of the model $\mathbf{m}$ and F can be written

$$P(\mathbf{m}, F|\mathbf{y}) \sim L(\mathbf{y}|\mathbf{m})P(\mathbf{m}|F)P(F). \tag{1}$$

For convenience, we choose the priors to be of multi-Gaussian form:

$$P(\mathbf{m}|F) \sim \frac{\exp[-\frac{1}{2}(\mathbf{m} - \bar{\mathbf{m}}(F))^T C_p^{-1}(F)(\mathbf{m} - \bar{\mathbf{m}}(F))]}{|C_p(F)|^{1/2}}, \tag{2}$$

where $\bar{\mathbf{m}}$ and C_p are assembled to capture the rock physics regressions for each rock type. Also, for convenience, this is written as a product over all lattice sites ("blocks"). The likelihood is also conveniently chosen to be of Gaussian form:

$$L(\mathbf{y}|\mathbf{m}) = L\big(\mathbf{y}|\mathbf{f}(\mathbf{m})\big) \sim \frac{\exp[-\frac{1}{2}(\mathbf{y} - \mathbf{f}(\mathbf{m}))^T C_d^{-1}(\mathbf{y} - \mathbf{f}(\mathbf{m}))]}{|C_d|^{1/2}}. \tag{3}$$

Here, C_d is a suitable noise covariance matrix, often diagonal with entries σ_j for the RMS noise level of the samples for stack j unrolled into $\mathbf{y}$.

The MRF for the facies labels we write as a standard pairwise Potts model

$$P(F) \sim \exp\left(-\frac{1}{2}\sum_i \sum_{j\sim i} \beta_{F_i,F_j} I(F_i, F_j)\right),$$

where $j \sim i$ denotes neighbors j of i, $I(F_1, F_2)$ is the indicator function ($I(F_1, F_2) = 1$, $F_1 = F_2$, otherwise 0), and the βs are a suitable set of penalty energies. We also use an equivalent form (though the parameter mappings are nontrivial)

$$P(F) \sim \prod_i \left[w_{F_i} \exp\left(-\frac{1}{2}\sum_{j\sim i} \beta_{h,v} I(F_i, F_j)\right)\right],$$

where the w_l, $l \in \mathcal{L}$ are pseudo-proportions and the β_h, β_v are horizontal/vertical facies discontinuity parameters shared by all labels for "permissible" transitions. Unphysical fluid state transitions are conceivable if the discrete classes include lithologies with mobile fluids. We embed heavy penalties in the appropriate β parameters to preclude these.

Hence, the maximization of the posterior (1) is equivalent to minimizing the energy-like exponent

$$E(\mathbf{m}, F) = \left(\mathbf{y} - \mathbf{f}(\mathbf{m})\right)^T C_d^{-1}\left(\mathbf{y} - \mathbf{f}(\mathbf{m})\right) + \left(\mathbf{m} - \bar{\mathbf{m}}(F)\right)^T C_p^{-1}(F)\left(\mathbf{m} - \bar{\mathbf{m}}(F)\right)$$
$$+ \log\left|C_p(F)\right| - \sum_i \sum_{j\sim i} \beta_{F_i,F_j} I(F_i, F_j) \qquad (4)$$

jointly over the spaces of $\mathbf{m}$ and F.

A simple toy limit of this problem is quite sobering: take the case that all rock types share a common prior dispersion C_p, and $\mathbf{f}(\mathbf{m}) = X\mathbf{m}$. Since the minimum of (4) in the subspace of $\mathbf{m}$ is available analytically, it can be substituted back in to leave a discrete minimization problem in the F-space that is quadratic in the term $\bar{\mathbf{m}}(F)$. In the two label case ($F_i \in [0, 1]$), this leaves a binary quadratic programming (BQP) problem for F, which is known to be NP-hard. In this limit, the matrix is also dense, which is known to be the most perverse case of BQP problems, in the sense of being most resistant to heuristics. The off-diagonal elements are also mainly negative, expressing anti-correlation, which also renders the energy function *non-submodular* and thus unsuitable for graph cutting approaches [7].

Nonetheless, summoning courage, we write down an optimization framework as an EM algorithm [4]. For the "E" step, the expectation of (4) with respect to F is required, for fixed $\mathbf{m}$. This is usually written

$$\begin{aligned} Q(\mathbf{m}|\mathbf{y}) &= \langle -2\log P(\mathbf{m}, F)\rangle_F \\ &= \left(\mathbf{y} - \mathbf{f}(\mathbf{m})\right)^T C_d^{-1}\left(\mathbf{y} - \mathbf{f}(\mathbf{m})\right) \\ &\quad + \sum_{\text{blocks } i} z_{ij}\left[(\mathbf{m}_i - \bar{\mathbf{m}}_{ij})^T C_{p,ij}^{-1}(\mathbf{m}_i - \bar{\mathbf{m}}_{ij}) + \log|C_{p,ij}|\right] + h(\beta, \mathbf{z}) \end{aligned} \qquad (5)$$

for some function $h()$ not directly dependent on $\mathbf{m}$. Here the z_{ij} are marginal probabilities ("memberships") of label j at grid-block i.

2.1 The Full EM Algorithm

For the "E" step, the memberships z_{ij} are estimable from a converged cycle of loopy Bayesian belief propagation (LBBP): see the Appendix for the details. The "M" step of the EM algorithm consists in minimization of (5) for fixed z_{ij} in the subspace of continuous $\mathbf{m}$. This amounts to a large scale nonlinear least squares problem which we solve using a Gauss-Newton method with line search backtracking. Specifically, the general nonlinear forward model is of form

$$\mathbf{y}(\mathbf{m}) = \begin{pmatrix} \mathbf{S}_1(\mathbf{m}) \\ \mathbf{S}_2(\mathbf{m}) \\ \vdots \end{pmatrix} = \begin{pmatrix} W_1 & 0 & 0 \\ 0 & W_2 & 0 \\ 0 & 0 & \ddots \end{pmatrix} \cdot \begin{pmatrix} \mathbf{R}_1(\mathbf{m}) \\ \mathbf{R}_2(\mathbf{m}) \\ \vdots \end{pmatrix}, \qquad (6)$$

where, for stack j, $\mathbf{S}_j$ is the synthetic seismic, W_j is a matrix representing convolution from the wavelet, and $\mathbf{R}_j(\mathbf{m})$ is the vector of reflectivities as a function of the model vector $\mathbf{m}$. The Jacobian is thus $X \equiv \frac{\partial \mathbf{y}}{\partial \mathbf{m}} = \text{diag}\{W_j\}\{J_1^T, J_2^T, \ldots\}^T$ where the Jacobian matrices $J_j = \partial \mathbf{R}_j / \partial \mathbf{m}$ are block matrices of form $J_j = [J_{jp}|J_{js}|J_{j\rho}]$, and each of these sub-blocks is a sparse (diagonal and superdiagonal only) matrix. These submatrix elements are routinely derived from various appropriate reflectivity models: we have implemented the Bortfield, Fatti, and isotropic Zoeppritz models, for catholicity. The Gauss-Newton point based on the current iterate $\mathbf{m}_0$ in the optimization is thus

$$\mathbf{m}' = \left(X^T C_d^{-1} X + C_p^{-1}\right)^{-1} \left(X^T C_d^{-1} \left(\mathbf{y} - \mathbf{y}(\mathbf{m}_0) + X\mathbf{m}_0\right) + C_{\text{eff},p}^{-1} \bar{\mathbf{m}}_{\text{eff}}\right). \quad (7)$$

Here the "effective" prior mean and dispersion have (blockwise over i) components given by

$$C_{\text{eff},p,i}^{-1} \equiv \sum_j z_{ij} C_{p,ij}^{-1}, \quad (8)$$

$$C_{\text{eff},p,i}^{-1} \bar{\mathbf{m}}_{\text{eff},i} \equiv \sum_j z_{ij} C_{p,ij}^{-1} \bar{\mathbf{m}}_{ij}, \quad (9)$$

which is a precision and membership re-weighted average of the facies means—an intuitively pleasing result. Since solving (7) is a large scale problem, computation of this point is performed (after some rescaling) using the conjugate gradient LSQR algorithm [12], since the coefficient matrix $X^T C_d^{-1} X + C_p^{-1}$ can participate in matrix-vector products using efficient sparse operations. The usual Armijo backtracking apparatus [6] is then applied in a line-search from $\mathbf{m}_0$ to $\mathbf{m}'$.

2.2 *Graph Cutting Algorithms*

A known greedy limit of the EM algorithm is the K-means algorithm, where, for the "E" step, the expectation in (5) is approximated by the value at the *mode*, i.e. $Q(\mathbf{m}|\mathbf{y}) \approx -2 \log P(\mathbf{m}, \hat{F})$, where $\hat{F}$ is the minimizer of (4) at the current, fixed $\mathbf{m}$. This corresponds to snapping memberships z_{ij} to 0 or 1. The mode of this discrete distribution is also computable using belief propagation algorithms (using the "max-product" variety). But for discrete energy functions with submodular character, like (4) with fixed $\mathbf{m}$, significantly faster and high quality solutions based on graph cutting (GC) methods [2, 7] are considered state-of-the-art. For a small number of facies, these algorithms scale almost linearly in the system size. From another point of view, the K-means approach consists in alternating minimizations in the F and $\mathbf{m}$ spaces, or a species of iterated conditional modes optimization.

EM algorithms produce parameter estimates at the mode of the marginal distribution obtained by integrating/summing out the latent variables. When we use LBBP, we form a pointwise facies-label estimate by greedy Bayesian classification based on the converged EM estimate in $\mathbf{m}$. Conversely, the graph cutting algorithm can be seen as a joint maximum likelihood estimate. In general these estimates are not the same, but are often qualitatively similar.

The EM algorithm is a local optimization method. Since the underlying problem is doubtless NP-hard, we are not entitled to expect an efficient global method. Our experience so far is that the LBBP method has a modestly wide basin of attraction around decent local minima, so reasonable choices of starting model appear to work well. Typically we set the starting model as an abundance weighted average of the facies means. The graph cutting algorithm, being greedier, is more vulnerable to local minima and starting choices can be rather delicate.

3 Examples

First we consider a 1D example. Figure 2 shows a typical example of how the EM algorithm evolves at a trace, for a 2-facies model aimed at inverting near-stack data produced by a sand slab embedded in a shale background.

Figure 3 shows a 3-facies example using a section across some NW-shelf Australian data (Stybarrow field). Here the MRF parameters $\beta_v = \beta_h = \beta$ are all lumped. This small cross-section ($\approx 3 \times 10^4$ parameters) is about a second on contemporary machines (about 3Gflop) using GC methods. The 3D case ($\approx 5 \times 10^5$) is about 5 minutes using LBBP. Usually we find LBBP about 5–10 times slower than GC for the E-step, and M-step/E-step work times about comparable using GC.

4 Discussion

It is prudent to recall that the principal sensitivity of the seismic data is to p-impedance Z_p. One should not expect a discrimination of facies very much better than the number of mixture components which are readily distinguishable in multi facies plots of the prior marginal distributions of Z_p. For example, at the loading crossover where shale and sand impedances match, one would require signals from the AVO shear terms comfortably larger than the noise levels to hope for facies discrimination on the basis of shear alone, and this is a demanding requirement.

Of course, with far-offset data, the sensitivity to shear improves somewhat, but it is still comparatively weak. Moreover, a number of other physical artifacts such as anisotropy commonly infect far-offset images (see [1] for an extended discussion), so the amplitudes in these images are much less reliable. We feel a sensible limit on the number of detectable facies is probably 2–4, depending on the data and prior distributions. We agree with Eidsvik et al. [5] that the dominant parameters in the problem are the seismic signal-to-noise ratios, and the coupling (β) plus proportions ratios in the MRF.

Belief propagation is known to be exact on trees, but approximate on graphs with cycles. The discrete model formulation clearly contains some artistry and liberty, and, intriguingly, LBBP is known to provide marginals for an effective energy

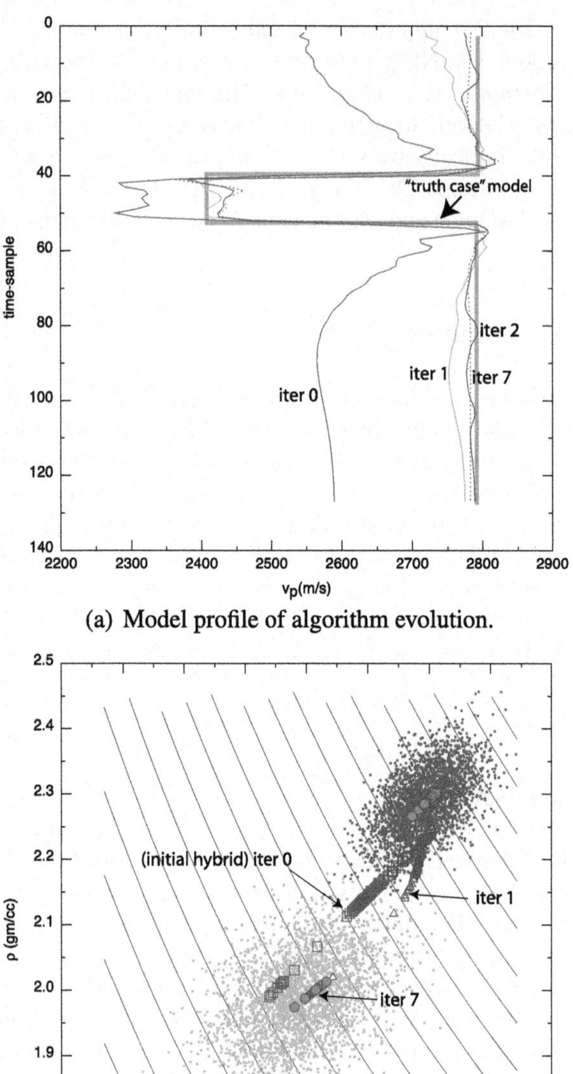

(a) Model profile of algorithm evolution.

(b) Scatterplots of mixture prior under algorithm evolution.

Fig. 2 *Top*: vertical profile of algorithm evolution with "truth case" model shown. Iteration 0 is the normal response from "pooled Gaussian" inversions, with low-frequency artifacts. *Bottom*: algorithm evolution down trace superposed on *red* and *green* clusters from rock physics prior for each facies. Contours are of p-impedance, so the forward model null space corresponds to translations of any model point along these contours

known as the Bethe free energy [15]. For weak or moderate couplings, the approximate marginals computed by LBBP are very good. For strong couplings, it may not converge at all, but this is also a regime highly resistant to all MCMC sampling tech-

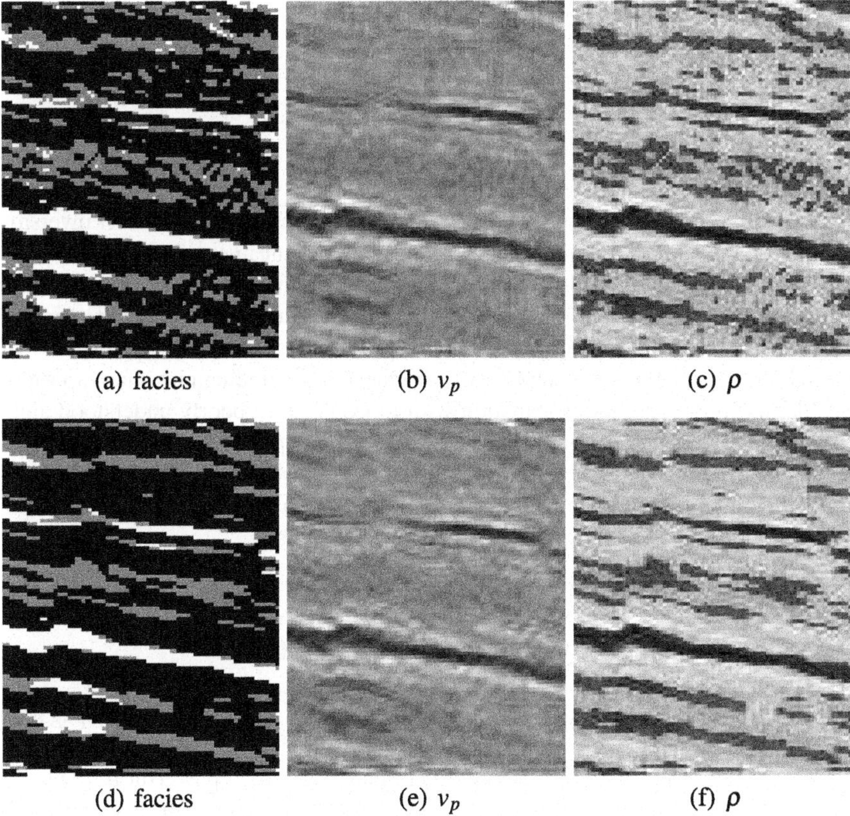

Fig. 3 Stybarrow inversion: cross section through well, showing importance of MRF smoothing of labeling: *top figures* are inversions with $\beta = 0$, *lower* with $\beta = 0.5$

niques except for special methods like Swendson-Wang [16] that depend strongly on certain symmetries in the model formulation that we cannot sustain in these models. It is also known [11] that for symmetric 2D Ising systems, the algorithm exhibits a phase transition at $\beta \approx 0.347$ (compared to the known actual phase transition at $\beta \approx 0.44$). Except on very small systems, significant convergence difficulties (e.g. starting point dependence) for stronger couplings than this can be expected in general if the data likelihood is weak (e.g. high noise level). However, perhaps because the seismic data is dense and informative, we find the algorithm converges well outside this range, albeit more slowly. The energy penalties for fluid ordering routinely cause more trouble for LBBP than graph cutting.

There are many open questions and possible extensions. It is clear that introduction of "hard data" values of F and $\mathbf{m}$ is a routine problem. So is the use of very low frequency kinematic velocity data routinely developed in seismic imaging. Extension of the prior spatial model for $\mathbf{m}$ to a continuous Gaussian MRF on the same graph as the facies model is also possible, for which pairwise marginals of the discrete MRF will be necessary. These can also be rendered from LBBP providing the pairs are neighbors, which will be the case.

5 Conclusion

Mixed categorical-continuous frameworks are an attractive idea for AVO inversion problems, since facies inferences are centrally important in reservoir characterization. MRFs are one of the few analytically amenable models for the discrete modeling, and lead naturally to hierarchical models on a undirected graph. The newer methods in computer vision such as belief propagation and graph cutting are extremely useful ingredients in an EM-like framework for finding maximum aposteriori like inversions. Belief propagation with loops is still a poorly understood algorithm, so the behavior of these schemes under strong spatial coupling is uncertain. But seismic data is dense and informative, so we believe useful calculations can be run with milder couplings, enabling the prior model of a discrete mixture with depth trends to infill the missing spectral content in seismic data.

Appendix: Loopy Bayesian Belief Propagation

Good introductions to these ideas can be found in [8, 10, 11, 15]. Belief propagation is known to embrace a variety of special algorithms, such as the Viterbi and forward-backward algorithm (which is used in [13]), and even broad powerful frameworks like the EM algorithm and Kalman filters. For our purposes, the basic "sum-product" algorithm is set out as follows. For probabilities of the form

$$P(F) = \prod_i g_i(F_i) \prod_{j \sim i} u_{ij}(F_i, F_j) \tag{10}$$

defined on an undirected graph where $j \sim i$ denotes "neighbors", messages m_{ij} are propagated along edges. Messages are updated using

$$m'_{ij}(F_j) = \sum_{F_i} g_i(F_i) u_{ij}(F_i, F_j) \prod_{\{k \sim i\} \setminus j} m_{ki}(F_i)$$

iterated over all sites in the lattice until convergence. Here $m_{ij}(F_j)$ is the message from vertex i to j about the state F_j, and $\{k \sim i\} \setminus j$ denotes neighbors of i excluding j. The null product (absence of any neighbors) defaults to 1. At initialization, all messages $m_{ij}(F_j)$ are initialized to 1. For stability, messages are often damped by a backtracking-like device like $m'_{ij}(F_j) \leftarrow m'_{ij}(F_j) + \alpha(m'_{ij}(F_j) - m_{ij}(F_j))$, $0 < \alpha < 1$, with e.g. $\alpha = \frac{1}{2}$. Messages are commonly normalized after each iterations such that $\sum_{F_j} m_{ij}(F_j) = 1$. Posterior marginals are then approximated by

$$P_i(F_i) \sim g_i(F_i) \prod_{k \sim i} m_{ki}(F_i),$$

where normalization is required such that the marginals sum to unity, $\sum_{F_i} P_i(F_i) = 1$. These marginals produce the z_{ij} memberships used in the EM algorithm. A variety of message-updating schedules are possible; we use a simple flooding scheme that updates messages simultaneously. A very closely related algorithm known as the Viterbi or "max-product" (all 'sum' operations above replaced with 'max') can find the modal configuration $\hat{F}$, but we usually find it is not competitive with graph cutting.

References

1. Avseth P, Mukerji T, Mavko G (2005) Quantitative seismic interpretation. Cambridge University Press, Cambridge
2. Boykov Y, Kolmogorov V (2004) An experimental comparison of min-cut/max-flow algorithms for energy minimization in vision. IEEE Trans Pattern Anal Mach Intell 26(9):1124–1137
3. Buland A, Kolbjornsen A, Omre H (2003) Rapid spatially coupled AVO inversion in the Fourier domain. Geophysics 68(3):824–883
4. Dempster AP, Laird NM, Rubin DB (1977) Maximum likelihood from incomplete data via the em algorithm. J R Stat Soc B 39(1):1–38
5. Eidsvik J, Avseth P, Omre H, Mukerji T, Mavko G (2004) Stochastic reservoir characterization using prestack seismic data. Geophysics 69(4):978–993
6. Fletcher R (1987) Practical methods of optimization. Wiley, New York
7. Kolmogorov V, Zabih R (2002) What energy functions can be minimized via graph cuts? In: Proceedings of the 7th European conference on computer vision—part III, ECCV '02. Springer, London, pp 65–81
8. Kschischang F, Frey B, Loeliger HA (2001) Factor graphs and the sum-product algorithm. IEEE Trans Inf Theory 47(2):498–519
9. Larsen AL, Ulvmoen M, Omre H, Buland A (2006) Bayesian lithology/fluid prediction and simulation on the basis of a Markov-chain prior model. Geophysics 71(5):R69–R78
10. Mackay DJC (2002) Information theory, inference & learning algorithms, 1st edn. Cambridge University Press, Cambridge
11. Mooij J, Kappen H (2007) Sufficient conditions for convergence of the sum-product algorithm. IEEE Trans Inf Theory 53(12):4422–4437
12. Paige CC, Saunders MA (1982) LSQR: an algorithm for sparse linear equations and sparse least squares. ACM Trans Math Softw 8:43–71
13. Ulvmoen M, Omre H (2010) Improved resolution in Bayesian lithology/fluid inversion from prestack seismic data and well observations: Part 1—methodology. Geophysics 75(2):R21–R35
14. Virieux J, Operto S (2009) An overview of full-waveform inversion in exploration geophysics. Geophysics 74(6):WCC1–WCC26
15. Wainwright MJ, Jordan MI (2008) Graphical models, exponential families, and variational inference. Found Trends Mach Learn 1:1–305
16. Winkler G (2003) Image analysis, random fields and Markov chain Monte Carlo methods: a mathematical introduction, 2nd edn. Springer, Berlin

Non-random Discrete Fracture Network Modeling

Eric B. Niven and Clayton V. Deutsch

Abstract Discrete fracture networks (DFNs) are commonly created as stochastic models of fractures in a rock mass. Most existing computer codes for creating DFNs generate fracture centroid locations randomly (with a Poisson process) and draw orientation independently of location. The resulting fracture networks do not have realistic spatial properties compared to the natural fracture networks they intend to model. DFNs generated in this manner commonly show fractures that are unrealistically close together and may have many more fracture intersections than are expected. This paper presents a new approach to DFN simulation that results in DFNs that are more geologically realistic in that target spatial statistics such as local fracture spacing, deviation in local fracture orientation and the number of fracture intersections are honored. The proposed algorithm relies on generating more fractures than are required and iteratively adding or removing fractures to find a subset that matches target input fracture network statistics.

1 Introduction

Research shows that as much as 60 % of the world's petroleum reservoirs are in naturally fractured reservoirs (NFRs) [1, 6, 9]. A NFR is a reservoir where the fractures in the reservoir rock have, or are predicted to have, a significant effect on reservoir fluid, either in terms of flow rates or flow anisotropy [3]. As production from convenient reserves continues to decline, producers are increasingly turning towards more challenging and non-traditional sources of petroleum such as oil sands or NFRs.

In order to optimize the management of NFRs, detailed information on the behavior, attributes and properties of the fracture network and rock matrix must be determined. The main challenge facing geologists and geostatisticians is that fracture

E.B. Niven (✉) · C.V. Deutsch
School of Mining and Petroleum Engineering, Department of Civil & Environmental Engineering, University of Alberta, 3-133 Markin/CNRL NREF Building, Edmonton, Alberta, Canada, T6G 2W2
e-mail: eniven@ualberta.ca

C.V. Deutsch
e-mail: cdeutsch@ualberta.ca

P. Abrahamsen et al. (eds.), *Geostatistics Oslo 2012*,
Quantitative Geology and Geostatistics 17,
DOI 10.1007/978-94-007-4153-9_22, © Springer Science+Business Media Dordrecht 2012

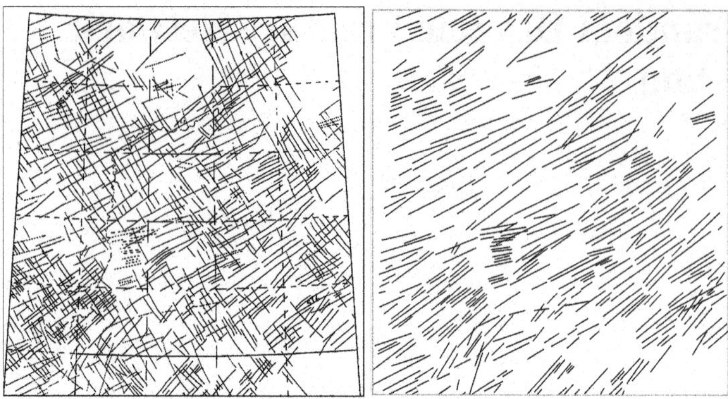

Fig. 1 On the *left*: A map of fractures from Northern Alberta [4]. On the *right*: The southwest-northeast fracture set digitized

networks in NFRs are extremely complex. It is impractical to gather enough data to make meaningful direct estimates of fracture location, size, orientation, permeability and flow response.

It is common practice to generate a discrete fracture network (DFN) as a model of the fractures in a reservoir. DFNs have the advantage of portraying both the fracture spatial distribution and details of individual fracture characteristics such as location, orientation, size, density and conductivity [8]. Large scale faults and some large scale fractures that show up on seismic surveys are explicitly specified in the DFN. Small and medium scale fractures (those that do not show up on seismic surveys and are referred to as being sub-seismic) are modeled probabilistically.

Many studies, and most commercially available DFN computer codes, use a Poisson process to randomly generate fracture centroid locations and draw fracture orientations independently of location. The Poisson process is characterized by an intensity function and is said to be homogeneous if the intensity function is constant over the area of interest and non-homogeneous if the intensity varies with location. Since Poisson process events (fracture centroids in this case) are independent, centroids may occur very close together regardless of whether a homogeneous or non-homogeneous intensity function is used.

Some research has focused on generating more geologically realistic DFNs by growing fractures from an initial seed location. Renshaw and Pollard [5] simulated fracture networks using geomechanical principles by propagating fracture tips when stress exceeds a critical threshold. Their approach was successful in yielding realistic images of fractures but was computationally prohibitive and limited to two dimensions. Srivastava's [7] work mimics that of Renshaw and Pollard in that fractures are propagated at their tips, but instead of being governed by geomechanical principles, his method is governed by statistical rules. Initial fracture locations are seeded and fracture traces are propagated in two dimensions at the surface by using sequential Gaussian simulation (SGS), which incorporates nearby data into a local distribution of possible azimuths for the next segment. Once the surface traces

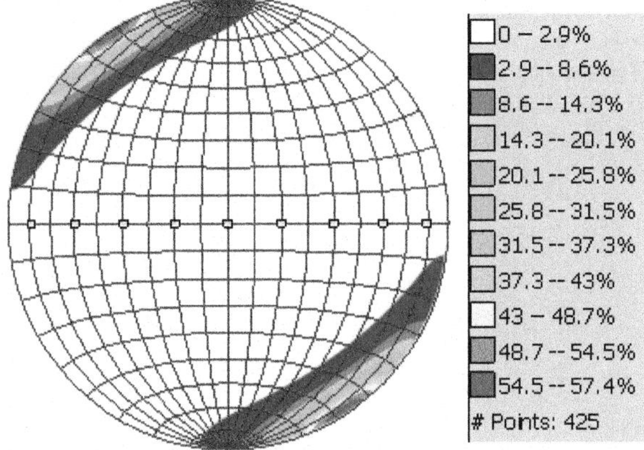

Fig. 2 Schmidt equal area stereonet for the fractures shown in Fig. 1

are simulated, they are propagated to depth, again using SGS, by simulating a dip angle from nearby data. Srivastava's approach was successful in yielding realistic three dimensional fracture networks; however, his method is mainly applicable for modeling fractures at the surface rather than at depth, such as in a petroleum reservoir.

This paper demonstrates the limitations of the typical Poisson process based techniques. A new algorithm for generating DFNs is presented. The central idea of the algorithm is to simulate more fractures than are required and iterate to find a subset DFN that closely matches target fracture network statistics such as local fracture spacing, deviation in local fracture orientation, fracture length, intensity and the number of fracture intersections. The strength of the proposed approach is that it handles any arbitrary histograms of local fracture spacing and local orientation while operating in three dimensions and maintaining the ability to simulate and optimize fracture networks with millions of fractures.

2 Motivation Demonstrated by an Example

Figure 1 shows a map of two lineament (or fracture) sets occurring in an area of Northern Alberta. The lineaments are inferred from satellite imagery and digital elevation models [4]. One set strikes southwest-northeast and the other strikes southeast-northwest. The southwest-northeast set was digitized for further analysis. In total, there are 425 fractures from the SW-NE set. A stereonet of the poles (Fig. 2) indicates that the fractures are from a single set. From a qualitative standpoint there appears to be some pattern to the fracture locations and orientations. There are 20 fracture intersections and fractures appear to be generally oriented similarly to their nearest neighbors.

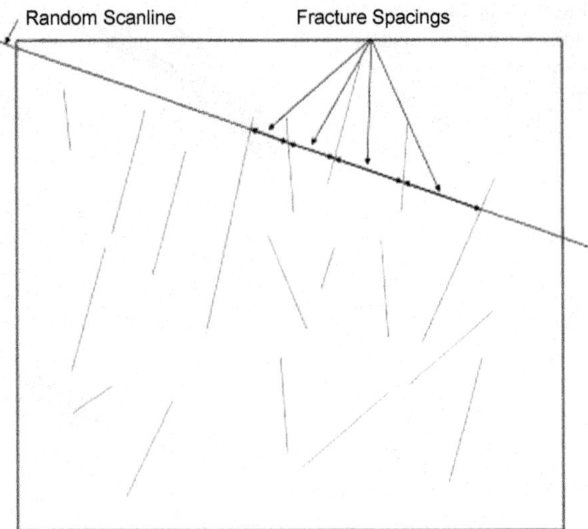

Fig. 3 Measuring fracture spacing along a scanline that is perpendicular to the average fracture orientation

2.1 Calculating the Fracture Spacing of a Fracture Network

Imaginary scan lines are drawn perpendicular to the average fracture orientation. The distance along the scanlines between intersections with fractures are individual measurements of fracture spacing (Fig. 3). One hundred scanlines seeded from random locations within the map area are used to build a relative histogram of fracture spacing for the digitized fracture map. The relative histogram is shown as the blue line on left side of Fig. 4.

Next, the similarity in the orientation of nearby fractures is assessed. Each fracture is visited and its nearest neighbor is identified in the direction that is perpendicular to the fracture plane (parallel to the fracture plane normal vector). The deviation in local fracture orientation is the angle between the normal vectors of a fracture and its nearest neighbor. A relative histogram of the deviation in local fracture orientation is shown for the digitized fractures on the right side of Fig. 4 in blue.

If the digitized fractures are taken as the truth, DFNs can be built to model those fractures and the spatial statistics of the digitized fracture network and the DFNs created to model them can be compared. The goal here is to determine whether the typical random DFN generation algorithms can effectively model the digitized fractures. 100 DFN realizations are generated to model the digitized fracture network. Fracture centroids are simulated randomly from within the area of interest. Fracture orientations are drawn from the same distribution as the digitized fractures and independently of centroid location. Histograms of fracture spacing and deviation in local fracture orientation for the 100 DFN realizations are shown in Fig. 4 in red.

The digitized fractures have an average spacing of 17.7 km with a standard deviation of 17.4 km. The randomly located DFN fractures have a larger average spacing of 24.1 km with a standard deviation of 23.4 km. Moreover, the fracture spacing histogram for the digitized fractures is narrower and more skewed than the for the

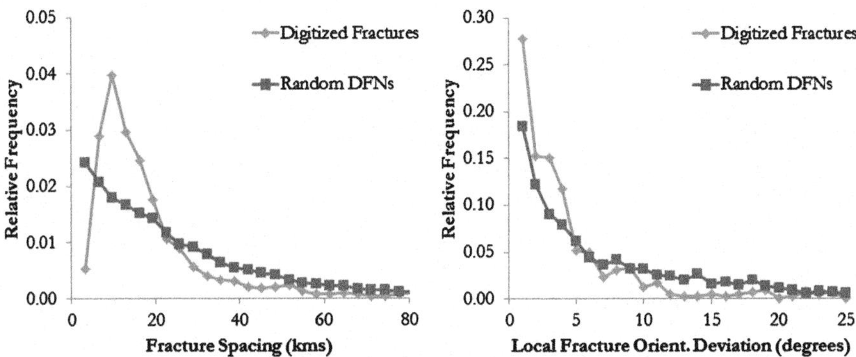

Fig. 4 On the *left*: Relative histograms of fracture spacing for digitized fractures, compared to 100 realizations of random DFNs. On the *right*: Relative histograms of the deviation in orientation between nearby fractures, compared to 100 realizations of random DFNs

100 DFN realizations. The bin with the highest frequency of fracture spacing for the digitized fractures is 13 km. By comparison, the most frequent bin for the 100 DFN realizations is the 3 km fracture spacing bin.

Upon examination of the histograms of deviation in local fracture orientation (right side of Fig. 4), it is apparent that almost all of the digitized fractures are oriented within 12 degrees of their nearest neighbor. However, for the 100 DFN realizations the histogram is much wider.

The number of fracture intersections is also an important parameter since well-connected fracture networks are more permeable than discontinuous fracture networks with few fracture intersections. There are 20 fracture intersections in the digitized fracture network. The number of fracture intersections is calculated for each of the 100 DFN realizations. The average number of fracture intersections per realization is 89.4 with a standard deviation of 8.4 intersections. The number of fractures in the digitized fracture network is 7.5 standard deviations less than the mean compared to the distribution of fracture intersections in the 100 DFN realizations.

Based on this analysis, the spatial statistics of the digitized fracture network shown in Fig. 1 cannot be modeled with a typical DFN (i.e. generated with random centroid locations and independently drawn fracture orientations). The DFNs are unable to honor the histograms of fracture spacing and deviation in local fracture orientation and showed far more intersections than were seen in the digitized fracture network.

3 A New Approach to DFN Simulation

A new approach to DFN simulation is proposed that relies on generating more fractures than are required and iteratively adding or removing fractures to find a subset that matches the input distributions of fracture intensity, length, the number of inter-

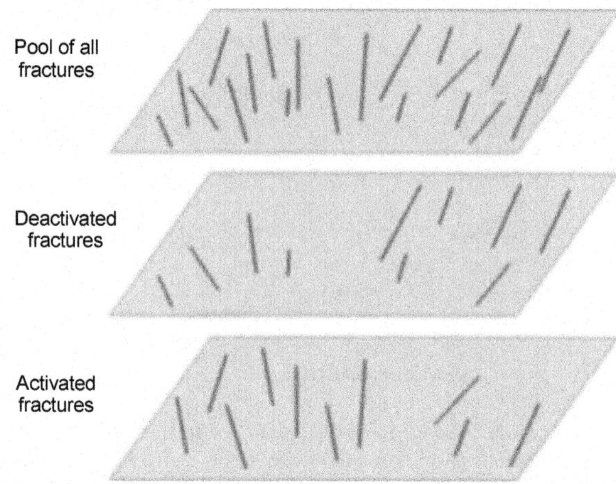

Fig. 5 A 2D illustration of the activated, deactivated and pool of fractures. There are 10 activated and 10 deactivated fractures. Thus, there are 20 fractures in the pool. The activated fractures are the DFN at any time during the iterative optimization process. The deactivated fractures are not used after the optimization is complete

sections, local fracture spacing and local fracture orientation. Note that the methodology is general enough to work in two or three dimensions.

The proposed methodology is as follows:

1. A pool of fractures is simulated using traditional means. The idea is to generate more fractures than are required.
2. Not all fractures are assigned to the DFN (Fig. 5). Some fractures will be assigned to the DFN and are termed activated while fractures that are not part of the DFN are termed deactivated.
3. An initial DFN is created by randomly visiting fractures in the pool and activating them to be part of the DFN. This process stops when the desired fracture intensity is achieved. Thus, the initial DFN has the target fracture intensity.
4. A search strategy is implemented to discover the locations of the fracture centroids with respect to all other fractures. The goal is to identify which fractures are close to each other and calculate the distances between them as a measure of fracture spacing. The angles between the normal vectors of nearby fractures are also calculated as a measure of the deviation in local fracture orientation.
5. An objective function is calculated for the initial DFN. The objective function measures the difference between actual and target histograms of local fracture spacing, deviation in local fracture orientation, fracture length, fracture intensity and the actual and target number of fracture intersections.
6. A random path to visit each fracture in the pool is determined.
7. The initial DFN is iterated upon by visiting a fracture on the random path and switching its activation. If the fracture is already activated and a part of the DFN, then the fracture is deactivated and removed from the DFN. If the fracture was deactivated and not in the DFN, the fracture is activated and assigned to the DFN.
8. The objective function is re-calculated for the modified DFN.
9. The change to the DFN is accepted if the objective function decreases.
10. The process repeats, visiting a new fracture each time (i.e. go to step 7) until the desired number of iterations is complete.

The methodology is statistical, but the end is a heuristic algorithm to generate DFNs that are not entirely statistical in nature by reproducing information that is calibrated from available measurements and analogue sites that are deemed representative. There are no geomechanical principles directly employed in the generation of the fracture networks.

Some steps in the methodology warrant further discussion.

3.1 Generating a Pool of Fractures and an Initial DFN

Generating a large pool of fractures, that exceeds the target fracture intensity, permits an optimization to find a suitable subset that comes close to matching target fracture network spatial statistics. The choice of pool size is subjective but internal testing has shown that a pool intensity that is two to three times the target intensity is usually sufficient.

The pool of fractures can be generated using the traditional approach. Fracture locations are simulated with a Poisson process (i.e. randomly from within an area of interest) and fracture orientations are drawn independently of fracture locations and from an appropriate input distribution determined from available data.

Every fracture in the pool (Fig. 5) starts out in the deactivated state (i.e. not part of the DFN). An initial DFN is created by randomly visiting fractures in the pool and activating them, which assigns them to the DFN. The initial DFN is complete when its intensity matches the target fracture intensity. Thus, the initial DFN always starts with the correct intensity. Fracture intensity could be specified as a fracture count or, more commonly, fracture area per volume of rock. Fracture count and fracture area per volume are commonly called the P10 and P32 fracture intensity, respectively.

3.2 Calculating the Spatial Statistics of the DFN

After the pool of fractures is created and a subset has been assigned to the initial DFN, the objective function is calculated. The objective function is based, in part, on the local fracture spacing and deviation in local fracture orientation.

A geologist or engineer might measure fracture spacing at an outcrop using scanlines (Fig. 3). However, with large fracture networks and many DFN iterations this approach is too computationally expensive. We propose to use the average perpendicular distance between a fracture and its nearest neighbor as a measure of local fracture spacing. Figure 6 shows the calculation of the perpendicular distance between a fracture and its neighbor. Each fracture is visited and its nearest neighbor is identified. Fracture spacing is usually measured normal to the plane of the fracture. Thus, the perpendicular distance between fractures is used. The perpendicular distances (d_{p2} and d_{p1}) between each fracture's centroid and the other fracture are determined. The average of d_{p2} and d_{p1} is taken to be the local fracture spacing. If the ray from one centroid does not intersect the fracture, the intersection is taken where it would have been if the fracture was infinite in extents.

In addition to local fracture spacing, the similarity in orientation between fractures and their single nearest neighbors (identified by the shortest perpendicular distance) is compared. The deviation in local fracture orientation is determined by finding the angle between the poles of a fracture and its nearest neighbor and is calculated from the dot product of the two normal vectors:

$$\theta = \arccos\left(\frac{\mathbf{a}\cdot\mathbf{b}}{|\mathbf{a}||\mathbf{b}|}\right), \tag{1}$$

where **a** and **b** are the normal vectors of the two fractures.

3.3 The Objective Function

Now that we have the local fracture spacing and deviation in local fracture orientation for each fracture we can calculate the objective function. The objective function measures the squared difference between the bins of target histograms and the histograms calculated from the simulated DFN. The objective function is

$$\begin{aligned} O &= \sum_{c=1}^{5} C_c O_c \\ &= C_{\text{spac}} \sum_{i=1}^{sbins} \left(S_i^{\text{target}} - S_i^{\text{DFN}}\right)^2 + C_{\text{or}} \sum_{i=1}^{orbins} \left(Or_i^{\text{target}} - Or_i^{\text{DFN}}\right)^2 \\ &\quad + C_{\text{len}} \sum_{i=1}^{lbins} \left(L_i^{\text{target}} - L_i^{\text{DFN}}\right)^2 + C_{\text{inter}} \left(Inter_i^{\text{target}} - Inter_i^{\text{DFN}}\right)^2 \\ &\quad + C_{\text{int}} \sum_{i=1}^{ibins} \left(I_i^{\text{target}} - I_i^{\text{DFN}}\right)^2, \end{aligned} \tag{2}$$

which consists of the following five components:

1. S_i^{target} and S_i^{DFN} are the target and DFN histograms of local fracture spacing, respectively. *sbins* is the number of histogram bins.
2. Or_i^{target} and Or_i^{DFN} are the target and DFN histograms of deviation in local fracture orientation, respectively. *orbins* is the number of histogram bins.
3. L_i^{target} and L_i^{DFN} are the target and DFN histograms of fracture length, respectively. *lbins* is the number of histogram bins. Note, each fracture has a length, which is measured horizontally through the centroid.
4. $Inter_i^{\text{target}}$ and $Inter_i^{\text{DFN}}$ are the target and DFN number of fracture intersections, respectively.
5. I_i^{target} and I_i^{DFN} are the target and DFN histograms of fracture Intensity, respectively. *ibins* is the number of histogram bins.
6. C_{spac}, C_{or}, C_{len}, C_{inter}, and C_{int} are coefficients, which allow each component to play an equally important role in the objective function and serve to make it unit-less in order to compare components with different original units.

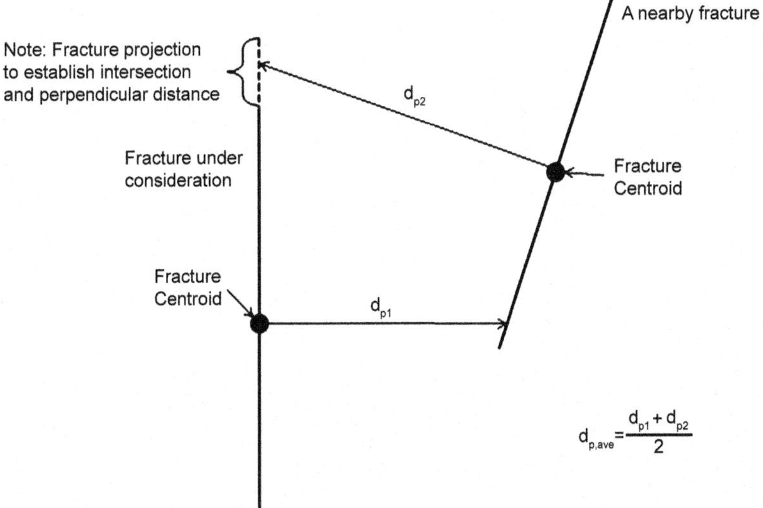

Fig. 6 The calculation of the perpendicular distance to the nearest fracture

The coefficients can be calculated automatically so that, on average, each component contributes equally to changes in the objective function. This can be achieved if each of the coefficients are inversely proportional to the average change of that component objective function [2].

The change in the objective function due to a fracture activation or deactivation is

$$\Delta O = O_{\text{new}} - O_{\text{old}} = \sum_{c=1}^{5} C_c \Delta O_c, \tag{3}$$

where c represents the five different components of the objective function. Then the coefficients are inversely proportional to the average change of that component objective function:

$$C_c = \frac{1}{\overline{|\Delta O_c|}} \quad \text{for } c = 1, \dots, 5. \tag{4}$$

The average change in each component of the objective function, ΔO_c, can be approximated by evaluating the average change of N changes (say $N = 1000$) to the DFN. The procedure is: (1) visit a fracture randomly, (2) change its activation, (3) observe the effect of the change on the objective function, (4) reverse the activation and (5) repeat, visiting a new fracture each time until N changes have been made. The average change for each component is

$$\overline{|\Delta O_c|} = \frac{1}{N} \sum_{i=1}^{N} \left[O_c^{\text{base}} - O_c^{i} \right] \quad \text{for } c = 1, \dots, 5, \tag{5}$$

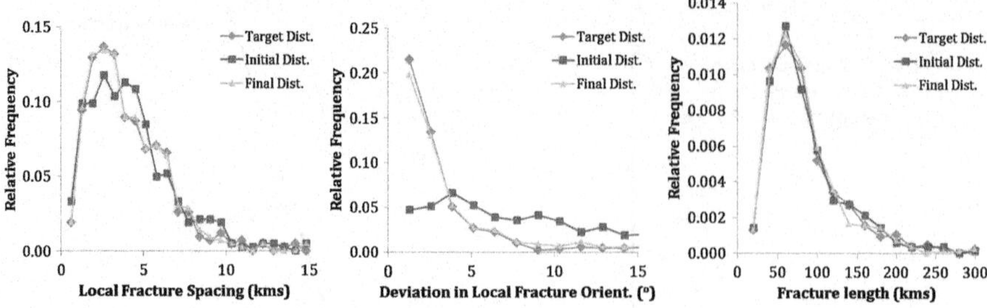

Fig. 7 Target, initial and optimized histograms of local fracture spacing, deviation in local fracture orientation and fracture length

where $\overline{|\Delta O_c|}$ is the average change for component c, O_c^{base} is the objective function for the initial DFN (i.e. the base case), and O_c^i is the objective function for the perturbed DFN. The base objective function, O_c^{base}, is calculated with the five coefficients equaling 1.

Once the coefficients, C_c, are determined, the algorithm would proceed to steps 6 and 7 (Sect. 3), where the initial DFN is iterated upon and changes to the DFN are accepted or rejected depending on whether or not the change results in a decrease in the objective function. The objective function is flexible and can optimize the fracture network on any or all of its five components. For example, information on fracture length is difficult to obtain in a petroleum reservoir setting where fractures are sampled by core. If there is great uncertainty in the fracture length histogram, the user could choose not to optimize on fracture length.

4 Example Application

The proposed approach is used to generate a DFN intended to model the digitized fractures in Fig. 1. Once again, the digitized fracture set is taken as the "truth".

Target fracture network statistics are determined from the digitized fracture network. Histograms of local fracture spacing, deviation in local fracture orientation and fracture length are calculated. There are 425 fractures in the digitized fracture network yielding 20 intersections, which are taken as the target fracture intensity and the target number of fracture intersections. Figure 7 shows the target histograms of local fracture spacing (i.e. perpendicular distance to the nearest fracture), deviation in local fracture orientation and fracture length that are calculated from the digitized fractures in Fig. 1. Our goal is to generate a DFN that matches these target histograms as well as the target intensity and number of fracture intersections.

The ratio of fracture pool size to target intensity is set at 2. This means that 850 fractures will be generated for the pool; half are assigned to the initial DFN.

The initial DFN is shown on the left side of Fig. 8. There are 425 fractures in the initial DFN. Visual inspection of the initial DFN shows many instances of fractures that are extremely close together compared to the digitized fracture map. There are

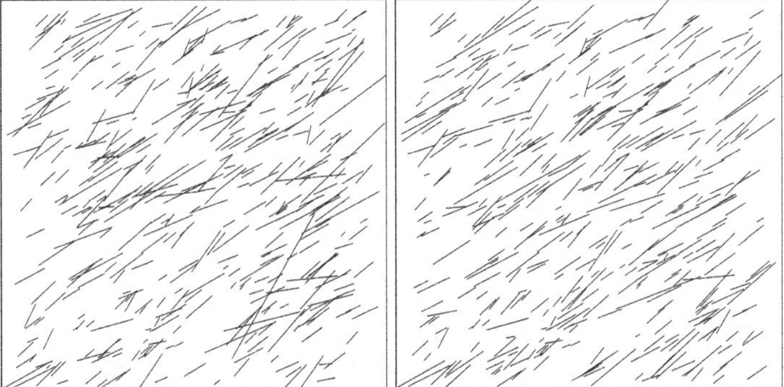

Fig. 8 On the *left*: An initial DFN generated to represent the fractures in the fracture map. On the *right*: The optimized DFN after 3400 iterations (visiting each of the 850 fractures 4 times)

86 fracture intersections in the initial DFN, which is more than four times as many as the target of 20.

The final DFN (right side of Fig. 8) is optimized over 3400 iterations (each of the 850 fractures in the pool is visited four times). Visual inspection of the optimized DFN shows that fractures are oriented more similarly to their nearest neighbor and has a more regular spacing, evidenced by fewer bare spots (areas without fractures), compared to the initial DFN.

Figure 7 also shows the target, initial DFN and optimized DFN histograms of local fracture spacing, deviation in local fracture orientation and fracture length. All three histograms for the optimized DFN are a very good match to the target histograms. The final optimized DFN has the correct fracture intensity (425 fractures) and the right amount of fracture intersections (20 intersections).

5 Conclusions

An algorithm for generating DFNs is proposed, which shows promise in that it allows for simulation of DFNs that match target input statistics on fracture spacing, relative orientation, length, intensity and the number of intersections. The algorithm works by simulating more fractures than are required and iterating to find a subset that best matches the target input histograms. An objective function is minimized to find the best quality fit between the target statistics and those from the DFN.

The algorithm is flexible and can be used in two or three dimensions. Our experience has been that tens of millions of fractures can be simulated and optimized in a reasonable computation time (less than a day on modern computers). One shortcoming of this approach is that it requires the user to define the target histograms and the number of fracture intersections. This may not be possible in cases where limited information on the fractures is available. However, if fracture information is available from core or borehole images in at least a few wells, or two dimensional aerial images of the fractures are available, the target histograms can be calculated.

References

1. Beydoun ZR (1998) Arabian plate oil and gas; why so rich and so prolific? Episodes 21(2):74–81
2. Deutsch CV (1992) Annealing techniques applied to reservoir modeling and the integration of geological and engineering (well test) data. PhD Thesis, Stanford University, California, 306 pp
3. Nelson RA (2001) Geologic analysis of naturally fractured reservoirs. Gulf Professional Publishing, Houston, 332 pp
4. Pana DI, Waters J, Grobe M (2001) GIS compilation of structural elements in northern Alberta. Release 1.0. 2001-01, Alberta energy and utilities board, Alberta geological survey
5. Renshaw C, Pollard D (1994) Numerical simulation of fracture set formation: a fracture mechanics model consistent with experimental observations. J Geophys Res 99(9):9359–9372
6. Roxar (2009) Naturally fractured reservoirs: an introduction to their appraisal and management [online]. Available from http://www.roxar.com/category.php?categoryID=2141. Cited April 22 2009
7. Srivastava RM (2006) Field verification of a geostatistical method for simulating fracture network models. In: Proceedings of the 41st US symposium on rock mechanics (USRMS): 50 years of rock mechanics—landmarks and future challenges, Golden, Colorado
8. Tran NH, Rahman MK, Rahman SS (2002) A nested neuro-fractal-stochastic technique for modeling naturally fractured reservoirs. In: Proceedings SPE Asia Pacific oil and gas conference and exhibition, pp 453–464
9. Waldren D, Corrigan AF (1985) An engineering and geological review of the problems encountered in simulating naturally fractured reservoirs. In: SPE Middle East oil technical conference and exhibition, Bahrain, pp 311–316

Part III
Mining

Kriging and Simulation in Presence of Stationary Domains: Developments in Boundary Modeling

Brandon J. Wilde and Clayton V. Deutsch

Abstract Perhaps the most critical decision in geostatistical modeling is that of choosing the stationary domains or populations for common analysis. The boundaries between the stationary domains must be modeled with uncertainty. The correlations and trends across these boundaries must be used in modeling. Interpolating a distance function is a useful method for modeling boundaries with uncertainty. The current implementation of distance function boundary modeling with uncertainty requires expensive calibration with simulated data and numerous reference models to ensure unbiasedness and fair uncertainty. A method for using the available data to calibrate the distance function is proposed which greatly reduces the calibration expense. The nature of the boundaries between stationary domains must be considered. Boundaries can be hard or soft. The boundary nature must be accounted for in the modeling. A contact plot is a useful tool for identifying the nature of the grade transition across a boundary. Guidelines for determining the nature of a boundary from a contact plot are suggested. With the location of the boundary defined and the nature of transition across the boundary determined, the next step is to model the grades in a manner that accounts for the boundary information. The resulting models should reproduce the boundaries of the stationary domains and reproduce the nature of the boundary at the boundary location.

1 Introduction

Interpolating a distance function has been shown to be a reasonable method for locating boundaries as it is simple and flexible [2, 4]. However, it needs a large amount of hard data and does not provide direct access to uncertainty. The definition of the distance function is related to the notion of distance to an interface separating two distinct domains. Distance is measured to the nearest unlike data location. Distance can be positive or negative depending on the location of the data inside or outside

B.J. Wilde (✉) · C.V. Deutsch
University of Alberta, Edmonton, Canada
e-mail: bwilde@ualberta.ca

C.V. Deutsch
e-mail: cdeutsch@ualberta.ca

P. Abrahamsen et al. (eds.), *Geostatistics Oslo 2012*,
Quantitative Geology and Geostatistics 17,
DOI 10.1007/978-94-007-4153-9_23,

the domain. The choice of sign for distance values inside or outside the domain is trivial, but consistency must be emphasized. The distance function varies smoothly between increasingly positive values outside and further away from the boundary interface to increasingly negative values inside and further away from the boundary surface. To determine the location of the boundary, the distance to the nearest unlike sample is calculated for all available samples. This distance function data is then used to condition the interpolation of distance function on a regular grid. The boundary is considered to lay at the transition between positive and negative interpolated distance function values.

There is uncertainty in the boundary location. Munroe and Deutsch [5, 6] propose assessment of the uncertainty by calibration of parameters C and β where C controls the width of the uncertainty and β controls the bias. These parameters are optimized to give appropriate uncertainty. Optimizing these parameters is an expensive operation requiring multiple reference models and two objective functions.

This work proposes a simpler, less expensive calibration. Only the C parameter is calibrated and this is done using only the data to calibrate in a relatively computationally inexpensive manner. A subset of data is removed prior to the calculation of the distance function at the data locations. The subset of data removed will hereafter be referred to as the jackknife data. This effectively creates two data sets: the distance function data and the jackknife data. The distance function data are used to condition the estimation of distance function at the jackknife data locations. A number of jackknife data that are coded as inside the domain will have positive distance function estimates (outside the domain) and a number of jackknife data that are coded as outside the domain will have negative distance function estimates (inside the domain). The C parameter is adjusted until the desired proportion of incorrectly classified jackknife data is correctly classified. Once the C parameter is determined, it is applied to the calculation of the distance function for all available sample data. All of the data are then used to condition the interpolation of the distance function.

In addition to identifying the location of the domain boundary, the nature of the grade transition across the boundary must be considered. Domain boundaries are typically referred to as hard or soft. Hard boundaries are found when there is an abrupt change in the mineralogy or grade. Contacts where the variable changes transitionally across the boundary are referred to as soft boundaries. Geological models should reproduce the boundary types indicated by the data.

2 Distance Function Formalism

The first requirement in the calibration of distance function uncertainty is a dataset where all data locations have been coded as either inside or outside the domain of interest:

$$i(\mathbf{u}_\alpha) = \begin{cases} 1 & \text{if domain of interest present at } \mathbf{u}_\alpha, \\ 0 & \text{otherwise,} \end{cases} \qquad \alpha = 1, \dots, n, \tag{1}$$

where $\mathbf{u}$ is a location vector, α is the sample index, and n is the number of samples. For each sample $\mathbf{u}_\alpha$, $\alpha = 1, \dots, n$, the nearest sample located in a different domain $\mathbf{u}_{\alpha'}$ is determined such that $i(\mathbf{u}_\alpha) \neq i(\mathbf{u}_{\alpha'})$. The Euclidean distance between these two locations is the distance function value at location $\mathbf{u}_\alpha : df(\mathbf{u}_\alpha)$. If $\mathbf{u}_\alpha$ is within the domain, the distance is set to negative; otherwise the distance is positively signed:

$$df(\mathbf{u}_\alpha) = \begin{cases} +(\mathbf{u}_\alpha - \mathbf{u}_{\alpha'}) & \text{if } i(\mathbf{u}_\alpha) = 0 \\ -(\mathbf{u}_\alpha - \mathbf{u}_{\alpha'}) & \text{if } i(\mathbf{u}_\alpha) = 1. \end{cases} \tag{2}$$

The calculation of distance would account for anisotropy, if present. Notice that this approach is designed for binary systems where locations are in or out of a particular domain. Multiple domains could be modeled hierarchically. The presence of many intermingled domains would not be possible with this approach. Also note that the distance function data correspond to the distance to the nearest *observed* contact; not the distance to the nearest *real* contact.

Once distance function has been calculated for each sample, this distance function data can be interpolated on a regular grid using a smooth estimator such as kriging or inverse distance estimation. A global estimator is particularly suited to the task as such an estimator is free from artifacts due to the search for local data. The boundary is considered to lay at the transition between positive and negative interpolated distance function values.

To illustrate, consider the sample locations coded as inside and outside the domain in Fig. 1a where black-filled bullets represent samples inside the domain and white-filled bullets represent samples outside the domain. The distance to the nearest outside sample is calculated for each inside sample and vice versa. These distances are shown in Fig. 1b. Samples outside the domain have positive distance function; samples inside the domain have negative distance function. These samples are interpolated yielding the map of values shown in Fig. 1b. Negative estimates are considered inside the domain and positive estimates are considered outside the domain yielding the map shown in Fig. 1c where black is inside the domain. An estimate of the boundary location falls at the transition between positive and negative distance function estimates (the black-white interface). There is, of course, uncertainty in the location of the boundary.

3 *C* Parameter

This work proposes a new method for quantifying distance function uncertainty. This method uses the data to calibrate a single additive factor, C, which modifies the distance function values calculated at the sample locations. This method is similar to the jackknife where a subset of the data is held back when estimation is performed and estimated values are compared with the true values at sample locations.

The C parameter modifies the distance function value at each sample location. C is an additive parameter, being added to the distance function when outside the domain and subtracted from the distance function when inside the domain:

$$\hat{df}(\mathbf{u}_\alpha) = \begin{cases} df(\mathbf{u}_\alpha) + C & \text{if } i(\mathbf{u}_\alpha) = 0 \\ df(\mathbf{u}_\alpha) - C & \text{if } i(\mathbf{u}_\alpha) = 1, \end{cases} \tag{3}$$

where $\hat{df}(\mathbf{u}_\alpha)$ represents the modified distance function at the sample locations.

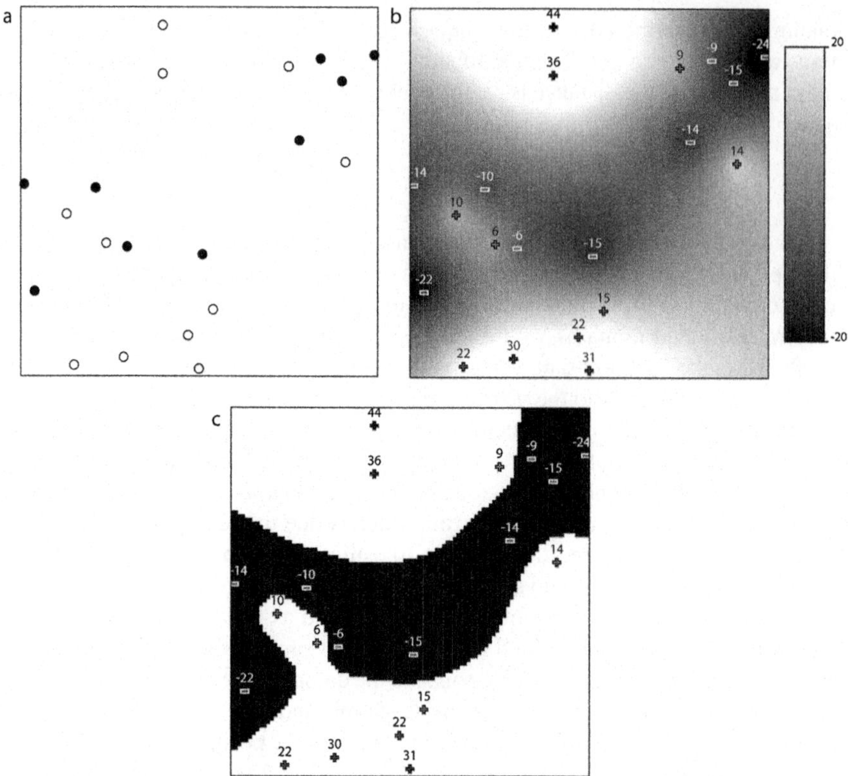

Fig. 1 (**a**) Locations coded as inside (*black*) and outside (*white*) the domain, (**b**) distance function calculated at each sample location and interpolated, (**c**) distance function less than zero considered inside the domain

The C parameter increases the difference between positive and negative distance function values. Once the C parameter has been applied to the data, the modified distance function is interpolated. Modified distance function estimates greater than C are considered outside the domain. Modified distance function estimates less than $-C$ are considered inside the domain. Any modified distance function estimates between $-C$ and C are within the range of boundary uncertainty; the boundary is located between modified distance function estimates of $-C$ and C.

The uncertainty band for different C values is illustrated in Fig. 2 for different values of C. The same data shown in Fig. 1 are used. A C value of 3 increases the positive and decreases the negative distance function values by 3. The modified distance function values are interpolated. Any modified distance function estimate greater than 3 is considered outside the domain (white) while any modified distance function estimate less than -3 is considered inside the domain (black). The grey areas have modified distance function estimates between -3 and 3 and represent the region within which the boundary may lay. This is repeated for C values of 5 and 7. As C increases, the data values change and the size of the grey boundary

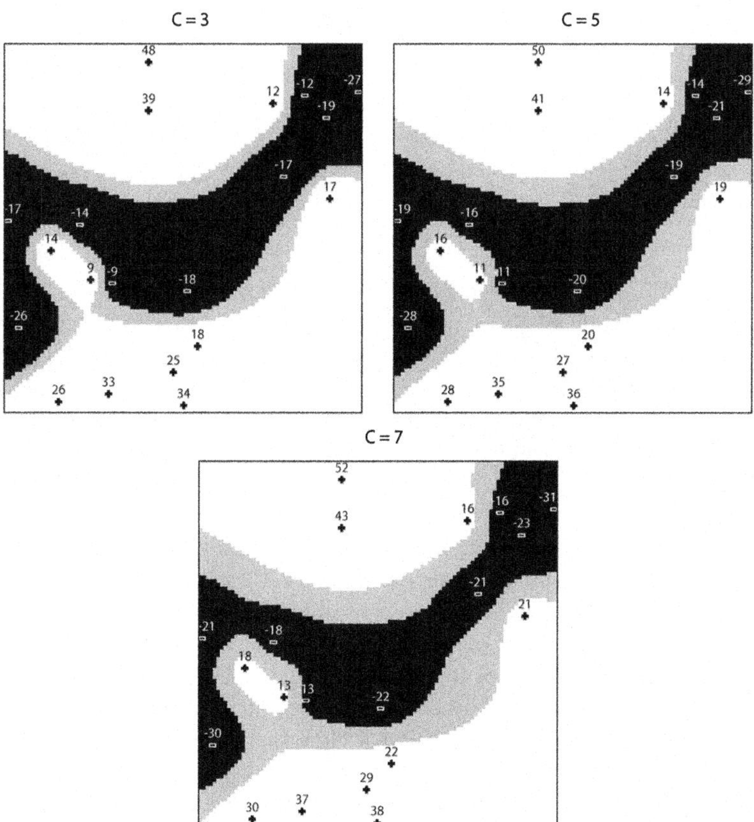

Fig. 2 Boundary uncertainty for C values of 3, 5, and 7. *Black*—surely inside domain, *white*—surely outside domain, *grey*—region of boundary uncertainty

uncertainty region increases. There is a need to infer a reasonable C value for each boundary.

4 *C* Calibration

The C parameter controlling the size of the boundary uncertainty is calibrated in a manner similar to the jackknife. A subset of the data is removed and the remaining data are used to estimate distance function at the jackknife locations. The number of jackknife data that fall on the wrong side of the boundary is reduced as C is increased.

The first step in the calibration of C is to remove a subset of the data. This can be done by randomly choosing drillholes to exclude. Distance function values are then calculated for the remaining data with an initial value of $C = 0$.

These distance function data are used to condition the estimate of distance function at each of the jackknife data locations. There are four possible outcomes re-

Fig. 3 Possible outcomes for distance function estimation at jackknife data locations

		Estimate	
		Out	In
Truth	In	Estimated Out Truly In	Correctly In
	Out	Correctly Out	Estimated In Truly Out

sulting from this estimation as shown in Fig. 3. The location could be: (1) correctly estimated to be outside the domain, (2) correctly estimated to be inside the domain, (3) incorrectly estimated to be outside the domain, and (4) incorrectly estimated to be inside the domain. We are interested in the number of data that fall on the wrong side of the boundary, that is, the number of times the estimate is positive but the data is coded as inside the domain and the number of times the estimate is negative but the data is coded as outside the domain. The number of times a data falls on the wrong side of the boundary for $C = 0$ is the base case.

C is then increased and the distance function values at the non-jackknife sample locations modified. This modified distance function is estimated at each of the jackknife locations. The boundary is now considered to fall between $-C$ and C. A data falls on the wrong side of the boundary when a jackknife location is coded as inside but has a modified distance function estimate greater than C or a jackknife location is coded as outside but has a modified distance function estimate less than $-C$. The number of data falling on the wrong side of the boundary decreases as C increases. C is increased until the number of data falling on the wrong side of the boundary is acceptable. The C value where this occurs is the calibrated C value which quantifies the boundary uncertainty.

This is illustrated in Fig. 4 using the same data as previously. There are jackknife data not used in the initial estimation of distance function that are now considered. The white-filled circles represent sample locations outside the domain while the black-filled circles represent sample locations inside the domain. For $C = 0$, there are two samples coded as outside the domain which fall inside (white circles in black region) and four samples coded as inside the domain which fall outside (black circles in white region) for a base case of six incorrectly classified data. Increasing the C parameter to 3 decreases the number of samples coded as outside the domain which fall inside from two to one (one white bullet that was in the black region now falls in the grey region) and decreases the number of samples coded as inside the domain which fall outside from four to three (one black bullet that was in the white region now falls in the grey region). Increasing C to 5 further decreases the number of black circles falling in the white region to two for a total of 3 incorrectly classified data, one half of the base case at $C = 0$.

The calibration of C is sensitive to which data are used as jackknife data. Using a different quantity and/or subset of the data will lead to a different number of incorrectly classified data for a given C value. Therefore, it is recommended that the C calibration be performed for a variety of jackknife subsets to ensure that the calibration is robust.

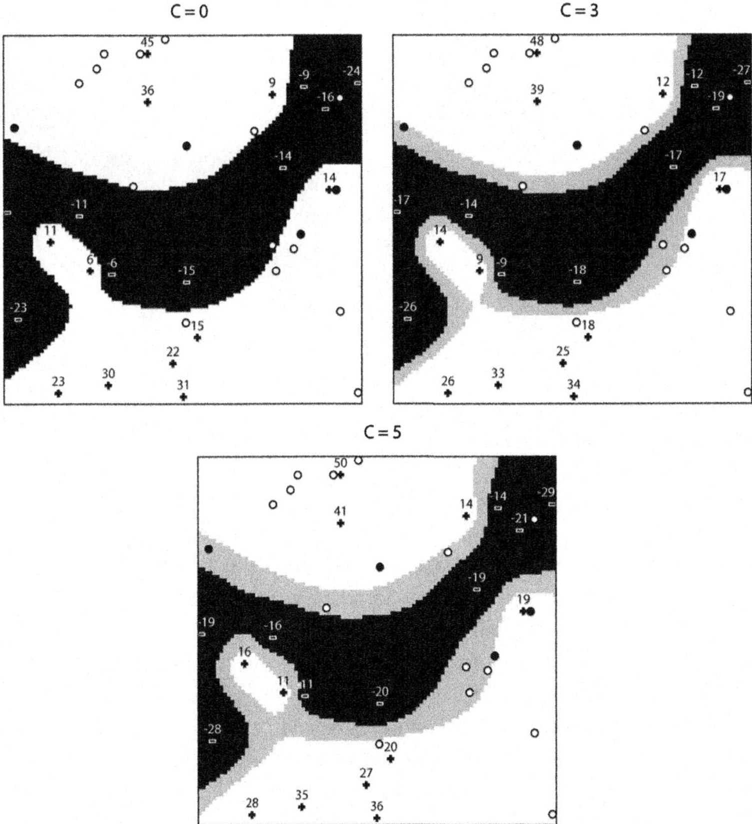

Fig. 4 Illustration of C calibration methodology for C values of 0, 3, and 5 respectively. The number of jackknife data on the wrong side of the boundary decreases as C increases

Once C has been determined it is used to calculate the modified distance function at all sample locations. The C parameter is applied as shown previously; it is added to positive distance function values and subtracted from negative distance function values. The resulting modified distance function is then interpolated. The boundary is considered to fall inside the distance function transition from $-C$ to C. Different boundaries can be extracted by applying a threshold between $-C$ and C. A threshold of C corresponds to the white-grey interface in Fig. 4 and leads to a dilated boundary that is big everywhere. A threshold near $-C$ corresponds to the black-grey interface in Fig. 4 and leads to an eroded boundary that is small everywhere. A threshold of zero is the base-case boundary.

5 Boundary Simulation

There can be cases where it is useful to have multiple realizations of the boundary such that it is neither big everywhere nor small everywhere, yet uncertainty in

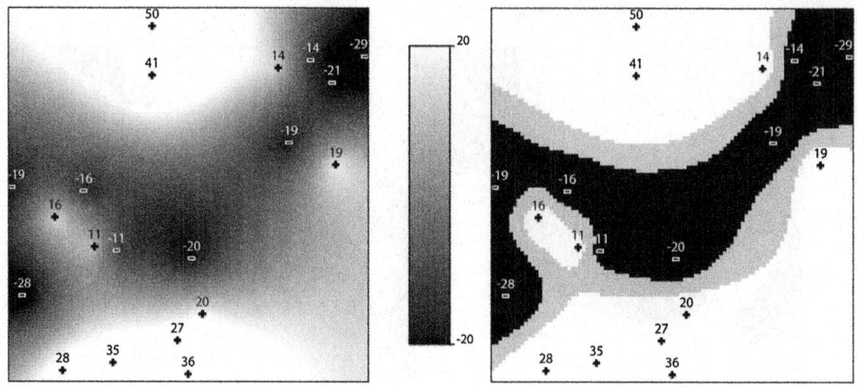

Fig. 5 The same two models of interpolated distance function. The *left map* is smoothly varying; the *right map* has thresholds applied at values of −5 and 5

its location is accounted for. This can be done within the calibrated distance function framework presented by simulating distance function values uniformly between $-C$ and C. Wherever the interpolated distance function is less than the simulated distance function is inside the domain and wherever the interpolated distance function is greater than the simulated distance function is outside the domain. The point where the interpolated distance function is equal to the simulated distance function is the boundary location.

To simulate L realizations of distance function between $-C$ and C, Gaussian deviates, $y^l(\mathbf{u}), l = 1, \ldots, L$, can be simulated and transformed according to:

$$df^l(\mathbf{u}) = 2CG^{-1}\big(y^l(\mathbf{u})\big) - C, \quad l = 1, \ldots, L, \tag{4}$$

where $df^l(\mathbf{u})$ is the simulated distance function value, $y^l(\mathbf{u})$ is an unconditionally simulated standard normal value and G^{-1} represents the determination of the standard normal CDF value corresponding to $y^l(\mathbf{u})$. Multiplying by $2C$ and subtracting C ensures that the values are between $-C$ and C. It is only necessary to simulate at locations where the interpolated distance function is between $-C$ and C. Interpolated values that are greater than C are surely outside the domain; interpolated values that are less than $-C$ are surely inside the domain. This can reduce the time required to simulate distance function realizations.

Determining whether a location, u, is inside or outside the domain requires comparing the interpolated distance function with the simulated distance function:

$$i(\mathbf{u}_\alpha) = \begin{cases} \text{inside} & \text{if } df^l(\mathbf{u}) > \hat{df}(\mathbf{u}) \\ \text{outside} & \text{if } df^l(\mathbf{u}) < \hat{df}(\mathbf{u}). \end{cases} \tag{5}$$

Since a boundary is being simulated, it is recommended that a Gaussian variogram be used in the simulation of the distance function as this model maintains high short-range continuity. A very small positive nugget is recommended for mathematical stability. There is no clear method for determining the range of the variogram model. A short range leads to a more rapidly changing boundary; a longer

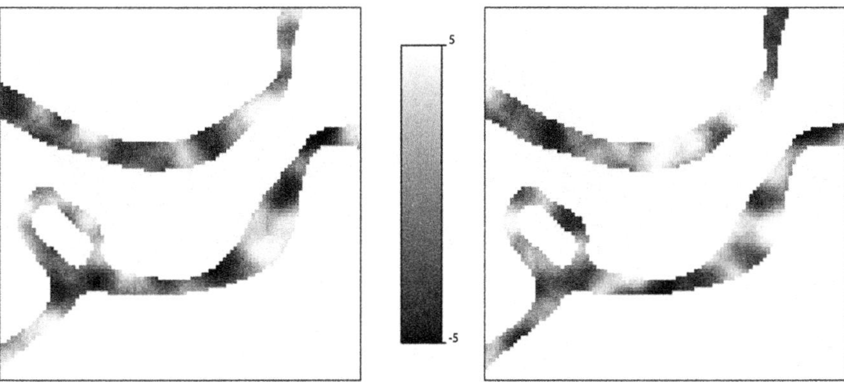

Fig. 6 Two realizations of distance function values simulated uniformly between −5 and 5 at the locations where interpolated distance function falls between −5 and 5

range creates a smoother boundary. This variogram range could come from geologic calibration from data or observations, visualization and expert judgment, or an analogue deposit. The variogram range can be too small leading to unrealistic transitions between domains. A short range variogram causes the boundary to change rapidly while a longer range leads to a smoother boundary. A too-long variogram range is similar to using a single distance function threshold to locate the boundary; the domain will be big or small everywhere.

An example of boundary simulation is shown using the same data as previously for $C = 5$. The smoothly varying interpolated distance function is shown in Fig. 5. Also shown in Fig. 5 is this same map with thresholds applied at −5 and 5. Those locations where the interpolated distance function falls between −5 and 5 are shaded grey. These are the locations where distance function values are simulated uniformly between −5 and 5. Two realizations of these simulated distance function values are shown in Fig. 6. The determination of whether each location falls inside or outside the domain is made by comparing the simulated distance function to the interpolated distance function. Those locations where the interpolated distance function is less than the simulated distance function are inside the domain and are shaded black in Fig. 7; the locations where the interpolated distance function is greater than or equal to the simulated distance function are outside the domain and are shaded white in Fig. 7. The two simulated boundary realizations are similar in a global sense in that the general shape of the black domain is about the same. However, locally the realizations can be quite different. In particular, the black domain in the first realization encompasses an island of the white domain where in the second realization this feature is not present. This relates well with the boundary uncertainty summarized by the grey shaded region in the right of Fig. 5.

6 Boundary Nature

In addition to identifying the location of domain boundaries, the nature of transition in the geologic properties across the domain boundaries should also be investigated.

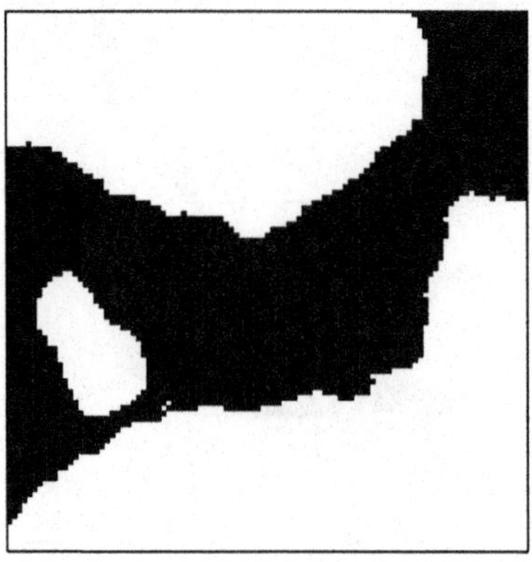
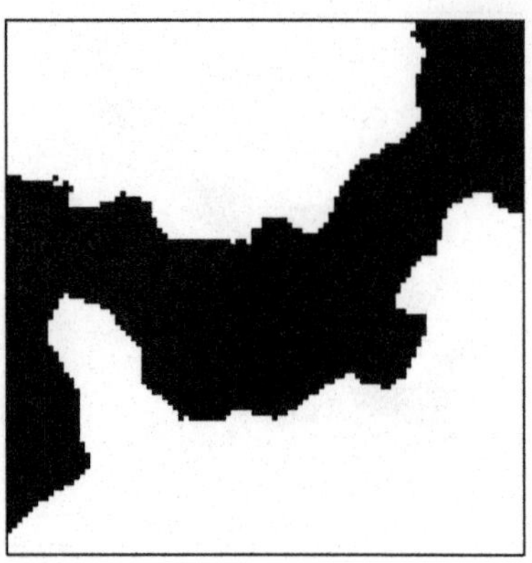

Fig. 7 Two boundary realizations based on the simulated distance function values in Fig. 6.

Domain boundaries are often referred to as either 'hard' or 'soft' [3]. Hard boundaries are found when there is an abrupt change in the mineralogy or grade without a transition at the scale of observation [7]. They do not permit interpolation or extrapolation across domains. Contacts where the variable changes transitionally across the boundary are referred to as soft boundaries. These allow selected data from either side of a boundary to be used in the estimation of each domain.

A contact analysis is undertaken to detect hard and soft boundary transitions as well as different types of hard boundary transitions. The corehole data with rocktype or facies information is required for this analysis. McLennan [4] proposes two types of contact analysis: expected value contact analysis and covariance function contact analysis. Cuba and Leuangthong [1] show how the variogram can be used to identify nonstationarity in the local variance in addition to the local mean.

A more qualitative contact analysis is performed by use of a contact plot. Figure 8 shows the general form of a contact plot for a soft and hard boundary. This type of plot is useful for determining the nature of a boundary. Sample data values z are plotted against their distance inside either the left domain, d_{12}, or right domain, d_{21}, from the boundary. Expected values are represented with the solid lines. Notice the transition zone bound by the vertical dotted lines on the left. The size of the transition zone may vary, but this zone will be present for soft boundaries. The stationary random function (SRF) to the left and right of the boundary are denoted by $Z_1(\mathbf{u})$ and $Z_2(\mathbf{u})$. In contrast to soft boundaries, the $Z_1(\mathbf{u})$ and $Z_2(\mathbf{u})$ SRFs are applicable all the way through their respective domains up until the hard boundary. There is no transition zone present.

In addition to plotting the expected value within each domain it is useful to plot the variability seen in each domain as shown in Fig. 9 where the grey shaded region represents the 90 % probability interval. The number of points within the specified distance is also shown for each domain.

Considering the expected value and relative spread allows inferences to be made regarding the nature of the boundary. Consider four boundary classifications: none, soft, hard stationary and hard nonstationary. The boundary illustrated by a contact

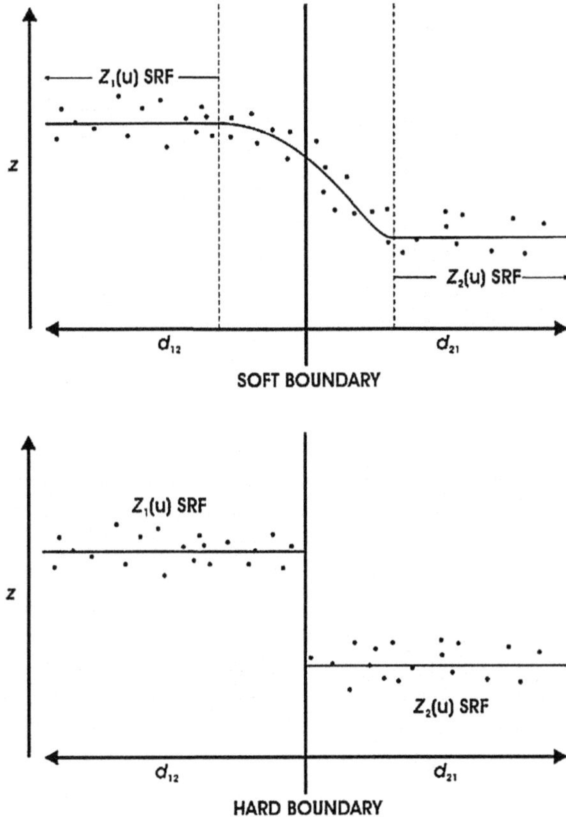

Fig. 8 The general form of a contact plot for both a soft and a hard boundary (modified from [4])

plot can be classified as one of these four types based on the slope of the expected value line and the width of the 90 % probability interval. For example, one could conclude that no boundary is present when the expected value at zero distance for each domain is within the range of variability at zero distance for the other domain. A hard boundary is likely present when the expected value at zero distance for at least one domain falls outside the range of variability for the other domain. This hard boundary is considered stationary when neither domain exhibits a strong trend near the boundary. The hard boundary would be considered nonstationary when a significant trend is present. A soft boundary is present when the grade within one or both domains exhibits a strong trend near the boundary with no significant change in grade at the boundary.

There are two general categories for modeling the nature of boundaries: implicit and explicit [4]. The first refers to the conventional method of pasting together domains predicted from separate stationary random functions and conditioning data sets. This is appropriate for hard boundaries. For soft boundaries, a near-boundary model describing how separate stationary random functions interact is needed to build in realistic geological transitions explicitly.

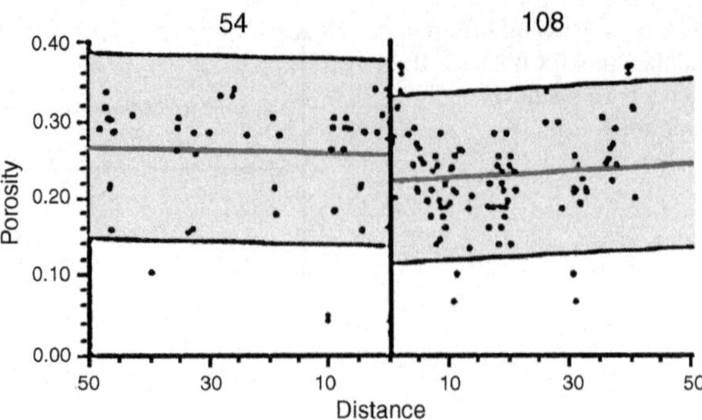

Fig. 9 Another form of contact plot showing the change in porosity between two different facies

7 Conclusions

Determining stationary domains is an important step in geostatistical modeling. The locations of the boundaries between stationary domains must be determined. There is uncertainty in the location of the boundary between stationary domains away from data. The uncertainty in boundary location should be accounted for and should be fair. The boundary uncertainty can be calibrated using the available data in a manner similar to the jackknife. An additive parameter, C, is calibrated to quantify the boundary uncertainty.

This calibration framework makes simulating boundary realizations straightforward. A uniform deviate between $-C$ and C is simulated. Comparing the simulated deviate to the interpolated distance function classifies each location as inside or outside the domain.

In addition to location, it is important to capture the nature of transition of the geological property across the boundary. The nature of this transition has important modeling implications. The boundary nature can be determined qualitatively by considering a contact plot.

References

1. Cuba M, Leuangthong O (2009) On the use of the semivariogram to detect sources of non-stationarity in mineral grades. In: Proceedings of APCOM
2. Hosseini AH (2009) Probabilistic modeling of natural attenuation of petroleum hydrocarbons. PhD Thesis, University of Alberta, 359 pp
3. Larrondo P, Deutsch CV (2004) Accounting for geological boundaries in geostatistical modeling of multiple rock types. In: Leuangthong O, Deutsch CV (eds) Geostats 2004: proceedings of the seventh international geostatistics congress. Springer, Berlin, pp 3–12
4. McLennan J (2008) The decision of stationarity. PhD Thesis, University of Alberta, 191 pp
5. Munroe MJ, Deutsch CV (2008) A methodology for modeling vein-type deposit tonnage uncertainty. Center for Computational Geostatistics Annual Report 10, University of Alberta, 10 pp
6. Munroe MJ, Deutsch CV (2008) Full calibration of C and in the framework of vein-type deposit tonnage uncertainty. Center for Computational Geostatistics Annual Report 10, University of Alberta, 16 pp
7. Ortiz JM, Emery X (2006) Geostatistical estimation of mineral resources with soft geological boundaries: a comparative study. J S Afr Inst Min Metall 106(8):577–584

Assessing Uncertainty in Recovery Functions: A Practical Approach

Oscar Rondon

Abstract Recovery functions provide one of the most important tools in the mining industry for summarizing mineral inventory information of a deposit as a function of cut-off grades. They are also used in several stages of the deposit evaluation, including mine planning, financial decision making and management. Recovery functions are commonly obtained from block grade estimates or by mean of change of support techniques. However, these practices do not allow for an assessment of the underlying uncertainty associated with them. This limitation can be overcome by generating multiple conditional simulations of the deposit from which the recovery functions are computed. However, this approach is time consuming and may not be feasible for deposits modeled with a large number of blocks. In this paper, a technique is proposed for simulating recovery functions and assessing the corresponding uncertainty without recourse to conditional simulation. Its application in a number of deposits has shown that if the sole purpose is to assess the uncertainty in the recovery functions then the technique can be used to eliminate the necessity of carrying out multiple conditional simulations.

1 Introduction

Assessment of an in-situ resource, i.e. the amount of mineral available in an area to be mined irrespective of whether this amount can be recovered economically or not [4], is usually carried out by reporting recovery functions such as ore tonnage, quantity of metal, mean ore grade and conventional profit from block grade estimates or change of support techniques as functions of cut-off grade. These traditional practices have the disadvantage of not providing any information for assessing the underlying uncertainty related to reported recovery figures.

O. Rondon
Golder Associates, Level 3, 1 Havelock Street, West Perth, Perth, WA 6005, Australia

Present address:
O. Rondon (✉)
Optiro, Level 4, 50 Colin Street, West Perth, Perth, WA 6005, Australia
e-mail: orondon@optiro.com

P. Abrahamsen et al. (eds.), *Geostatistics Oslo 2012*, Quantitative Geology and Geostatistics 17,
DOI 10.1007/978-94-007-4153-9_24,

Conditional simulation can be used to obtain multiple equiprobable realizations of the ore body from which recovery functions at a given block support are reported to assess the corresponding uncertainty. In practice, this approach is time consuming and may not be worth doing it if the resulting simulations are not used for further purposes, such as evaluation of recoverable reserves or mine planning optimization.

Recovery functions provide global figures which do not depend on the spatial location of estimated or simulated block grades but on their statistical distribution at the corresponding block support [8]. Therefore, if the sole purpose is to assess the uncertainty in the recovery functions, a change of support model can be used to derive the statistical distribution at a given block support. This approach is described in this paper by using the discrete Gaussian model for change of support. Through a number of case studies it has been shown that results obtained compare favorably with the outcomes from conditional simulations.

2 The Discrete Gaussian Model

Let $Z(x)$ and $Z(v)$ be the grade at point support and block support v respectively. The discrete Gaussian model introduced by Matheron [7] expresses $Z(x)$ and $Z(v)$ as functions of two standard Gaussian variables $Y(x)$ and Y_v as

$$Z(x) = \phi\big(Y(x)\big), \tag{1}$$

where ϕ is the point anamorphosis function derived from the point support data [8] and

$$Z(v) = \phi_v(Y_v), \tag{2}$$

where ϕ_v is the block anamorphosis function which is derived via Cartier's relation by further assuming that the two standard Gaussian variables $Y(x)$ and Y_v have joint Gaussian distribution with correlation coefficient $r > 0$ [5, 8]. This implies that the block anamorphosis function ϕ_v is given by

$$\phi_v(y) = \int_{-\infty}^{+\infty} \phi\big(ry + u\sqrt{1-r^2}\big) g(u)\,du, \tag{3}$$

where g denotes the standard Gaussian density [5, 8]. The coefficient r corresponds to the variance correction factor from point to block support and is chosen so as to respect the variance of $Z(v)$ by inverting

$$\mathrm{Var}\big(Z(v)\big) = \sum \phi_n^2 r^{2n}, \tag{4}$$

where the coefficients ϕ_n correspond to the expansion of the point anamorphosis function ϕ in terms of Hermite polynomials [5, 8]. An alternative method has been proposed by Emery [2] which allows to compute r without inverting (4).

It is worth noting that even if the point support distribution is considered known, the problem of determining the distribution of $Z(v)$ is still undetermined and the choice of an appropriate change of support model is of crucial importance [6]. In what follows, it is assumed that the discrete Gaussian model is a plausible model for approximating the distribution of block grades and therefore, the methodology presented here is largely limited by the suitability of the discrete Gaussian model to the data being modeled.

3 Variability in Recovery Functions

Computation of ore tonnage,

$$T(z) = E[1_{Z(v)\geq z}], \tag{5}$$

and quantity of metal,

$$Q(z) = E\big[Z(v)1_{Z(v)\geq z}\big], \tag{6}$$

above cut-off z delivers only average values from which the average grade,

$$M(z) = \frac{Q(z)}{T(z)}, \tag{7}$$

and conventional profit

$$B(z) = Q(z) - zT(z), \tag{8}$$

above cut-off are obtained. Therefore, it is not possible to access the full range of variability in the corresponding recovery functions. To overcome this limitation the discrete Gaussian approach can be used in any of the following two ways: First, one may compute theoretical limits based on percentiles of grades above cut-off or second, one may compute a number of simulated block grades which are used to report the recovery functions.

Theoretical limits on grades above cut-off can be obtained by noting that if z_τ stands for the τ-th percentile of block values $Z(v)$ above cut-off z then

$$\begin{aligned}\tau &= P\big(Z(v) \leq z_\tau \,\big|\, Z(v) \geq z\big)\\ &= \frac{P(z \leq Z(v) \leq z_\tau)}{P(Z(v) \geq z)}\\ &= \frac{P(y \leq Y_v \leq y_\tau)}{P(Y_v \geq y)},\end{aligned} \tag{9}$$

where $y = \phi_v^{-1}(z)$ and $y_\tau = \phi_v^{-1}(z_\tau)$ correspond to the equivalent Gaussian cut-off and τ-th percentile respectively. Since Y_v has standard Gaussian distribution with cumulative distribution G one has

$$\tau = \frac{G(y_\tau) - G(y)}{1 - G(y)}, \tag{10}$$

and therefore $y_\tau = G^{-1}(\tau + G(y)(1-\tau))$ from which $z_\tau = \phi_v(y_\tau)$ is obtained. By varying the percentile τ it is then possible to obtain theoretical limits on block grades above cut-off.

To simulate block grades, one should note that a key property of the discrete Gaussian model is that Y_v has standard Gaussian distribution regardless of the block support v. Therefore, by simulating a number N of standard Gaussian variables $Y_{v_1}, \ldots, Y_{v_N}$, it is possible to obtain a number N of simulated block grades $Z_1(v), \ldots, Z_N(v)$ via the block anamorphosis function. Simulated Ore Tonnage $\widetilde{T}(z)$ and metal $\widetilde{Q}(z)$ are then computed as

$$\widetilde{T}(z) = \frac{1}{N}\sum_{i=1}^{N} 1_{Z_i(v)\geq z}, \tag{11}$$

$$\widetilde{Q}(z) = \frac{1}{N}\sum_{i=1}^{N} Z_i(v) 1_{Z_i(v)\geq z} \tag{12}$$

for each cut-off z, from which the simulated average grade $\widetilde{M}(z) = \widetilde{Q}(z)/\widetilde{T}(z)$ and conventional profit $\widetilde{B}(z) = \widetilde{Q}(z) - z\widetilde{T}(z)$ are obtained [4]. By considering another set of input standard Gaussian simulated values and repeating the process, a number of simulated recovery functions can be obtained to assess their variability.

This approach is extremely fast and does not demand the computational resources required to carry out and store the conditional simulations.

3.1 Accounting for the Information Effect

The information effect refers to the fact that a pay block v is defined according to whether its estimate $Z^*(v)$ and not the real block grade $Z(v)$ is above an economic cut-off z or not. Therefore, some ore blocks will be misclassified as waste and vice versa.

Modeling of the information effect is carried out by assuming that

$$Z^*(v) = \phi_v^*(Y_v^*) \tag{13}$$

can be written as a weighted linear combination of samples with positive weights summing up to one [1, 5], where Y_v^* stands for a standard Gaussian variable and ϕ_v^* corresponds to the anamorphosis function obtained from the point anamorphosis function but with correction factor s chosen so as to respect the variance of $Z^*(v)$ by inverting (14).

$$\mathrm{Var}\big(Z^*(v)\big) = \sum \phi_n^2 s^{2n}. \tag{14}$$

Note that this allows to model the distribution of $Z^*(v)$ from which effective ore tonnage,

$$T^*(z) = E[1_{Z^*(v)\geq z}], \tag{15}$$

can be computed as well as theoretical limits on estimated block grades as shown in (9). Moreover, by further assuming that Y_v^* and Y_v have a joint Gaussian distribution with correlation coefficient $\rho > 0$ satisfying

$$\mathrm{Cov}\big(Z^*(v), Z(v)\big) = \sum \phi_n^2 r^n s^n \rho^n \tag{16}$$

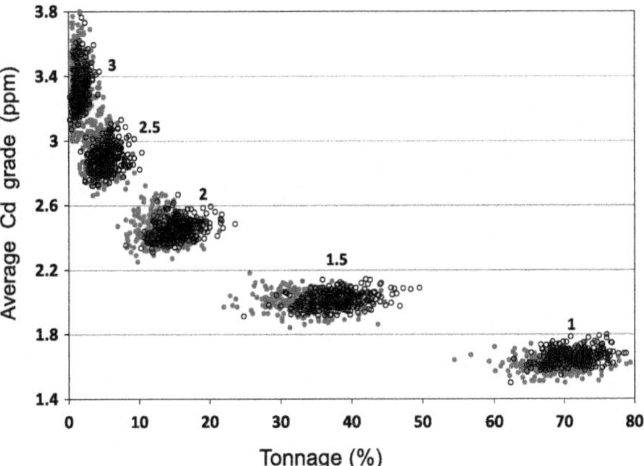

Fig. 1 Comparison of average grades and tonnes above indicated cut-offs obtained from conditional simulations (*black*) and the discrete Gaussian model (*red*)

the effective metal can be computed as [1, 5]

$$Q^*(z) = E\left[Z(v)1_{Z^*(v)\geq z}\right]. \tag{17}$$

Simulation of recovery functions taking into account the information effect can be carried out by considering

$$\widetilde{T}^*(z) = \frac{1}{N}\sum_{i=1}^{N} 1_{Z_i^*(v)\geq z}, \tag{18}$$

$$\widetilde{Q}^*(z) = \frac{1}{N}\sum_{i=1}^{N} Z_i(v)1_{Z_i^*(v)\geq z}, \tag{19}$$

where $Z_i(v) = \phi_v(Y_{v_i})$ as before and $Z_i^*(v) = \phi_v^*(Y_{v_i}^*)$. Here $Y_{v_i}^*$ is given by

$$Y_{v_i}^* = \rho Y_{v_i} + \sqrt{1-\rho^2}U_i \tag{20}$$

with U_i an independently simulated standard Gaussian variable. This step ensures the correlation ρ between $Y_{v_i}^*$ and Y_{v_i} is preserved.

4 An Illustrative Example

To illustrate the technique, 300 conditional simulations of cadmium (Cd) grades (ppm) at a support of 10 m by 10 m are computed from 259 declustered Cd grades from the well-known Jura data set [3]. The simulations are then reblocked to a support of 150 m by 150 m and corresponding recovery figures compared to the outcomes of the proposed approach using the same number of simulations.

Figure 1 summarizes the comparison of tonnes and average block grades above cut-off. The results show that variations in these figures computed from simulations obtained via the discrete Gaussian model are within the range of variability of corresponding figures computed from conditional simulations of Cd. This example

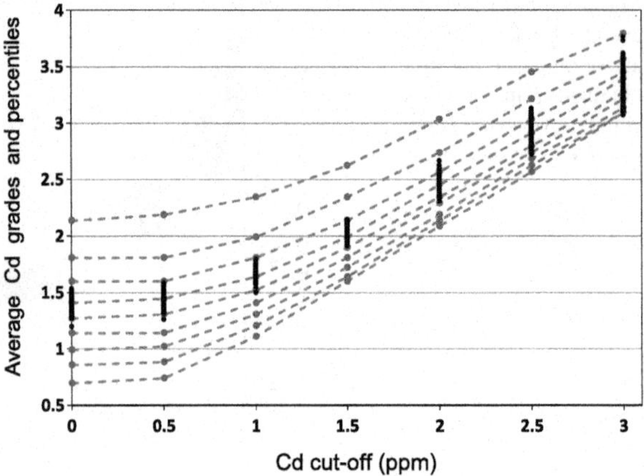

Fig. 2 Average Cd block grades obtained from conditional simulations (*black*) and theoretical percentiles on block grades above cut-off (*red*) derived from the discrete Gaussian model. Percentiles are reported by increments of 10 starting at the 10th percentile

shows that the proposed approach may be used to assess the variability in recovery functions without carrying out the conditional simulations.

Computation of theoretical percentiles above cut-off allows the assessment of variations not on average block grades but on block grades above cut-off. The variability is assessed by analyzing the percentiles at each cut-off. For instance, one can conclude from Fig. 2 that there is an 80 % probability of having block grades less than 3.03 ppm Cd and greater than 2.08 ppm Cd, which correspond to the 90th and 10th theoretical percentiles respectively.

5 An Iron Ore Deposit

Iron (Fe) grades (%) from an iron ore deposit with a nominal drill spacing of 100 m by 100 m are used in this case study to compare the variability in recovery functions obtained from conditional simulations and the proposed approach. The experimental variogram computed using all available data was fitted with a variogram model comprising up to three structures, a nugget and two spherical structures.

The comparison is carried out by computing recovery functions from 300 direct block conditional simulations at a block support of 25 m by 25 m by 2 m and comparing the results to those obtained via the discrete Gaussian model using a variance correction factor of 0.84. Figure 3 displays tonnes, metal, average grade and conventional profit as functions of the cut-off grade. The results show that simulated recovery functions obtained via the discrete Gaussian model reasonably reproduce the corresponding variability of the recovery functions obtained with the conditional simulations. However, for this deposit, setting up and running the direct block conditional simulations required a lot more effort and computational resources.

Modeling the distribution of the estimates $Z^*(v) = \phi_v^*(Y_v^*)$ as described before allows the assessment of the variability in estimated tonnes and average Fe grades above cut-off for an ordinary kriging model with a block support of 25 m by 25 m

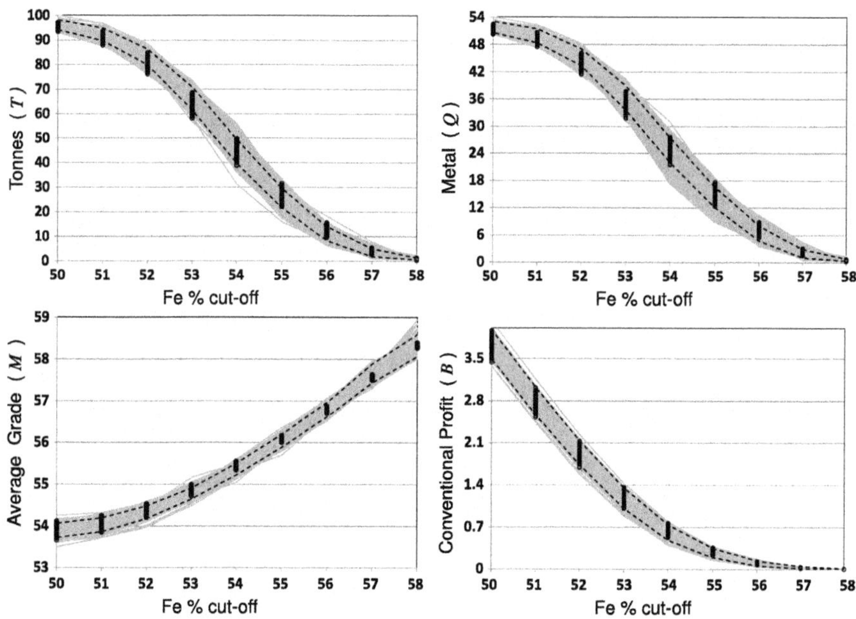

Fig. 3 Recovery functions at a block support of 25 m by 25 m by 2 m derived from direct block conditional simulations of Fe grades (*black*) and the discrete Gaussian model (*grey lines*) as functions of the cut-off grade. *Dashed lines* represent the 10th and 90th percentiles from the distribution of corresponding simulated recovery functions

by 2 m available for this deposit. Figure 4 displays the estimated tonnes and grades and their corresponding variability.

Variations on tonnes show what it is expected from an ordinary kriging model. Tonnes are overestimated for low Fe cut-offs and underestimated for high Fe cut-offs. However, it is now possible to assess the extent to which the ordinary kriging model could be overestimating or underestimating tonnes. With respect to grades, simulations suggest one could expect to have slightly higher estimated average grades above cut-off.

6 Conclusions

Traditional in-situ recovery functions are reported without any information regarding to potential variability in the reported figures. In this article, a simple and practical approach based on the discrete Gaussian model for change of support was described to assess the uncertainty in such recovery functions. Comparison to outcomes from actual conditional simulations in a number of case studies suggest the technique can be used to quickly assess the corresponding variability without carrying out a number of conditional simulations.

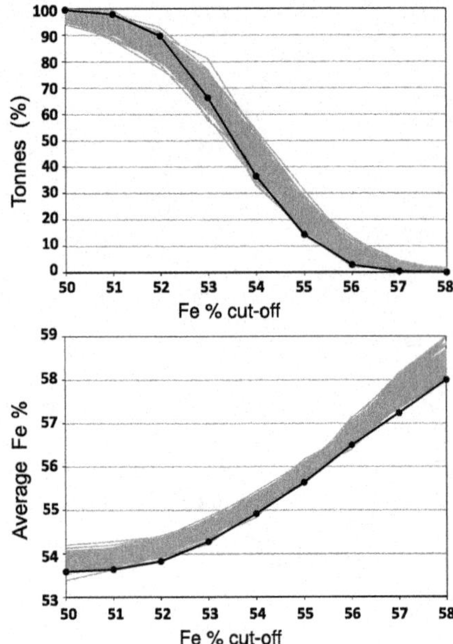

Fig. 4 Tonnes (*top*) and grades (*bottom*) as function of different Fe cut-offs obtained from ordinary kriging estimates (*black line*) and simulated values (*grey lines*) after modeling the distribution of the estimates at a block support of 25 m by 25 m by 2 m

Acknowledgements The author wishes to thank professor Ute Mueller for her valuable help during the preparation of the manuscript.

References

1. Deraisme J, Roth C (2001) The information effect and estimating recoverable reserves. In: Kleingeld WJ, Krige DG (eds) Proceedings geostatistics 2000, Cape Town, Geostatistical Association of Southern Africa
2. Emery X (2007) On some consistency conditions for geostatistical change of support models. Math Geol 39(2):205–223
3. Goovaerts P (1997) Geostatistics for natural resources evaluation. Oxford University Press, New York, 483 pp
4. Journel A, Kyriakidis P (2004) Evaluation of mineral reserves. A simulation approach. Oxford University Press, London, 216 pp
5. Lantuéjoul Ch (1994) Cours de Selectivité. Ecole des Mines de Paris, Centre de Geostatistique, Fontainebleau
6. Lantuéjoul Ch (1998) On the importance of choosing a change of support model for global reserves estimation. Math Geol 20(8):1001–1019
7. Matheron G (1976) Forecasting block grade distributions: the transfer functions. In: Guarascio M et al (eds) Advanced geostatistics in the mining industry, pp 151–239
8. Rivoirard J (1994) Introduction to disjuntive Kriging and non-linear geostatistics. Clarendon, Oxford, 180 pp

Comparative Study of Localized Block Simulations and Localized Uniform Conditioning in the Multivariate Case

Jacques Deraisme and Winfred Assibey-Bonsu

Abstract The general indirect estimation technique of recoverable resources during long-term planning derives the unknown Selective Mining Unit (SMU) distribution from the modeled distribution within large kriged blocks (panels). The Multivariate Uniform Conditioning (MUC) technique provides a consistent framework to achieve this task with the practical advantage, in that no specific hypothesis on the correlation between the respective secondary elements is required. The Localized MUC (LMUC) technique has been developed to enhance the indirect MUC by localizing the results at the SMUs scale. The paper investigates the possibility of improving the LMUC estimates through multivariate block simulations which incorporate all the correlations of the secondary and main elements. The tonnages and metals represented by the simulated grade tonnage curves are used to derive probable tonnages and metals, which are decomposed and distributed into the SMUs referred to as Localized Multivariate Simulated Estimates or LMSE. After a review of MUC, Direct Block Simulations (DBS), LMUC and LMSE a comparative case study based on a porphyry copper gold deposit in Peru is presented.

1 Introduction

Uniform Conditioning (UC) consists of estimating the grade distribution on selective mining unit (SMU) support within a panel, conditioned to a panel grade, usually estimated by Ordinary Kriging to circumvent non-stationarity issues. The general framework which forms the basis of Uniform Conditioning is the Discrete Gaussian Model (DGM) of change of support, based in particular on the correlation between Gaussian-transformed variables. Rivoirard [6] and Emery [3] discuss some of the tests to determine if UC is a suitable methodology for a particular type of mineral-

J. Deraisme (✉)
Geovariances, 49 Ave Franklin Roosevelt, Avon 77212, France
e-mail: deraisme@geovariances.com

W. Assibey-Bonsu
Geostatistics & Evaluation, Gold Fields Ltd., PO Box 628, West Perth, WA 6872, Australia
e-mail: Winfred.AssibeyBonsu@goldfields.com.au

P. Abrahamsen et al. (eds.), *Geostatistics Oslo 2012*,
Quantitative Geology and Geostatistics 17,
DOI 10.1007/978-94-007-4153-9_25,

ization. They have been performed as part of the case study and confirm the diffusive nature of the mineralization compatible with the application of the DGM. The DGM applies to the multivariate case, where the correlations between all variables on any support can be calculated after transformation into Gaussian space. More precisely, the model only requires specifying the correlations between any variable and the primary or main variable, cut-offs being applied to the main economic element of the deposit; while the correlations between the secondary variables are not modeled explicitly.

Moreover when the selection of SMUs is based on a combination of different elements, Multivariate Uniform Conditioning as applied in this paper, uses a new variable or *equivalent* grade for the block estimate expressed by means of a net smelter returns (NSR). This NSR variable is calculated as a linear combination of grade variables, the weights being derived from the economic valuation of the respective elements. It is then declared as the main variable. The drawback is that the formula used to calculate the NSR variable depends on economic parameters that may vary during the mine life. Besides, the correlations between the different element grades are not used in the model and can only be checked after the estimates have been derived....

In order to appreciate the impact of such a decision (i.e., of cut-off being applied to the main economic element of the deposit); the paper investigates an alternative approach based on block co-simulations. In this latter case, the linear model of coregionalization is fully incorporated in the analyses, i.e. all correlations between any pair of variables are modeled. The expected values of tonnages and metals after applying a cut-off within panels from a series of simulated SMUs are derived and compared with the corresponding grade tonnage curves calculated from the MUC approach.

To compare the alternative techniques, at a local scale, the grade tonnage curves obtained by both approaches have been assigned to each SMU by using a localization post-processing method (see Deraisme [7]). These localized estimates are referred to as Localized Multivariate Uniform Conditioning (LMUC) or Localized Multivariate Simulated Estimates (LMSE) in the paper.

To optimize the simulations process a direct block simulation (DBS) method has been used that is also based on the properties of the DGM. In the DBS method the change of support coefficient is calculated from the multivariate variogram model of Gaussian variables (see Emery [3]), which is different from the change of support coefficient calculated for MUC.

2 Models for Nonlinear Geostatistics

2.1 Basis of the Discrete Gaussian Model

Let v be the generic selection block (SMU) and $Z(v)$ its grade, that will be used for the selection at the future time of exploitation (we assume that this grade will then be perfectly known, i.e. there is no information effect). The recoverable resources above cutoff grade z for such blocks are:

$$\begin{aligned} \text{Ore:} \quad & T(z) = \mathbb{1}_{Z(v)\geq z}, \\ \text{Metal:} \quad & Q(z) = Z(v)\mathbb{1}_{Z(v)\geq z}. \end{aligned}$$

We use here the discrete Gaussian model for change of support (e.g. Rivoirard, 1994). A standard Gaussian variable Y, with pdf g, is associated with each raw variable Z. Let $Z(x) = \Phi(Y(x))$ be the sample point anamorphosis. The block model is defined by its block anamorphosis $Z(x) = \Phi_r(Y_v)$, given by the integral relation:

$$\Phi_r(y) = \int_{\mathbb{R}} \Phi\big(ry + \sqrt{1-r^2}u\big) g(u)\,du, \tag{1}$$

where the change of support coefficient r is obtained from the variance of blocks. Then the global resources at cutoff z are

$$\text{Ore:} \quad E\big[T(z)\big] = E[\mathbb{1}_{Z(v)\geq z}] = E[\mathbb{1}_{Y_v\geq y}] = 1 - G(y), \tag{2}$$

$$\text{Metal:} \quad E\big[Q(z)\big] = E\big[Z(v)\mathbb{1}_{Z(v)\geq z}\big] = E\big[\mathbb{1}_{Y_v\geq y}\Phi_r(Y_v)\big] = \int_y \Phi_r(u)g(u)\,du, \tag{3}$$

where G is the standard Gaussian cdf, and y is the Gaussian cutoff related to z through $z = \Phi_r(y)$.

In the multivariate case indices are added to distinguish the variables. Let Z_1 be the metal grade used for the main variable, and let Z_2 be one of the secondary metal grades. In addition to the univariate case seen above, we now wish to estimate the other metals, for instance:

$$Q_2(z) = Z_2(v)\mathbb{1}_{Z_1(v)\geq z}.$$

Its global estimation is given by:

$$\begin{aligned} E\big[Q_2(z)\big] &= E\big[Z_2(v)\mathbb{1}_{Z_1(v)\geq z}\big] \\ &= E\big[\mathbb{1}_{Z_1(v)\geq z} E\big[Z_2(v)\big|Z_1(z)\big]\big] \\ &= E\big[\mathbb{1}_{Y_{1v}\geq y} E\big[\Phi_{2,r_2}(Y_{2v}\big|Y_{1v})\big]\big] \\ &= E\big[\mathbb{1}_{Y_{1v}\geq y} \Phi_{2,r_2\rho_{1v,2v}}(Y_{1v})\big] \\ &= \int_y \Phi_{2,r_2\rho_{1v,2v}}(u)g(u)\,du, \end{aligned} \tag{4}$$

where r_2 is the change of support coefficient for Z_2, and Y_{1v} and Y_{2v} are bi-Gaussian, with a correlation $\rho_{1v,2v}$.

2.2 UC in the Multivariate Case

Multivariate UC [5] involves estimating the recoverable resources of blocks v in panel V from the sole vector of panel estimates $(Z_1(V)^*, Z_2(V)^*, \ldots)$ The problem is simplified by assuming that:

- $Z_1(v)$ is conditionally independent of the auxiliary metal panel grades given $Z_1(V)^*$, and so the UC estimates for the selection variable correspond to the univariate case.
- Similarly, $Z_2(v)$ is conditionally independent of $Z_1(V)^*$ given $Z_2(V)^*$.
- $Z_1(v)$ and $Z_2(v)$ are, conditionally independent of the other metal panel grades given $(Z_1(V)^*, Z_2(V)^*)$. It follows that the multivariate case reduces to a multi-bivariate case. In particular we have:

$$[Q_{2V}]^* = E\big[Z_2(v)\mathbb{1}_{Z_1(v)\geq z}\big|Z_1(V)^*, Z_2(V)^*\big]. \tag{5}$$

We further impose, for the metal at cut-off 0:

$$E\big[Z_2(v)\big|Z_1(V)^*, Z_2(V)^*\big] = Z_2(V)^* = E\big[Z_2(v)\big|Z_2(V)^*\big]. \tag{6}$$

This is similar to the univariate case, so that $Z_2(V)^*$ must be conditionally unbiased. The model is entirely specified by the anamorphosis, the different change of support coefficients, and the correlations between the Gaussian variables $(Y_{1v}, Y_{1V^*}, Y_{2v}, Y_{2V^*})$. The correlation between Y_{1v} and Y_{2v}, and that between Y_{1V^*} and Y_{2V^*} allow completing the correlations by using the conditional independence relationships. For more details on the equations see [2].

2.3 Direct Block Simulations (DBS)

The DGM relies on the partition of the domain into small blocks v. Then each sample point is considered as random within its block, and conditioned to its block value (here the multivariate value of the different elements), the point (multivariate) value does not depend on any other variable, whether they are values of other blocks or other points, even in the same block. This allows the deduction of the point-point and point-block covariances from the block-block covariances. After anamorphosis, all Gaussian values are considered as multi-Gaussian, allowing conditional simulation to be implemented.

The methodology [3] considers that the Gaussian transform on block support is nothing but the regularized point Gaussian variable, normalized by its variance, i.e. the square of the change of support coefficient r. It leads to another determination of the change of support coefficient, using the variogram and variance of the Gaussian variable, instead of the raw variable variograms used in the previous method:

$$r^2 = \operatorname{var} Y(v) = \operatorname{var} Y(x) - \overline{\gamma_Y(v, v)} \simeq 1 - \overline{\gamma_Y(v, v)}.$$

Developing that methodology leads to a consistent method for simulating directly the block grades conditioned to *point* data. The method is based on the use of the turning bands simulation technique (see Lantuejoul [4]), characterized by the fact that the simulations are achieved in two steps: first non-conditional simulations to reproduce the variogram model and second conditioning of the simulations by co-kriging from the data.

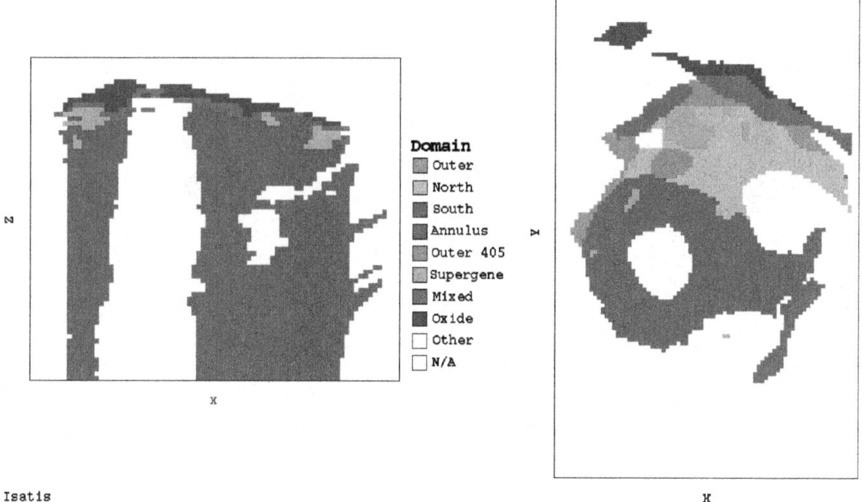

Fig. 1 Example of West East and plan sections with colored estimation domains

3 Case Study

3.1 Geology of the Deposit

The case study is based on a porphyry copper gold deposit in Peru. The mineralization is found in intrusive rocks within sedimentary rocks. Oxidation, weathering, leaching and subsequent secondary enrichment has led to the formation of four mineral domains with distinct metallurgical behaviors (Fig. 1). All of the ore beneath the Oxide Domain comprises parts of the sulphide zone, which is separated into three main domains on the basis of degree of oxidation and consequent change in sulphide mineralogical composition. The sulphide zone has three main domains, which from top to bottom are the Mixed Domain, the Supergene Domain and the Hypogene Domains. The Supergene Domain is an enriched copper blanket comprising chalcocite-covellite-chalcopyrite. The study presented in this paper was conducted in one of the Hypogene Domains. The variables studied are total gold (AUTOT), total copper (CUTOT) and a combination of gold and copper grades giving an economic value, Net Smelter Return (NSR).

3.2 Data Analysis

More than 5000 composited data on 2 m cores, later considered as point supports, have been used to perform LMUC and LMSE on $10 \times 10 \times 10$ m^3 SMU support from:

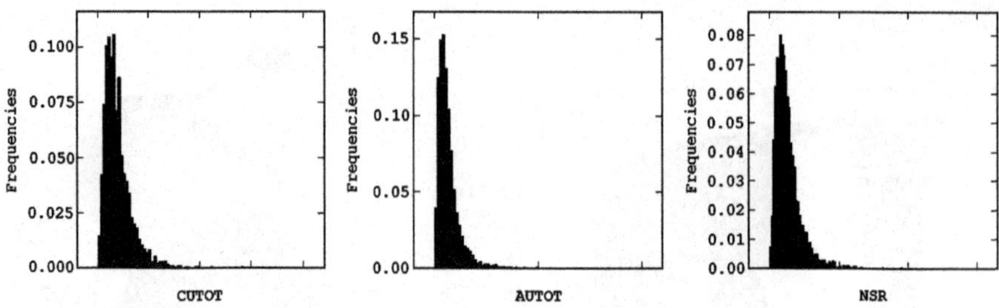

Fig. 2 Histograms of the 2 m composites for CUTOT, AUTOT and NSR

Table 1 Matrix of coefficients of correlation between 3 variables on 2 m composites

	CUTOT	AUTOT	NSR
CUTOT	1	0.68	0.90
AUTOT	0.68	1	0.94
NSR	0.90	0.94	1

- MUC calculated on 50 m × 50 m × 10 m panels with NSR as the main variable.
- 50 blocks co-simulations of CUTOT and AUTOT.

The three variables have a positively skewed distribution (Fig. 2) with coefficients of variation from 0.7 to 0.9. Besides the correlations are highly significant (see Table 1).

3.3 Variographic Analysis

Two different variographic analyses were performed in order to carry out the analyses for both approaches. To sum up there are two important differences:

- Firstly, for MUC a variogram model of raw grades is used while a variogram model of Gaussian grades is used for the simulations.
- Secondly, for MUC the variogram model concerns the 3 variables NSR-CUTOT-AUTOT, while for the simulations only the 2 original grades variables CUTOT-AUTOT have to be modeled. In both cases the models are multivariate linear model of co-regionalization. The Net Smelter return is calculated only after having obtained the simulated block values of grades.

MUC requires the calculation of change of support coefficients on the SMU support and cokriging of the panels for the 3 variables NSR-CUTOT-AUTOT, NSR being the main variable. The variograms have been calculated from the raw variables and modeled by a small nugget effect and two spherical variograms with longer ranges vertically than horizontally (Fig. 3).

For simulating CUTOT and AUTOT the process is the following:

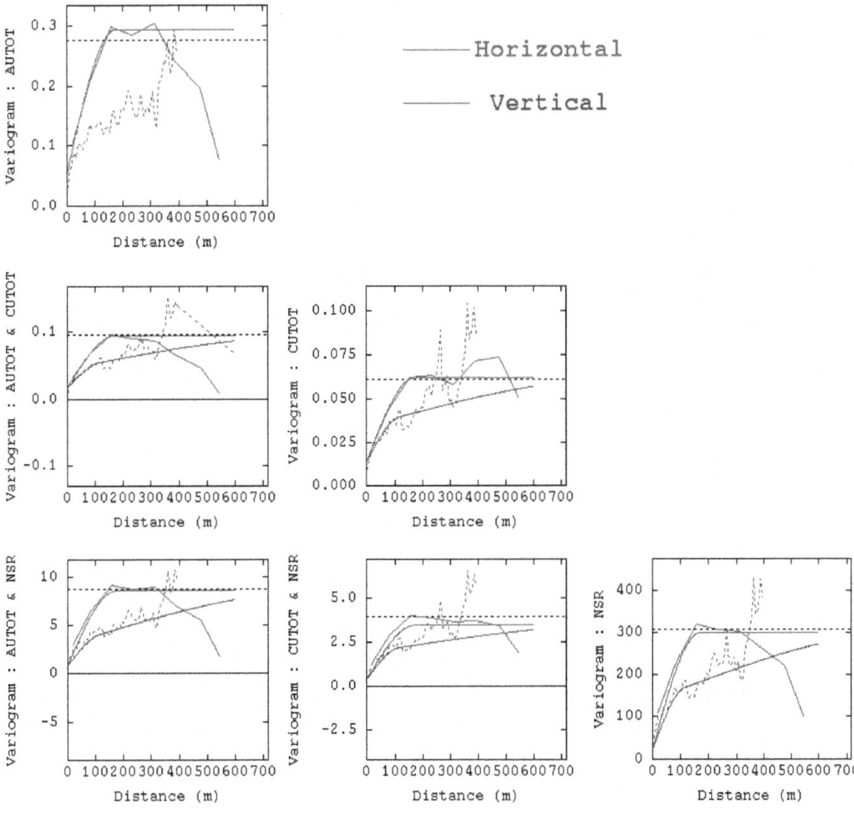

Fig. 3 Experimental (*dotted line*) and modeled (*single line*) variograms and cross-variograms for NSR, CUTOT and AUTOT in horizontal and vertical directions

- normal score transforms of both variables,
- variograms of the normal variables are calculated and modeled,
- the variogram model is then regularized on the SMU support,
- a variogram model on the Gaussian variables regularized on the SMU support was then fitted. On the block support the nugget effect is almost non-existent.

3.4 Change of Support

The change of support coefficients are calculated in two different ways for MUC (Table 3) and DBS (Table 2). The change of support used in MUC is calculated:

- For the SMU support, from the variance computed from each variogram model of raw data.
- For the kriged panels, from the theoretical dispersion variance of the cokriging of the panels. In order to account for the heterogeneity of the cokriging configura-

Table 2 Change of support coefficients of SMUs, calculated from the variogram of raw data and used for MUC

	NSR	CUTOT	AUTOT
Punctual Variance (Anamorphosis)	307.934	0.061	0.277
Variogram Sill	300.000	0.062	0.295
Gamma(v, v)	36.685	0.017	0.062
Real Block Variance	271.549	0.045	0.214
Real Block Support Correction (r)	0.951	0.876	0.908
Main-Secondary Block Support Correction	–	0.978	0.985

Table 3 Change of support coefficients for SMUs, calculated from the variogram of Gaussian transforms of the data

	CUTOT	AUTOT
Real Block Variance	0.044	0.201
Real Block Support Correction (r)	0.874	0.886
Correlation between Gaussian variables		0.860

tions, the panels can be classified according to the variance of the main variable, with a value of the change of support coefficient that depends on the class.

For the simulations the change of support coefficients result from the regularization of the Gaussian variogram model on the SMU support. The change of support coefficients are rather close to the ones obtained in the previous calculations.

3.5 *Results*

After having performed the calculations of MUC and DBS, the results are analyzed at the global then at the local scale.

3.5.1 Comparison of the Global Grade Tonnage Curves Obtained by MUC and DBS

The grade tonnage curves are immediately derived from the MUC results. For calculating the same curves from 50 simulations the following procedure was used:

- The NSR values were calculated for each SMU and each simulation using the same formula as used for the composites.
- For the SMUs regrouped into panels 50 m × 50 m × 10 m, the tonnages and metals for each cut-off is calculated for each SMU.
- The 50 possible tonnages and metals values are then averaged in order to get an estimate of the recovered tonnage and metals.

Figure 4 shows that globally, and for the entire domain, both curves are similar.

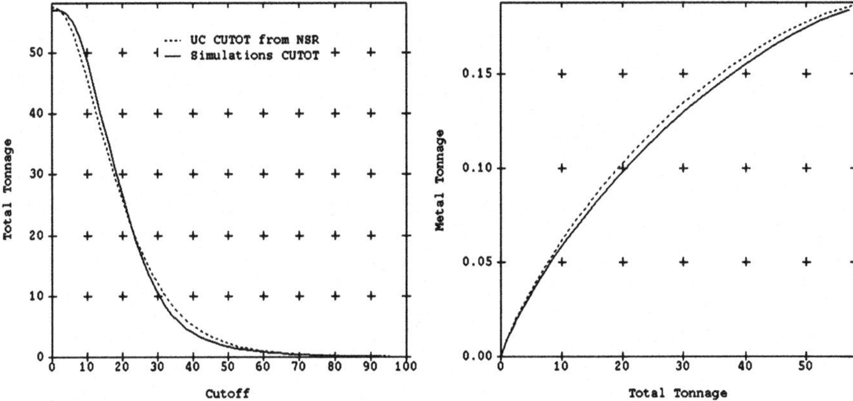

Fig. 4 Grade-tonnage curves on SMUs CUTOT calculated from MUC or DBS

3.5.2 Comparison of the Local Grade Tonnage Curves Obtained by MUC and DBS

The grade tonnage curves calculated at the scale of each panel by both methods have been localized on the SMU by using as a guide the same kriged estimate of the NSR variable (see Abzalov [1]). In other words the highest grades from the grade tonnage curves are assigned to the SMU whose kriged NSR is the highest and so on. The results are called respectively LMUC and LMSE.

An example of a bench with LMUC grades assigned from MUC or LMSE from simulations is shown in Fig. 5 and Fig. 6.

The similarity between both techniques can be compared by plotting the scatter diagrams between SMUs grades assigned by both approaches (Fig. 7). The comparison of both techniques was aimed at investigating the impact of not incorporating the correlation between the secondary variables when using the MUC approach. The linear coefficient of correlation resulting from both approaches is quite similar (Table 4), but as shown by Fig. 8, the scatter diagram between CUTOT and AUTOT grades shows a correlation closer to bi-Gaussian when using the simulations approach than when using the MUC approach.

This observation is probably a consequence of the strong multi-Gaussian property of the simulations achieved in the frame work of the Gaussian model. This interpretation is confirmed by the slightly lesser standard deviation of the grades distribution observed for the simulations compared to the MUC results.

4 Conclusions

When the orebody mineralization follows the conceptual framework of diffusive models, multi-Gaussian models provide a practical solution for calculating nonlinear quantities such as tonnages and metals after cut-off.

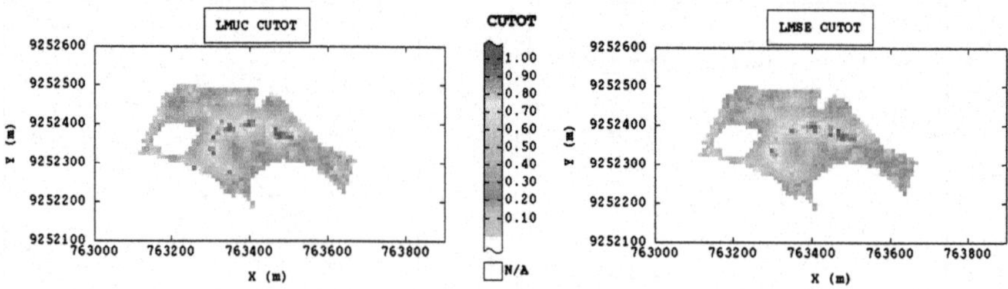

Fig. 5 Plan section of the CUTOT grades assigned to SMUs from LMUC or from LMSE, the cut-off being applied on NSR

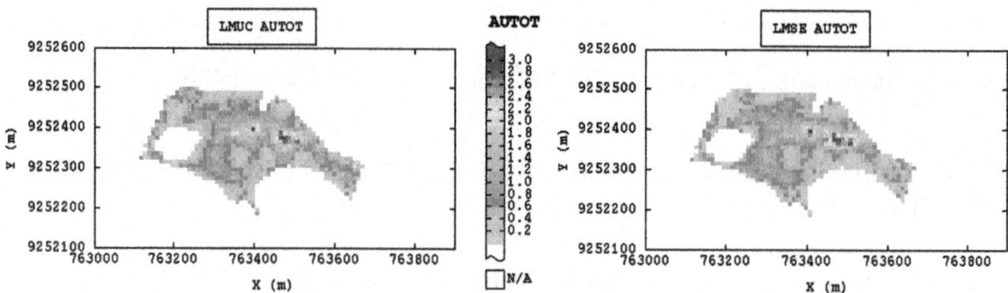

Fig. 6 Plan section of the AUTOT grades assigned to SMUs from LMUC or from LMSE, the cut-off being applied on NSR

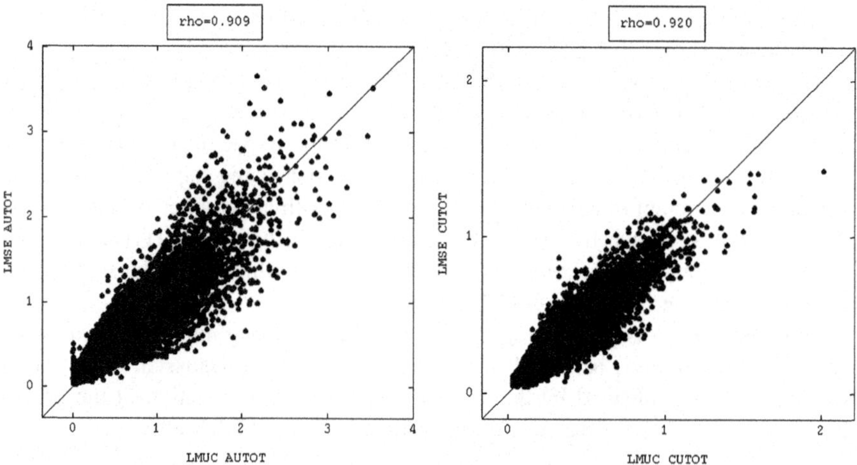

Fig. 7 Scatter diagrams between SMUs grades obtained via LMUC and LMSE for AUTOT (*left*) and CUTOT (*right*)

Two different approaches of these models are available and have been compared on a real case study using Multivariate Uniform Conditioning and Block co-simulations.

Table 4 Statistics on the grades assigned to the SMUs by LMUC and LMSE (The variance of the variable is put on the diagonal, and the coefficients of correlation of two variables out of the diagonal)

VARIABLE	LMUC CUTOT	LMUC AUTOT	LMSE CUTOT	LMSE AUTOT
LMUC CUTOT	0.175	0.893	0.920	0.859
LMUC AUTOT	0.893	0.372	0.801	0.909
LMSE CUTOT	0.920	0.801	0.157	0.898
LMSE AUTOT	0.859	0.909	0.898	0.343

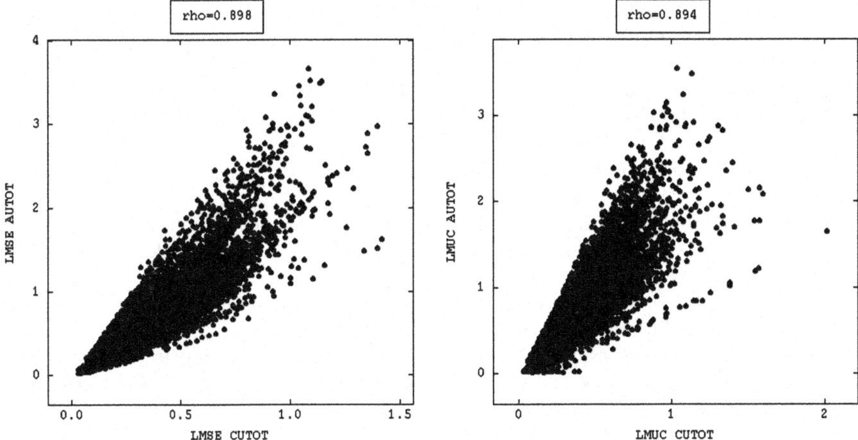

Fig. 8 Scatter diagrams between AUTOT and CUTOT grades of SMUs obtained via LMSE (*left*) and LMUC (*right*)

Preliminary conclusions can be drawn from this example: indeed other applications will help in confirming their generality.

The MUC approach does not use explicitly the correlations between the secondary variables. Nevertheless, these correlations are in practice well reproduced as confirmed by the simulation approach.

Block co-simulation are developed in a full multi-Gaussian approach, it results in making the correlations between pairs of block variables tending towards bi-gaussianity.

The comparison has been made both on the global and local scales for grade, tonnage and metal using LMUC and LMSE techniques. The study shows that both approaches lead to similar results. However, the MUC approach has the advantage of being straightforward and less time consuming.

Working with simulations is also practically feasible as soon as a dedicated algorithm for block simulations is used. Two important benefits come out of the simulation approach. The first is that all models are obtained prior to any transform by economic parameters. The second is to have access to the quantification of the uncertainty allowing for example the selection of scenarios corresponding to different risk levels or the building of localized confidence intervals.

Acknowledgements The authors are grateful to Gold Fields for permission to publish this paper based on a case study of the Group's Cerro Corona mine.

References

1. Abzalov MZ (2006) Localised uniform conditioning (luc): a new approach for direct modelling of small blocks. Math Geol 38(4):393–411
2. Deraisme J, Rivoirard J, Carrasco P (2008) Multivariate uniform conditioning and block simulations with discrete Gaussian model: application to Chuquicamata deposit. In: VIII international geostatistics congress, GEOSTATS 2008, Santiago, Chile, 1–5 December 2008. Gecamin, Santiago, pp 69–78
3. Emery X, Ortiz JM (2004) Internal consistency and inference of change-of-support isofactorial models. In: Geostatistics Banff 2004, pp 1057–1066
4. Lantuejoul C (2010) Geostatistical simulation: models and algorithms. Springer, Berlin, pp 192–204
5. Rivoirard J (1984) Une méthode d'estimation du récupérable local multivariable. Technical report N-894, CGMM Fontainebleau, Mines Paris-Tech, 1984. Edited under http://www.cg.ensmp.fr/bibliotheque
6. Rivoirard J (1994) Introduction to disjunctive kriging and non-linear geostatistics. Spatial information systems. Clarendon, Oxford
7. Deraisme J, Assibey-Bonsu W (2011) Localised uniform conditioning in the multivariate case—an application to a porphyry copper gold deposit. In: APCOM 2011 conference, Wollongong, Australia

Application of Stochastic Simulations and Quantifying Uncertainties in the Drilling of Roll Front Uranium Deposits

Gwenaële Petit, Hélène De Boissezon, Valérie Langlais, Gabrielle Rumbach, Askar Khairuldin, Thomas Oppeneau, and Nicolas Fiet

Abstract Uranium mineralization in roll front deposits mined in Kazakhstan is due to the circulation of uranium bearing oxidized fluids. Deposition associated with permeability and redox properties of the host rocks, results in very complex deposit geometries, and highly heterogeneous uranium grades. In these deep and low grade deposits, uranium is extracted by acid in situ recovery (ISR or ISL), via a network of injector and producer wells. Stochastic simulation methods allow for the characterization of ore deposit uncertainties (global and local). This paper describes a case study of uncertainty assessments in drilling campaigns. On completion of exploration drilling, the deposit is modeled in 3 dimensions with plurigaussian simulations to characterize the variability of lithology, redox context and ore body geometry. Two approaches are used and compared. The first is based on drill hole data and vertical proportion curves, whereas the second integrates the geological interpretation of the deposit through a 3D deterministic model. Uranium grades are then simulated in permeable mineralized rocks. Simulation results are summarized into probability maps. They are used to propose additional drilling for mineralization delineation. When ore limits are defined, borehole spacing inside exploitation limits is dictated by the uncertainty level accepted for resource estimation and mining plan.

G. Petit (✉) · H. De Boissezon · V. Langlais · G. Rumbach · T. Oppeneau
AREVA, La Défense, France
e-mail: gwenaele.petit@areva.com

H. De Boissezon
e-mail: helene.deboissezon@areva.com

V. Langlais
e-mail: valerie.langlais@areva.com

T. Oppeneau
e-mail: thomas.oppeneau@areva.com

A. Khairuldin · N. Fiet
Katco, Almaty, Kazakhstan

A. Khairuldin
e-mail: askar.khairuldin@areva.com

N. Fiet
e-mail: nicolas.fiet@areva.com

P. Abrahamsen et al. (eds.), *Geostatistics Oslo 2012*,
Quantitative Geology and Geostatistics 17,
DOI 10.1007/978-94-007-4153-9_26,

1 Introduction

After a short introduction to the geological environment, the mining technique and the objectives of the study, the methodology based on plurigaussian simulations [1] and direct block simulation [2] is presented. It is applied to a case study on which the results are discussed.

Roll front uranium deposits are developed by precipitation of uranium contained in oxidized bearing fluids transiting through a highly permeable formation. Changes in the formation characteristics, permeability loss, shale content, or presence of organic matter, allow trapping of the uranium. Resulting deposit geometry is usually very complex and can have a sinuous shape over several kilometers. Uranium grades are also highly heterogeneous, making difficult 3D modeling and resources estimation. Genesis of roll-front deposits leads to a strong zonation: upstream oxidized lixiviated rocks, mineralization, and reduced zone downstream.

The *In Situ Recovery* (ISR) technique comprises a network of injector and producer wells that inject a solution (acid or alkaline) into the aquifer. The uranium is then dissolved enriching the solution that is pumped out. After uranium extraction at the processing facilities, the solution is reconditioned prior to its re injection in the system.

Two important features need to be evaluated when assessing potential recoverable resources. One is to define the limits of the mineralized zone; the other is to evaluate the recoverable uranium quantity contained within these limits and its related uncertainties. One should note that the potentially recoverable resources are not only defined by a grade criterion, but should also account for the permeable character of the rock. The leaching solution will have no access to the uranium contained in non permeable rocks. The mineralization boundaries will determine the area to be equipped with the injector/producer well network. The accepted uncertainty will be dictated by the targeted resources level and/or the need for mine planning.

A practical case has been chosen to fine tune and assess the methodology. The first section will present the available data. Then the methodology will be detailed and finally its practical application to the case study is discussed.

2 Available Data and Information

The studied zone (1100 m by 1000 m by 70 m) has been densely drilled through several campaigns. Three main vintages can be identified in Fig. 1: 68 exploration drill holes that correspond to the initial data set (spacing 200 m by 50 m), 49 mining development drill holes aiming at mineralized contour definition and recoverable resource assessment (spacing 100 m by 50 m), and 98 exploitation wells dedicated to ore production (spacing 40 m by 40 m).

Most of the exploration drill holes (54 out of 68) have been cored within the productive interval. A detailed geological description in terms of lithology and redox facies is thus achieved.

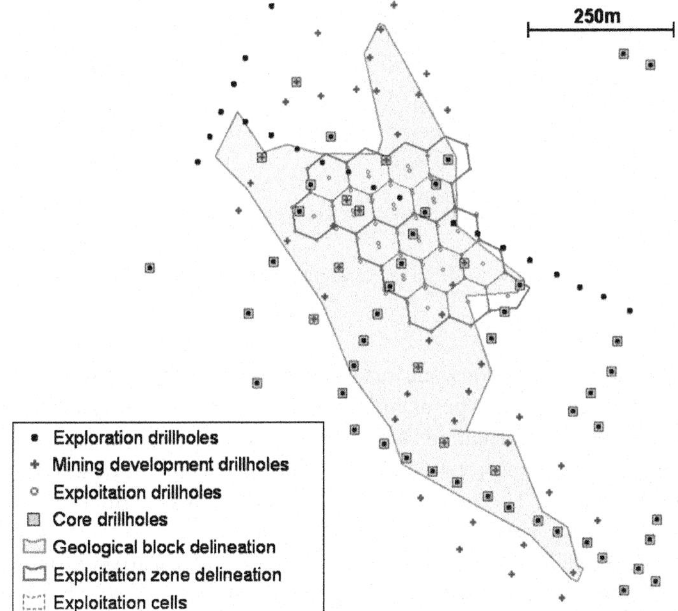

Fig. 1 Map presenting available drill hole data and information identified by drilling campaigns, geological block and exploitation limits

For all the drill holes, radioactivity gamma and resistivity logs are acquired. Accounting for various parameters related to the hole and to the rock, the gamma radiometry is transformed into equivalent uranium grade. One should note that an average disequilibrium is used between radium and uranium. The resistivity log is interpreted into different lithologies from coarse sands to shales.

3 Methodology

The objective is to define a method for the evaluation of recoverable uranium resources, which also quantifies associated uncertainties in the context of ISR technique. Simulation techniques are selected for this purpose. Simulations are run on a 10 m by 10 m by 1 m oriented grid (N116, mineralization orientation at exploration stage knowledge), leading to 770,000 cells. The methodology is based on the 117 holes available at the end of the mining development stage and is presented in the next sections.

3.1 Surfaces and Cross Sections—Deterministic Model

In this area, the roll front is developed across two major formations. Upper formation is interpreted as a juxtaposition of deltaic channels and bars, without shale

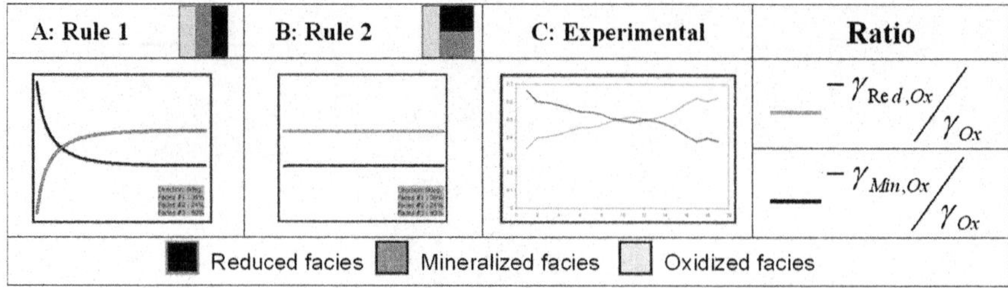

Fig. 2 Variogram ratios for the oxidized facies according to the facies rules and the experimental curves

contacts, forming a continuous, uniform sand body. Lower formation is associated with a meandering fluvial type environment, with heterogeneous, low continuity sand bodies. Those two formations are separated by a discontinuous lacustrine shale deposit. Stratigraphic interpretation of the area is based on core descriptions and resistivity interpretation. The top and base surfaces of these formations are used to define the simulation grid.

From the drill hole information, cross sections are constructed describing the three domains: (1) the oxidized one that contains no uranium; (2) the mineralized one that has economic interest; (3) the reduced one up stream.

Based on these cross sections a 3D envelope of the mineralized volume is built. It provides a reference volume to which the simulated ones can be compared.

3.2 Stochastic Model

In order to calculate the amount of leachable uranium and its spatial distribution, a three step methodology has been developed, with the distribution in 3D of: (1) oxidized, mineralized and reduced rock; (2) uranium within the defined mineralized volume; (3) permeable/impermeable rocks.

3.2.1 Mineralized Volume

This section is dedicated to mineralized volume modeling with plurigaussian simulations [1]. Exploration and mining development drill holes information is used to subdivide material into the three following rock categories: oxidized (36 %), mineralized (24 %) and reduced (41 %). The three rock qualities are simulated as the outcome may be used for resource calculation, but also for reactive transport simulation, where the redox quality needs to be considered.

The first step is to decide upon the "facies rule", that represents the existing relationship between the three facies categories: oxidized, mineralized, and reduced. In this context two possibilities may be considered: do we have an ordered relationship between the three facies going upstream (oxidized facies) to downstream (reduced

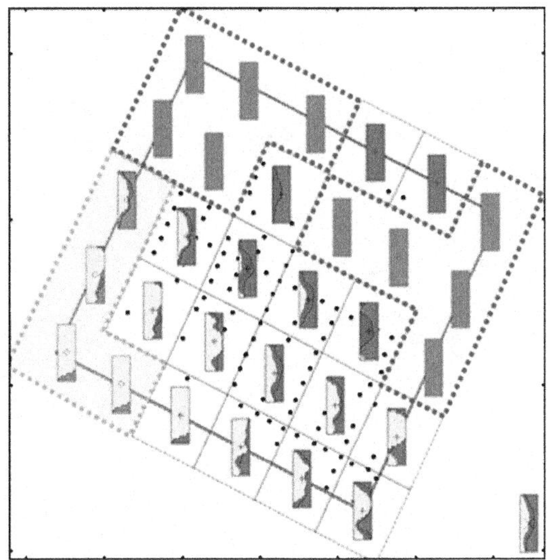

Fig. 3 VPC computed on regular polygons for oxidized, mineralized, and reduced facies. User added VPC's are contoured with dashed line

facies) as depicted in Fig. 2.A (Rule 1), or do we have a split first between oxidized and non oxidized facies, and then between the mineralized facies and the barren reduced one as shown in Fig. 2.B (Rule 2).

According to the rules, the ratios between cross variograms and simple ones [3] have an expected behavior illustrated in Fig. 2.A and 2.B. The experimental curves (Fig. 2.C) advocate clearly for an ordered relationship between the three facies (Rule 1) from upstream to downstream, modeled with one Gaussian.

According to the roll front description, stationarity of the "redox" facies proportions cannot be expected. This feature is clearly depicted in Fig. 3, where the oxidized zone lies in the South West corner. The purely reduced zone is poorly recognized as it has no economical interest.

The local vertical proportion curves (VPC) are computed according to a regular polygon pattern (180 m by 200 m) as shown in Fig. 3. Most of the VPC are computed from the drill hole information available within the block dotted polygon. The other VPC, circled with dashed lines, are user added based on geological information, to better constrain the final 3D proportion grid. Within the poorly drilled northern zone (light blue polygon, Fig. 3), VPC with 100 % reduced facies are added, as we know that this upstream zone has not been crossed by the roll. In the south east corner the closest VPC have been duplicated as no major changes are expected (light yellow polygon, Fig. 3). The 3D proportion interpolation is performed by kriging the VPC with a 2000 m range spherical variogram model.

An alternative proportion matrix has been constructed using the deterministic geological model. This was performed converting the 3D mineralized envelope and the oxidized limit surfaces to proportion of simulation grid cells for being mineralized or oxidized. The proportion of reduced facies is deduced from the two others. Initial deterministic proportions are then smoothed with a moving average using a structural element of 100 m by 100 m by 5 m thick.

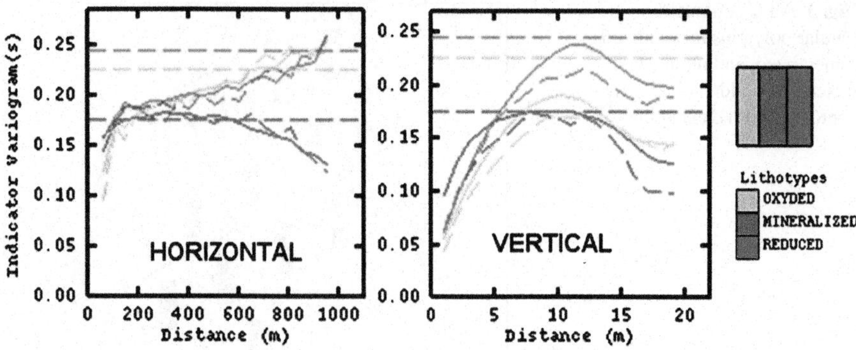

Fig. 4 Variogram model for the underlying Gaussian in roll front geometry simulation

Finally, the experimental variograms are calculated and the variogram model of the Gaussian function is adjusted with an isotropic spherical model (Spherical [100 m, 100 m, 13 m]) (Fig. 4).

3.2.2 Grade Simulation

The grade distribution is simulated with the direct block simulation technique [2]. First, uranium grade is regularized at a 1 m core length. Regularized data distribution is transformed in a Gaussian distribution through an anamorphosis model, using Hermite polynomial functions. The experimental Gaussian variogram is adjusted with 22 % nugget effect and a spherical model with 78 % sill, ranges of 100 m in N140, 50 m in N230 and 5 m vertically.

To perform direct block simulation, corrections of support are necessary. Variogram model is transformed in a block variogram, and anamorphosis of the grades is corrected to the block support (10 m by 10 m by 1 m).

3.2.3 Permeable/Impermeable Volume

Only uranium within permeable rocks can be reached by the leaching solution. Therefore the distribution of these rocks is a key factor for recoverable resource evaluation and later on for reactive transport simulation.

From the core description and the resistivity interpretation a total of 13 lithologies are defined, some in very minor occurrences. They are grouped into 5 main classes, taking into account the different environment, continuity and petrophysical properties. Three lithotypes are permeable and will contribute to the resources: coarse, medium and fine sands. Two lithotypes are not permeable: shale plus silty lithotypes, and carbonated cemented ones. Despite the fact that carbonates consume a significant amount of injected reagent, they are not simulated here as they represent only 0.4 % of the different lithotypes in the area.

Table 1 Lithotype proportions and rules for the two formations

	Coarse sand	Medium sand	Fine sand	Shale & silt
Upper formation	11 %	69 %	14 %	6 %
Lower formation	2 %	31 %	37 %	30 %

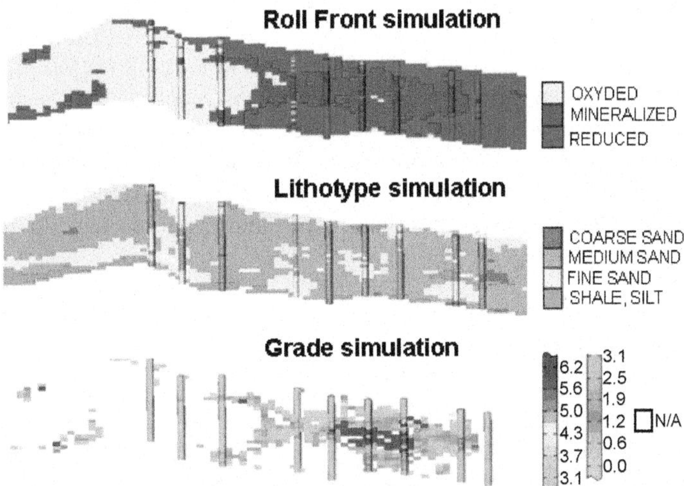

Fig. 5 Simulation results for lithotypes, roll front geometry, and grades, on a vertical section of the grid. Conditioning drill holes (9) are represented in vertical lines

As presented earlier, the roll front is developed within two formations which have distinct depositional environments, with proper orientation and continuities of sedimentary objects. Therefore, upper and lower formations are simulated independently. Lithotype analysis leads to the use of 2 underlying Gaussians in the plurigaussian approach [1]: one for impermeable/permeable rocks and another one for sand quality within permeable rocks. Lithotype proportions and rules are summarized in Table 1. The proportion matrix is constructed from interpolation of local VPC derived from drill holes information.

3.3 Outcome

This three step simulation is illustrated in Fig. 5, where cross sections of lithotype, redox and grade realization are shown.

Uncertainties regarding mineralization contours and metal quantities can be assessed, through the calculation of histograms, statistical maps or probability maps.

In the following, mineralization is restricted to permeable rock. Uranium within non permeable rock is disregarded, as it is not leachable. From the 500 simulations various maps can be derived for analysis purposes. In the case study, the probability maps of having a grade thickness (GT or accumulation) greater than 0.06 m % (denoted $P\_GT > 0.06$ m %) are used as a guideline for delineating hole locations. The cut off is based on economical considerations. The average GT maps (GT_mean) are presented to reveal synthetically the uranium grade heterogeneities.

4 The Case Study

Two main objectives are pursued. The first one is to use a probabilistic approach to help position mining development holes. Only this recognized geometry will be equipped with exploitation wells at a later stage. The second one focuses on density of infill drilling versus resource precision. The aim is to define the optimal drill spacing required to reach a particular degree of resource confidence.

As the method is run "post mortem", drill holes positions have to be chosen among existing ones (Fig. 1).

4.1 Delineation

The strategy here is, from the exploration final stage, to examine how uncertainty evaluation can help identify potential areas of interest that need to be tested. Can such a method help reduce inherent ore geometry uncertainty, and at the same time reducing the number of delineation drill holes?

The mineralization is simulated with exploration drill holes only, spacing is 200 m by 50 m. The 500 simulation results are summarized into the probability map $P\_GT > 0.06$ m %. This map is used to help positioning additional holes. Two zones are tested: zones potentially mineralized outside of the actual limit (black contour in Fig. 6, Maps A to E) and zones inside the actual limit, potentially not mineralized.

First of all when comparing simulation results based on exploration drill holes only (Fig. 6.B) with the ones obtained with the deterministic geological model and the exploration holes (Fig. 6.A); one clearly sees areas still open in the north and in the south-east, when the deterministic geological model is not considered. When deterministic geological model is used, the rather quick probability transition from 1 to 0 is mainly model driven.

Based on these observations, additional drilling is proposed close to the 50 % probability line on the $P\_GT > 0.06$ m % map (Fig. 6.B), where the probability slope is smooth and keeping the existing 200 m or 100 m sections alignment of the drill holes. The 10 new boreholes proposed are illustrated with white dots in Fig. 6.B. Adding these 10 holes gives the probability map shown in Fig. 6.C. It reveals an interesting extension zone in the NW and confirms the ore body continuity in the SE. The mineralization contour is largely precised all over.

From the previous delineation stage (Fig. 6.C), 7 additional holes are proposed to confirm the extension to the NW, either the mineralization can be connected to the actual block limit or not, and to precise SE borders as the probability gradient is still

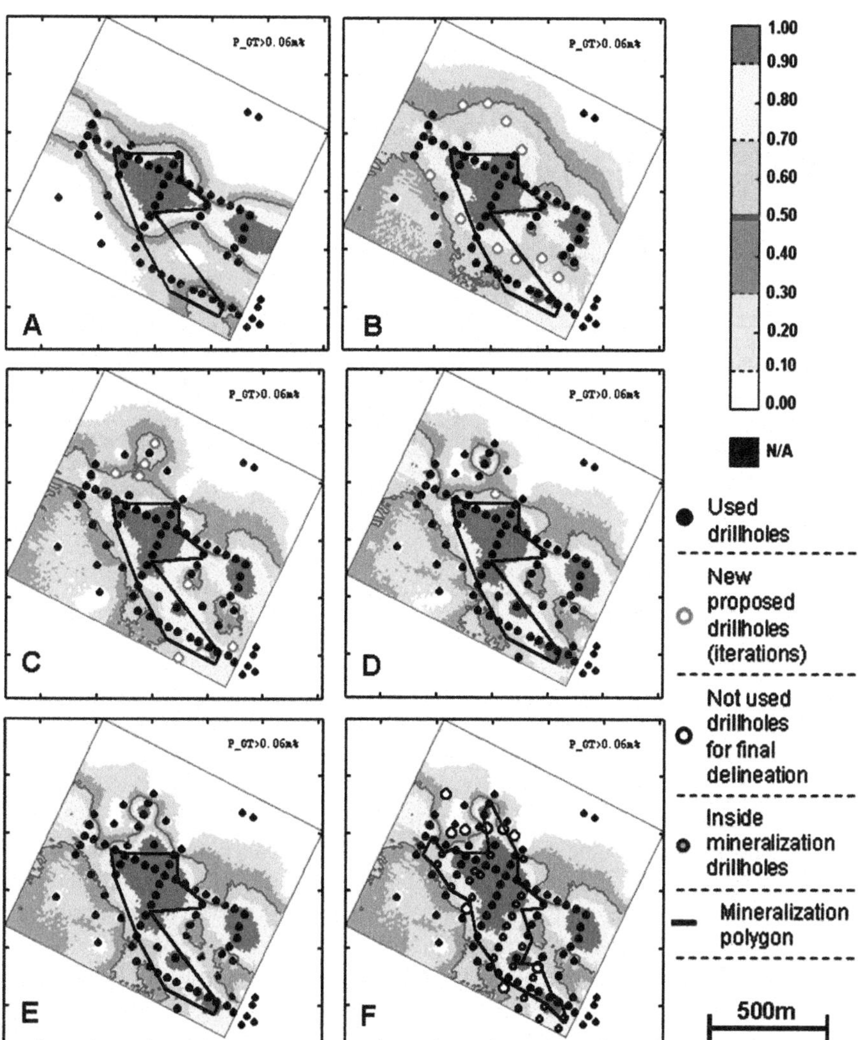

Fig. 6 Delineation results: Probability maps of having a grade thickness greater than 0.06 m %. Drill holes used for the simulations are illustrated in *black dots*; new proposed holes in *white dots*. *Blue line contour* represent 50 % of probability ($P\_GT > 0.06$ m %). Map [**A**]: Results obtained with exploration holes and the deterministic geological model. Map [**B**]: with exploration drill holes only. Maps [**C**], [**D**], [**E**]: updated results for delineation iterations with 10, 7, and 1 additional holes. Map [**F**]: final mineralization delineation and comparison of the delineation approach with real development

very smooth. They are illustrated in white dots in Fig. 6.C. Figure 6.D shows the new $P\_GT > 0.06$ m %. From boreholes added in the north, one hole confirms the extension of the ore, and the 3 others are barren, reducing drastically the possibility of a mineralized extension. In the SE, the 3 additional boreholes precise slightly the contour.

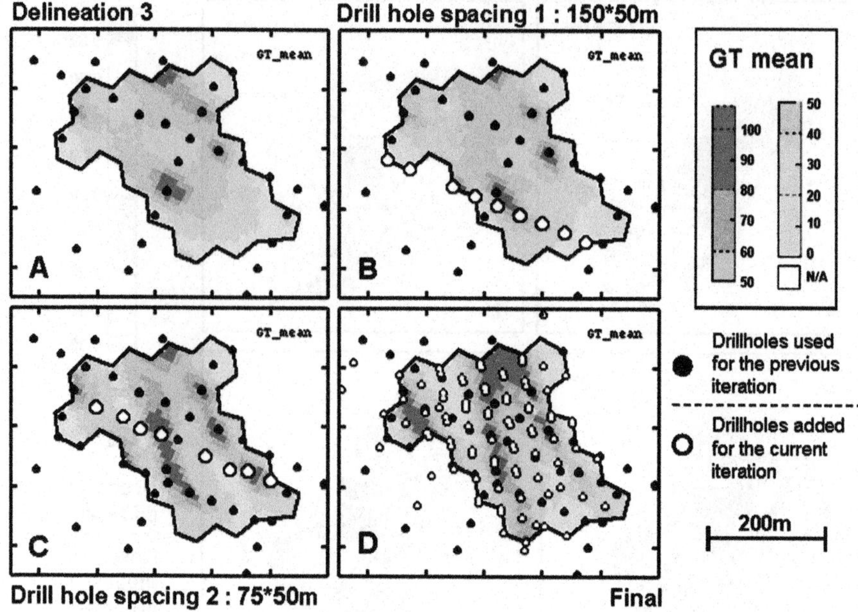

Fig. 7 Maps of average grade thickness from 500 simulations. Map [**A**]: final delineation holes. Map [**B**]: infill 1, well spacing 50 m by 150 m. Map [**C**]: infill 2, well spacing 50 m by 75 m. Map [**D**]: final reference, calculated with all available drill holes. The GT have been rescaled to 100

The ultimate iteration expects to test the continuity of the extension NW with only one hole. The outcome is shown in Fig. 6.E. The connection is confirmed and mineralization geometry precision is enough to design the final ore extraction limits.

At exploration drill hole spacing, a large extension was missed on the north west of the block. Stochastic simulation and probability maps allows for the contour of the block to be precised, thereby saving on drillings. In Fig. 6.F, all development drill holes have been reported in white and grey dots. In the north area extension, at least 6 drill holes could have been saved using this method as a guide (illustrated with white dots). To the south, probably 3 drill holes could have been saved.

4.2 Impact of Drill Hole Spacing on Uncertainty Assessment Within the Mineralized Zone

Once the contour of the mineralization is defined, with stochastic approach, recoverable resources and associated uncertainty can be calculated within that limit. This section is dedicated to evaluating the drill hole spacing required to reach an acceptable level of confidence for resources and exploitation planning.

Starting from the final delineation stage (Fig. 7.A), lines of infill drill holes are added in two iterations. The first one adds 9 holes and gives a 50 m by 150 m drilling

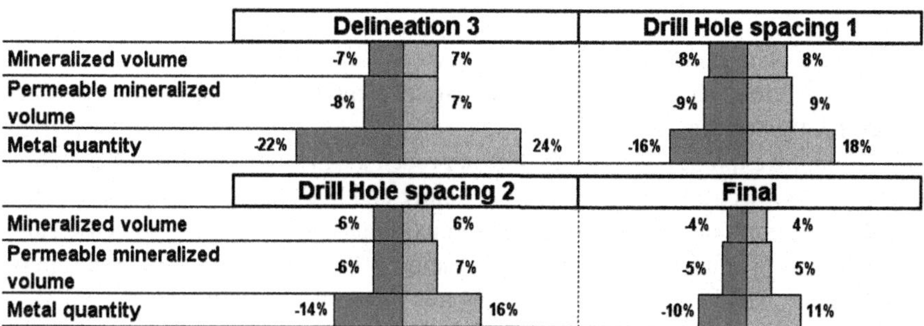

Fig. 8 P5–P95 expressed in percentage of variation around the mean for mineralized volume, permeable mineralized volume and metal quantities

pattern. The second adds 8 holes, giving a 50 m by 75 m drilling pattern. They are represented by white dots in Fig. 7.B and Fig. 7.C.

Average grade thickness maps, from 500 simulations are shown Fig. 7 for the different recognition stages. The higher hole density, the better heterogeneities are identified. The first infill line (Fig. 7.B) brings very low precision improvement compared to the initial state (Fig. 7.A). The second one (Fig. 7.C) identifies heterogeneity of the mineralization and results are very close to the final state (Fig. 7.D), using all drill holes. In this area, 75 m by 50 m spacing for the holes appears sufficient to precise mineralization heterogeneities for production cells design.

The histogram of ore tonnage, recoverable ore tonnage and metal tonnages are calculated for the 4 iterations. Results are summarized in Fig. 8, showing P5 and P95 expressed as percent of the mean for total and recoverable mineralized volume, and metal quantity. Uncertainty is reduced as drill hole density increases. Ore volume variability is low (maximum 10 %) as the area is within the mineralized zone. Impact of lithology distribution is almost none as impermeable rock proportion is low in this part of the deposit. Maximum uncertainty is on metal quantity, which is directly related with high grade heterogeneity. If metal uncertainty is decreasing from 23 % to 17 % when adding the first infill line, it stays very high at 15 %, when adding the second infill line (Fig. 8).

This case study shows reducing drill hole spacing results in a decrease on ore and metal tonnages uncertainties. The 50 m by 150 m spacing is not enough to properly characterize heterogeneities of the zone, whereas 50 m by 75 m spacing identifies mineralization heterogeneities enough to improve exploitation wells design. But this spacing does not characterize grades enough to reduce uncertainties on metal.

5 Conclusions and Perspectives

A method for recoverable resources and associated uncertainties evaluation has been set up in the context of roll front deposit exploited by ISR. It is based on plurigaussian simulations for discrete variables and direct block simulations for grade. An application on a uranium deposit is discussed.

For mineralization delineation, probability maps bring guidelines for drill holes positioning. The mineralization contour can be defined saving some holes. After the first iteration (10 holes), a large extension on the North West is identified, and mineralization contour is precised.

Within the exploitation contour, this approach gives some criteria to define hole spacing versus acceptable resource uncertainty and heterogeneity definition. These factors are important to mine planning and exploitation cell design. The case study on drill hole spacing shows that infill spacing 50 m by 75 m identifies properly spatial heterogeneities but uncertainty on metal quantity is still high (±15 %).

In later perspective, these models will be injected into a reactive transport simulator in order to reproduce the process of uranium extraction by ISR as shown in [4], to evaluate resources uncertainty impact on production plans.

Acknowledgements The authors would like to thank Katco for supporting this work. The authors are particularly grateful to Anselme Diracca from Katco for all valuable and constructive discussions.

References

1. Armstrong M, Galli A, Beucher H, Le Loc'h G, Renard D, Doligez B Eschard R, Geffroy F (2011) Plurigaussian simulations in geosciences. Springer, Berlin
2. Deraisme J, Rivoirard J, Carrasco P (2008) Multivariate uniform conditioning and block simulations with discrete Gaussian model: application to Chuquicamata deposit. In: Ortiz JM, Emery X (eds) 2008 GEOSTATS, vol 1, pp 69–78
3. Rivoirard J (1988) Modeles à résidus d'indicatrices auto-krigeables. Sci Terre Inform Geol 28:303–326
4. Nos J (2011) Modèle conceptuel d'une exploitation d'uranium par récupération in situ—interprétation des données de production et apport de la modélisation du transport réactif en milieu hétérogène. PhD thesis, Mines ParisTech

Multivariate Estimation Using Log Ratios: A Worked Alternative

Clint Ward and Ute Mueller

Abstract Common implementations of geostatistical methods, kriging and simulation, ignore the fact that geochemical data are usually reported in weight percent and are thus compositional data. Compositional geostatistics is an approach developed to ensure that the constant sum constraint is respected in estimation and simulation. The compositional geostatistical framework was implemented to test its applicability to an iron ore mine in Western Australia. Cross-validation was used to compare the results from ordinary cokriging of the additive log ratio variables with those from conventional ordinary cokriging. Two methods were used to back-transform the additive logratio estimates, the additive generalized logistic back transformation and Gauss-Hermite Quadrature approximation. Both the Aitchison distance and the Euclidean distance were used to quantify the error between the estimates and original sample values. The results follow the required constraints and produce better estimates when considering the Aitchison distance. When the Euclidean errors are considered the conventional ordinary cokriging estimates are less biased but the distribution of errors for the additive logratio estimates appear to be superior to the conventional ordinary cokriging estimates.

1 Introduction

Geochemical analyses of sampled drill cuttings or drill core are used to evaluate the chemical composition of mineral deposits. The interdependence between the geochemical analytes is studied in a variety of situations. Iron ore mining is an example; specifically where sufficiently high concentrations of deleterious elements such as Silicon oxide (SiO_2) and Aluminum oxide (Al_2O_3) can affect the saleability of the product. Where economic concentrations of Iron (Fe) also have levels of deleterious

C. Ward (✉)
Cliffs Natural Resources, Level 12, 1 William Street, Perth, WA 6000, Australia
e-mail: Clint.Ward@CliffsNR.com

U. Mueller
Edith Cowan University, 270 Joondalup Drive, Perth, WA 6027, Australia
e-mail: u.mueller@ecu.edu.au

P. Abrahamsen et al. (eds.), *Geostatistics Oslo 2012*,
Quantitative Geology and Geostatistics 17,
DOI 10.1007/978-94-007-4153-9_27,

elements close to the upper limit, understanding the interdependence between the Fe and the deleterious elements is of critical importance to ensure profitability and avoid contractual penalties. This interdependence is the result of multiple factors including mineralogy, weathering, and host rock composition. The mining value chain encompasses exploration, estimation of the in-situ mineral resources and ore reserves, mining of the ore reserve, and finally sales of the product to a market. Product specifications are evaluated at all stages of this value chain and financial models are generated from them, often heavily dependent upon the interdependence or correlation between the analytes.

Geostatistical methods, kriging and simulation, are commonly used to estimate or simulate the analytes in space. The common implementations of these estimation techniques ignore the fact that the analytes are compositional in nature, meaning that they represent some part of a whole, and the outputs from these techniques do not honor this constraint. As a result, the interdependence can also be misrepresented.

Appropriate interpolation methods for regionalized compositions should comply with the non-negativity and constant sum constraints of compositional data. Examples of methods that do so are nearest neighbor, inverse distance, triangulation and local sample mean interpolation. However, the spatial covariance structure is not taken into account. Compositional kriging [5] and compositional cokriging [2] take into account the spatial covariance structure, but in order to satisfy the constant sum constraint arising from the compositional nature of the data, an additional constraint is incorporated in the kriging systems. Compositional kriging has two drawbacks: firstly the cross-correlations are not taken into account, omitting potentially valuable information, and secondly in the derivation of the compositional kriging system, the prediction error variance is considered only as an objective function in a minimization problem, rather than as a measure for constructing confidence intervals. The inclusion of an additional constraint in the simple cokriging system also has drawbacks. Negative estimates can still result, and the addition of a constraint to the cokriging system may lead to bias [2].

Compositional data techniques pioneered in the 1980's sought to develop a simple and appropriate methodology to analyze compositional data [1]. These techniques were extended to cover the case of compositional data with location and location dependent attributes using geostatistics [4]. This study will use compositional techniques to estimate the values of spatially located analytes such that the outputs both adhere to the compositional nature of the input data, and honor the interdependencies between the different analytes. Specifically estimates will be computed based on the ordinary cokriging of the alr transform (see Sect. 2) of the analytes. The alr transform allows the use of standard geostatistical techniques and so unbiasedness of the transformed data is guaranteed. In contrast to the approach taken in [4] where the Fast Fourier Transform was used to compute variograms and cross-variograms we will use the linear model of co-regionalization to construct variograms and cross-variograms.

2 Theoretical Framework

In this section we outline the compositional data approach that was be taken in this study. The description largely follows [4].

2.1 *The alr Transform*

Compositional data are multivariate data where each component represents some part of a whole and which carry only relative information. When D components are to be accounted for, then the corresponding composition is a vector $\mathbf{z} \in \mathbb{R}^D$ of positive components which add up to a constant value $c > 0$:

$$\sum_{i=1}^{D} z_i = c. \tag{1}$$

Analogously, a *regionalized composition* is a vector random function $\mathbf{Z} : A \longrightarrow \{\mathbf{X}(\mathbf{u}) : \mathbf{u} \in A\}$ such that for each $\mathbf{u} \in A$ the vector $\mathbf{X}(\mathbf{u})$ is a composition and the sample space for $\mathbf{Z}(\mathbf{u})$ is the D-simplex $S^D = \{\mathbf{Z} \in \mathbb{R}_+^D : \sum_{i=1}^D z_i = c\}$. The *additive logratio* (alr) transformation alr : $S^D \longrightarrow \mathbb{R}^{D-1}$ given by

$$\operatorname{alr}\big(\mathbf{Z}(\mathbf{u})\big) = \mathbf{X}(\mathbf{u}) = \left(\ln\left(\frac{Z_1(\mathbf{u})}{Z_D(\mathbf{u})}\right), \ldots, \ln\left(\frac{Z_{D-1}(\mathbf{u})}{Z_D(\mathbf{u})}\right)\right) \tag{2}$$

embeds the D-simplex into $(D-1)$-dimensional space and removes the constant sum constraint. The *additive generalized logistic* (agl) transform may be used to recover the values in S^D:

$$\operatorname{agl}\big(\mathbf{X}(\mathbf{u})\big) = \mathbf{Z}(\mathbf{u}) = c \times \frac{(\exp(X_1(\mathbf{u})), \ldots, \exp(X_{D-1}(\mathbf{u})), 1)}{1 + \sum_{i=1}^{D-1} \exp(X_i(\mathbf{u}))}. \tag{3}$$

2.2 *Spatial Covariance Structure*

In [4] the spatial covariance structure for r-compositions is discussed and it is shown that D^4 covariance functions are needed for a complete characterization and in their calculation all possible logratios are accounted for. In the case of the alr transform only a subset of all possible logratios are considered (namely those using Z_D as the divisor) and the alr cross-covariance function is given by

$$\boldsymbol{\Sigma}(\mathbf{h}) = \big[\operatorname{Cov}\big(X_i(\mathbf{u}), X_j(\mathbf{u}+\mathbf{h})\big)\big]_{i,j=1,2,\ldots,D}. \tag{4}$$

If the covariance structure of $\mathbf{Z}(\mathbf{u})$ is symmetric, then the number of covariance terms required reduces to $(D-1)D/2$. In the case of an intrinsic alr random function which is symmetric, the corresponding semivariograms are required and they are defined by

$$\psi_{ij}(\mathbf{h}) = \frac{1}{2}\operatorname{Cov}\left[\ln\left(\frac{Z_i(\mathbf{u})}{Z_D(\mathbf{u})}\right) - \ln\left(\frac{Z_i(\mathbf{u}+\mathbf{h})}{Z_D(\mathbf{u}+\mathbf{h})}\right), \ln\left(\frac{Z_j(\mathbf{u})}{Z_D(\mathbf{u})}\right) - \ln\left(\frac{Z_j(\mathbf{u}+\mathbf{h})}{Z_D(\mathbf{u}+\mathbf{h})}\right)\right]. \tag{5}$$

As in classical geostatistics, the covariance function and the semivariogram function $\boldsymbol{\Psi}$ are related via the equation

$$\boldsymbol{\Psi}(\mathbf{h}) = \boldsymbol{\Sigma}(\mathbf{0}) - \boldsymbol{\Sigma}(\mathbf{h}), \tag{6}$$

so that fitting of a linear model of coregionalization (LMC) using standard variography tools is possible.

2.3 Cokriging

Ordinary co-kriging (OCK) cross-validation was used to compute estimates of alr($\mathbf{Z}(\mathbf{u})$) at the sample locations. The cross-validation technique employed is the common implementation in geostatistical studies whereby each sample point is sequentially removed from the dataset and the value at the location estimated using the remaining data. We recall the matrix formulation of the estimator below. The ordinary co-kriging estimate at location $\mathbf{u}_0$ is given by

$$\mathbf{X}(\mathbf{u}_0) = \sum_{\alpha=1}^{n} \mathbf{B}_\alpha \mathbf{X}(\mathbf{u}_\alpha), \tag{7}$$

where the vector of weights $\mathbf{B}_\alpha$ at location $\mathbf{X}(\mathbf{u}_\alpha)$ is determined from the ordinary co-kriging system

$$\sum_{\alpha=1}^{n} \mathbf{B}_\alpha \mathbf{C}(\mathbf{u}_\beta - \mathbf{u}_\alpha) + \Lambda = \mathbf{C}(\mathbf{u}_\beta - \mathbf{u}_0) \tag{8}$$

$$\sum_{\alpha=1}^{n} \mathbf{B}_\alpha = I_{D-1}. \tag{9}$$

The matrix Λ is a matrix of Lagrange multipliers of size $D-1$. The ordinary co-kriging variance-covariance matrix is given by

$$\Sigma_0\big(\mathbf{X}(\mathbf{u}_0)\big) = \mathbf{C}(\mathbf{0}) - \sum_{\alpha=1}^{n} \mathbf{B}_\alpha \mathbf{C}(\mathbf{u}_\alpha - \mathbf{u}_0) - \Lambda \tag{10}$$

and the matrix $\mathbf{C}(\mathbf{0})$ is estimated from the sills of the coefficient matrices of the LMC fitted to the data.

2.4 Back-Transformation of Co-kriging Estimates

The OCK estimates derived in the previous section are estimates in (unconstrained) $D-1$ dimensional real vector space and need to be back transformed to the D-

simplex prior to comparison with the input data. Two methods are available for doing so:

a. a crude back transform of the results via the application of the agl transformation;
b. the use of Gauss-Hermite quadrature to compute an estimate of the expected value of the r-composition $\mathbf{Z}(\mathbf{u}_0)$

Given the vector of expected values $\mathbf{X}(\mathbf{u}_0)$ and corresponding variance covariance matrix $\Sigma_0(\mathbf{X}(\mathbf{u}_0))$ the expected composition $\mu_{\mathbf{Z}}(\mathbf{u}_0)$ is given by

$$\mu_{\mathbf{Z}}(\mathbf{u}_0) = \int_{\mathbb{R}^{D-1}} \pi^{-\frac{D-1}{2}} \operatorname{agl}\big(\sqrt{2}R^T\mathbf{Y} + \mathbf{X}^*(\mathbf{u}_0)\big) \exp\big(-\mathbf{Y}^\mathbf{T}\mathbf{Y}\big)\, d\mathbf{Y}. \tag{11}$$

Here the matrix R denotes the Cholesky decomposition of $\Sigma_0(\mathbf{X}(\mathbf{u}_0))$. The integral can be approximated numerically using k Gauss-Hermite quadrature points: Setting $g(\mathbf{Y}) = \pi^{-\frac{D-1}{2}} \operatorname{agl}(\sqrt{2}R^T\mathbf{Y} + \mathbf{X}^*(\mathbf{u}_0))$

$$\int_{\mathbb{R}^{D-1}} g(\mathbf{Y}) \exp\big(-\mathbf{Y}^\mathbf{T}\mathbf{Y}\big)\, d\mathbf{Y} = \sum_{i_1=1}^{k} \sum_{i_2=1}^{k} \cdots \sum_{i_D=1}^{k} \prod_{\ell=1}^{D} w_{i_\ell} g(Y_{i_1}, Y_{i_2}, \ldots, Y_{i_D}). \tag{12}$$

2.5 *Performance Measures*

The cross validation results will be assessed using firstly the errors in the estimates of the individual attributes and secondly the traditional Euclidean distance as well as the *Aitchison distance* defined as

$$\delta_a\big(Z(\mathbf{u}), Z^*(\mathbf{u})\big) = \sqrt{\sum_{i=1}^{D} \left(\ln \frac{Z_i(\mathbf{u})}{g(Z(\mathbf{u}))} - \ln \frac{Z_i^*(\mathbf{u})}{g(Z^*(\mathbf{u}))} \right)^2}. \tag{13}$$

As a further measure the STRESS (standardized residual sum of squares) will be used. STRESS compares the dispersion of the data with the dispersion of the estimates derived from OCK and is defined as

$$\text{STRESS} = \sqrt{\frac{\sum_{\alpha<\beta} (\delta(\mathbf{Z}(\mathbf{u}_\alpha), \mathbf{Z}(\mathbf{u}_\beta)) - \delta(\mathbf{Z}^*(\mathbf{u}_\alpha), \mathbf{Z}^*(\mathbf{u}_\beta)))^2}{\sum_{\alpha<\beta} (\delta(\mathbf{Z}(\mathbf{u}_\alpha), \mathbf{Z}(\mathbf{u}_\beta)))^2}}. \tag{14}$$

Here δ denotes either the Aitchison or the Euclidean distance and $\mathbf{Z}^*(\mathbf{u})$ denotes the estimate at location $\mathbf{u}$ obtained from conventional OCK or OCK of the logratios.

3 Application

This study used a single bench of 6 m long blast hole (BH) samples from an iron ore mine located in the central Yilgarn, Western Australia (Fig. 1).

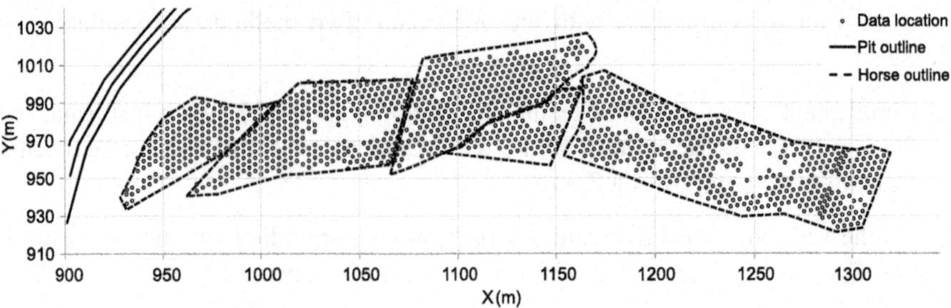

Fig. 1 Location map of BH collars and fault imbricate outlines in relation to pit outline (mine grid)

Table 1 Descriptive statistics of BH analytes

Variable	n	Min	Max	Mean	Std. Dev.	CoV
Al_2O_3	1296	0.08	6.87	1.25	1.06	0.85
Fe	1296	52.43	71.74	64.09	2.43	0.04
SiO_2	1296	0.34	15.48	2.24	1.79	0.80
r	1296	25.98	40.11	32.42	1.33	0.04

Eighteen analytes measured in weight percent were available within the dataset, although only the three main elements of interest for iron ore mining (Fe, SiO_2 and Al_2O_3) were examined. As these analytes on their own form a subcomposition, a filler variable r was introduced in order to satisfy the constant sum constraint. As this study is focused on the method rather than the data, the decision was taken to extract a single bench and examine the data in two dimensions only. The bench consists of five discrete rotated fault imbricates. The longest strike distance within each imbricate (constrained by the angular tolerance) is 60 m.

3.1 Data description

The data exhibit the typical hematite enriched iron ore distributions; negatively skewed Fe, positively skewed SiO_2 and Al_2O_3 distributions (Table 1). The scatter-plots also display typical behavior: strong negative correlation both between Fe and Al_2O_3 and Fe and SiO_2, and strong positive correlation between SiO_2 and Al_2O_3 (Fig. 2). Comparing the statistics in Table 1 and Table 2 highlights the differences between the original and alr transformed variables.

The reduction in asymmetry in the histogram due to the alr transformation is illustrated in Fig. 3 where the histograms of the Al_2O_3 and the corresponding transformed factor ($X3$) are juxtaposed. In practical terms, the alr transformation has a normalizing effect (Fig. 3), in some cases making the spatial covariance more tractable to model.

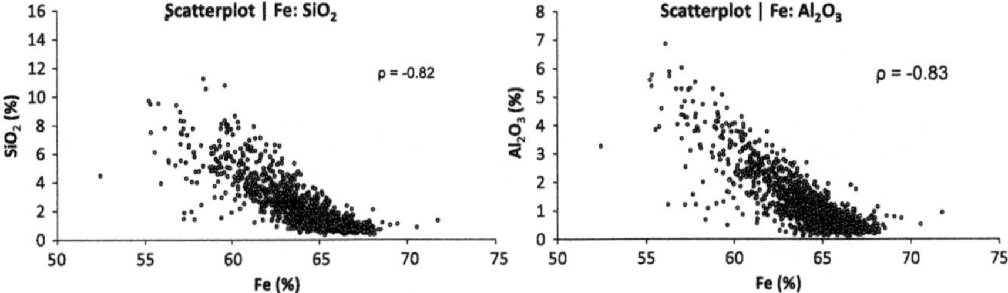

Fig. 2 Scatterplots Fe vs SiO_2 (*left*) and Fe vs Al_2O_3 (*right*)

Table 2 Descriptive statistics of alr transformed variables

Variable	*n*	Min	Max	Mean	Std. Dev.	CoV
$X1$	1296	0.28	1.02	0.681	0.058	0.085
$X2$	1296	−6.03	−1.45	−3.562	0.800	−0.225
$X3$	1296	−4.61	−0.56	−2.926	0.719	−0.246

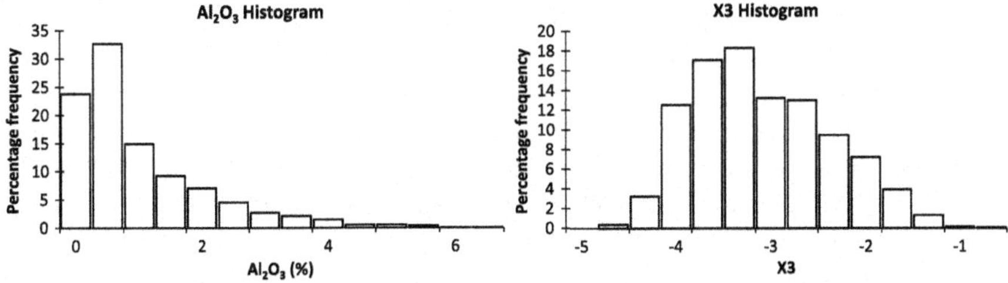

Fig. 3 Histogram of Al_2O_3 (*left*) and histogram of $X3$ (*right*)

An LMC was fitted for the untransformed Fe, SiO_2, and Al_2O_3 variables. This model consists of a nugget together with a two spherical structures with ranges along strike of 30 and 450 m, while the range across strike is 20 m. As the covariance structure is symmetric, or at least locally symmetric (Fig. 4 for the cross-covariance function for $X1$ and $X3$), the use of an LMC is justified for the alr spatial structure also. In the case of the LMC of the transformed data a nugget and two spherical structures provide an adequate fit for the variogram in the strike direction, while across strike an additional zonal spherical structure is required. The ranges of the alr LMC are 20 m and 56 m in the strike direction and 6 m and 20 m across strike. The model fit the variogram of variable $X3$ in strike direction is shown in Fig. 4 to illustrate the quality of the fit.

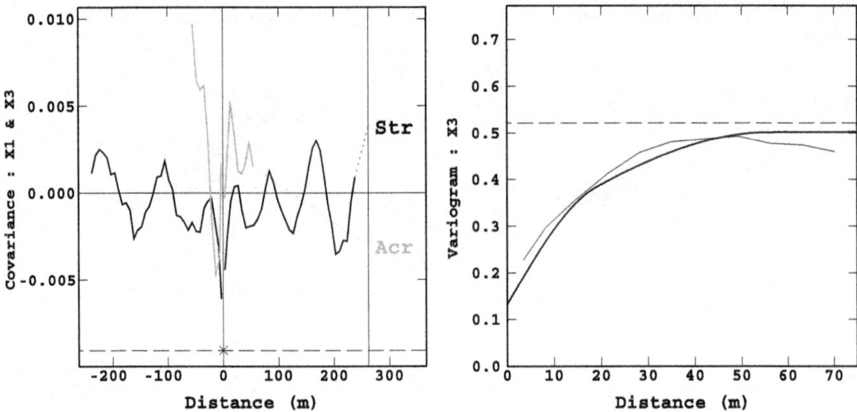

Fig. 4 Strike direction and across strike cross-variogram $X1$–$X3$ (*left*) and fitted LMC—$X3$ strike direction (*right*)

3.2 Analysis of Results

The purpose of the study was to analyze the effectiveness of the log ratio approach with iron ore data using OCK cross-validation. The effectiveness was measured by computing the errors both between the sample data and the back transformed logratio estimates as well as with estimates from a 'standard' technique; conventional OCK. The error measurements were computed using Euclidean and compositional distance measures. Back transformation of the estimated factors $X1$ to $X3$ was performed using the agl function directly, and with the theoretically supported, but cumbersome to implement Gauss-Hermite quadrature method, as outlined in Sect. 2. In both the conventional OCK and logratio cases the same search strategy was applied.

The mean of the OCK estimates and the Gauss-Hermite quadrature estimates are similar to the sample mean; however the estimates derived using Gauss-Hermite quadrature integration possess a larger range (Table 3). The coefficients of variation are generally highest for the agl back-transformed estimates, while all three sets of estimates display lower variation than those of the sample (see Table 1). This is caused by the inherent smoothing effect of the interpolation method. However; the Gauss-Hermite quadrature approximations best reproduce the spread of the estimates as evidenced by the standard deviation values.

Based on the means of the univariate errors, that is the average difference between the sample value and corresponding estimate for each of the attributes, one would rank the OCK equal with the G-H approximation, and superior to the agl (Table 4). The variability of these errors is approximately the same. The statistics alone do not fully describe the relationships however; an examination of the error histograms (Fig. 5) shows that the distribution of Gauss-Hermite approximation errors for Al_2O_3 and SiO_2 appear more centered on zero and more peaked than the OCK and agl error distributions; desirable qualities when examining error distributions.

Table 3 Summary statistics of estimates

Variable	n	Min	Max	Mean	Std. Dev.	CoV
Al_2O_3–agl	1296	0.01	4.57	1.07	0.67	0.63
Al_2O_3–G-H	1296	0.26	4.60	1.24	0.69	0.56
Al_2O_3–OCK	1296	0.10	3.87	1.24	0.62	0.50
Fe–agl	1296	57.65	68.38	64.36	1.67	0.03
Fe–G-H	1296	58.42	67.23	64.11	1.65	0.03
Fe–OCK	1296	59.06	66.95	64.11	1.54	0.02
SiO_2–agl	1296	0.01	8.60	2.02	1.31	0.65
SiO_2–G-H	1296	0.63	11.59	2.21	1.43	0.65
SiO_2–OCK	1296	0.67	8.35	2.22	1.31	0.59

Table 4 Univariate error statistics

Variable	n	Min	Max	Mean	Std. Dev.	Skew
Al_2O_3–agl	1296	−1.90	5.02	0.18	0.79	1.68
Al_2O_3–G-H	1296	−1.92	4.63	0.01	0.80	1.64
Al_2O_3–OCK	1296	−2.19	4.12	0.01	0.82	1.54
Fe–agl	1296	−10.14	5.61	−0.27	1.76	−0.94
Fe–G-H	1296	−8.85	5.75	−0.03	1.76	−0.97
Fe–OCK	1296	−9.69	6.22	−0.03	1.79	−0.99
SiO_2–agl	1296	−3.11	9.17	0.22	1.11	1.45
SiO_2–G-H	1296	−4.02	5.59	0.03	1.08	0.98
SiO_2–OCK	1296	−3.25	8.01	0.02	1.10	1.24

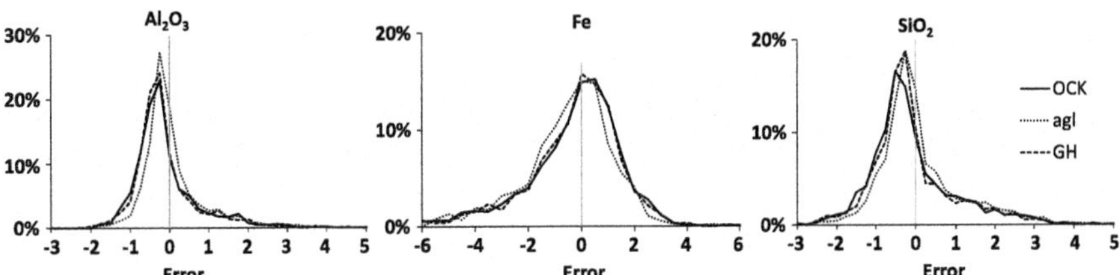

Fig. 5 Histogram of errors for Al_2O_3 (*left*), Fe (*center*) and SiO_2 (*right*)

As is the case with the univariate errors, the smaller the magnitude of the compositional error, the more accurate the estimate. When the deviations of the estimates from the sample values (Table 5) are measured jointly via either the traditional Euclidean distance or the Aitchison distance, the agl back-transformed estimates have the smallest mean error followed by the Gauss-Hermite quadrature approximations.

Table 5 Compositional error statistics

Variable	n	Min	Max	Mean	Std. Dev.	CoV
Euclidean OCK error	1296	0.041	9.776	1.750	1.426	0.81
Euclidean agl error	1296	0.030	10.971	1.673	1.513	0.90
Euclidean GH error	1296	0.063	9.677	1.700	1.42	0.84
Aitchison OCK error	1296	0.011	1.777	0.423	0.278	0.66
Aitchison agl error	1296	0.005	3.598	0.380	0.279	0.73
Aitchison GH error	1296	0.006	1.595	0.398	0.262	0.66

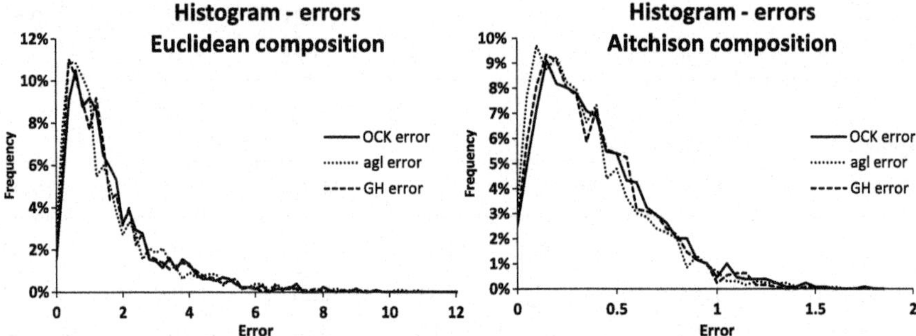

Fig. 6 Histogram of compositional errors; Euclidean (*left*), Aitchison (*right*)

The OCK estimates are more biased than the Gauss-Hermite quadrature estimates when judged by these distance measures. Overall, the estimates derived from compositional techniques are more accurate and precise than the corresponding conventional OCK estimates. This is evidenced by the statistics and the shape of the error distributions (Fig. 6).

The STRESS measure provides a single number which can be interpreted as a normed measure of loss of information when observations are substituted by estimates; thus, the smaller the STRESS the better the estimates [3]. The description of the results above has consistently ranked the compositional techniques highly in all categories. When the STRESS measure is used, the increased complexity of the compositional techniques pays dividends and the overall compositional accuracy of the estimates is superior to that of the conventional approach (Table 6).

4 Conclusions and Further Work

Experience by other practitioners has shown that Euclidean measures will favor non-compositional techniques and Aitchison type measures will favor the logratio methods (*pers. comm.*, R. Tolosana-Delgado). This has been proven partially invalid with the results of this study; the univariate distance measure shows conventional ordinary co-kriging to result in similarly biased estimates to the results achieved with

Table 6 STRESS statistics

Distance measure (δ)	agl	Gauss-Hermite	OCK
Euclidean	0.745	0.816	1.225
Aitchison	0.641	0.698	0.760

the use of the cokriging of the alr transformed data. The compositional distance measures specifically support the compositional methods when using compositional distance measures and generally support the agl back-transformed results over the Gauss-Hermite quadrature approximation. There are alternative distance measures, such as the Mahalanobis distance which are advocated in the literature for theoretical reasons; applying examples of these alternatives will be explored in later studies, as will the use of simulation techniques. The study has demonstrated that the use of logratio techniques in mining geostatistics is a valid alternative and should be explored further.

References

1. Aitchison J (2003) The statistical analysis of compositional data. Blackburn Press, Caldwell, 416 pp (reprint)
2. Koushavand B, Deutsch C (2008) Constraining the sum of multivariate estimates. In: Twelfth annual report of the centre for computational geostatistics. University of Alberta, pp 308-1–308-12
3. Martin-Fernandez JA, Olea-Meneses RA, Pawlowsky-Glahn V (2001) Criteria to compare estimation methods of regionalized compositions. Math Geol 33(8):889–909
4. Pawlowsky-Glahn V, Olea RA (2004) Geostatistical analysis of compositional data. Oxford University Press, New York, 181 pp
5. Walvoort D, de Gruijter J (2001) Compositional kriging: a spatial interpolation method for compositional data. Math Geol 33(6):951–966

Measuring the Impact of the Change of Support and Information Effect at Olympic Dam

Colin Badenhorst and Mario Rossi

Abstract The change of support and information effect concepts are fundamental in every resource model. It underpins all aspects of resource estimation in every deposit worldwide, yet is poorly understood, rarely taught, and even more rarely applied. This paper describes the practical implications of these concepts using conditional simulation, by deriving a recoverable resource estimate for the first 11 benches of the proposed Olympic Dam open cut mine. The Olympic Dam deposit is one the world's largest polymetallic resources at 9 billion tonnes grading 0.8 % Cu, 270 ppm U_3O_8, 0.32 g/t Au and 1.5 g/t Ag. BHP Billion is currently undertaking a feasibility study of a large open cut operation with an estimated mine life in excess of 100 years. The resource estimation practices at Olympic Dam comprise of a combination of linear and non-linear techniques to estimate 16 different grade variables critical to the resource. In the southern portion of the deposit, at the site of the proposed open cut, the current resource estimate data spacing is insufficient to predict the recoverable tonnage and grade that will be selected using closely spaced grade control blast holes once mining commences. Conditional simulation has been used to generate a recoverable resource estimate by quantifying the tonnage and grade uplift resulting from the change of support and the information effect that occurs at the time of mining. Ten realizations of Cu, S, U_3O_8 and Au were generated using Sequential Indicator Simulation. The simulations were validated visually and statistically, and a single realization was then chosen to represent reality. Several grade control databases were constructed by sampling the realization at the expected blast hole spacing. Each database was used to estimate the first few pushbacks of the proposed open cut mimicking future grade control estimates. Variations were quantified and grade tonnage curves at the smaller grade control support were compared to the larger blocks of the resource. This information has been used to optimize the

C. Badenhorst (✉)
Olympic Dam, BHP Billiton, Level 8, 55 Grenfell St., Adelaide, SA 5000, Australia
e-mail: Colin.Badenhorst@bhpbilliton.com

M. Rossi
GeoSystems International, Inc., 2385 NW Executive Center Dr., Suite 100, Boca Raton, FL 33431, USA
e-mail: mrossi@geosysint.com

P. Abrahamsen et al. (eds.), *Geostatistics Oslo 2012*,
Quantitative Geology and Geostatistics 17,
DOI 10.1007/978-94-007-4153-9_28,

predictions of expected tons and grades fed to the mill, adjusting the recoverable resource estimate to control its smoothing. This information is critical for optimal mine planning. The results and conclusions of this work unequivocally demonstrate why every resource geologist should have a deep understanding of the change of support and information effect, and how it can be applied in their resource models using conditional simulation.

This publication includes information on Mineral Resources which have been compiled by S. O'Connell (MAusIMM). This is based on Mineral Resource information in the BHP Billiton 2010 Annual Report which for can be found at http://www.bhpbilliton.com. All information is reported under the 'Australasian Code for Reporting of Mineral Resources and Ore Reserves, 2004' (the JORC Code) by Shane O'Connell who is a full-time employee of BHP Billiton and has the required qualifications and experience to qualify as Competent Person for Mineral Resources under the JORC Code. Mr. O'Connell verifies that this report is based on and fairly reflects the Mineral Resources information in the supporting documentation and agrees with the form and context of the information presented.

	Measured Resource (Mt)	Indicated Resource (Mt)	Inferred Resource (Mt)	BHP Billiton interest (%)
Olympic Dam	1,280 @ 1.08 % Cu	4,725 @ 0.86 % Cu	3,222 @ 0.74 % Cu	100

1 Introduction

The Olympic Dam deposit, situated approximately 570 km NNW of Adelaide and 16 km north of the township of Roxby Downs in South Australia, is one the world's largest polymetallic resources containing 75 million tonnes of copper, 2.4 million tonnes of uranium oxide and 97 million ounces of gold. The deposit was discovered by WMC Resources in 1975, and mining using sub-level open stopping techniques commenced in mid-1988 at a rate of 1.3 million tonnes per annum. Several expansion phases in 1992, 1995 and 1996 culminated in annual ore production of 9.8 million tonnes per annum by 1999. In June 2005, BHP Billiton purchased WMC Resources, and embarked on an in-depth feasibility study aimed at establishing a large open cut operation with an estimated mine life in excess of 100 years.

The open cut expansion would be in the southern half of the deposit, which has been defined using both vertical and angled surface drilling on an average spacing of 50×35 m, allowing a resource to be estimated to a block support of $10 \times 10 \times 5$ m. Five-meter long composites were used. Whilst the spacing is sufficient for a global resource estimate, it is insufficient to provide a prediction of the reserves recovered at the time of mining. During mining, closely spaced grade control drillhole samples will be used to define the ROM tonnage and grade. This close spacing of holes (information effect) and smaller blocks (support effect) will result in a tonnage and grade different to that estimated using the currently available wider resource drillhole spacing.

Conditional simulation [3] was undertaken on the deposit, and a single realization corresponding to the median copper metal content was chosen to represent "reality". A scripting technique was employed to sample this realization at a spacing that mimics the likely grade control blasthole spacing that will be employed at the time of mining. These simulated blastholes were then used to produce a grade control estimate over the first 11 benches of the open cut. These estimates represent the recoverable resource at the time of mining, and their predictions allow mine planners to optimize the open cut design and better predict the variations in the expected tonnage and grade feed to the mill.

Conditional simulation was performed over the entire deposit for Cu, U_3O_8 and Au. For the purposes of this paper, only the Cu simulation was considered. In order to comply with BHP Billiton confidentiality requirements, all tonnage and grade information have been appropriately factored.

2 The Resource Model

Current resource estimation practices at Olympic Dam are complex and on a scale rarely observed in the industry. There is a rich dataset available, comprising of more than 1 600 surface HQ (63.5 mm) and NQ (47.6 mm) size drillholes sampled at 1 m intervals, and 7 900 underground BQ (36.5 mm) and LTK48 (35.6 mm) size drillholes sampled at 2.5 m intervals. Most surface drillhole depths average 1 000 m below surface.

Copper sulphide abundance and type (chalcopyrite, bornite and chalcocite), and haematite abundances are the primary controls on ore mineralization, and modeled using combinations of deterministic and probabilistic methods. These models are used to construct over 68 unique estimation domains for Cu, which is estimated using Ordinary Kriging (OK) with local sample sharing between domains tailored to honor the observed grade characteristics and relationships across unique domain boundaries. Density is estimated using OK, within domains defined by haematite abundance estimated using Multiple Ordinary Kriging (MIK).

Estimates are validated by comparing composites to block estimates both statistically and also visually through means of cross- and longitudinal sections and plan views. Discrete Gaussian (DG) change of support modeling is used to assess the amount of smoothing obtained during estimation, and the estimates tuned to DG models as required.

In the area of the open cut, the average drillhole spacing is 50×35 m, with some spacing as close as 35×35 m locally. The resource model in this area is estimated at a $10 \times 10 \times 5$ m support, and then regularized to $10 \times 10 \times 15$ m for mine planning purposes.

3 The Simulation Model

The Olympic Dam deposit has been simulated several times in recent years. Both Sequential Gaussian Simulation [2] and Sequential Indicator Simulation [1] techniques have been used, as well as a variant of SGS using Bayesian Updating [4].

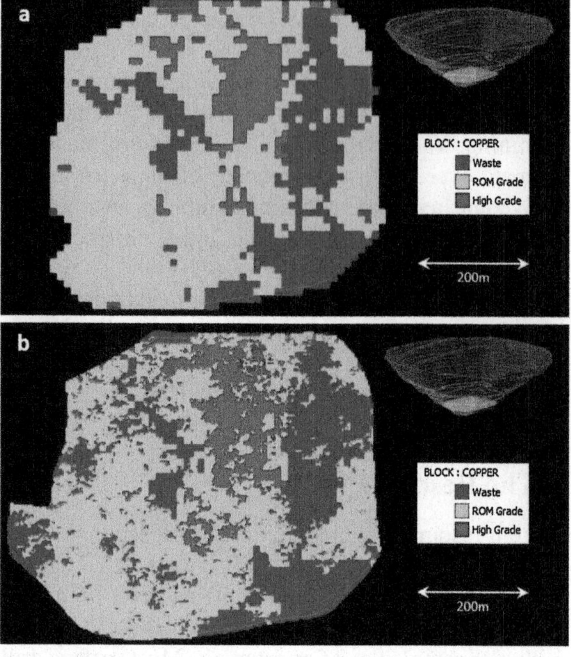

Fig. 1 Plan view of bench 6 at the −355 m RL showing (**a**) Cu estimate at a 10 × 10 × 15 m support in the resource model and (**b**) Cu realization at a 2.5 × 2.5 × 15 m support in the simulation model. The *white line* in the perspective view of the open cut shows the elevation at which mining exposes the first ore bench after having removed 350 vertical meters of waste overburden. *Blue* = 0–0.5 % Cu, *Green* = 0.5–2 % Cu, *Red* > 2 % Cu

The simulation models have been used to:

1. Characterize the uncertainty in the resource estimates at several scales;
2. Predict recoverable resources and reserves (change of support and information effect);
3. Test the resource classification for different production rates;
4. Provide daily, weekly, monthly, and quarterly profiles of ore feed to the plant;
5. Provide input into the optimization of underground stope design.

Data for this paper was drawn from a recent conditional simulation aimed specifically to characterize the selectivity required for the proposed open cut operation.

A SIS technique using 25 cut-offs and a median indicator variogram was used to generate 10 realizations of Cu, U_3O_8 and Au at a 2.5 × 2.5 × 2.5 m node spacing. The basic inputs to the simulation were 425 000 samples, composited to 5 m lengths, from 9 500 drillholes. Simulation was performed within the primary copper sulphide domains (chalcopyrite, bornite and chalcocite) to mimic the domain control employed in resource estimation.

For the purpose of this paper, a subset of the simulation covering the first 11 benches of the open cut footprint was extracted, and regularized to a 2.5 × 2.5 × 15 m support. The 15 m vertical dimension corresponds to the proposed bench height of the open cut, and only the Cu realization corresponding to the median metal content of the simulation was retained.

Figure 1 shows a comparison between the Cu estimate and Cu realization in bench 6 at the −330 m RL. Note that in order to expose ore, the open cut must first move 350 vertical meters of un-mineralized overburden.

4 The Grade Control Model

The conditional simulation model is used to bridge the gap between a global resource estimate and the locally recovered reserves by predicting the change of support and information effect. This is done by mimicking the mining process and simulating the grade control process.

A script was developed to sample the simulation at a 10×10 m spacing, and these samples were converted into a base case grade control database designed to simulate the likely production blast hole drilling at the time of mining.

Grade control at the time of mining was mimicked by estimating Cu grade into $5 \times 5 \times 15$ m blocks for each of the 11 benches using the simulated grade control drillholes. An Inverse Distance Weighting technique was employed using the same domain control/sharing as the resource estimate. The Inverse Distance technique was chosen in this exercise for simplicity and expediency, and does not impact the final results and conclusions. On the other hand, the simulated blast holes and indeed the simulation model itself will only quantify realistic variations if they themselves are a reasonable representation of reality.

The grade control estimation plan consisted of using 2 passes for each Estimation Domain (Non-Sulphides, Chalcopyrite, Bornite, and Chalcocite). The first pass used $12.5 \times 12.5 \times 15$ m search radii in the Easting, Northing, and Elevation directions, with a minimum of 4 and a maximum of 5 simulated samples. The second pass increased the search radii to $25 \times 25 \times 15$ m search radii in the Easting, Northing, and Elevation directions, with a minimum of 3 and a maximum of 5 samples. In most cases, samples were shared across boundaries:

1. Bornite and Chalcocite were estimated sharing as a single unit, sharing samples in both passes;
2. Chalcopyrite was estimated using also Bornite samples in both passes;
3. Non-sulphides were estimated with Chalcopyrite samples only in the second pass.

Figure 2 shows a visual comparison between the resource and grade control estimates for a few benches. Note the effect of the increased data density and smaller block size. There is a sharper definition of the waste/stock/high grade contacts.

5 Comparing the Resource and Grade Control Estimates

The resource estimate and grade control estimate for each bench was compared by using classic grade-tonnage analysis and summarized in Table 1. The material is classified according to their destination:

1. Waste material (Cu < 0.5 %);
2. ROM (run-of-mine) material ($0.5\ \% < \text{Cu} < 2\ \%$);
3. High grade (Cu > 2 %).

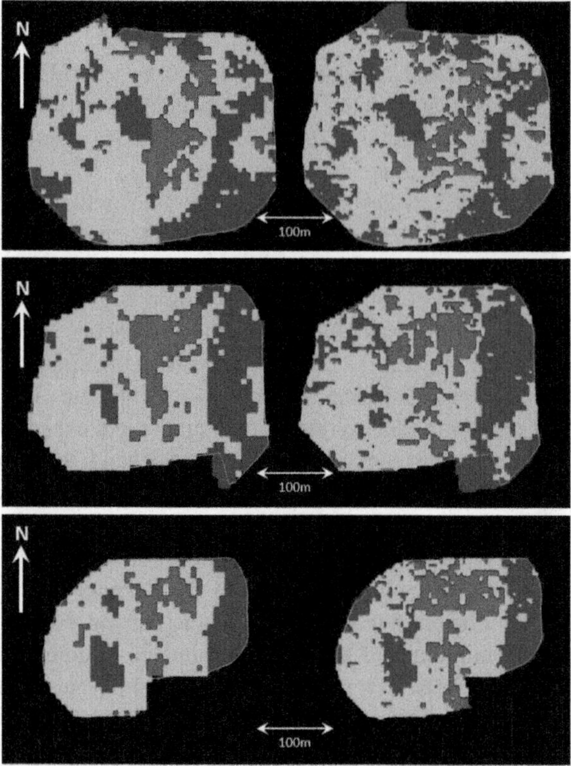

Fig. 2 Plan view of bench 3, 9 and 10 showing a comparison between the resource model (on the *left*) and the grade control model (on the *right*). The *legend color* ranges are the same shown in Fig. 1

In Table 1, a positive differential indicates more copper metal predicted by the grade control model, whereas a negative differential indicates more copper metal predicted by the resource model.

At higher cut-offs, the tonnage and grade undergoes a significant uplift (Fig. 3). This is directly attributable to the resolution achieved through the change in support (smaller blocks) and the information effect (many closely spaced blast holes).

Bench 2 is considered an anomaly, in that for the high grade bench the grade control model suggests that the resource model overstates the copper metal at cut-offs less than 3 % Cu, probably due to the presence of locally high grade Cu composites.

6 Real-World Grade Control Model

The free-selection basis described earlier is a theoretical exercise that illustrates the metal uplift achieved if each block could be extracted against a specific cut-off on a block-per-block basis. In real-world mining scenarios, blasting and digging practices do not allow the blocks to be selected on this basis. Typically, dig lines are generated to derive practical mineable shapes that attempt to balance the maximum tonnage and grade with the production schedule and available mining equipment.

Table 1 Comparison between resource model and grade control model on a free selection basis, showing the tonnage, grade and metal content for the Waste, ROM and High Grade classifications. Note that the change of support and information effects are more dramatic as the cut-off increases

Bench No.	Grade Category	Resource Model			Free Selection Model			% Diff. Metal
		MTons ($\times 10^6$)	Grade %	Cu Metal ($\times 10^3$)	MTons ($\times 10^6$)	Grade %	Cu Metal ($\times 10^3$)	
1	Waste	9.43	0.07	7	9.78	0.11	11	63 %
	ROM	4.58	1.04	48	4.20	1.03	43	−9 %
	High Grade	0.70	2.54	18	0.73	2.84	21	17 %
2	Waste	5.59	0.17	9	5.83	0.21	12	29 %
	ROM	6.29	1.08	68	6.39	1.05	67	−1 %
	High Grade	1.52	2.71	41	1.17	2.84	33	−20 %
3	Waste	3.01	0.26	8	3.24	0.26	8	7 %
	ROM	6.13	1.10	67	5.89	1.10	65	−4 %
	High Grade	1.29	2.72	35	1.30	2.72	35	1 %
4	Waste	2.91	0.25	7	3.26	0.26	8	17 %
	ROM	5.96	1.07	64	5.57	1.07	60	−7 %
	High Grade	1.27	2.71	35	1.31	2.68	35	2 %
5	Waste	2.18	0.27	6	2.54	0.29	7	25 %
	ROM	5.02	1.05	53	4.55	1.04	47	−10 %
	High Grade	1.10	2.72	30	1.21	2.83	34	14 %
6	Waste	1.90	0.26	5	2.17	0.28	6	23 %
	ROM	4.92	1.06	52	4.35	1.01	44	−16 %
	High Grade	0.79	2.59	21	1.09	2.67	29	41 %
7	Waste	1.57	0.25	4	1.72	0.28	5	23 %
	ROM	3.68	1.08	40	3.42	1.08	37	−7 %
	High Grade	0.75	2.58	19	0.86	2.64	23	16 %
8	Waste	1.13	0.22	2	1.43	0.26	4	49 %
	ROM	3.26	1.10	36	2.97	1.13	34	−6 %
	High Grade	0.82	2.64	22	0.81	2.81	23	5 %
9	Waste	0.68	0.17	1	0.86	0.23	2	71 %
	ROM	2.22	1.16	26	1.96	1.16	23	−12 %
	High Grade	0.64	2.62	17	0.72	2.64	19	13 %
10	Waste	0.61	0.22	1	0.77	0.27	2	55 %
	ROM	2.10	1.12	24	1.74	1.16	20	−14 %
	High Grade	0.32	2.96	9	0.51	2.77	14	52 %
11	Waste	0.43	0.23	1	0.45	0.24	1	7 %
	ROM	1.13	1.09	12	0.98	1.13	11	−10 %
	High Grade	0.16	3.12	5	0.30	2.93	9	73 %
Total Waste		29.44	0.18	51.93	32.06	0.21	66.29	28 %
Total ROM		45.28	1.08	488.69	42.03	1.07	450.30	−8 %
Total High Grade		9.37	2.68	251.40	10.00	2.75	274.94	9 %

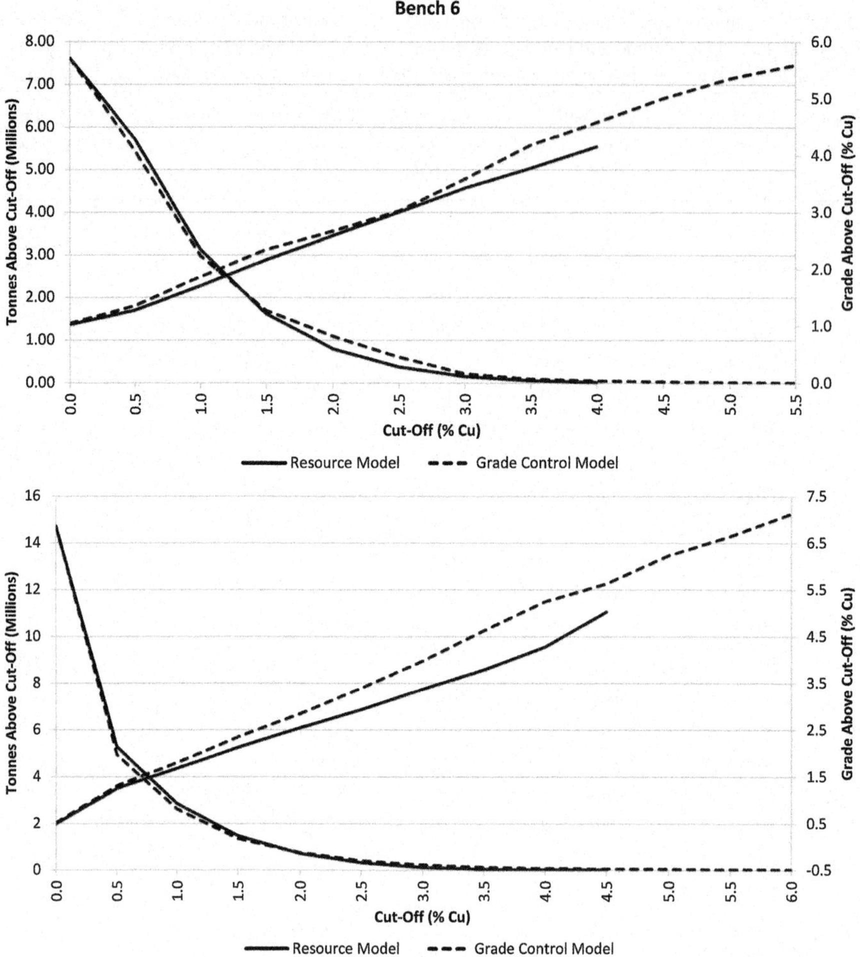

Fig. 3 Grade-tonnage curves for Bench 1 and Bench 6 showing tonnage and grade uplift achieved at higher cut-offs due to the change of support and information effect in the grade control model

Depending on the selectivity possible from the mining equipment, the dig-lines will be generated in such a way that some ore blocks are allocated to the waste category, and some waste blocks will be allocated to the ore category.

Practical mining shapes (dig-lines) have been generated for the 11 benches of the open pit. In practice, dig-lines like these would be used by the operation to select, load, and haul the material to the different destinations. Figure 4 shows an example of the dig lines generated for bench 2 of the open cut. Note that some of the shapes may be minable, or not, depending on the direction of mining. Still, it is a realistic example of what would be expected during the ore/waste selection process.

Note from Fig. 4 that there can be significant additional ore loss and dilution when the non-free selection scenario is considered. In this example, almost all miss-

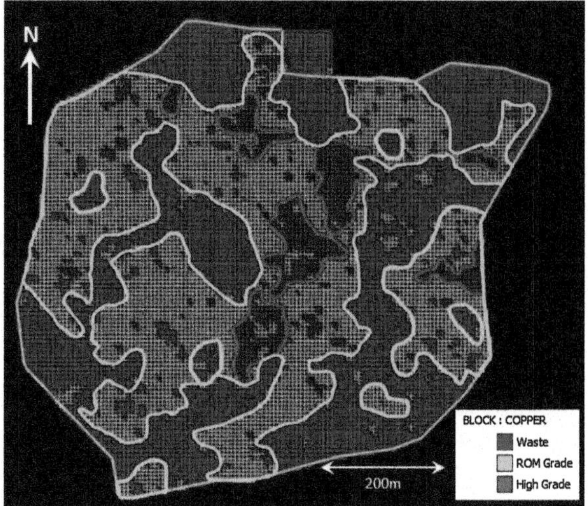

Fig. 4 Plan view of bench 2 showing the dig lines and dilution incurred at the time of mining

classifications that can occur do occur: waste to ROM (blue blocks inside green outlines), waste to high grade (blue blocks inside red dig lines), ROM to high grade (green blocks inside red dig lines), ROM to waste (green blocks inside blue dig lines, or with no dig lines at all), and high grade to ROM (red blocks insider green outlines). For this bench, the only exception is that there is no high grade sent to waste as this represents the more typical practice of sending more waste material to the ore (ROM and high grade) destinations. In the case of the small high grade dig-line to the NW in Fig. 4, the high grade dig line butts against waste material, and some unplanned ore loss is possible.

Another critical issue to consider is the significant difference between the free selection and the "real" grade control model. While both are based on the same samples and grade control estimates, the process of adjusting the selection to what is likely to be mineable implies a significant loss in recovered grade and ore.

Table 2 compares the resource model with the ("real") grade control model after dig lines have been defined. Table 3 compares the grade control model without dig lines (free selection) with the "real" grade control model after dig lines have been defined.

The resource model (Table 2) in effect is not such a bad predictor of the "real" grade control model (i.e., recoverable resources/reserves) in some benches. But indeed, it is difficult to calibrate the resource model to accurately predict what is going to be recovered at the time of mining.

Comparing the free selection grade control with the "real" grade control (Table 3), it is evident that dilution and ore loss is significant at the time of mining. This is to the point that, for several benches, the significant uplift observed from the resource model to the free selection model is lost when defining dig lines. But this does not occur for all benches, with some gains in metal realized through gains in tonnages.

Table 2 Comparison between resource model and "real" grade control model (using the dig-lines), showing tonnage, grade and metal content for Waste, ROM and High Grade classifications. Note that the change of support and information effects are more dramatic as the cut-off increases

Bench No.	Grade Category	Resource Model			Grade Control Model			% Diff. Metal
		MTons ($\times 10^6$)	Grade %	Cu Metal ($\times 10^3$)	MTons ($\times 10^6$)	Grade %	Cu Metal ($\times 10^3$)	
1	Waste	9.43	0.07	7	8.72	0.05	4	−39 %
	ROM	4.58	1.04	48	5.28	1.01	53	11 %
	High Grade	0.70	2.54	18	0.71	2.45	17	−5 %
2	Waste	5.59	0.17	9	5.60	0.24	14	47 %
	ROM	6.29	1.08	68	6.71	1.04	70	3 %
	High Grade	1.52	2.71	41	1.08	2.69	29	−30 %
3	Waste	3.01	0.26	8	32.93	0.31	9	15 %
	ROM	6.13	1.10	67	6.31	1.10	70	4 %
	High Grade	1.29	2.72	35	1.19	2.50	30	−14 %
4	Waste	2.91	0.25	7	3.11	0.30	9	24 %
	ROM	5.96	1.07	64	5.73	1.08	62	−3 %
	High Grade	1.27	2.71	35	1.30	2.50	32	−7 %
5	Waste	2.18	0.27	6	2.41	0.33	8	36 %
	ROM	5.02	1.05	53	4.62	1.02	47	−11 %
	High Grade	1.10	2.72	30	1.27	2.67	34	13 %
6	Waste	1.90	0.26	5	2.13	0.34	7	42 %
	ROM	4.92	1.06	52	4.30	0.99	43	−18 %
	High Grade	0.79	2.59	21	1.18	2.47	29	41 %
7	Waste	1.57	0.25	4	1.59	0.31	5	27 %
	ROM	3.68	1.08	40	3.38	1.04	35	−12 %
	High Grade	0.75	2.58	19	1.03	2.37	24	23 %
8	Waste	1.13	0.22	2	1.33	0.29	4	55 %
	ROM	3.26	1.10	36	2.89	1.10	32	−11 %
	High Grade	0.82	2.64	22	0.98	2.48	24	13 %
9	Waste	0.68	0.17	1	0.82	0.29	2	106 %
	ROM	2.22	1.16	26	1.89	1.11	21	−18 %
	High Grade	0.64	2.62	17	0.83	2.46	20	21 %
10	Waste	0.61	0.22	1	0.69	0.30	2	58 %
	ROM	2.10	1.12	24	1.79	1.14	20	−14 %
	High Grade	0.32	2.96	9	0.54	2.61	14	51 %
11	Waste	0.43	0.23	1	0.43	0.22	1	−5 %
	ROM	1.13	1.09	12	0.96	1.12	11	−13 %
	High Grade	0.16	3.12	5	0.34	2.68	9	83 %
Total Waste		29.44	0.18	51.93	29.77	0.22	65.32	26 %
Total ROM		45.28	1.08	488.69	43.86	1.06	463.94	−5 %
Total High Grade		9.37	2.68	251.40	10.45	2.52	263.20	5 %

Table 3 Comparison between free selection grade control model and the "real" grade control model (using the dig-lines), showing tonnage, grade and metal content for Waste, ROM and High Grade classifications

Bench No.	Grade Category	Resource Model			Free Selection Model			% Diff. Metal
		MTons ($\times 10^6$)	Grade %	Cu Metal ($\times 10^3$)	MTons ($\times 10^6$)	Grade %	Cu Metal ($\times 10^3$)	
1	Waste	9.78	0.11	11	8.72	0.05	4	−60 %
	ROM	4.20	1.03	43	5.28	1.01	53	22 %
	High Grade	0.73	2.84	21	0.71	2.45	17	−19 %
2	Waste	5.83	0.21	12	5.60	0.24	14	14 %
	ROM	6.39	1.05	67	6.71	1.04	70	4 %
	High Grade	1.17	2.84	33	1.08	2.69	29	−13 %
3	Waste	3.24	0.26	8	2.93	0.31	9	7 %
	ROM	5.89	1.10	65	6.31	1.10	70	8 %
	High Grade	1.30	2.72	35	1.19	2.50	30	−15 %
4	Waste	3.26	0.26	8	3.11	0.30	9	6 %
	ROM	5.57	1.07	60	5.73	1.08	62	4 %
	High Grade	1.31	2.68	35	1.30	2.50	32	−9 %
5	Waste	2.54	0.29	7	2.41	0.33	8	9 %
	ROM	4.55	1.04	47	4.62	1.02	47	−1 %
	High Grade	1.21	2.83	34	1.27	2.67	34	−1 %
6	Waste	2.17	0.28	6	2.13	0.34	7	15 %
	ROM	4.35	1.01	44	4.30	0.99	43	−2 %
	High Grade	1.09	2.67	29	1.18	2.47	29	0 %
7	Waste	1.72	0.28	5	1.59	0.31	5	4 %
	ROM	3.42	1.08	37	3.38	1.04	35	−5 %
	High Grade	0.86	2.64	23	1.03	2.37	24	6 %
8	Waste	1.43	0.26	4	1.33	0.29	4	4 %
	ROM	2.97	1.13	34	2.89	1.10	32	−5 %
	High Grade	0.81	2.81	23	0.98	2.48	24	7 %
9	Waste	0.86	0.23	2	0.82	0.29	2	20 %
	ROM	1.96	1.16	23	1.89	1.11	21	−7 %
	High Grade	0.72	2.64	19	0.83	2.46	20	8 %
10	Waste	0.77	0.27	2	0.69	0.30	2	2 %
	ROM	1.74	1.16	20	1.79	1.14	20	1 %
	High Grade	0.51	2.77	14	0.54	2.61	14	−1 %
11	Waste	0.45	0.24	1	0.43	0.22	1	−11 %
	ROM	0.98	1.13	11	0.96	1.12	11	−4 %
	High Grade	0.30	2.93	9	0.34	2.68	9	6 %
Total Waste		32.06	0.21	66.29	29.77	0.22	65.32	−1 %
Total ROM		42.03	1.07	450.30	43.86	1.06	463.94	3 %
Total High Grade		10.00	2.75	274.94	10.45	2.52	263.20	−4 %

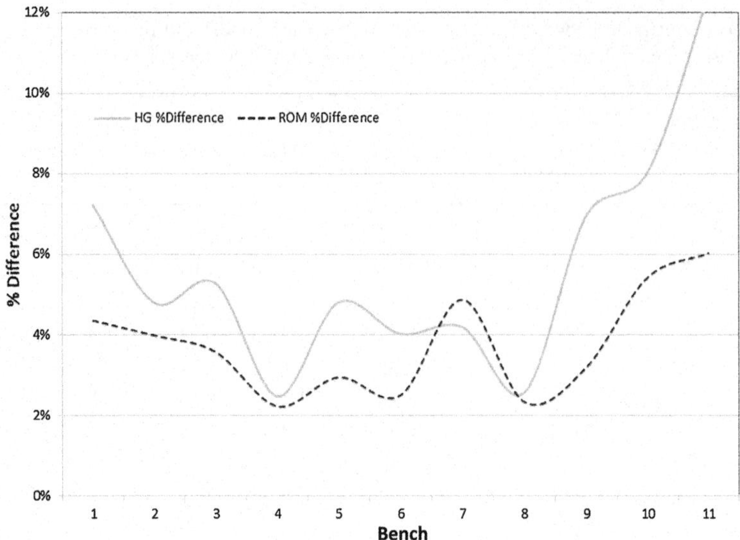

Fig. 5 Relative difference of average Cu grades at 0 % cut-off for all 15 databases, and for each Bench (1 through 11)

Additionally, the simulation sampling script was also used to create variations of the base case grade control database by varying the Northing and Easting collar positions to mimic real-world production scenarios. A total of 15 grade control database variations were created by offsetting the Northing and Easting collar positions from the base case scenario by 2.5 m, 5 m and 7.5 m.

Figure 5 shows the changes in average Cu grades for each bench and at 0 % cut-off for the 15 databases. The variance is shown as a relative difference between the maximum and minimum average grades with respect to its mean.

Figure 5 shows that the sampling variance for benches with lower ROM and High Grade tonnages (Benches 1, 2, 9, and 11) is larger. Even for the middle benches, the sampling variance is always greater than 2 %. This is a spatial variance, in the sense that it quantifies the consequence of shifts in the position of the samples; it is equivalent to the stochastic variations that can be observed from one simulation to the next.

The conditional simulations developed reproduce the original variability of the drill hole data. In reality, the variances observed in Fig. 5 can only be expected if the variability of the drill hole data, and the blast hole data used to select ore and waste, are similar.

7 Conclusions

The results of this work unequivocally demonstrate why every resource and mine geologist should have a deep understanding of the change of support and information effects, and how these concepts need to considered in their resource models. The main conclusions that can be extracted are:

1. The impact of the change of support and information effect at Olympic Dam is significant. Conditional simulation allows us to characterize them by themselves or as a combined effect, as presented in this paper.
2. The resource model should be built as a recoverable resource model, trying to incorporate future dilution and ore loss. This allows better prediction of the metal recovered at the time of mining.
3. The amount of dilution and ore loss to be incorporated into the resource model can be calibrated using conditional simulations and grade control. It is generally more significant for higher cut-off grades and at depth, where the high-grade ore zones are more continuous compared to the ore at shallower elevations. Grade control practices for the shallower benches will require more diligence than for benches deeper down into the deposit.
4. To adequately predict dilution and ore loss at the time of mining, it is necessary to develop the full grade control process. Assuming free selection would be dangerous and naive, in the sense that it predicts significantly higher metal recovered than what is actually achievable. This may raise unreasonable expectations, and leads to poor reconciliation.

Acknowledgements The work presented here is a culmination of five years worth of team effort. In particular, the authors would like to acknowledge the contributions of Anthony Bottrill and Shane O'Connell. We would also like to acknowledge BHP-Billiton Uranium for their permission to publish this work.

References

1. Alabert F (1987) Stochastic imaging of spatial distributions using hard and soft information. MSc thesis. Stanford University, Stanford, CA, 197 pp
2. Isaaks EH (1990) The application of Monte Carlo methods to the analysis of spatially correlated data. PhD thesis, Stanford University, Stanford, CA, 213 pp
3. Journel AG (1988) Fundamentals of geostatistics in five lessons. Stanford Center for Reservoir Forecasting, Stanford
4. Rossi ME, Badenhorst C (2010) Collocated co-simulation with multivariate Bayesian updating: a case study on the olympic dam deposit. In: Castor R et al (eds) Proc. of the 4th international conference on mining innovation (MININ 2010), Santiago, Chile

Comparative Study of Two Gaussian Simulation Algorithms, Boddington Gold Deposit

Michael Humphreys and Georges Verly

Abstract Simulation results depend on many parameters such as: modeling assumptions (e.g. Gaussian or Indicator simulation); implementation (e.g. point or block simulation); and case-specific parameters (e.g. top-cut values). The user is often in the dark when it comes to the impact on the results of modeling assumptions and software implementation. Not much literature is available and checks are difficult to complete. This paper is a comparative study of two point and block Gaussian related simulations. The study shows that careful calibration/validation is necessary in both cases to avoid significant biases. Point simulation is easy to validate against primary data because the support does not change; block simulation is more difficult to validate. Both software/algorithms were able to provide what they have been designed for, i.e. conditional simulated values that reproduce the required grade point or block distribution and variogram. Differences, however, were noted between the re-blocked point simulation and direct block simulation results. Differences in application and results of the methods together with advantages and disadvantages are discussed.

1 Introduction

Simulations of Au grade have been run as a measure of risk and for other investigation work (e.g. bench height) at Newmont Mining's Boddington Gold Mine located in the south-west corner of Western Australia. Various simulation methods have been used by Newmont personnel and consultants—all based on a Gaussian approach. There has been past inconsistency with the simulations overcalling the grade compared to estimations—Multiple Indicator Kriging, Uniform Conditioning and Ordinary Kriging (OK).

M. Humphreys (✉)
Newmont, Level 1, 388 Hay St, Subiaco 6009, Australia
e-mail: Michael.Humphreys@Newmont.com

G. Verly
AMEC Americas, 111 Dunsmuir St, Vancouver, BC, Canada, V6B 5W3
e-mail: georges.verly@amec.com

P. Abrahamsen et al. (eds.), *Geostatistics Oslo 2012*,
Quantitative Geology and Geostatistics 17,
DOI 10.1007/978-94-007-4153-9_29,

Limited production data from around two years operation also supported the case of simulation overcalling the grade. A detailed investigation into simulation at Boddington was undertaken [4, 5] to resolve the continuing uncertainty from previous work around choice of parameters (e.g. capping) and methodology (e.g. point vs. block simulation).

Papers are available concerning simulation but those that describe the important practical details of implementation, the reasoning behind choice of parameters and their effects are rare. It is also easy to use just what is obvious or available in a given software without asking questions or taking that extra validation step.

This paper delves into practical details and hopes to promote discussion concerning choices and effects. The objectives of the paper are: (1) to compare two different algorithms for simulation, though both Gaussian based, from a practical point of view; (2) to suggest a series of validations to perform when comparing two simulations; (3) to provide a list of important steps to follow when simulating; (4) to suggest a series of validations for a grade simulation with the expectation that commercial software packages should be able to perform them obviously and inherently in time.

2 Boddington Deposit

Boddington is a large tonnage, low grade deposit covering an area of approximately 2 km by 6 km. Gold and copper are hosted in the bedrock mineralization. Current published reserves (2011) are approximately 969 Mt @ 0.65 g/t Au (20.3 Moz Au) and 0.11 % Cu. Boddington occurs in the Wells formation consisting of felsic to intermediate volcanics and related intrusives. Barren dolerite dykes run through the orebody. There are over 40 domains with over 200 000 composites of 4 m length. The selective mining unit (SMU) size is 20 by 20 by 12 m. Milling plans for more than 30 Mt of ore per annum.

3 General Approach

Different types of simulation are available. In this case, previous test work had chosen a Gaussian based approach. The simulation runs described herein followed this choice. However, the overall size of the Boddington deposit raised the question of point vs. block simulation—for computing time and file size reasons. Both simulation methods were run and compared for this work.

The general outline of the approach implemented here is presented in [8]. After capping [7], a restricted envelope (an Exploratory Data Analysis or EDA envelope) was defined to limit composite selection initially to where there was a reasonable density of data. The EDA envelope defines an area of confidence for calibrating the simulation parameters and validating the simulation results against the original data. Using data within the EDA envelope, variograms and distributions were modeled and normal score transforms were performed. Simulations were then run. Validation at each step of the simulation within the EDA envelope was performed—checking re-production of input variograms, expected statistics, local mean (by swath plots) both before and after back-transform.

The last step was a calibration of the simulations to the local mean. For various reasons—in particular lack of stationarity at the local scale—the simulations may produce grades departing from a mean of the local area. The simulations can be adjusted or calibrated to reflect this local mean (using a moving average or Ordinary Kriged model for example). Nowak and Verly [8] describe this as "progressive correction". If this adjustment is large, then it is an indicator that there is likely to be a problem.

Once this calibration/validation process was completed with an acceptable result, the simulations are normally re-run for the entire domain (not just the EDA envelope), using the parameters defined from the well informed EDA domain. Validation is again performed, keeping in mind that the results are impacted by the simulated values located outside the EDA envelope. Results presented in this paper are limited to the calibration/validation work of the simulations within the EDA envelope. For this study, the effect of uncertainty of the underlying geological models and initial parameter choice are not considered.

4 Project Specifics

Only one of the significant gold domains was thoroughly investigated for this work. The focus was gold as it provides the bulk of the revenue. The domain size is approximately 500 m by 900 m by 900 m with a NW strike. There are over 18 500 four meter composites for this domain with significant clustering close to the surface.

A Gaussian based approach was used for simulation as prior testwork had indicated this was appropriate. These tests included the indicator cross variogram ratio and indicator variogram reproduction from transform, as well as point-block bigaussianity tests (H-scatter plot and ratio of variogram to madogram [1, 6]). Most recent internal tests with more data indicate that at high cutoff grades the Discrete Gaussian model approach might not be as appropriate. This is not an issue at the much lower life-of-mine cutoff grade level.

Although it could be attempted, no detrending option or non-stationary models were implemented. These create their own array of issues when applied. A past trial of detrending prior to simulation for Boddington was not successful.

Two simulation methodologies were compared—a point-based and a block-based approach. The 4 m composites were used for the basis of both point and block simulations. The same variogram model (from the Gaussian-transformed composite grades) was used as the starting point for both methodologies. The point-based simulation was a sequential Gaussian simulation (SGS) implementation. Various SGS programs are available in commercial, public or private software. AMEC's in-house

SGS code was used. This is similar to the public GSLib software but has been optimized for speed and file access. No significant differences in results are expected with other GSLib based codes (e.g. usgsim) but this will require later checking.

The block simulation method has been applied at Boddington in two ways. Originally the method in [3] was used. The method applied here follows [2] using the Isatis software. Both methods use the name DBSim (Direct Block Simulation) but are different in their mathematics. Unless otherwise specified, DBSim in this paper refers to [2], which is essentially implementing a Discrete Gaussian change of support to the input data and parameters and working in that space. This DBSim also runs a turning bands (TB) methodology.

Simulations are run in the Gaussian space, leaving the simulated distributions to be checked in that space before the back transform is applied. For the 20 by 20 by 12 m blocks, grades are simulated, either directly by DBSim after back transforming, or after back transforming and reblocking the point SGS simulation (5 by 5 by 4 m nodes to 20 by 20 by 12 m blocks).

5 Practical Application

Details from the application are presented in this section. As mentioned previously, once an acceptable run through is achieved using the EDA envelope, the entire domain is simulated using all data within the domain but using the variogram and distribution defined from the well informed area.

Exactly the same dataset of over 18,500 composites was used for both simulation approaches. Previous work had determined a capping of 8 g/t Au for estimation, even though there was a maximum grade of over 200 g/t Au. This removed about 6 % of the metal, which is acceptable in practice for a gold deposit. Capping of the grades was implemented after reviewing the grade distribution in multiple ways (histograms, log probability plots, indicator correlation, spatial location, etc.). Previous simulation work—with less strict capping—had shown an over-calling of the grade compared to resource model estimates and to the reconciliation from mining. It is believed that implementing the same capping here for the simulations and estimates is the root cause of better matching the mean grade of the two (and mine reconciliation). In Boddington's case, the high-grade tail of the domain considered seems built on isolated, individual grades with little volume impact that over influence the simulation result and the modeled underlying distribution if not heavily restricted in some fashion.

5.1 EDA Envelope

The goal was to define a reasonably well drilled area where the domain dominated the block considered, i.e. an area where reasonable calibration and validation can be achieved. For our 20 by 20 by 12 m block model, a depth where the drilling started to become significantly sparser was used as a restriction. The other restriction was to use blocks that were greater than 80 % within the domain considered. This was to avoid any issue with weighting when reviewing results. These were relatively subjective calls which could be implemented in alternative and perhaps more objective approaches, but the general principle was achieved. The impact of "fringe" mineralization on statistics is reduced for the calibration of the simulation. As expected, the EDA restriction gave a reduction in number of data but a more significant reduction in domain volume considered.

5.2 Declustering

There was grouping of data from various drilling directions and variously spaced drilling programs; thus a declustering was applied within the EDA envelope. Some of the techniques available for declustering are open to inaccuracy due to the waste dolerite dykes present in the domain and using the edge of the domain as a hard boundary. Weighting where grades have been removed (due to waste or other domain) should be different to weighting where grade samples do not exist. For example, data next to the dolerite dyke could have an artificially inflated weighting due to including the volume of the waste dyke as part of its influence because the data in the dyke has been removed. Therefore in this case, a nearest neighbor (NN) estimate of small blocks was done to create the weighting for declustering and to avoid fringe and void effects. Differences in declustered statistics for various methods were not grossly different in this case, but choice of a small enough block for the NN is significant. Other methods have sensitivity to size of cell/ellipse and origin.

5.3 Gaussian Transform and Variogram

The EDA envelope data were transformed to Gaussian space for the simulations using NN declustering weights separately within the two software approaches. These gave very similar results. The normal score transform for SGS did not use a tail extrapolation as in the standard GSLib and was thus more similar to the polynomial approach used for DBSim. Different methods of transform can introduce variation, and there are many options. Those used here are well known and tested. As mentioned earlier, various confirmatory test work on the multigaussian assumption had been previously run internally and by external auditors.

The Gaussian (normal score) variogram was modeled for the EDA envelope data. This model was used for both simulation methods. An alternative fitting was trialled, which had a zonal aspect (very long ranges). However, it later proved difficult for the DBSim approach to produce simulations with normal score statistics using this zonal aspect without putting a large factor/adjustment to the model. Therefore, the zonal aspect model was abandoned for these trials.

5.4 Simulation (EDA Envelope)

At this stage the two methodologies diverge in their application. Even the choice of the local number of conditioning data differs with 16 composites or previously simulated values for the point simulation and 32 composite values to condition the block simulation. The search ellipse size was taken directly from previous estimation work for DBSim. For the SGS multi-grid approach, larger searches were used for the first passes and smaller searches for the last passes. Note that DBSim does not have a multi-grid option since it is based on the turning bands method.

Point simulation proceeded directly from here, with all parameters set. However, the DBSim required regularization of the variogram model in the Gaussian space to the Selective Mining Unit (SMU) block size—20 by 20 by 12 m—as well as incorporation of change-of-support into a model of distribution and the Gaussian transform function.

5.4.1 Validation of Gaussian Value Simulations

Fifty simulations were run within the EDA envelope and statistics were calculated on the results. In all cases the statistics did not come back exactly as $N(0, 1)$. The mean differed slightly from 0.0 but the variances differed enough to warrant review (the mean of the 50 simulation means and variances was used as the measure). This is a most important check because the variance greater than 1.0 for this positively skewed raw gold grade population will result in a positive bias for the back transformed grades.

The option of using SK or OK for the simulation conditioning made a difference. For example, exactly the same parameters for everything else in the DBSim except the SK or OK conditioning, produced a variance of 1.04 for SK and 1.14 for OK. It is understood the higher variance from the OK approach is a common problem. Due to this the SK conditioning was applied for both DBSim and SGS. Later work showed that using the SK conditioning (applying a global mean of 0.0) is a possible issue where there is a high weight on the mean.

With the means being close enough to 0.0, the input variogram model structure sills were proportionally adjusted to reproduce a variance closer to 1.0. Opposite sill adjustments were made for SGS and DBSim. SGS required an increase in sill (factor 1.05); DBSim required a decrease in sill (factor 0.95 for SK, 0.80 for OK conditioning).

With the $N(0, 1)$ distribution now well reproduced, other validations in the Gaussian space were performed. Variograms of the simulations were validated against the input model (Fig. 1a) along with a visual check of the Gaussian values in 3D. Some swath plots (plots of average value by coordinate) and histogram/distribution checks were carried out on the simulated point Gaussian values. Swath plots cannot be directly compared for the DBSim simulation because both point and block distributions are $N(0, 1)$. In other words, a Gaussian block value is not an average

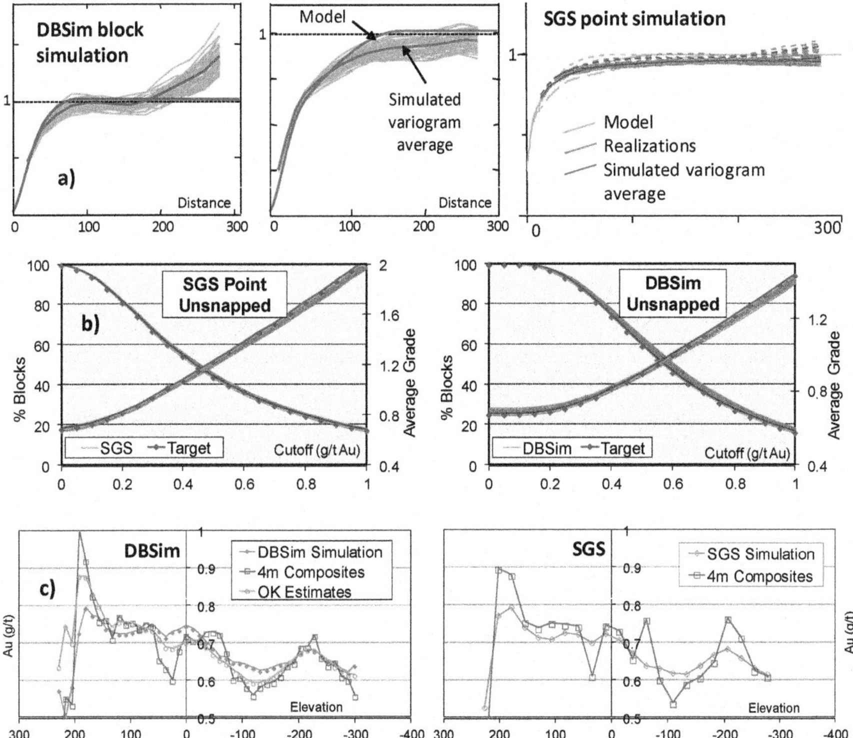

Fig. 1 Validating the simulations. (**a**) reproduction of variogram model in the Gaussian space. (**b**) reproduction of *grade-tonnage curves* before adjustment to a kriged model. (**c**) swath plots before adjustment to a kriged model

of Gaussian point values. For DBSim, the focus of the validations was on the back-transformed values. Note that for SGS, all the validations were also repeated on the back-transformed values.

5.4.2 Validation of Grade Simulations

The simulated Gaussian values were back-transformed to simulated grades. The DBSim grade simulation does not need to be re-blocked. The SGS point grade simulation is re-blocked to SMU size. In practice, the validation often stops at this stage. For this study, however, a full suite of validations on the simulated grades was applied: variograms, swath plots, grade-tonnage curves, statistics and visual. Excepting swath plots, these were reviewed looking at all individual simulations.

Table 1 lists the percentage difference in statistics for the back-transformed data. As expected, the differences are small. If no correction were made for the variance of the simulated Gaussian values, the result would be a positive bias in raw grade of 4–5 % and higher for variance [5].

Table 1 Percentage difference of back transformed statistics to input data

Type	Mean	Variance	CV
Point SGS	0.30 %	−2.70 %	−1.60 %
Block DBSim	1.50 %	−4.60 %	−3.90 %

Variograms in raw grade space are acceptably reproduced (almost identical in shape to the Gaussian reproduction in Fig. 1a) as are the grade-tonnage curves (Fig. 1b). For SGS, the point grade variogram was reconstructed from the Gaussian variogram. For DBSim, the point grade reconstructed variogram was regularized from point to 20 by 20 by 12 m SMU size (it was difficult to back transform the Gaussian SMU variogram model used for the simulation).

Swath plots show the mean of the simulations reproduces the global trend well but varies locally from the average declustered grades of data (derived from the NN block model) and the local averages produced by OK estimates (Fig. 1c). The local variation of simulations from local means and OK estimates could arise from a number of sources. The use of SK for conditioning is considered the likely cause, with its associated attraction to the global mean. Figure 1c swath plots show relatively good results.

5.4.3 Validation of OK Snapped Simulations

Nowak and Verly [8] discussed different ways to improve the trend and local mean reproductions. They suggested a progressive correction as a simple and practical way to better respect the local mean while preserving the local CV's. In this case, the mean of the simulations was simply "snapped" to a local mean provided by a kriged model as per

$$\text{AuSim}_{\text{New}} = \text{AuSim}_{\text{Old}} \times \frac{\text{Au}_{\text{OK}}}{\text{AuSim}_{\text{avg.}}}. \tag{1}$$

Multiple options for the number of samples used to create the OK estimate were trialled, varying from 24 to 64 points. Estimates were created in 10 by 10 by 12 m blocks; then these values were copied into the simulation models. The general principle is that a local area mean is required; thus a more smoothed estimate is likely to be most appropriate. Too few samples can result in variability perhaps calibrated too high (to SMU/block variability). It is best to trial a few cases.

Snapping alters the swath plot to identically match that from the local mean being snapped to—in our case the OK model. The greater smoothing of the OK used for snapping (more points) results in lowering the variance (Table 2) and the variogram sills. The shape of the variograms is generally well preserved.

Grade tonnage curves do change after snapping, although the effect isn't obviously significant. Figure 2 shows percentage differences to the target distribution—metal, tonnes, and grade—for the point SGS simulation. The average difference

Table 2 Percentage difference of snapped simulation statistics to data input (point grade variance for SGS; SMU grade dispersion variance for DBSIM)

	Point SGS			SMU DBSim		
	Un-snapped	OK24 snapped	OK64 snapped	Un-snapped	OK24 snapped	OK64 snapped
Mean	0.30 %	−0.40 %	−0.60 %	1.50 %	0.90 %	0.70 %
Variance	−2.70 %	−6.10 %	−13.20 %	−4.60 %	6.30 %	−4.20 %
CV	−1.60 %	−2.70 %	−6.30 %	−3.80 %	2.20 %	−2.80 %

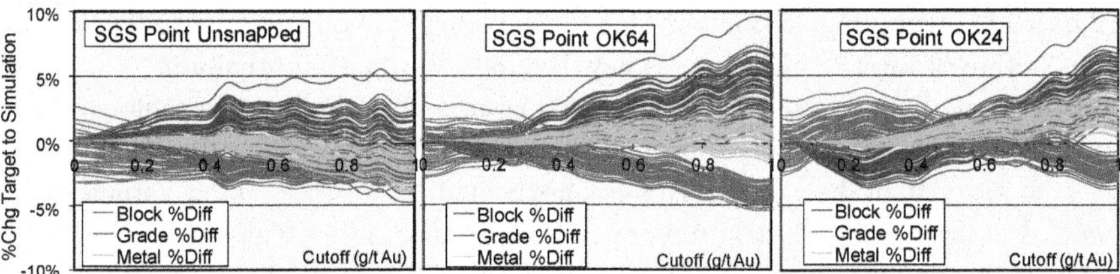

Fig. 2 Validating the reproduction of point *grade-tonnage curves* % difference after adjustment to a kriged model

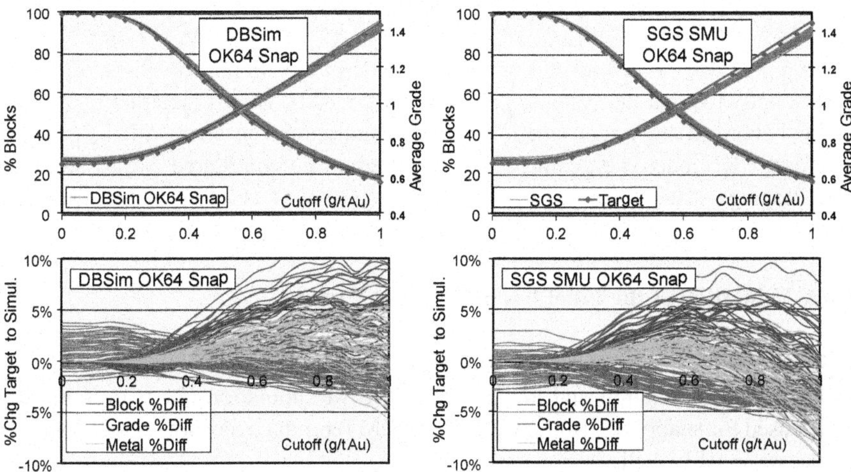

Fig. 3 Validating the SMU grade simulations after adjustment to a kriged model. The target is the SMU grade distribution obtained by Discrete Gaussian change of support

from the target was less than two relative percents at the lower cutoffs corresponding to the mining practice. The differences from the target for the DBSim simulation are slightly larger but still well within acceptable limits (Fig. 3).

The choice was made to proceed with the snapped simulations to the OK64 estimated model. The point simulations were then re-blocked to the 20 m by 20 m by

Table 3 Percentage difference of snapped SMU simulation statistics to data input (SMU grade dispersion variance)

	SMU SGS			SMU DBSim		
	Unsnapped	OK24 snapped	OK64 snapped	Unsnapped	OK24 snapped	OK64 snapped
Mean	−0.70 %	−0.30 %	−0.40 %	1.50 %	0.90 %	0.70 %
Variance	−23.20 %	−4.80 %	−15.00 %	−4.60 %	6.30 %	−4.20 %
CV	−11.70 %	−2.20 %	−7.40 %	−3.80 %	2.20 %	−2.80 %

12 m SMU (averaging 48 points). There is the question whether to snap the simulations before or after re-blocking. This will require further investigation.

The immediate concern is that the un-snapped point SGS when reblocked lost too much variance (Table 3) in comparison with the theoretical target (CV was 11.7 % lower than the target). This is unexpected because the point variance (Table 2) and variogram are well reproduced. Different origin offsets were considered for rescaling the SGS simulations. The variances/CV's were always lower than expected. However, the snapped simulations re-blocked with a more acceptable variation (CV was 7.4 % lower than the target). The sills of the re-blocked, snapped simulation variograms were low but the shape validated acceptably. The grade tonnage curves validated acceptably as did the corresponding swath plot and 3D visualization.

Snapping does alter statistics and variography of the simulated values, often not significantly [5], but the benefit is obvious for the SMU SGS simulation in Table 3 giving better statistical reproduction. Here, a simulation consistent with the primary data and OK model was required. Snapping could be considered as just stronger conditioning.

5.4.4 Validation of the Final Grade Tonnage Curves

Figure 3 shows the simulated SMU grade tonnage curves for the snapped, reblocked SGS simulation and the snapped DBSIM simulation (64 point OK estimates used for snapping). Both SGS and DBSIM reproduce the target grade tonnage curves reasonably well.

6 Conclusions

A summary of the two methodology practices discussed in this paper is presented in Table 4. There are advantages and disadvantages to both.

The DBSim methodology application has worked well for this domain at Boddington, with very similar results to that obtained from re-blocking the SGS point

Table 4 Comparison of methodology practice for DBSim and SGS point simulations

DBSim	Theory is more complex. Simple co-kriging behind the scene, Discrete Gaussian model, (global) theoretical change of support. Use turning bands.
SGS Point	Theory simpler. Change of support (local) by simple re-blocking.
DBSim	Fast and less disk space. Using large numbers of turning bands (1 000) slowed the calculation.
SGS Point	Not as fast but can be optimized. The software utilized here only took 8.5 minutes for 50 realizations of 1 million nodes out of 4.5 million on a four year old laptop.
DBSim	Internally consistent (SMU level).
SGS Point	Internally consistent (SMU level).
DBSim	No re-blocking issue since the change of support is one parameter of the simulation.
SGS Point	Issues when re-blocking. Block variance was not as predicted by the theoretical calculation, leading to block CV and variograms (sills) not as predicted.
DBSim	Inflexibility with block sizes. Cannot re-block to varied, non-multiple block sizes. Need a percent block model to account for blocks straddling domainal/geological boundaries.
SGS Point	More flexible allowing re-blocking to various block sizes and across domainal/geological boundaries.
DBSim	More limited validation with primary data due to change of support. No validation directly of simulated Gaussian values to swath plots of Gaussian composites or comparison to composite Gaussian variograms or 3D visualization. On the back transformed grades there is no option for comparison to composite grade variograms.
SGS Point	Full validation possible.

simulation. In some cases, the DBSim approach was able to produce better validation (e.g. block variances and block variograms) than the SGS approach. This may well be due to the global change of support being inherently part of the simulation process for DBSim but not for SGS.

The SGS methodology worked very well at the point level. However, after re-blocking the un-snapped simulations resulted in lower variances than that predicted by theoretical global support volume-variance relationship. This was unexpected since the point variance and variogram were well reproduced by the simulation.

Multiple validations plus adjustments through the process of simulation are considered essential and perhaps not as widely implemented nor openly discussed as they should be. Validation via statistics, visual, swath plots and grade-tonnage curves is important at various stages through the simulation process on both the Gaussian and back transformed values. Looking specifically at the better informed areas (the EDA envelope) is material. Topcutting can be critical. This work has shown the ready departure of Gaussian simulation statistics from $N(0,1)$ and the option for variance adjustment with a proportional variogram. The use of SK (global or local) or OK for conditioning is something that could be followed further publicly, considering the effect on simulated variance which in turn can bias the overall grade mean, and matching of the local mean or grade trend. Snapping the simulation to a local mean value to better represent trends or local grade variation offers promise, although various options need to be investigated further as to how it can be done or whether it can be better incorporated into the simulation itself (perhaps utilizing trends or non-stationary models) rather than as an adjustment. Should it be done at a point or block level? It is uncertain whether there would be a difference.

Acknowledgements Many thanks to Newmont and AMEC for the time and permission for this paper and the work involved in it. Also, thanks to Harry Parker for his time and advice.

References

1. Chiles JP, Delfiner P (1999) Geostatistics: modelling spatial uncertainty. Wiley series in probability and mathematical statistics, 695 pp
2. Emery X (2009) Change-of-support models and computer programs for direct block simulation. Comput Geosci 35:2047–2056
3. Godoy M (2002) The effective management of geological risk in long-term production scheduling of open pit mine. Unpublished PhD thesis, WH Bryan Mining Geology Research Centre, University of Queensland
4. Humphreys M (2010) Newmont Boddington Orebody, two methods applied—simulation and uniform conditioning. Presented at 2010 SME annual meeting, Phoenix, Arizona, Feb 28–March 3
5. Humphreys M (2011) Learning from simulation at Boddington Gold Mine. In: Proceedings 35th APCOM symposium
6. Lajaunie C (1993) Lestimation géostatistique non linéaire. In: Cours C-152, Centre de Géostatistique, École des Mines de Paris
7. Marinho M, Machuca M (2009) Capping and outlier restriction: state-of-art. In: Proceedings APCOM 2009, pp 337–345
8. Nowak M, Verly G (2005) The practice of sequential Gaussian simulation. In: Leuangthong O, Deutsch C (eds) Proceedings geostatistics Banff 2004. Springer, Berlin, pp 387–398

Non-multi-Gaussian Multivariate Simulations with Guaranteed Reproduction of Inter-Variable Correlations

Alastair Cornah and John Vann

Abstract Stochastic modeling of interdependent continuous spatial attributes is now routinely carried out in the minerals industry through multi-Gaussian conditional simulation algorithms. However, transformed conditioning data frequently violate multi-Gaussian assumptions in practice, resulting in poor reproduction of correlation between variables in the resultant simulations. Furthermore, the maximum entropy property that is imposed on the multi-Gaussian simulations is not universally appropriate. A new Direct Sequential Cosimulation algorithm is proposed here. In the proposed approach, pair-wise simulated point values are drawn directly from the discrete multivariate conditional distribution under an assumption of intrinsic correlation with local Ordinary Kriging weights used to inform the draw probability. This generates multivariate simulations with two potential advantages over multi-Gaussian methods: (1) inter-variable correlations are assured because the pair-wise inter-variable dependencies within the untransformed conditioning data are embedded directly into each realization; and (2) the resultant stochastic models are not constrained by the maximum entropy properties of multi-Gaussian geostatistical simulation tools.

A. Cornah (✉) · J. Vann
Quantitative Group, PO Box 1304, Fremantle, WA 6959, Australia
e-mail: ac@qgroup.net.au

J. Vann
Centre for Exploration Targeting, The University of Western Australia, Crawley, WA 6009, Australia

J. Vann
School of Civil Environmental and Mining Engineering, The University of Adelaide, Adelaide, SA 5000, Australia

J. Vann
Cooperative Research Centre for Optimising Resource Extraction (CRC ORE), The University of Queensland, St. Lucia, Qld 4067, Australia

P. Abrahamsen et al. (eds.), *Geostatistics Oslo 2012*,
Quantitative Geology and Geostatistics 17,
DOI 10.1007/978-94-007-4153-9_30,

1 Introduction

The advantages of stochastic modeling of in-situ mineral grade attribute variability and geometric geological variability through conditional simulation are increasingly recognized within the mining industry. Typical applications include studies for recoverable resource estimation, mining selectivity, drill hole spacing, multivariate product specification, quantification of project risk through the value chain and many others. Recent applications for conditional simulations involve feeding multiple realizations of key attributes (grade or geometallurgical) through a virtual mining and processing sequence in order to analyze a range of different project scenarios under real variability conditions. The development of such approaches means that the demand for realistic multivariate simulations is increasing.

Multi-Gaussian conditional simulation algorithms, mainly the Turning Bands Method (TBM) [14] and Sequential Gaussian Simulation (SGS) [7] are now widely used to simulate continuous attributes in the minerals industry. Both these algorithms require that conditioning data honor the properties of the multi-Gaussian model [11] and consequently a prior, single point, univariate transformation of the data to their normal scores equivalents is therefore generally required. This ensures Gaussianity of the transformed point attribute data, but higher order properties of the multi-Gaussian distribution must also be met [8, 11, 20]. This is usually assumed to be the case, but in practice that assumption is often violated.

For most mining projects, multiple continuous interdependent attributes are of essential interest. Examples include bulk commodity mining (iron, manganese, coal, bauxite, phosphate, etc.); multi-element metalliferous deposits (base metals, or base-precious metals), etc. In many mining operations deleterious components and geometallurgical attributes are as important to financial viability and ultimate operational performance as the revenue variable(s). Because of this, reproduction of inter-variable dependencies in realizations is as important as the reproduction of marginal histograms for each individual variable. The traditional multivariate extension of the multi-Gaussian simulation approach uses the Linear Model of Coregionalization (LMC) to enforce inter-variable correlations between simulated attributes [9]. The n inter-related random variables $z^{1\ldots n}$ are transformed into their normal scores equivalents $y^{1\ldots n}$ through monotonic, univariate, single point normal scores transforms. Multivariate pair-wise dependencies in the untransformed sample data are also transferred into their normal scores equivalents, potentially violating the multi-Gaussian assumption. Because the inverse of the normal scores transform ϕ^{-1} is also monotonic, applying the back transforms (pertaining to non-multi-Gaussian sample normal score equivalents) to vector of n simulated multi-Gaussian distributions $\phi^{-1}(y^{s1\ldots sn})$ will result in multivariate pairwise dependencies in the simulated data that are not found in the untransformed sample data; often these pairings are *mineralogically impossible*. In addition, the limitations of modeling direct and cross variograms through the LMC is well documented [12, 18, 23].

Figure 1 presents the bivariate distributions between four grade variables for sample data and a back transformed multi-Gaussian TBM simulation. The black polygons represent areas of bi- and multivariate pairings in the simulated values that are

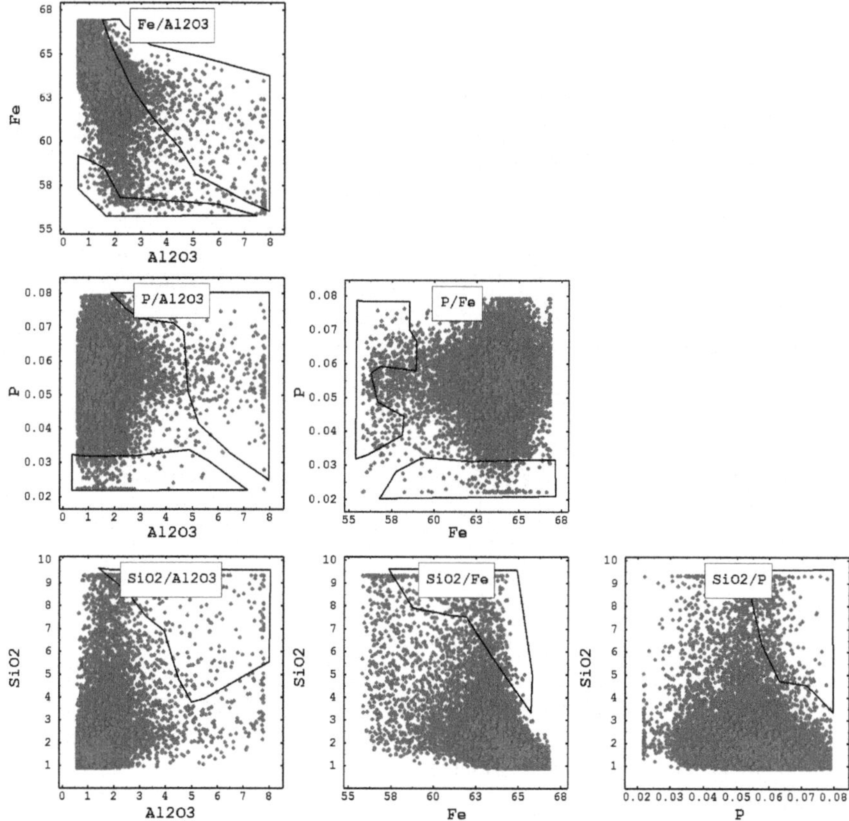

Fig. 1 Comparison of pairwise dependencies in the sample dataset (*blue*) for Al_2O_3, Fe, P and SiO_2 against simulated values (*red*) generated using the multi-Gaussian LMC approach. The *black polygons* denote invalid simulated bivariate pairings

impossible due to mineralogical constraint. While the realization generally honors the marginal distributions of each of the variables in the conditioning data and the general shape of the bivariate relationships, there are clearly a significant number of mineralogically impossible pairings present in the simulation.

Existing solutions to this problem within the multi-Gaussian framework aim to first decorrelate the conditioning data so that it is free from multivariate non-multi-Gaussian behaviors. Simulation by TBM or SGS can then be carried out independently and the original inter-variable correlations are reconstructed post-simulation. The Stepwise Conditional Transform (SCT) was first used in geostatistics in [17] and involves the transformation of multiple variables to be univariate and multivariate Gaussian, with no cross correlations, under an assumption of intrinsic correlation. However, the method requires relatively large datasets to be effective, can be sensitive to ordering and loses efficacy for subsequent variables. Min-Max Autocorrelation Factors (MAF) [3, 6, 22] is a principal components based decomposition which is extended to incorporate the spatial nature of data. However, non-linear dependencies between attributes are problematic [2] and will also result in unlikely or impossible attribute value pairings.

Nevertheless, even overlooking this issue, a well known feature of the multi-Gaussian approach is the imposition of maximum entropy upon stochastic realizations [16]. This may be pragmatically compatible with the real deposit grade architecture for some mineralization styles, depending on the scale considered, but in many cases it is at least questionable. For example in Banded Iron Formation hosted iron ore deposits, such as those in the Pilbara region of Australia and elsewhere, from which the example dataset is drawn, the lower extreme of the Fe grade distribution and upper extremes of the SiO_2 and Al_2O_3 distributions generally comprise shale bands. These shale bands are thin relative to standard mining selectivity and are cannot be practically separated from the rest of the material during mining, but are far from spatially disordered. In fact individual shale bands are *continuously traceable over hundreds of kilometers* and are used by exploration and mining geologists as marker horizons to aid geological interpretations. Consequently, imposition of maximum disorder in the simulated extreme low Fe and extreme high SiO_2 and Al_2O_3 values in these deposits will not only result in poor spatial representations, it is also utterly inconsistent with important features of the real grade architecture and known geological continuities.

Sequential Indicator Simulation [1, 10, 11] provides a potentially lower entropy alternative to the multi-Gaussian approach for the simulation of a single continuous attribute but no practical multivariate extension is available. The requirement for a conditional simulation algorithm which avoids the maximum entropy property and also ensures proper reproduction of inter-variable dependencies is therefore clear. It is also clear that these two properties (maximum entropy and poor inter-variable dependency reproduction) are coupled. In summary, whilst the multi-Gaussian assumption provides congenial properties for the simulation of an individual variable, it imposes maximum entropy on the resulting realizations and causes difficulty in replicating non-multi-Gaussian inter-variable dependencies.

2 Direct Sequential Simulation

Direct Sequential Simulation (DSS) [4, 13, 15, 21] has been proposed to avoid the requirement for a multi-Gaussian assumption and also therefore offers the possibility of lower entropy stochastic images. DSS is based upon a concept that was introduced in [15] who submitted that sequential simulation will honor the covariance model whatever choice of local conditional distribution (cdf) that simulated values are drawn from, provided that distribution is informed by the Simple Kriging (SK) mean $z(x_u)^*$ and SK variance $\sigma^2_{sk}(x_u)$, where

$$z(x_u)^* = m + \sum_{\alpha} \lambda_\alpha(x_u)\big[z(x_\alpha) - m\big],$$

and x_α are the conditioning data (including sample data and previously simulated nodes). One critical drawback was that, unless the cdf is fully defined by its mean and variance (which is true only in the case of a few parametric distributions such as the Gaussian), the realization does not reproduce the input (target) histogram. Recognizing this, further development of DSS was carried out by [21] who, instead of using $z(x_u)^*$ and $\sigma^2_{sk}(x_u)$ to define a local cdf from which to draw z^s (given a parametric assumption), $z^s(x_u)$, drew directly from the (untransformed) global cdf $F_z(z)$. The draw uses a Gaussian transform ϕ of the original $z(x)$ values. The SK estimate $z(x_u)^*$ is converted into its Gaussian equivalent $y(x_u)^*$ and with the standardized estimation variance this defines the sampling interval in the Gaussian global conditional distribution:

$$G\big(y(x_u)^*, \sigma^2_{sk}(x_u)\big).$$

The drawn Gaussian value y^s is then back transformed to a simulated value $z^s(x_u)$ using the inverse of the transform ϕ^{-1}:

$$z^s(x_u) = \phi^{-1}\big(y^s\big).$$

This development allowed DSS realizations to honor the target histogram as well as the variogram. The method was extended to incorporate a secondary variable, using Collocated Cokriging [24] to define the sampling of the global conditional distribution of the secondary variable [21]. An improvement was subsequently proposed in [13] whereby the second variable is drawn conditionally on the ranked Gaussian equivalents for the first variable. This apparently results in improved reproduction of dependency between variables; however, the method could be progressively more problematic as the number of variables involved increases. Furthermore, we suggest that this method still does not capitalize on a critical potential benefit of the direct simulation approach, which is to allow pair-wise dependencies in the drillhole data to flow through the simulation.

3 Direct Sequential Cosimulation (DSC)

By avoiding the constraints of trying to meet multi-Gaussian assumptions, the direct approach opens the possibility of circumventing the need to destroy and then subsequently attempt to reconstruct inter-variable dependencies. The proposed Direct Sequential Co-simulation (DSC) algorithm, which is built upon DSS, extracts pair-wise dependencies from the experimental data and embeds them directly into the realization, thus, by construction, guaranteeing the reproduction of inter-variable dependencies.

The DSC algorithm follows the traditional methodology for sequential simulation: visiting a random sequence of nodes, estimating the local (multivariate) cumulative distribution function, drawing (multivariate) values and repeating until all nodes have been visited. As with other sequential simulation algorithms, the kriging weights that are used to determine the local (multivariate) cdf must incorporate sample data locations and previously simulated points. The drawn values are sequentially incorporated into the simulation until all nodes have been simulated. In keeping with the existing DSS algorithms outlined in [13] and summarized above, simulated data values are drawn directly from the (untransformed) global conditional distribution. However, two key differences exist: (1) the simulated data values are drawn without an intermediate Gaussian step and (2) pair-wise multivariate simulated values for all attributes of interest are drawn directly and simultaneously from the experimental multivariate distribution. Correlations between variables inherent within the experimental data thus become embedded within the realizations because of the pair-wise draw and inter-variable dependencies in the resultant simulations are *assured.*

The proposed DSC approach firstly requires that all variables in the sample dataset are collocated. Secondly, an assumption of intrinsic correlation between the attributes of interest is required. Under this assumption, direct and cross covariance functions of all variables are proportional to the same basic spatial correlation function. In practice this is a reasonable assumption for key variables in many iron and base metal deposits. These two requirements permit the following: at each location to be simulated, a single set of OK weights (based upon the intrinsic spatial covariance function and the geometry of data locations as well as previously simulated nodes) can be used to define a probability mass weighting for each multivariate pair-wise data location in the surrounding neighborhood (simulated and sample data) to be drawn into the simulation. This can then be utilized for the draw of pair-wise simulated values from the experimental multivariate distribution. OK weights are effectively used to represent the conditional probability for a pairwise value to be drawn from the experimental multivariate distribution. This is consistent with the interpretation of OK as an E-type (conditional expectation) estimate [19] and consistent with the same author's usage of OK weights to model local conditional distributions.

Because OK (with unknown mean) is used instead of SK the algorithm simulates the discrete multivariate distribution and cannot generate a value different from the original data. In other words, in their point form, the output point simulations are not continuous. The DSC algorithm progresses as follows.

1. Define a random path through all nodes to be simulated.
2. For each node:
   a. Determine local OK weights $\lambda_\alpha^{OK}(u)$ for the surrounding experimental data locations $z(x^i)$ and previously simulated locations $z^s(x^i)$.
   b. Sort the OK weights for each experimental and simulated data location $\lambda_\alpha^{OK}(u)$ by magnitude and calculate the cumulative frequency weighting value $C(0, 1)$ for each $\lambda_\alpha^{OK}(u)$.
   c. Draw a p value from a uniform distribution $U(0, 1)$ and match to the cumulative frequency $C(0, 1)$; assign $z^{s1 \ldots n}(x_u)$ and add the multivariate pair-wise values at the drawn experimental data or simulated data location to the conditioning dataset.
3. Proceed to the next node along the random path and repeat steps 2a–2d until every node has been simulated. As with other sequential simulation algorithms

another realization is generated by repeating the entire procedure with a different random path.

Note that OK is required in DSC because applying a probability weighting to the mean as required by SK is impossible in the pair-wise draw sense; consequently negative weights $\lambda_\alpha^{\mathrm{OK}}(u) < 0$ are a potential outcome. Because p is drawn from a bounded uniform distribution any pair-wise data location within the neighborhood which is assigned a negative weight is excluded from being drawn at the simulated location and the remaining weighting is rescaled to equal 1. The implication of this is that in the presence of negative weights, unbiasedness in the expectation of the simulation is not explicitly guaranteed:

$$E\{z^s(x_u)\} \neq E\{z(x_u)^*\}.$$

Other approaches, such as adding a constant positive value to all OK weights if there is one or more negative weight, followed by restandardization to 1 [19] would also fail to guarantee unbiasedness. Our testing has shown the impact of this to be minimal provided that estimation neighborhood parameters are chosen to minimize negative weights. Assuming this is the case; at each grid location simulated values for each variable of interest are centered upon the local OK estimate $z_{\mathrm{OK}}^*(x_u)$:

$$z_{\mathrm{OK}}^*(x_u) = \sum_{\alpha=1}^{n(u)} \lambda_\alpha^{\mathrm{OK}}(u) z(u_\alpha) \quad \text{with} \sum_{\alpha=1}^{n(u)} \lambda_\alpha^{\mathrm{OK}}(u) = 1.$$

The range of simulated values is provided by the range of experimental and simulated data values within the search neighborhood and the variance is the weighted variance of the data values within the neighborhood:

$$\sigma_{\mathrm{OK}}^2(x_i) = \sum_{\alpha=1}^{n} \lambda_{\mathrm{OK}} \times \left[z(x_i) - z(x_u)^*\right].$$

In the extreme case of no spatial continuity within the underlying geological process, and a neighborhood covering all of the sample data, simulated data pairs at non-data locations will be drawn at random from the multivariate global conditional distribution.

4 Example for an Iron Ore Deposit

A two dimensional iron ore dataset which is sourced from a single geological unit and comprises 290 exploration sample locations within a 1.2 km by 1.5 km mining area is used to demonstrate the algorithm. Four grade attributes (Fe, SiO_2, Al_2O_3 and P) are fully sampled across the area at an approximate spacing of 75 m by 75 m. For demonstration purposes, 25 point realizations of the four attributes (Fe, SiO_2, Al_2O_3 and P) at 19 026 nodes have been generated using the proposed DSC algorithm. Experimental direct variograms and cross variograms were generated

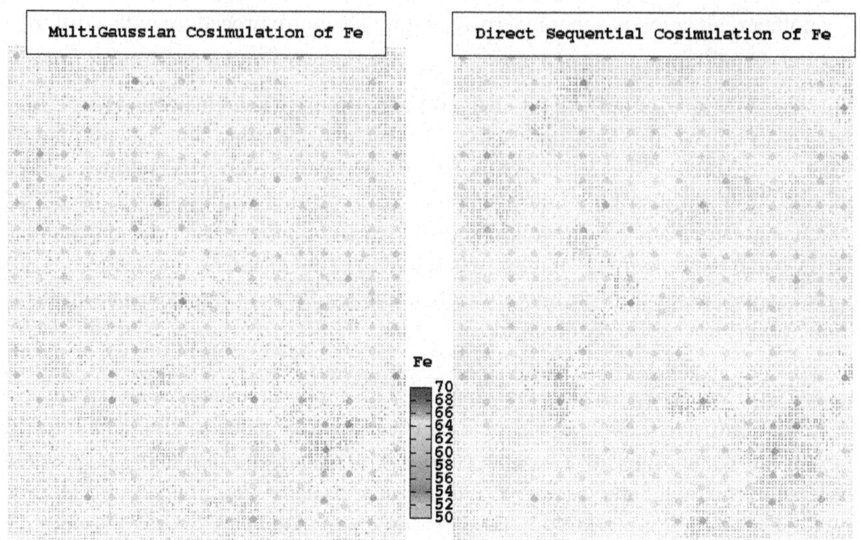

Fig. 2 A comparison between multi-Gaussian and DSC realizations of Fe

for the four variables; the direct experimental variogram for Fe was modeled using a 20 % nugget contribution and single spherical structure with 80 % contribution and 200 m range. This model represents a reasonable fit to the direct SiO_2, Al_2O_3 and P experimental variograms when the total sill is appropriately rescaled, thereby satisfying the intrinsic correlation requirement for the proposed DSC algorithm.

A comparison between a DSC realization and a multi-Gaussian realization is shown in Fig. 2. The grade architecture generated by DSC is more like the mosaic model [10, 20] and is thus appealing in cases like the one presented here, where maximum entropy is clearly unacceptable. A DSC realization of the four attributes is shown in Fig. 3 with the bivariate distributions between the four simulated attributes also shown. The two key advantages of the DSC algorithm are apparent from the figure, specifically (1) the inter-variable relationships in the simulated attributes match those observed in the sample dataset; and (2) the architecture of the grades in the realization is driven by the sample data itself and does not have the appearance of maximum entropy. In addition, by construction the realizations honor the declustered sample histograms of the variables of interest. Experimental direct and cross variograms for the 25 DSC realizations of the four grade attributes are compared against their sample data equivalents in Fig. 4. A close comparison is noted for both the direct and cross experimental variograms. Ergodicity of the simulations around the sample data experimental variograms is approximately consistent with that which might be expected from multi-Gaussian algorithms.

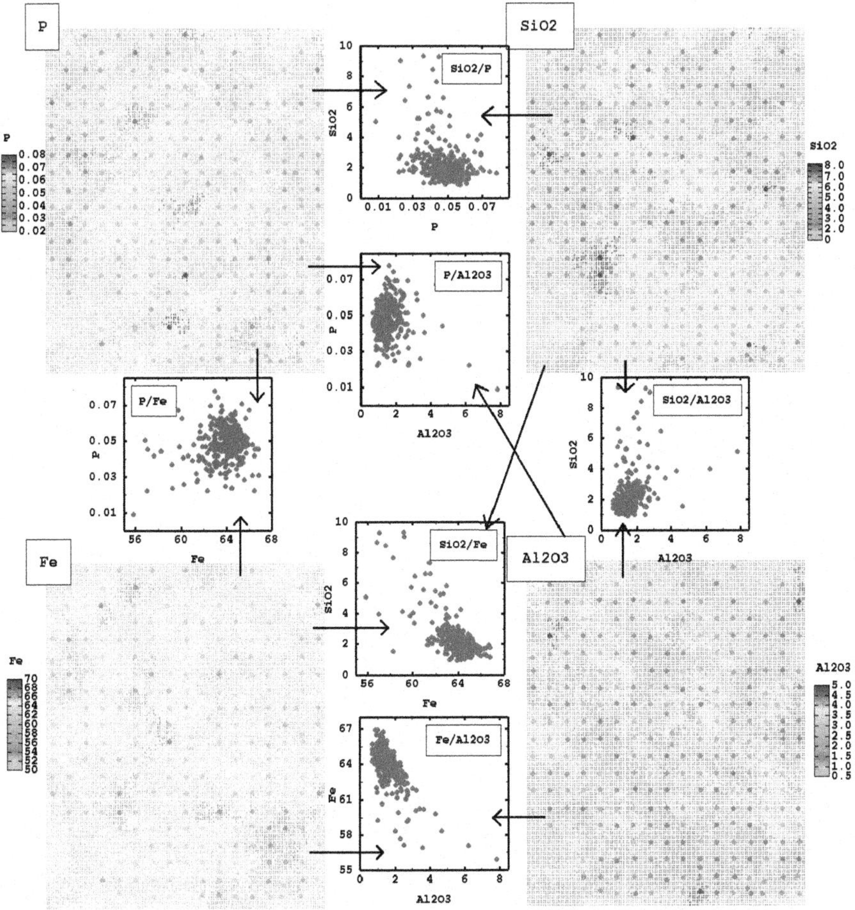

Fig. 3 A DSC realization of Al_2O_3, Fe, P and SiO_2 with inter-simulated attribute dependencies

5 Limitations

Because the proposed DSC method simulates the discrete multivariate distribution, simulated values cannot deviate from those in the experimental dataset. This is both a strength of the algorithm, because it allows the pair-wise dependencies in the sample data to be embedded directly into the realizations; but it is also a weakness because potential for simulation of adjacent nodes with the exact same value exists. This is particularly likely to be the case if the modeled spatial continuity is high. Consequently re-blocking to change support is a critical pre-requisite to any per realization use of the simulations with the assumption being that the averaging process will reproduce a realistic continuous distribution at the support re-blocked to. Prior kernel smoothing of the input data would result in a reference distribution from which to draw continuous multivariate simulated values, however this could

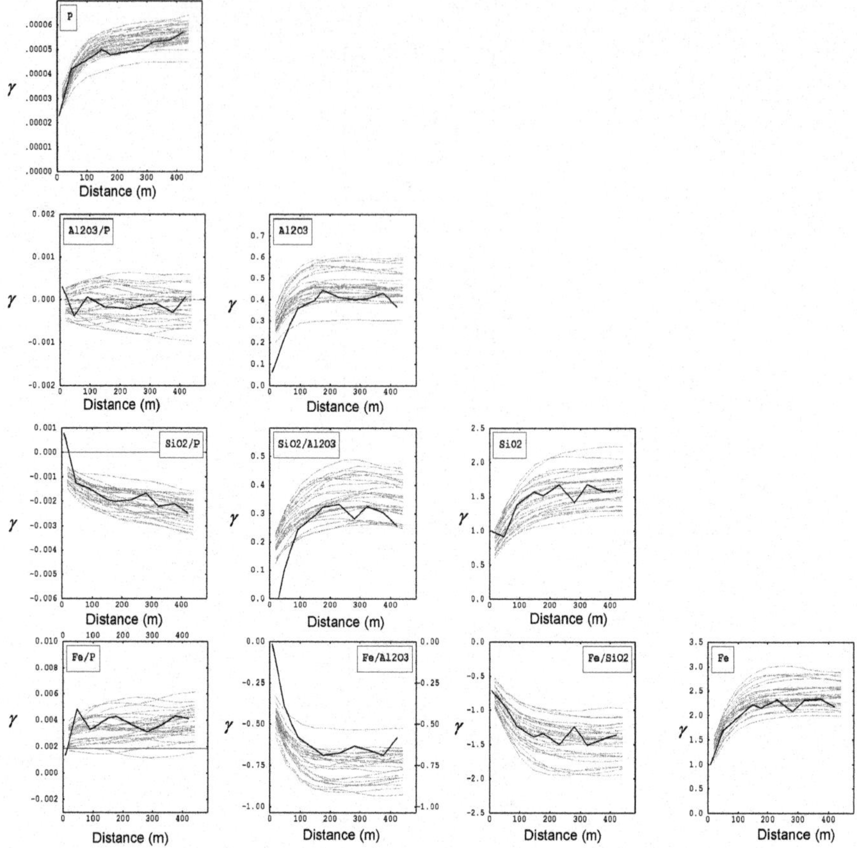

Fig. 4 Comparison of experimental direct and cross variograms for 25 DSC realizations of Al_2O_3, Fe, P and SiO_2 and the sample dataset

introduce invalid data pairings which are avoided in the algorithm's proposed implementation. A further disadvantage of the DSC approach over the multi-Gaussian approach is that unlike the latter [5], no direct block equivalent is available. This is not a drawback from small simulations, but is a potential restriction for large scale mining problems.

6 Conclusions

Applications of conditional simulation within the minerals industry rely directly upon the integrity of the simulated architecture of those in situ attributes. Adequately representing the connectivity of extremes of attribute distributions is critical because very often these are the key drivers of value or causes of interaction with mining, processing, or product specification constraints; adequate representation of inter-attribute dependencies is equally important. The authors believe that these requirements demand the ongoing development of conditional simulation algorithms for continuous multivariate attributes outside the of the multi-Gaussian framework. The DSC algorithm presented is intended to contribute to this effort by providing a low entropy alternative to the multi-Gaussian approach for the conditional simulation of multiple continuous attributes that also guarantees that inter-variable dependencies are honored.

The intrinsic correlation, full sampling and collocation of sampling requirements of DSC are no more onerous than those of SCT and in general the practical implementation of DSC is far more straightforward than the multivariate multi-Gaussian approaches. No LMC modeling is required and the additional transformations and associated validation steps that are required in MAF and the SCT are also avoided.

Acknowledgements Professor Julian Ortiz of the University of Chile and our Quantitative Group colleagues, in particular Mike Stewart, are thanked for feedback and discussions about the proposed method prior to the writing of this paper. Any remaining deficiencies are entirely the responsibility of the authors.

References

1. Alabert F (1987) Stochastic imaging of spatial distributions using hard and soft information. Unpublished master's thesis, Department of Applied Earth Sciences, Stanford University, Stanford, California
2. Boucher A, Dimitrakopoulos R (2004) A new joint simulation framework and application in a multivariate deposit. In: Dimitrakopoulos R, Ramazan S (eds) Orebody modelling and strategic mine planning, Perth, WA, 2004
3. Boucher A, Dimitrakopoulos R (2009) Block simulation of multiple correlated variables. Math Geosci 41(2):215–237
4. Caers J (2000) Direct sequential indicator simulation. In: Proceedings of 6th international geostatistics congress, Cape Town, South Africa
5. Deraisme J, Rivoirard J, Carrasco Castelli P (2008) Multivariate uniform conditioning and block simulations with discrete Gaussian model: application to Chuquicamata deposit. In: Ortiz J, Emery X (eds) Proceedings of the eight international geostatistics congress, pp 69–78
6. Desbarats A, Dimitrakopoulos R (2000) Geostatistical simulation of regionalized pore-size distributions using min/max autocorrelation factors. Math Geol 32(8):919–942
7. Deutsch C, Journel A (1992) GSLIB—geostatistical software library and user's guide. Oxford University Press, New York
8. Deutsch C, Journel A (1998) GSLIB: geostatistical software library and user's guide, 2nd edn. Oxford University Press, New York
9. Dowd P (1971) The application of geostatistics to No. 20 level, New Broken Hill Consolidated Ltd. Operations Research Department, Zinc Corporation, Conzinc Riotinto of Australia (CRA), Broken Hill, NSW, Australia
10. Emery X (2004) Properties and limitations of sequential indicator simulation. Stoch Environ Res Risk Assess 18:414–424
11. Goovaerts P (1997) Geostatistics for natural resources evaluation. Oxford University Press, New York
12. Goulard M, Voltz M (1992) Linear coregionalization model: tools for estimation and choice of cross-variogram matrix. Math Geol 24(3):269–285

13. Horta A, Soares A (2010) Direct sequential co-simulation with joint probability distribution. Math Geosci 42(3):269–292
14. Journel A (1974) Geostatistics for conditional simulation of ore bodies. Econ Geol 69(5):673–687
15. Journel A (1994) Modeling uncertainty: some conceptual thoughts. In: Dimitrakopoulos R (ed) Geostatistics for the next century. Kluwer Academic, Dordrecht, The Netherlands, pp 30–43
16. Journel A, Alabert F (1989) Non-Gaussian data expansion in the earth sciences. Terra Nova 1:123–134
17. Leuangthong O, Deutsch C (2003) Stepwise conditional transformation for simulation of multiple variables. Math Geol 35(2):155–173
18. Oliver D (2003) Gaussian cosimulation: modeling of the cross-covariance. Math Geol 35(6):681–698
19. Rao S, Journel A (1997) Deriving conditional distributions from ordinary kriging. In: Baafi E, Schofield N (eds) Geostatistics—Wollongong 96. Kluwer Academic, London, pp 92–102
20. Rivoirard J (1994) Introduction to disjunctive kriging and non-linear geostatistics. Clarendon Press, Oxford
21. Soares A (2001) Direct sequential simulation and co-simulation. Math Geol 33(8):911–926
22. Switzer P, Green A (1984) Min/max autocorellation factors for multivariate spatial imagery. Stanford University, Department of Statistics
23. Wackernagel H, Petigas P, Touffait Y (1989) Overview of methods for coregionalisation analysis. In: Armstrong M (ed) Geostatistics, vol. 1, pp 409–420
24. Xu W, Tran T, Srivastava RM, Journel A (1992) Integrating seismic data in reservoir modeling: the collocated cokriging alternative. SPE 24742

Field Parametric Geostatistics—A Rigorous Theory to Solve Problems of Highly Skewed Distributions

Rochana S. Machado, Miguel Armony, and João Felipe Coimbra Leite Costa

Abstract Linear kriging methods is not suited to estimate local grades and local reserves for highly skewed variables. This paper presents a new framework, the Field Parametric Geostatistics (FPG) that transforms noisy variograms into well-behaved variograms and proposes a mathematical model which justifies empirical procedures commonly used, such as trimming or capping arbitrarily very high values. A consistent theory is built with solid premises and rigorous mathematical procedures. The model is based upon two underlying properties associated to grade continuity and data representativeness. The method deals with both distribution function and spatial arrangement, consequently all the available information is used simultaneously. When dealing with quasi-point variables such as grade, there is an underlying assumption that all the samples have the same influence or representativeness, thus yielding non-parametric variograms, as the grade distribution is not taken into account. In FPG these point variables are transformed and replaced by macroscopic robust variables. The method, when applied to non-skewed variables, yields similar results to classic kriging. In this paper a summary of FPG estimation theory is presented and its evaluation techniques are shown through examples, which results are compared to results obtained by standard techniques.

1 Introduction

Gold grades estimation requires an understanding of the underlying phenomenon and should not be based on blind use geostatistical techniques. Mineral resources analysts working with gold deposits must always adapt the use of the mathemat-

R.S. Machado (✉) · J.F.C.L. Costa
Federal University of Rio Grande do Sul (UFRGS), Porto Alegre, Brazil
e-mail: rochana.machado@gmail.com

J.F.C.L. Costa
e-mail: jfelipe@ufrgs.br

M. Armony
MAIM Ltda, Rio de Jeneiro, Brazil
e-mail: miguelarmony@gmail.com

P. Abrahamsen et al. (eds.), *Geostatistics Oslo 2012*,
Quantitative Geology and Geostatistics 17,
DOI 10.1007/978-94-007-4153-9_31,

ical methods to provide a satisfactory result. The presence of outlier grades, very common in gold mineralization, introduces significant variability on statistical parameters and experimental semivariogram calculation. Linear kriging methods, such as ordinary or simple kriging, is not suited in these situations [4]. The main problem during gold grade estimation is the overestimation of high grades; the needed to avoid it has motivated professionals over decades to develop a variety of ways to control this effect. Cutting or capping the high values to limit their influence during grade estimation is a widespread technique in the mining industry. The specific concern is that very high values should not be assigned too much weight, or they contribute to an apparent tonnage of high grade ore that does not exist [10]. Generally, the choice of an upper limit is based on previous experiences or statistical parameters related to the cumulative distribution function, such as the 95th or 98th percentile of data. These empirical practices are very common in gold grade estimation. Indicator kriging (IK) [8] was developed initially to avoid both overestimations by using fully the very high grades and underestimation by cutting or abandoning these values. It allows attributing to blocks the probability to have grade values above a chosen cutoff. Indicator transform is an effective way of limiting the effect of very high values [6], but an overestimation is expected just because at the highest percentile the grades remain very high and their values are used during the estimation. The fundamental question we ask is why the rigorous mathematical theory in kriging fails to estimate local grades and local reserves for highly skewed distributions. The target is to provide a mathematical model which captures the physical basis of gold deposits. This methodology explain and justify the practice of resource analysts of lowering high grades, and opens a new research frame: the Field Parametric Geostatistics.

2 Problems Regarding Grade Estimation Using Kriging

Standard kriging methods are not able to deal with highly skewed distribution due to two main reasons. First note that when the mineral resource analyst lowers a high grade, for instance a 100 ppm sample, indeed it is assumed that this grade cannot be extended to one tonne. The cutting grade procedure can be mathematically justified assuming that the mineral resource analyst is reducing the spatial influence instead of cutting the grade. The expression 100 ppm contains an a priori wrong inference. Indeed, if the sample weights 10 kg its real grade would be 1 g in 10 kg. This initial mistake will cause error propagation in ore reserve estimation if the mineral resource analyst does not cap the grade. The question is for which amount of space (volume) the ratio 100 g/t remains true. In a given deposit there exist information about this property. This information is not captured by a standard kriging approach.

Next by observing the way data collection two desired model properties can be inferred. Every sample represents a spatial portion larger than itself (Representativeness). The portion represented by a sample fills some volume around it (Continuity). In kriging the continuity is given by the semivariogram. Note in particular that the range of the semivariogram is independent of the sample value. In the polygons method the representativeness is given by the area of influence. For the polyhedra there is a volume or a tonnage of influence, and the same for estimation using vertical sections and declustered cells. They are all piecewise continuous. FPG call any area or volume of influence by a generic designation—the extension. This mineral resource estimation method works with two variables. This pair, i.e., grade and extension are herein called the field of sample, as for an electron in Physics. Kriging methods treats all samples as if they have the same influence (range), hence it does not have a sufficient flexibility.

In addition, for kriging approaches samples are considered to be point observations with infinite precision. However, the samples have a size, therefore they are not point observations and the observations have an accuracy limited by the measuring equipment. If the grade ranges from 0.00 g/t to 100.00 g/t there are only 10001 possible values. It is therefore also possible to use a discrete spectrum. This is essential to the mathematical formulation of FPG theory.

3 The Theory of Field Parametric Geostatistics

Following the above analysis, a theory was developed, based on a variable transformation known as Field Parametric Geostatistics (FPG), [1–3, 9]. The principal aim of FPG is to build a variable that fulfills all the conditions pointed out. The mathematical procedure to transform grades into new variables is as follows:

- Assign to each sample a declustering weight, on the presence of preferential sampling. Declustering weights can be obtained using *declus* from GSLIB [5].
- Sum up all the declustering weights of the samples with the same grade—the result is the extension of the grade.
- Assign to each grade a value representing the representativeness of this grade, according to its occurrence in a dataset. For instance, if there are five samples of 2 ppm and one sample of 50 ppm, the 2 ppm samples are considered five times more representative of the deposit than sample of 50 ppm.
- Build the cumulative extension of the grades.
- Standardize the cumulative extension of the grades to range from zero to one.

When dealing with tonnages as extensions and the density is not constant, the declustering weight must be multiplied by density. In the following sections we discuss the new variable, show a simple example and discuss some of the features of the methodology.

3.1 Standardized Cumulative Extension

Standardized cumulative extension of the grades is the new chosen variable (called τ). It fulfills the needs required for the variable to be modeled: uses all

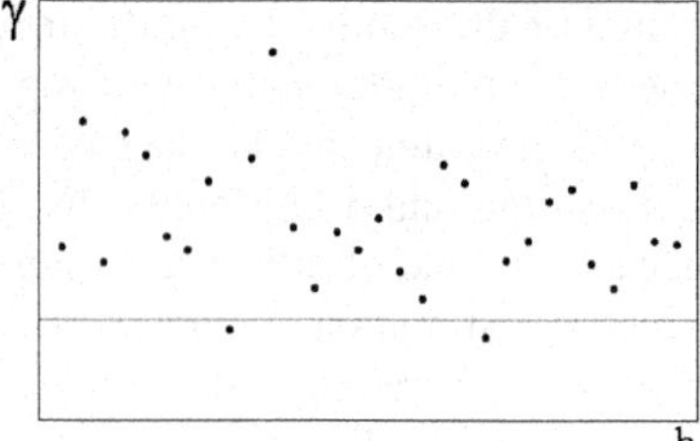

Fig. 1 Variogram shape of the 140 samples of GSLIB synthetic data

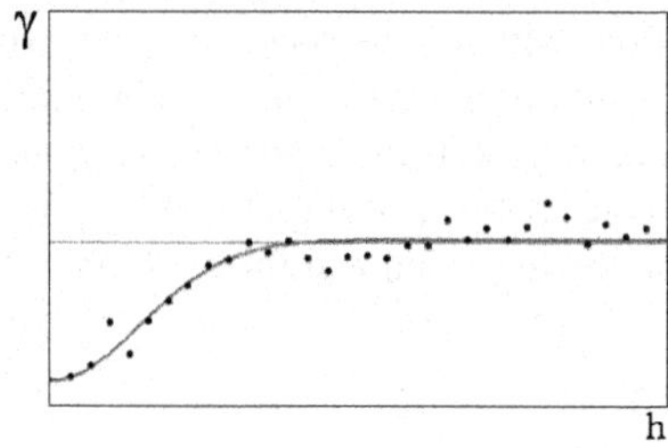

Fig. 2 Variogram shape of the FPG transform of the 140 GSLIB synthetic data

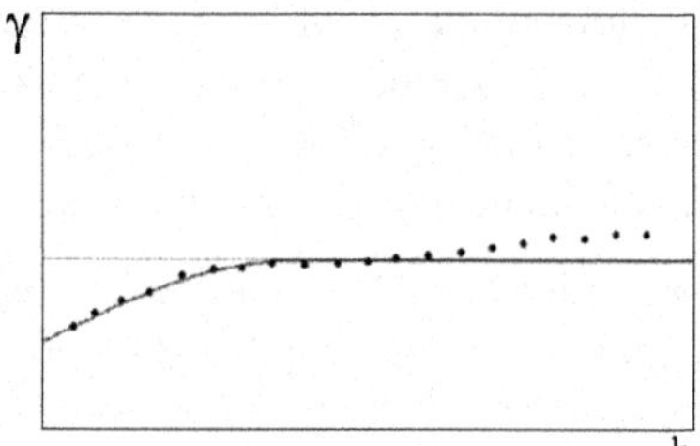

Fig. 3 Variogram shape of the thickness of a kaolin layer in the Amazon state, Brazil

the available information (grade and extension values), comprehend both desired properties and is a discrete variable. Standardized cumulative extension is called the extension function **g**. It is a totally discrete function because it has a finite number of points. The extension function is not a simple uniform transform nor a probability function nor a distribution function. Instead, it represents a real physical entity.

The standardized cumulative extension of the grades (τ), has the following properties:

1. It transforms bad or noisy variograms into well-behaved variograms as it can be seen in Figs. 1 and 2 where variograms of the synthetic 140 points presented in GSLIB [5] are shown. When the distribution is symmetric or almost symmetric, the variograms have the same shape, as shown in Figs. 3 and 4. Moreover, in the last case kriging results are compatible.
2. It reduces automatically the influence of the high grades without manual intervention.

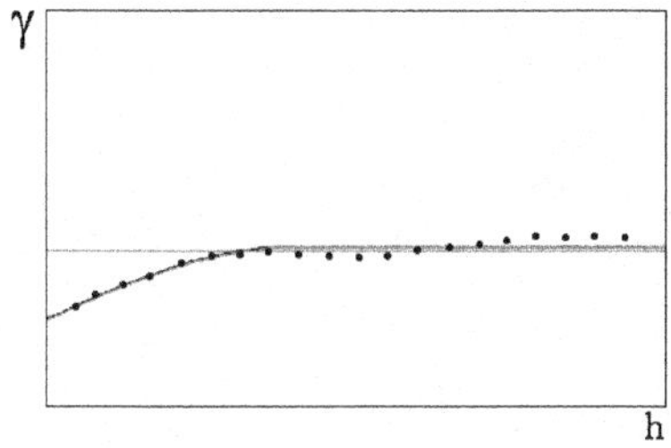

Fig. 4 Variogram shape of the FPG transform of thickness of the same kaolin layer

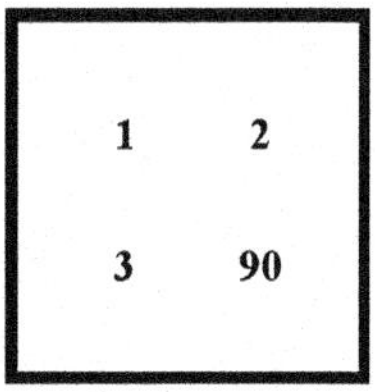

Fig. 5 A squared mining unit with four blast holes

3.2 A Simple Example of Using the Discrete Spectrum

In a small squared mining unit four blast holes are drilled in an almost symmetric way as shown in Fig. 5. The assayed samples have as results the grades 1.00 g/t, 2.00 g/t, 3.00 g/t and 90.00 g/t of gold. The precision of the grades is two decimals. Figure 6 shows an extension function conditioned to the four values. Note that although the lines segments seem to be continuous they are sequences of points associated with the given precision. From 1.00 g/t to 2.00 g/t there are 100 points. From 2.00 g/t to 3.00 g/t there are 100 points. From 3.00 g/t to 90.00 g/t there are 8700 points. As the holes are inside a small block it can be assumed that all the points of the block are associated to a unique random variable and the extension function can be treated as a probability function. The probability to obtain a new sample with grade between 1.00 g/t and 2.00 g/t, between 2.00 g/t and 3.00 g/t, and between 3.00 g/t and 90.00 g/t has the same value 0.25. But the probability to get a new sample with exactly 90.00 g/t is 87 times lower than to get a sample with grade 3.00 g/t, for example. In the continuous spectrum this conclusion would be very difficult to reach.

3.3 The Inflexion Point

When the resource analyst lowers the grades above a given cutoff value, there is an underlying assumption that these grades belong to a different population which is distinct from the grades below the cutoff. The grade distribution is a result of several physical and chemical phenomena occurring along the time; there must be some reason, detectable or not, why these samples behave in a different way.

A simple example where two different populations can be detected is a mineralization on fractured zones; there is a population of disseminated gold over the area,

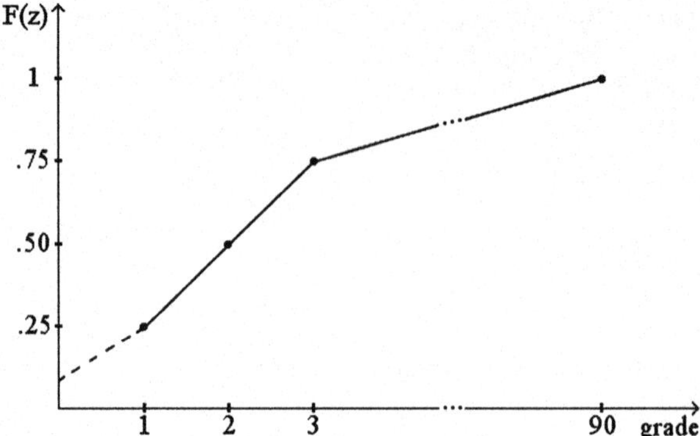

Fig. 6 CCDF of the grades inside the mining unit

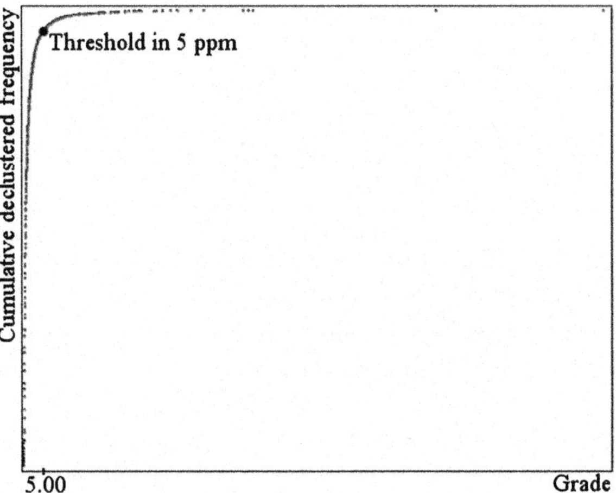

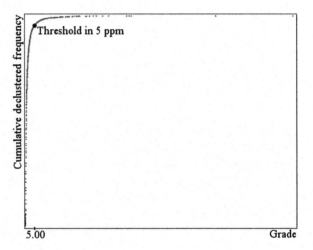

Fig. 7 Shape of the cumulative frequency of gold grades in the Amapari deposit (two populations)

and a population of high grade from the aggregation of the element on fractures. It seems like there are two different mineralization processes, but it is a very rare case where the reasons of the existence of two populations can be detected. One way to separate the populations is to draw the cumulative frequency of the grade and search for an inflexion point. In Calculus an inflexion point is the point where the first and second derivatives are null. Here this term is used as in the common language: the inflexion point is the point where the angle of the curve with the vertical axis grows significantly. Figure 7 shows the cumulative frequency of gold grades from Amapari deposit. It can be easily seen that the inflexion point is some value around 5 g/t.

Sometimes there is not a clear inflexion point and FPG treats the whole set of grades as a single population, although the analyst cuts the grades above a chosen value. The analysis of the cumulative frequency is not always the best way to separate populations; there are other factors which can be used for this purpose. The inflexion point is treated as limiting threshold, where the existence of this inflection determines that only the grades above this point should be estimated by FPG technique; in earlier study [1] was verified that its implementation for whole dataset would lead to underestimation. The knowledge of the nature of the data and, in the mineral frame, the knowledge of the genesis of the deposit can be very important and sometimes crucial to apply efficiently the techniques of FPG.

3.4 Point or Derived Variables and Global or Primitive Variables

A variable referring to a macroscopic portion associated to a point or derived variable is called a global or a primitive variable. Examples: area, volume or tonnage associated to a grade, or time associated to instantaneous velocity. They are called primitive because they are intuitive and need no definitions: a volume or an area can be seeing, a mass can be touched, and the weight can be felt. A variable obtained as a ratio of two primitive variables, measured in a portion that can be considered as a point is called a point variable or a derived variable. Examples: instantaneous velocity in m/sec, a sample grade in %, g/t among others.

3.5 Absolute Natural Classes and Conditional Natural Classes

In FPG theory $N(d) = (t_{\max} - t_{\min})/d$ is called the number of *absolute natural classes* associated to the precision d, where $t_{\max}$ and $t_{\min}$ are respectively the maximum and minimum grade values taking into account the possible extreme values. Consequently, the number of natural classes is the same as the possible values. A *conditional natural class* is the interval between two sequential measured values. The number of conditional natural classes is the same as the number of different measured values. It is conditioned to the set of values of the samples.

4 Summary of Mathematical Formulation

4.1 General Formulation

Let $\{\theta_j\}$, $j = 1, \ldots, m$ be an order statistic of a variable with quasi punctual support taking values into the natural classes of precision $d, m \leq N(d)$, and let $\theta_0 < \theta_1$ be the minimum chosen value to take care of the tail effect. Let $\theta_{\min} = \theta_0$ and $\theta_{\max} = \theta_m$. Then

$$\theta_{j-1} < \theta_j, \quad j = 1, \ldots, m. \tag{1}$$

Making the correspondence of $\{\theta_j\}$ to the m classes:

$$(\theta_{j-1}, \theta_j], \quad j = 1, \ldots, m, \tag{2}$$

then each class will contain one or more values. Then define a mapping that makes the correspondence of each class $(\theta_{j-1}, \theta_j]$ to one and only one extension $T_j \in \mathcal{R}$. Next, define

$$T(\theta_i) = \sum_{j=1}^{i} T_j. \tag{3}$$

$T(\theta_i)$ is the summation of all field extensions whose punctual variable value is less or equal to θ_i. Obviously we have $T(\theta_{i-1}) < T(\theta_i)$. Denote the upper end point T:

$$T = T(\theta_m), \tag{4}$$

and define a set of weights $\{\omega_j\}$ such that

$$T_j = \omega_j T, \quad 0 \leq \omega_j \leq 1. \tag{5}$$

A standardized cumulative global variable is now defined as

$$\tau_i = T(\theta_i)/T. \tag{6}$$

Let $g(\theta)$ be a discrete non-decreasing function, defined over the whole interval $[\theta_0, \theta_m]$, so that

$$g(\theta_i) = \tau_i. \tag{7}$$

The values τ can be understood as the values of $g(\theta)$ in the points chosen to represent the interval [1]. Then $g(\theta_0) = 0$, $g(\theta_m) = \tau_m = 1$, and we may define $\tau_0 = 0$. The function g will be called the extension function. The cumulative global variable τ will be used for modeling and estimation as the application $\{\theta_i\} \to \{\tau_i\}$ is isomorphic.

4.2 Application to Natural Conditional Classes

Let $\{t_i(\boldsymbol{u})\}$ be an order statistic with n samples taking values on the set of rational numbers, where $\mathbf{u}$ means the sample location. Then, $t_i(\mathbf{u}_\alpha) \leq t_{i+1}(\mathbf{u}_\beta)$ $\forall i = 1, \ldots, n-1$. To the set $\{t_i(\mathbf{u})\}$ corresponds the set $\{\theta_j\}$, $j = 1, \ldots, m$, $m \leq n+1$. The set $\{\theta_j\}$ is an order statistic of m different values of the set $\{t_i(\mathbf{u})\}$ plus $\theta_{\max} = \theta_m$, the maximum value chosen for tail effect correction. Then $\theta_{j-1} < \theta_j$ $\forall j = 1, \ldots, m$, where $\theta_0 = \theta_{\min}$. Such a set containing all the measured values with a given precision d, and only measured values will be called the set of natural conditional classes associated to a precision d. The natural conditional classes are given by (2): $(\theta_{j-1}, \theta_j]$, $j = 1, \ldots, m$. So, all the expressions from (3) to (7) apply to this case.

The weight ω_j defined in (5) is composed of three factors. The first factor concerning the number of samples with value θ_j in other words, the frequency of θ_j. The second factor referring to spatial arrangement that expresses the extension to be associated to each sample with a measured value θ_j. For example, the factor related to the declustering process or the area of influence. The third is an additional factor for representativeness. For example, the specific gravity, when tonnage is taken as the global variable, or sample size or type. The weight ω_j can be written as

$$\omega_j = \frac{1}{n} \sum_{k=1}^{\mu_j} \alpha_k \beta_k,$$

where μ_j is the number of samples with value θ_j, α_k is a spatial weight, β_k is an internal factor for representativeness and n is the total number of samples.

5 Using FPG Techniques

The FPG techniques can be applied in various settings. Here we provide the guideline for the workflow in two situations. Procedure guidelines for mineral resources estimation assuming a single population:

- Data analysis.
- Declustering, if necessary.
- FPG transform: create a piecewise linear extension function.
- Variography of the transformed variable.
- Kriging the transformed variable.
- Back-transform the results to grades by linear interpolation in the extension function.

Procedure guidelines for mineral resources estimation assuming two populations:

- Data analysis.
- Assign carefully the inflexion point (see section "The inflexion point").
- Lower grades: use Classical Geostatistics.
- Higher grades: use FPG transform.
- For both low and high grades: variography and kriging.
- Back-transformation of the FPG results.
- Indicator kriging of the high grade population, assuming that resulting values represent the grade contribution of high grade samples on each block that will be estimated.
- The final grade of each block is obtained by weighting, using the proportions given by indicator kriging.

6 Case Studies

In order to demonstrate the FPG principles, three datasets with high-skewed distributions had been chosen. They are: the Amapari gold deposit in Brazil and the U and V Walker Lake datasets. For an FPG application, a previous analysis of frequencies

Table 1 Statistical parameters for U and V

	Count	Mean		Count	Mean
U	275	604.08	V	470	435.30
U (declustered)	275	555.00	V (declustered)	470	408.53

is carried out, in order to determine if there is any inflexion in the declustered cumulative frequency curve. For Walker Lake variables, there is no inflexion point and the FPG transformation is performed to all data and the estimated values are back transformed to grades. In the case study of the gold deposit, there are evidences of two populations, and an inflexion point could be assigned. This point is treated as limiting threshold, determining that only the grades above it should be estimated by the FPG technique. The first population, with low grades, is estimated by ordinary kriging of the original grades. The second population, with high grades, has their values processed by FPG, estimated by ordinary kriging and back transformed to the original grades. The final grade of each block is obtained by weighting, using the proportions given by indicator kriging. For the sake of comparison the grade estimation will also be made by conventional techniques, including ordinary kriging of the original values and after capping the extreme values. For Walker Lake the estimated values are compared against the reference values, derived from the exhaustive dataset.

6.1 Walker Lake Datasets U and V

The Walker Lake dataset [7] is derived from a digital elevation model. Its numerical variables are U and V, which will be used in this study, and both present only one population. The exhaustive set has 78000 measured values. The statistical parameters for U and V are presented in Table 1.

Different methods have been tested and their results were compared with a reference model, U (Reference) and V (Reference), using block size of 5 m $\times$ 5 m. Table 2 shows the statistical parameters for U and V estimates. U (OK) and V (OK) are the results of ordinary kriging of the original values; U (capping Q97) and V (capping Q90) are the results of ordinary kriging after capping the extreme values. The results of median indicator kriging were post processed with *postik* from GSLIB: U (E-type 1.5) and V (E-type 1.5) are the results with a hyperbolic upper tail extrapolation of parameter $\omega = 1.5$ and U (E-type 3.0) and V (E-type 3.0) are the results with a hyperbolic upper tail extrapolation of parameter $\omega = 3.0$. For both variables the global average of deposit was best approximated by the FPG technique, especially for the U variable. The correlation coefficients are used as a comparative parameter, as it can be seen in Table 3. FPG produces better results, and this improvement is evident when comparing the results with ordinary kriging of the original values.

Table 2 Statistical parameters for the U and V estimates

	Count	Mean		Count	Mean
U (Reference)	3120	266.04	V (Reference)	3120	277.98
U (OK)	3101	542.88	V (OK)	3120	300.44
U (capping Q97)	3120	543.20	V (capping Q90)	3120	304.92
U (E-type 1.5)	3120	464.56	V (E-type 1.5)	3120	302.10
U (E-type 3.0)	3120	428.82	V (E-type 3.0)	3120	296.64
U (FPG)	3042	255.22	V (FPG)	3120	261.84

Table 3 Correlation coefficients for U and V

	U (FPG)	U (OK)	U (capping Q97)	U (E-type 1.5)	U (E-type 3.0)
Reference	0.510	0.383	0.375	0.501	0.502

	V (FPG)	V (OK)	V (capping Q90)	V (E-type 1.5)	V (E-type 3.0)
Reference	0.867	0.832	0.808	0.855	0.859

6.2 Amapari Gold Deposit

The area containing the deposit is located in the state of Amapá, Brazil. The colluvia domain of the deposit was chosen to develop this study. A division of colluvia domain into northern and southern portions was performed, whose statistics are presented in Table 4. An analysis of the distribution and spatial arrangement shows that two different populations can be recognized: the first ranging from 0.00 ppm to about 5 ppm and the second, equal or greater than 5 ppm, to be treated by FPG using cumulative extensions.

Different methods have been tested and their results were compared. Table 5 shows the overall statistics of block estimates in northern and southern portions. AU (OK) is the result of ordinary kriging of the original values; AU (FPG) is the result obtained by FPG technique; AU (capping Q95) and AU (capping Q98) are the results of ordinary kriging after capping the extreme values above the 95th and the 98th percentile of the distribution (AU = 5.03 ppm and AU = 9.20 ppm, respectively).

The method yields similar local results to classic kriging in the northern part of the deposit. The results show a higher variation in the south, where the variance is greater. The correlation coefficient is used as a comparative parameter as it can be seen in Table 6. In this case FPG produces results similar to ordinary kriging with a capping in the extreme values.

7 Conclusions

The approach trough Field Parametric Geostatistics, provides a consistent framework for grade estimation. In the case study of Walker Lake data the improvement

Table 4 Statistics for AU

Variable	Region	Count	Mean	Region	Count	Mean
AU, ppm	North	1720	1.47	South	4127	1.50
AU (declus), ppm	North	1720	0.90	South	4127	0.97

Table 5 Statistics for AU estimates

Variable	Region	Count	Mean	Region	Count	Mean
AU (OK), ppm	North	213979	0.72	South	267061	0.81
AU (capping Q98), ppm	North	213979	0.72	South	267061	0.75
AU (capping Q95), ppm	North	213979	0.70	South	267061	0.69
AU (FPG), ppm	North	213979	0.59	South	267061	0.62

Table 6 Correlation coefficients for Amapari

	Region	AU (OK)	AU (capping 98)	AU (capping Q95)
AU (FPG)	North	0.877	0.864	0.846
AU (FPG)	South	0.716	0.861	0.865

is observed both globally and locally. In the case of Amapari, Field Parametric Geostatistics yields results comparable to ordinary kriging with capping. In the FPG approach the choice of top cut is however not required. The approach of Field Parametric Geostatistics is thus a robust alternative for grade estimation.

References

1. Armony M (2000) Geoestatística paramétrica de campo. PhD thesis (unpublished), Federal University of Rio de Janeiro, Rio de Janeiro
2. Armony M (2001) Field parametric geostatistics. Int J Surf Min Reclam 15(2):100–122
3. Armony M (2005) Logical basis for field parametric geostatistics. Int J Surf Min Reclam 19(2):144–157
4. Costa JFCL (2003) Reducing the impact of outliers in ore reserves estimation. Math Geol 35(3):323–345
5. Deutsch CV, Journel AG (1998) GSLIB: geostatistical software library and user's guide, Oxford University Press, New York, 369 pp
6. Glacken MI, Blackney PA (1998) A practitioners implementation of indicator kriging. In: Proceedings of a one day symposium: beyond ordinary kriging—non-linear geostatistical methods in practice. Geostatistical Association of Australasia, Perth, pp 26–39
7. Isaaks EH, Srivastava RM (1989) An introduction to applied geostatistics. Oxford University Press, New York, 547 pp
8. Journel AG (1982) The indicator approach to estimation of spatial distributions. In: Proceedings of 17th APCOM. Society of Mining Engineering, Golden, pp 793–806

9. Machado RS, Armony M, Costa JFCL, Koppe JC (2011) FPG—geostatistical rigorous solutions for highly-skewed distributions. In: Proceedings of GEOMIN 2011. Gecamin Ltd, Santiago, Chap. 4
10. Sinclair AJ, Blackwell GH (2004) Applied mineral inventory estimation. Cambridge University Press, Cambridge, 381 pp

Multiple-Point Geostatistics for Modeling Lithological Domains at a Brazilian Iron Ore Deposit Using the Single Normal Equations Simulation Algorithm

Hélder Abel Pasti, João Felipe Coimbra Leite Costa, and Alexandre Boucher

Abstract Orebody modeling is critical for the evaluation and engineering of mineral deposits. The building of the 3D geometry is conventionally based on vertical and horizontal sections interpreted by a mine geologist. In more advanced cases, geostatistical methods are used such as indicator kriging and/or simulations, truncated gaussian or plurigaussian simulations, which allows to automate the modeling process. These methods are probabilistic and use the variogram to represent the geological heterogeneity. Multiple-point geostatistics (MPG) is an alternative to traditional variogram-based geostatistical modeling, whereas a fully explicit representation of the geological patterns (a training image) is used in place of variograms. Although it is now routinely used in modeling of oil and gas reservoirs, there are few studies showing application of this technique in mineral deposits. The advantages of the MPG approach are to provide a more realistic representation of the geology through a more accessible parameterization (the visual training image instead of the analytic variogram). This paper presents initial results of MPG with the SNESIM algorithm applied to multiple lithological domains at a Brazilian iron ore deposit. Additionally, the steps involved in dataset preparation for adequate use of the algorithm are discussed.

1 Introduction

Modeling mineral deposits involves building 3D representations of lithotypes or geochemical classes. Such models are widely used in the mining industry to evaluate

H.A. Pasti (✉) · J.F.C.L. Costa
Federal University of Rio Grande do Sul, Av. Bento Gonçalves, 9500, Bloco IV Prédio 75, Sala 104, Agronomia, Porto Alegre, RS 91501-970, Brazil
e-mail: hapasti@yahoo.com.br

J.F.C.L. Costa
e-mail: jfelipe@ufrgs.br

A. Boucher
Advanced Resources & Risk Technology LLC, 902A Suntree Ct, Sunnyvale, CA 94086, USA
e-mail: aboucher@ar2tech.com

P. Abrahamsen et al. (eds.), *Geostatistics Oslo 2012*,
Quantitative Geology and Geostatistics 17,
DOI 10.1007/978-94-007-4153-9_32,

mineral resources and ore reserves, being the basis of later stages of planning and extraction along the mine life. The challenge of assessment of future recoverable reserves is critical to the mining industry [5] in that ore reserves condition both investment and profitability associated with any mining venture.

Kriging or sequential simulation methods are generally used to generate 3D geological models to quantify uncertainty associated with the estimation and modeling processes [1, 3]. These techniques are based on a variogram model [2] and, although they are sufficient to solve various grade and lithological geometries, they are poor to characterize deposits with sparse sample data, high variability or very complex formation processes.

Multiple-point geostatistics (MPG) is an alternative to traditional variogram-based geostatistical modeling, because it uses a fully explicit representation of the geological patterns, the training image (TI), in place of the variogram [8] resulting in a more realistic representation of the geology. An advantage of MPG is the ability of generating a wide variety of high resolution 3D structural models simply by modifying the training image.

This paper presents some initial results of multiple-point simulation with the SNESIM algorithm [9] applied to model the main lithological domains of a Brazilian iron ore deposit. The simulated models were compared against the geological model generated traditionally to contrast both techniques and evaluate

- the reproduction of the features of mineral deposit,
- the effect of data conditioning on geological features reproduction.

Additionally, this paper shows that it is possible to build a satisfactory geological model (definition of the mineralized envelop) during preliminary exploration stages without significant manual inputs, saving time and decreasing costs.

2 Brief MPG Introduction

The basis of MPG is to replace the building of a kriging system by looking up a conditional distribution extracted from a training image. The first implementation came from Guardiano and Srivastava [4] who proposed to scan the training image and record the frequency of a central node given a neighboring data patterns. The SNESIM algorithm [9] provided a computational breakthrough by storing all conditional distributions contained in the training image within a search tree data structure. This data structure allowed for fast retrieval of probabilities associated with data patterns. The results is an algorithm that combines the flexibility and ease of conditioning of pixel-based algorithms with some of the capacity that object-based algorithms (Boolean-type) have to reproduce complex shapes.

It is now recognized [6] that MPG can reproduce complex spatial statistics involving two or more points at a time, allowing capturing of information beyond the reach of a simple variogram model.

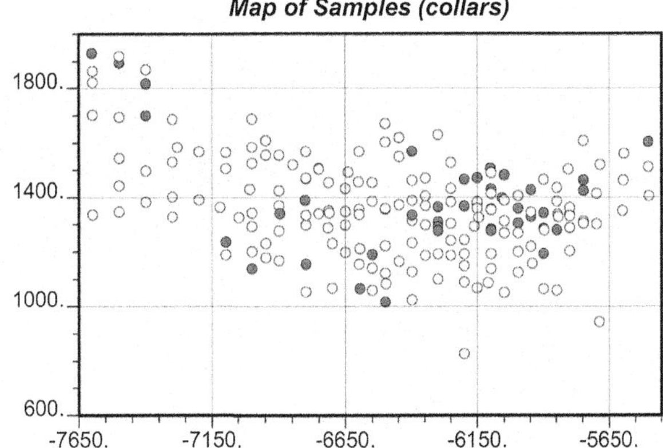

Fig. 1 The samples were assigned with numerical codes (indicators): Itabirites and Hematite samples have the code 0 (*white circles*), and the remaining samples (Waste) have the code 1 (*gray circles*)

3 Deposit and Data Preparation

The mineral deposit under study is an iron ore mine from Quadrilátero Ferrífero (Brazil), located on the eastern limb of the Moeda Syncline. Mineral deposits in that region are composed by BIFs (Banded Iron Formations) of enriched, oxidized, metamorphosed and heterogeneously deformed itabirites [7]. The quartzitic itabirite rock is predominant in the region. Iron is present as hematite, magnetite and martite.

The geometry is controlled by the presence of second-order folds of the Moeda Syncline trending NE-SW. Transverse faults from major continuity direction demarcate the different ore bodies that make up the mineral complex in the studied region.

For these studies, the structural elements were simplified in two main lithologies groups: Ore (Itabirites, Hematite) and Waste.

3.1 Dataset

There are a total of 2 176 samples from 195 boreholes, all with chemical analysis; 2 159 of them have a lithological descriptions. The sampling grid is approximately regular (100 m by 100 m in XY plane and composites of 10 m along Z axis) with a few clusters (Fig. 1).

In addition to boreholes, 40 vertical sections regularly spaced at 50 m are available. These sections provide information about the contacts between lithologies. They were generated combining expert interpretation and indicator kriging. Figure 2 shows a typical deposit sections, interpreted and built in the traditional way.

Fig. 2 Vertical section along the relative coordinate-5 800 along X axis. Itabirites are *dark gray*, Hematite are gray and Waste are *light gray*. In the subsequent studies, Hematites and Itabirites were grouped

4 Methodology

This study used the SGeMS [8] implementation of the SNESIM algorithm. The Hematite and Itabirites were grouped and only the large scale contact between ore and waste was simulated, thus reproducing the general shape of deposit.

The conditioning data consist of boreholes samples and 40 interpreted vertical sections. Another option would have been to integrate them as a probability field. Although it would have been preferable to use a very high resolution model (blocks of $2 \times 2 \times 2$), a coarser block model was built from each wireframe section, filling the volume with cells of $10 \times 10 \times 10\ m^3$. This support is close to the actual support adopted for mining in the studied area.

To properly focus on the simulation of the contact between ore and waste, the large simulation grid covering the entire deposit was separated into two minor grids, parallels along X axis. Each subgrid contains one side of the ore waste contact and will be referred as the left simulation grid (containing cells with small Y coordinate values) and the right simulation grid (cells with larger Y coordinates). These grids are typical Cartesian grids, each one with approximately 500 000 cells of $10 \times 10 \times 10\ m^3$.

The training image (TI) was built using a training image generator available in the latest version of SGeMS. The TI is a sinusoid object along X axis and a half circle along the YZ plane, to mimic the general shape of the orebody. The parameters used for the sinusoid function are: wavelength = 375, width = 80, thickness = 35, amplitude = 8 and length = 700. The TI is constructed at the same support of the simulation grid and represents the general geological features of the deposit to be reproduced. Note that the TI does not need to be geo-referenced to the deposit coordinates but must be defined on the same support volume.

The generated TI was merged with the topographic surface (Fig. 3) and finally divided in two parts: one representing the left ore-waste contact (which is used for simulating within the left simulation grid) and one for the right ore-waste contact. This decision made the simulation process easier and faster, since the algorithm have to scan a smaller volume to retrieve information and build the search tree.

The simulation of the ore and waste blocks was restricted to the parts of the deposit that were deemed uncertain; i.e. along the ore-waste contact zones. That zone

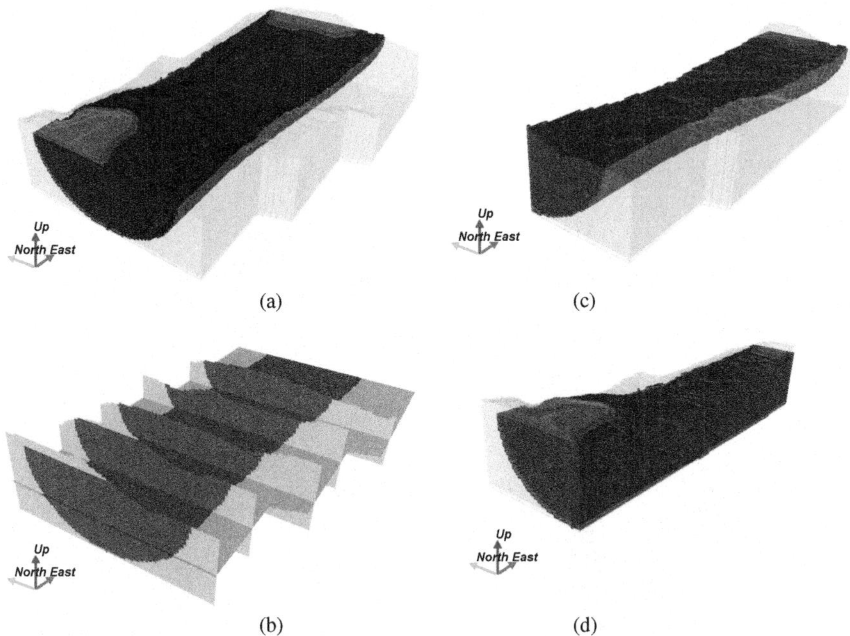

Fig. 3 Perspective view of (**a**) the 3D TI used in the simulation process, (**b**) cross section of TI, (**c**) *left side* and (**d**) *right side* ore-waste general contact in TI. *Dark gray cells* are the Orebody and *light gray cells* are the Waste

of uncertainty was determined by running a moving window between consecutive sections, if that moving window contained both ore and waste all the pixels within the window were set as uncertain and will be simulated. When the pixels on the sections were all waste (ore) than it was assumed that the pixel within the moving window were also waste (ore) and do not need to be simulated. In order to check the sensitivity of the method, three sets of vertical sections were used (40, 20 and 10 vertical sections, regularly spaced one each other from 50 m, 100 m and 200 m, respectively). Note that each set yields a different simulation domain. The search window of the algorithm was set as an ellipsoid with azimuth of 90, rake 0 and dip 0 degrees, 30 m of length along minor, 50 m along median axis and a variable major axis length, according to the spacing in vertical sections: 60 m, for sections regularly spaced at 50 m, 110 m when using sections spaced in 100 m and length of 210 m for the sections spaced at 200 m. The resulting volume characterizing the uncertainty zone of the ore-waste contact is shown in Fig. 4. These scenarios allow to investigate the effect of increasing the amount of information on the numerical models.

The values within the vertical sections that were located inside the uncertainty zones were not used as conditioning data (see Fig. 5), their only function is to delineate the potential geometry of the zone of uncertainty of the ore-waste contact. Only the boreholes were used for conditioning the simulations inside the

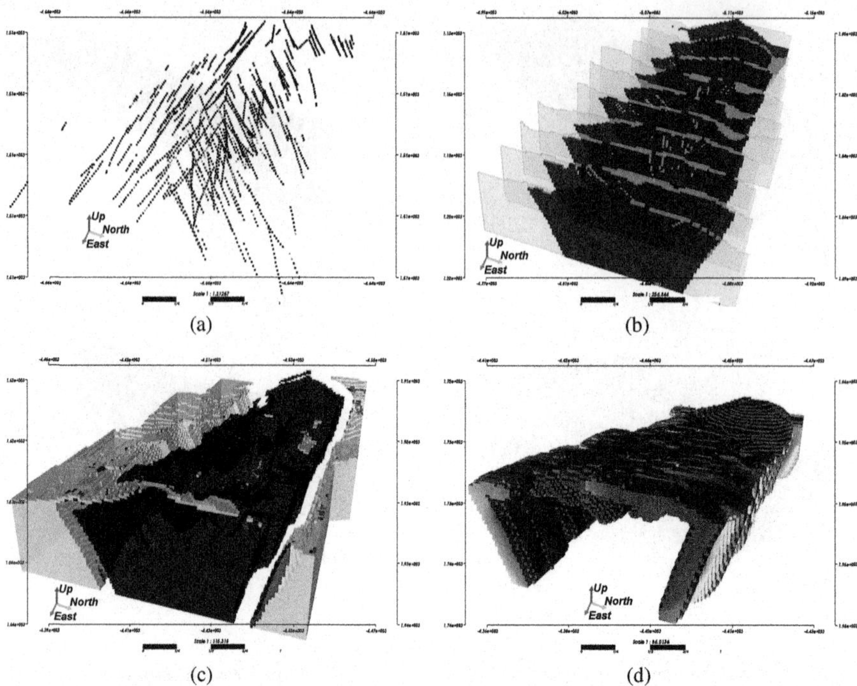

Fig. 4 (**a**) Borehole samples; (**b**) set of 10 sections regularly spaced at 200 m which were used to create an uncertainty zone; (**c**) volume outside the uncertainty zone, and (**d**) volume inside the uncertainty zone

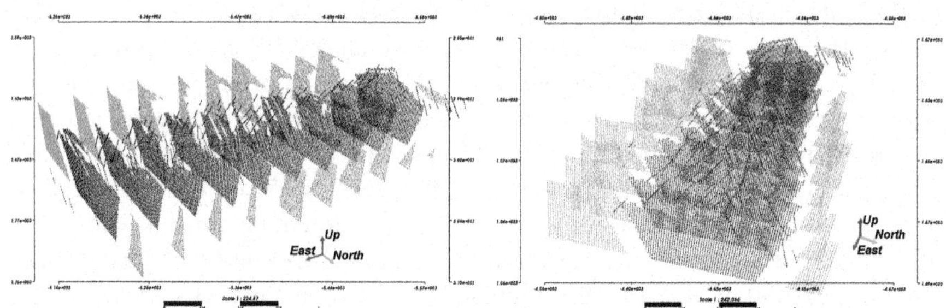

Fig. 5 Data used for conditioning the simulations. Note that the sections are only used as hard data outside the uncertainty zone

uncertainty zones. Being interpreted, the information within the vertical section were not deemed to be reliable enough to be used as hard data. The simulations of both uncertainty zone (left and right) are done independently then merged back into the full grid model.

The results derived from MPS were compared against a geological interpreted model, which was the result of expert interpretation of vertical and horizontal

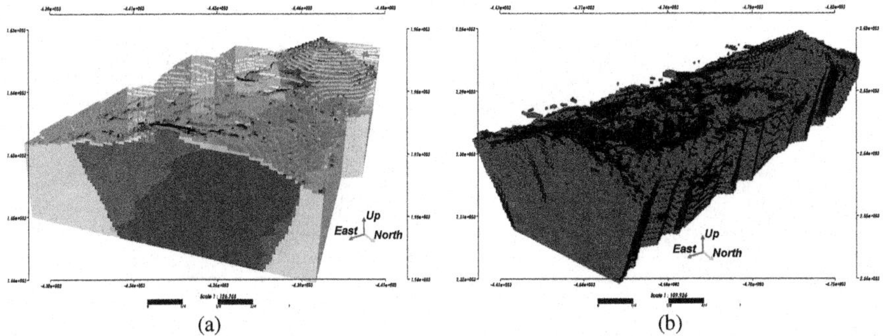

Fig. 6 Perspective view of interpreted geological model, showing (**a**) the general shape of deposit and (**b**) only the ore cells. *Dark gray* are ore cells and *gray* are waste cells

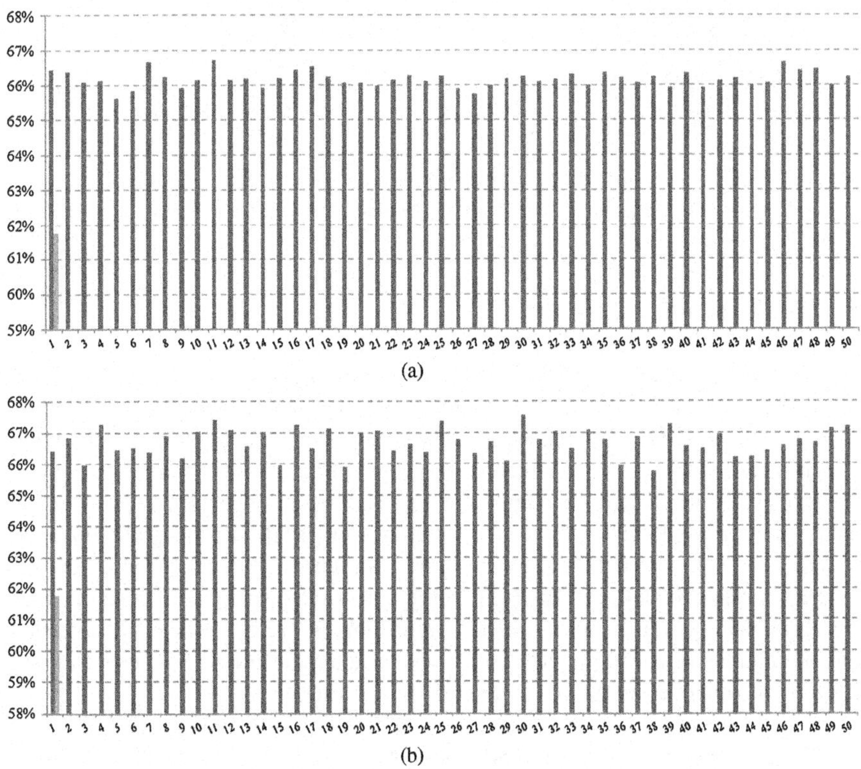

Fig. 7 Proportions for each of the 50 simulations conditioned by (**a**) 40 sections and (**b**) 10 sections. *Dark gray columns* are proportions of ore on each deposit geomodel. The *single light gray column* refers to interpreted model

sections combined with categorical indicator kriging. This interpreted model has regular blocks of $10 \times 10 \times 10\ \text{m}^3$ and its lithological information was simplified to Ore and Waste, as shows Fig. 6.

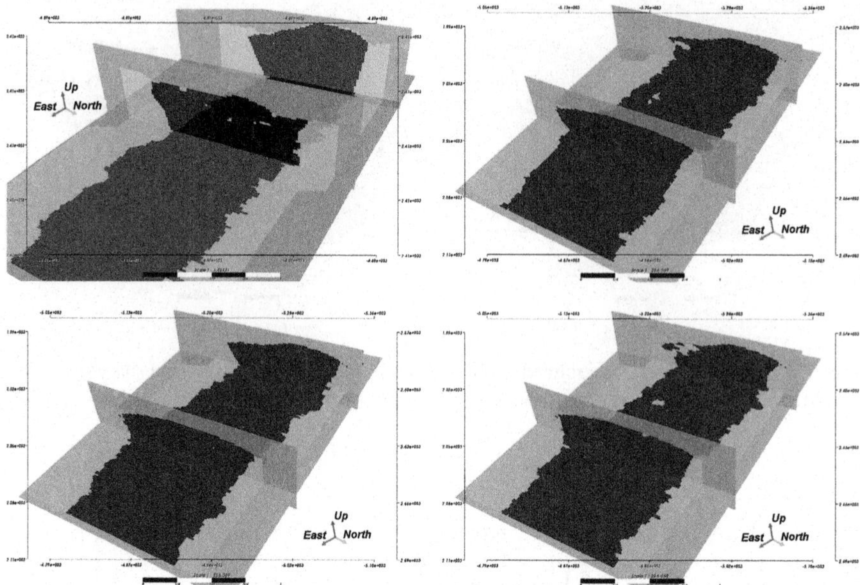

Fig. 8 Cross sections from ore cells of the interpreted geological model (*top left*) and three outcomes from the multiple-point simulations conditional to the boreholes and 20 sections regularly spaced at 100 m

5 Results and Discussion

Figure 7 shows the global proportions of the ore lithotype for two of the three evaluated scenarios. The ore-waste proportion for each simulation fluctuates around 66 %. All the simulated models show higher ore content than the interpreted model, which is around 62 %. The discrepancy between the simulations and the interpreted model can be partially explained by specific the training image used and by the hard data conditioning. Recall that the interpreted model was deterministically obtained, based on sections and data interpretation, and should not be seen as the true model or as a reference model. A possible interpretation of the discrepancy is that the interpreted model may have underestimated the volume of ore in the deposit. If needed, the simulations could be repeated with several alternate training images to better quantify the uncertainty on the volume of the deposit.

Figure 8, shows a few isometric views of the simulated models with the interpreted geological model. As expected there is a greater variability within the simulated models than with the interpreted model.

Figure 9 shows the probability for each pixel to be ore or waste, the values were obtained by averaging the set of 50 simulations (e-type map). It can be seen that the actual uncertainty zone in the ore waste classification is smaller than the one originally derived from the sections. This indicates that the volume of the section-derived uncertainty zone was sufficient to include the transition from one lithotype to the next. The pattern information from the training image combined with the

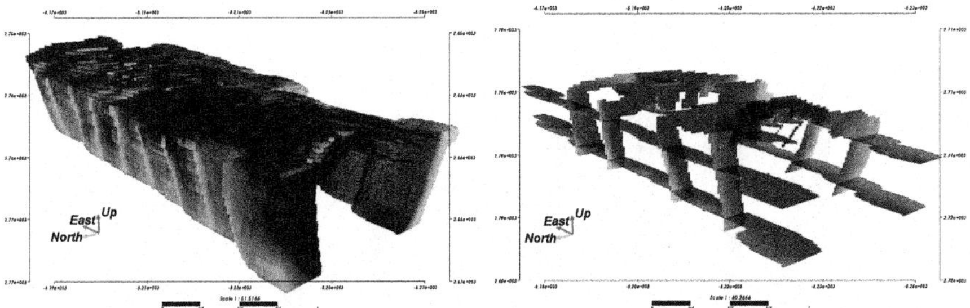

Fig. 9 Perspective view of the ore-waste probability map for the 200 m spacing scenario. *Black blocks* were assigned as ore (coding of 0) and *gray* as waste (coding of 1). *Blocks with other colors* (ranging from *black* to *gray*) are within the transition zone between ore and waste

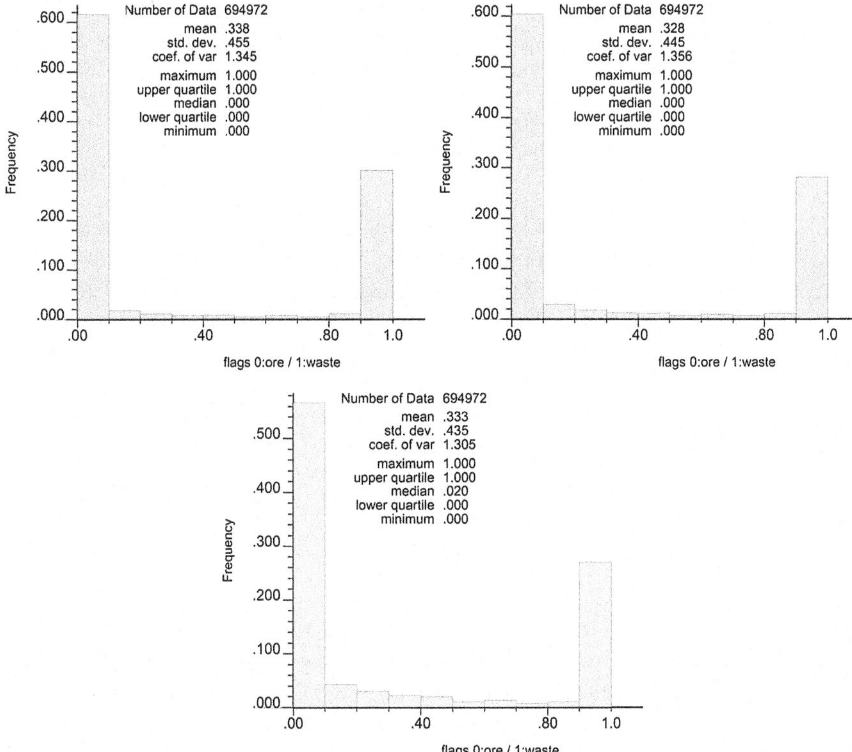

Fig. 10 Histogram of E-Type map, conditioned by 40, 20 and 10 sections (spaced respectively at 50 m, 100 m and 200 m), considering the global area. The X axis gives the probability of a block to be assigned as waste (code 1) or ore (code 0). Y axis gives the fraction of the total quantity of blocks which were flagged with that respective probability value

drillhole data helped reduced that zone of uncertainty. Note that when the number of vertical sections increases, there is a tighter control on the final geometry, resulting in many cells having a probability of 0 or 1 to be ore (Fig. 10).

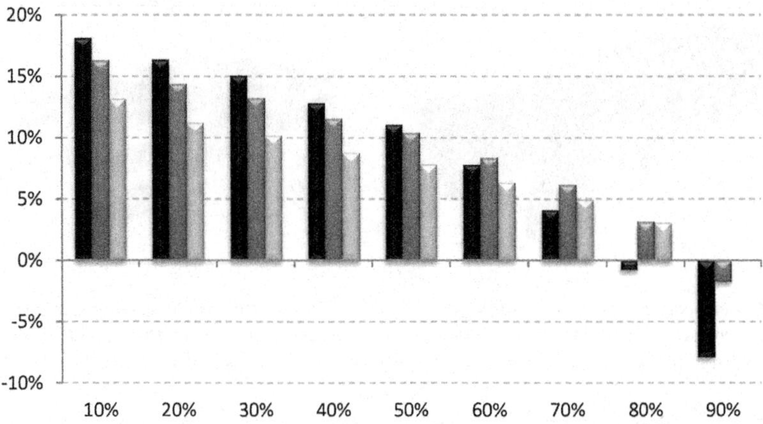

Fig. 11 Evaluation of the ore volume range. The *X* axis gives the probability of a block be assigned as ore. The *Y* axis gives the total ore volume deviation when compared against the interpreted model. *Black columns* are for conditioning using 10 vertical sections (spaced at 200 m). *Dark gray columns* are for conditioning using 20 vertical sections (regularly spaced at 100 m) and *light gray* columns are for conditioning with 40 sections

Figure 11 shows the difference in ore volume between the interpreted model and the volume resulting by the application of a threshold to the probability models (e-type). For instance, a threshold of 50 % indicates that all cells with at least 50 % probability to be ore are considered as being ore. A 10 % threshold would provide a very optimistic tonnage while a 90 % will be more pessimistic. Figure 11 shows those results for both the sections at 50 m (40 vertical sections), 100 m (20 vertical sections) and 200 m (10 vertical sections). The interpreted model is equivalent to applying a 60 % threshold to the simulations.

6 Conclusions

This study shows that it may be possible to build a geological model during preliminary exploration stages without long and demanding manual modeling processes with multiple-point geostatistics. The simulated models demand less effort (in time and cost) than a traditional approach. Additionally, the methodology applied to a 3D iron ore deposit allowed to access the uncertainty on the volumes for each facies modeled. This entire process minimizes the deterministic character of traditional modeling approach adding a probabilistic interpretation (uncertainty zone) about the actual shape and lithological contacts of the orebody. If there is a significant uncertainty related to the potential shape of the orebody, several scenarios can be created using alternate training images.

The simulation of this deposit did not require a large amount of interpreted sections. A few sections combined with borehole samples were sufficient to simulate the mineralized envelop of the deposit. From a broader perspective, the number of sections needed for a particular deposit would depend on the geometrical complexity of the contact patterns, the borehole density and the risk tolerance given the development phase of the deposit (early exploration versus production). Integrating more sections into the early delineation of the zone of uncertainty predictably decreases the uncertainty on the geometry of the deposit. The next step of this project is to consider more lithotypes leading to greater complexity of the orebody geometry.

References

1. Deutsch CV (2002) Geostatistical reservoir modeling. Oxford University Press, New York, 376 pp
2. Deutsch CV, Journel A (1998) Geostatistical software library and user's guide: version 2.0, 2nd edn. Oxford University Press, New York, 384 pp
3. Gómez-Hernandéz JJ (2005) Geostatistics. In: Rubin Y, Hubbard SS (eds) Hydrogeophysics. Water science and technology library series, vol 50. Springer, Netherlands, pp 59–83
4. Guardiano F, Srivastava RM (1993) Multivariate geostatistics: beyond bivariate moments. In: Soares A (ed) Geostatistics-Troia, vol 1. Kluwer Academic, Dordrecht, pp 133–144
5. Journel AG, Kyriakids PC (2004) Evaluation of mineral reserves: a simulation approach. Oxford University Press, New York, 216 pp
6. Liu Y, Harding A, Gilbert R, Journel A (2005) A workflow for multiple point geostatistical simulation. In: Leuangthong O, Deutsch CV (eds) Geostat Banff 2004. Quant geology and geostat book series, vol 14. Springer, Netherlands, pp 245–254
7. MBR, Minerações Brasileiras Reunidas Ltda (2005) Internal report. Gerência Geral de Geologia e Planejamento Operacional—GGGPO, Gerência de Geologia de Longo Prazo—GEGLP, 249 pp
8. Remy N, Boucher A, Wu J (2009) Applied geostatistics with SGeMS: a user's guide. Cambridge University Press, New York, 264 pp
9. Strebelle S (2002) Conditional simulation of complex geological structures using multiple-point statistics. Math Geol 34(1):1–21

Practical Implementation of Non-linear Transforms for Modeling Geometallurgical Variables

Ryan M. Barnett and Clayton V. Deutsch

Abstract Evaluation of process performance within mining operations requires geostatistical modeling of many related variables. These variables are a combination of grades and other rock properties, which together provide a characterization of the deposit that is necessary for optimizing plant design, blending and stockpile planning. Complex multivariate relationships such as stoichiometric constraints, non-linearity and heteroscedasticity are often present. Conventional covariance-based techniques do not capture these multivariate features; nevertheless, these complexities influence decision making and should be reproduced in geostatistical models. There are non-linear transforms that help bridge the gap between complex geologic relationships and practical geostatistical modeling tools. Logratios, Min./Max. Autocorrelation Factors, Normal Scores, and Stepwise Conditional transformation are a few of the available transforms. In many circumstances these transforms are used in sequence to model the variables for a given deposit. As each technique possesses its own limitations, challenges may arise in choosing the appropriate transforms and the order in which they are applied. These practical challenges will be examined, with a new technique named conditional standardization introduced as a potential solution to address non-linear and heteroscedastic multivariate features. A generalized workflow is proposed to aid in the selection and ordering of multiple transformations. Common problems such as bias and poor reproduction of spatial correlation are illustrated in a geometallurgic case study, along with a demonstration of the corrective measures. Although these transforms are presented within a mining context, they are equally suited to any petroleum or environmental application where multiple variables are being considered.

R.M. Barnett (✉) · C.V. Deutsch
University of Alberta, 3-133 NREF Building, Edmonton, Canada, T6G 2W2
e-mail: rmbarnet@ualberta.ca

C.V. Deutsch
e-mail: cdeutsch@ualberta.ca

P. Abrahamsen et al. (eds.), *Geostatistics Oslo 2012*,
Quantitative Geology and Geostatistics 17,
DOI 10.1007/978-94-007-4153-9_33,

1 Introduction

The necessity for modeling all aspects of a resource that are deemed critical to mine project design and economic feasibility studies is well documented [3]. When modeling multiple related variables under these circumstances, the multivariate Gaussian distribution is commonly implemented due to its simplicity and mathematical tractability. Unfortunately, continuous variables occurring in geologic media are rarely Gaussian in nature, due to the presence of complex features such as non-linearity, heteroscedasticity and constraints. There are a number of transformation techniques that are available for the removal of these complex features, producing well behaved distributions that approach Gaussianity. As the dimensionality of the data to be modeled may render co-simulation frameworks impractical, there are additional transformations for the decorrelation of variables, allowing independent simulation to proceed without the need for cross-variograms.

With certain transforms overlapping in their purpose, confusion may arise for practitioners in selecting the most applicable technique. Compounding this issue is the common need for 'chained' transformation sequences, in order to address a range of multivariate complexities. This paper introduces the more popular transformations that are currently used in practice, with a brief discussion of potential limitations for each method.

A shared limitation for multiple transforms is the poor handling of non-linear and heteroscedastic features. This motivated the development Conditional Standardization, which is proposed as a simple transform to augment the established methods. We also propose that a logical order exists for the application of multiple transformations in sequence. As this ordering relates to broad complex multivariate features which must be universally addressed, it is believed to be independent of considerations that are specific to each deposit. Although different mineralogies may only demand that certain complexities and associated transforms be applied, this generalized ordering should remain valid as an initial framing for transformation workflows.

A nickel laterite dataset will be used to demonstrate a chained transformation workflow. A concern for applying several transforms in sequence, may be the potential for redundancy in certain techniques and whether measurable value is gained from the additional effort. Individual transforms will therefore be sequentially excluded in order to gauge their relative impact on major model features. These diagnostic features will include the reproduction of histograms, multivariate relationships and spatial cross-correlations.

2 Common Transformations

Beginning with the most familiar and well established technique in geostatistics, the normal score transform allows for any distribution to be converted to a univariate standard Gaussian form. Achieved through matching quantiles of a variable's

CDF with quantiles of the standard normal CDF [2], the transformation guarantees a univariate normal distribution. It is important to recognize, however, that it is highly unlikely to produce a *multivariate* normal distribution. Although facilitating the application of Gaussian based modeling tools is its most obvious application, the normal score transform also standardizes a distribution and removes outliers. This makes it a very useful tool when used in conjunction with other transforms that benefit from such well behaved features.

Principal Component Analysis (PCA) is a dimension reduction and decorrelation technique, which seeks to transform a correlated multivariate distribution into orthonormal linear combinations through spectral decomposition of the covariance matrix [5]. These linear combinations are guaranteed to be uncorrelated at a zero lag distance, with associated eigenvalues revealing the relative variability which each component contributes to the multivariate system. This allows for components that provide no effective information to be discarded from subsequent simulation, with the back-transformation restoring the original dimensionality and correlation. PCA may therefore be particularly attractive in situations where a large number of variables are being considered. An immediate extension of PCA, Min/Max Autocorrelation Factors (MAF) involves a two-step spectral decomposition of the covariance matrix at zero and non-zero lag distances. If the variables are fully described by a two structure linear model of corregionalization (LMC), then MAF will decorrelate the variables at all lags, producing excellent reproduction of the cross-correlations. The technique was first introduced by [9] in the field of spatial remote sensing, and was successfully applied in geostatistics by [1]. A limitation of both PCA and MAF, is that only linear and homoscedastic distributions will be fully captured by spectral decomposition of the covariance matrix. As a result, execution of these techniques on complex multivariate distributions may result in poor reproduction of the original relationships following simulation and back-transformation.

The stepwise conditional transformation attempts to remove complex multivariate features while simultaneously decorrelating the variables at a zero lag distance to form an uncorrelated multivariate Gaussian distribution. Applying this technique, the first variable is simply normal score transformed. The second variable is then partitioned according to the probability class of the first variable, before having the normal score transformation independently executed on each discretized bin. The third variable is transformed conditional to probability classes of the first and second variables, and so on. Back-transformation reintroduces all of the original complex features. Introduced by Rosenblatt in 1952 [8], it was popularized in the field of geostatistics by Leuangthong and Deutsch [6]. While possessing many attractive features, limitations of stepwise include the data intensive nature of the technique and consideration of only the zero lag distance for decorrelation.

Logratios are a commonly applied family of transformations for the removal of compositional and ratio constraints. The additive logratio transform (ALR) [7] is the most commonly applied of these transforms in the earth modeling realm. Applying it transforms the variables to logarithmic ratios of their original form, effectively removing the constraints (e.g. compositional components are no longer bound between 0 and 100 %) and reducing the dimension of the system by one.

Back-transformation explicitly reinforces the targeted constraints and the original dimensionality of the system. The largest challenge for the ALR transform revolves around the potential for zeros values to enter the divisor of the ratio fractions, although a variety of documented methods exist to mitigate this issue [7].

While practitioners should pursue histogram reproduction through exploration and sensitivity analysis of the modeling workflow, a final transformation may ultimately be required to enforce adherence to the original distribution. This is of particular importance in the case of the primary resource variable, since it will have a first order impact on reserve estimates. Through a quantile-to-quantile transform between the model and original declustered distribution, perfect reproduction may be achieved. A more robust application of this method, however, will more heavily apply this correction to simulated values in locations that are relatively uncertain. The GSLIB implementation *trans* [2] achieves this by weighting the transformation based on the associated kriging variance of each simulated location.

3 Conditional Standardization

Conditional standardization is proposed as a technique for the removal of complex multivariate features, transforming non-linear and heteroscedastic data to approach linearity and homoscedasticity. In doing so, well behaved distributions are produced that are more suitable for either co-simulation frameworks or linear decorrelation transformations such as PCA and MAF.

The theory is attractively simple; consider a bivariate distribution consisting of two variables X and Z for n number of observations:

$$X_{1xn} = [x_1 \cdots x_n] \quad \text{and} \quad Z_{1xn} = [z_1 \cdots z_n].$$

Suppose that the relationship between these two variables is non-linear and heteroscedastic in nature, such as the schematic bivariate distribution displayed in Fig. 1. Observe that subtracting the Z values by a function which describes the mean of Z conditional to the value of X (red line in Fig. 1), will yield a residual distribution that is effectively linear. Likewise, if the values are divided by a function which describes the standard deviation Z conditional to the value of X, then a homoscedastic distribution will be produced. The bivariate form of conditional standardization is therefore given by

$$Z' = \frac{Z - E\{Z|X\}}{\sqrt{\mathrm{Var}\{Z|X\}}}. \tag{1}$$

The derivation of these conditional mean and standard deviation functions may be determined either parametrically through forms of regression or non-parametrically by discretizing the distribution based on the value of the conditioning variable (blue lines in Fig. 1). This concept may also be extended to higher dimensions, where a variable is transformed conditional to two or more variables. The generalized form of the transformation for p number of conditioning variables is given by:

$$Z' = \frac{Z - E\{Z|X_1, \ldots, X_p\}}{\sqrt{\mathrm{Var}\{Z|X_1, \ldots, X_p\}}}. \tag{2}$$

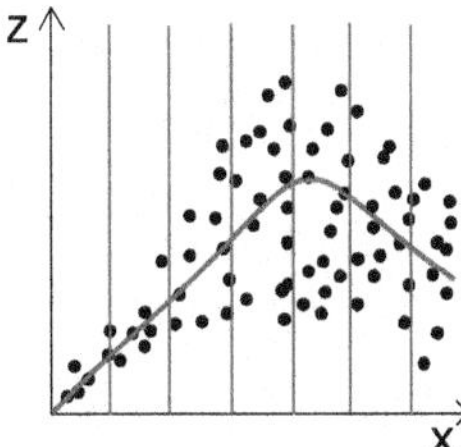

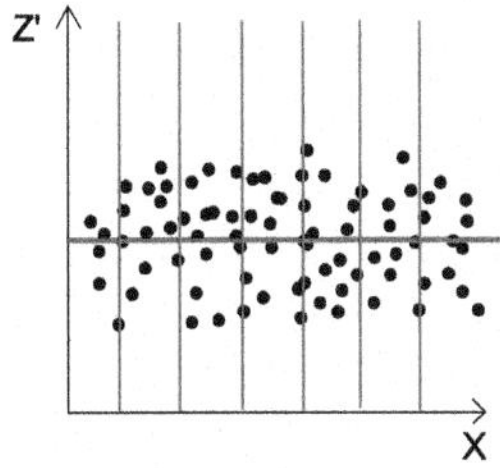

Fig. 1 Schematic of a non-linear and heteroscedastic bivariate distribution that has been partitioned according to conditional probability classes of X (*left*). Subtraction of the conditional mean and division of the conditional standard deviation yields a linear and homoscedastic distribution (*right*)

The back-transformation of the data or simulated values is achieved by a rearranging of the forward transformation, producing:

$$Z = Z' \times \sqrt{\mathrm{Var}\{Z|X_1,\dots,X_n\}} + E\{Z|X_1,\dots,X_p\}. \tag{3}$$

The success of this transform is largely dependent on the calculation of conditional mean and standard deviation functions which accurately describe the non-linearity and heteroscedasticity of the distribution. The non-parametric approach is generally expected to produce superior results, since no assumptions of the functional form of the distribution must be made. Parametric application may still be considered as a viable option in cases where a low number of data, or high dimensionality makes the non-parametric discretized approach impractical.

Should the non-parametric approach be chosen, there is no strict rule regarding the number of classes that are required for partitioning the conditioning variable, or the number of data that are required in each bin for the subsequent calculations of mean and standard deviation. The fewer the classes, the more likely that complex features will remain within the partitioned bins following transformation. Conversely, increasing the number of classes risks reducing the number of data in each bin to below a threshold that would permit the stable calculation of the conditional mean and standard deviation. Smoothing algorithms where data searches pass beyond the class partitions in sparsely populated regions of a distribution may help, but based on observation it is unlikely that greater than three conditioning variables will be viable in practice. In these cases practitioners may choose between either using a parametric calculation of the conditional functions, or a 'nested' application of the non-parametric approach. Nested application refers to using only one, two or three conditioning variables, to remove complex features from the higher order conditioned variables. Addressing these selected relationships will oftentimes resolve the majority of the complexity between variables that are not directly transformed conditional to one another. This is not guaranteed, however, and careful decision making must take place regarding the ordering of these variables. Considerations may include:

- Reproduction of the multivariate relationships which the primary resource variable holds with all secondary variables (resource variables becomes the first conditioning variable for all transformations)
- Reproduction of a multivariate relationship between secondary variables where the ratio or correlation between them is of critical interest (one secondary variable must condition the other)
- Conditioning of variables that demonstrate dramatic bivariate complexity. Under these circumstances, it is unlikely that a well behaved distribution will be produced unless one variable directly conditions another

The above considerations will often lead to difficult decision making, as not all of them may be satisfied by a nested conditional standardization application.

4 Chained Transformations

While a variety of transforms have been introduced, no single technique addresses all of the complexities which may exist between the variables of a mineral deposit. Transforms will often be used in combination as a result, but which techniques should be selected and in what order should they be applied? The more direct question may relate instead to the order in which multivariate complexities should be removed and reintroduced by the forward and back transformations. Given a complex multivariate observation vector Z, Fig. 2 provides the general order that multivariate features may be addressed in order to obtain an uncorrelated distribution Y that approaches multivariate Gaussianity. Major points of emphasis include:

- Back-transformations are executed in the reverse order of which they were applied going forward
- Should logratios be required to remove compositional constraints, it must be applied as the first forward transformation. This will ensure that the fundamental constraints it targets will be explicitly honored, since no subsequent transforms will follow it when back-transforming
- Should Logratios be required for both compositional and ratio/stoichiometric constraints, the former must be addressed before the later. When specific constituent variables are converted to ratios of one another before having logratios applied, the relative proportion which they previously represented in the greater composition will be lost (preventing the application of logratios to the compositional constraint)
- Removal of non-linear and heteroscedastic features precede decorrelation for two major reasons: (i) any transformations following decorrelation have the potential to reintroduce a measure of correlation and (ii) decorrelation methods such as PCA and MAF more accurately decompose linear and homoscedastic distributions
- A final normal score transformation should be applied to all variables prior to simulation in order to insure marginal Gaussianity

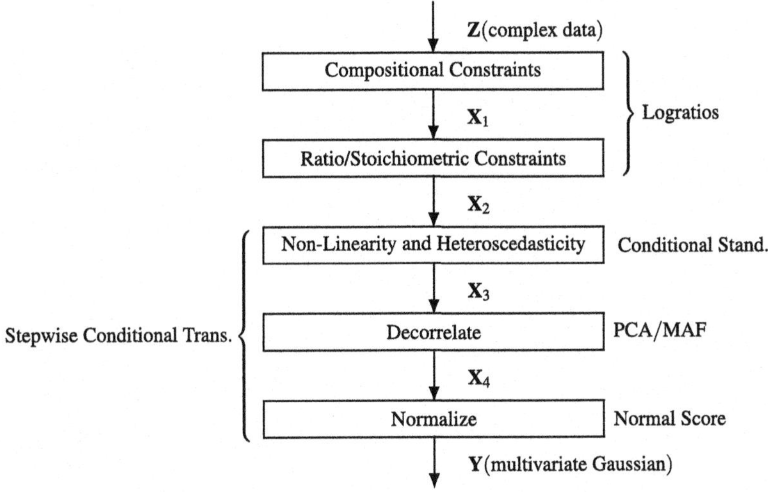

Fig. 2 Order in which multivariate complexities should be addressed to form an uncorrelated multivariate Gaussian distribution. Transformations are given for the removal of each complexity

- Should a histogram correction be required, it is applied following all back-transformation steps

The stepwise conditional transform may drastically simplify the transformation workflow according to Fig. 2, but bear in mind that various benefits may be gained by the alternate series of transforms that appear on the right side this figure.

5 Case Study

A chained transformation workflow will be demonstrated using a nickel laterite dataset. The workflow will follow the series of techniques that appear on the right hand side of Fig. 2, with a few additions and modifications:

- Logratios will be applied only once to remove the compositional constraint
- A normal score transform will precede conditional standardization to correct for the very skewed distributions that logratios create. These skewed distributions would otherwise create instability in the calculation of the mean and standard deviation functions
- MAF will be used rather than PCA for the more robust spatial decorrelation
- A normal score transform will precede MAF, in order to center the data at zero (required) and remove outlier values (recommended)

Following all transformations, the resultant well behaved multivariate distribution will be used to model the deposit through sequential Gaussian simulation (SGS) [2, 4]. Simulated values will then be back-transformed in the reverse order of the

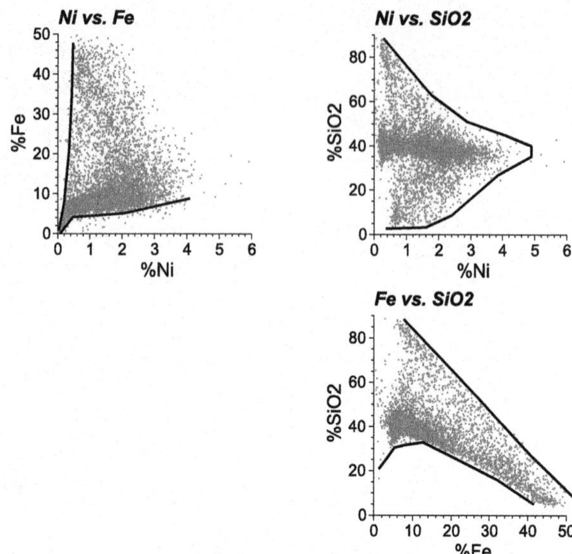

Fig. 3 Scatter-plots between the Ni, Fe and SiO_2 with complex features outlined

forward transforms, with a final histogram correction applied to the nickel resource. To gain understanding of each transformation's effect on final reproduction, parallel modeling workflows will be executed with specific transforms excluded from the process. Results with and without these transforms will then be displayed using key diagnostic plots.

The nickel laterite dataset is composed of 7740 homotopically sampled assays. A typical mining model for a nickel laterite deposit will require the Nickel (Ni) resource, as well as iron (Fe) and silica dioxide (SiO_2) which exert a critical influence on smelting extraction. Additional variables are likely to be modeled in practice, but will be excluded from this study due to presentation space constraints. As highlighted in the declustered data scatter plots in Fig. 3, complex multivariate features play a major role in this multivariate system.

To demonstrate the new conditional standardization technique, an intermediate step in the transformation workflow appears in Fig. 4. The prior distribution in this figure has had the logratio and normal score transform applied. Marginal histograms on the bivariate scatter plots reveal this univariate Gaussianity, reinforcing that the normal score transform does not guarantee a multivariate normal distribution (as heteroscedastic and non-linear features clearly remain). It is these complex features that will not be captured by the subsequent MAF transform. Observe in Fig. 4 that unlike the prior distribution, the conditional standardized multivariate distribution is both homoscedastic and linear.

Following the remainder of the modeling workflow, back-transformed simulated values are displayed in Fig. 5, along with the declustered data distributions. Simulated values without the use of conditional standardization in an otherwise identical workflow are also shown for comparison. To more clearly demonstrate the relative density of data in each bivariate relationship, Gaussian kernel density plots accom-

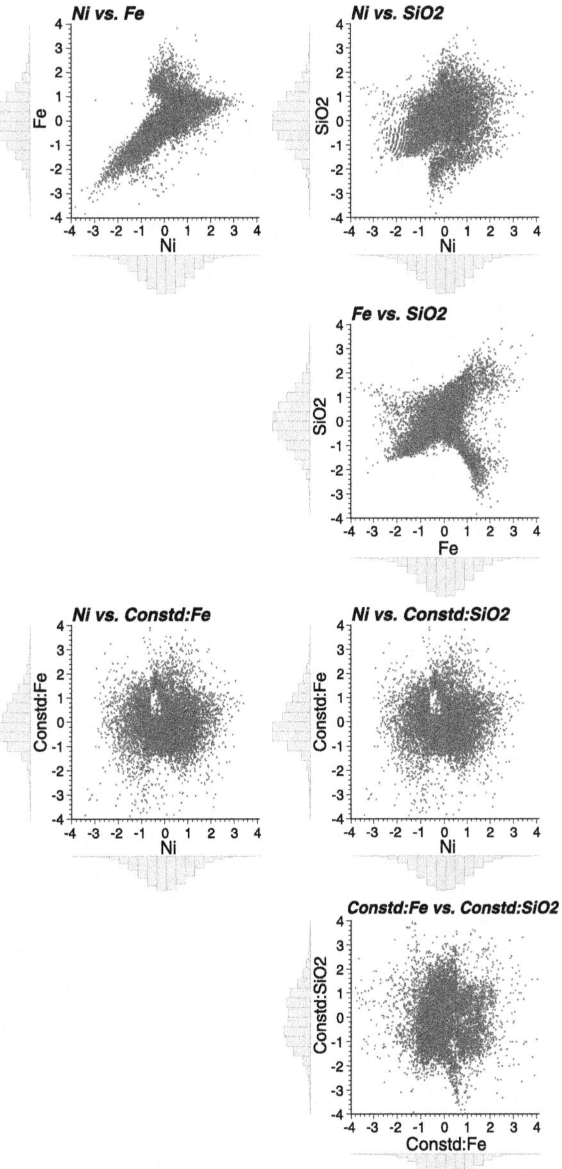

Fig. 4 Scatter-plots of Ni, Fe and SiO_2 following logratio and normal score transforms (*top*). Non-linear and heteroscedastic features of this distribution have been removed following conditional standardization (*bottom*)

pany the scatter plots in this figure. A clear improvement in the reproduction of the original multivariate relationships is seen for the workflow that utilizes conditional standardization.

As was seen in Fig. 4, following conditional standardization the distributions are largely uncorrelated. To determine whether the subsequent MAF transform is necessary given the little remaining correlation, Fig. 6 displays the experimental cross-

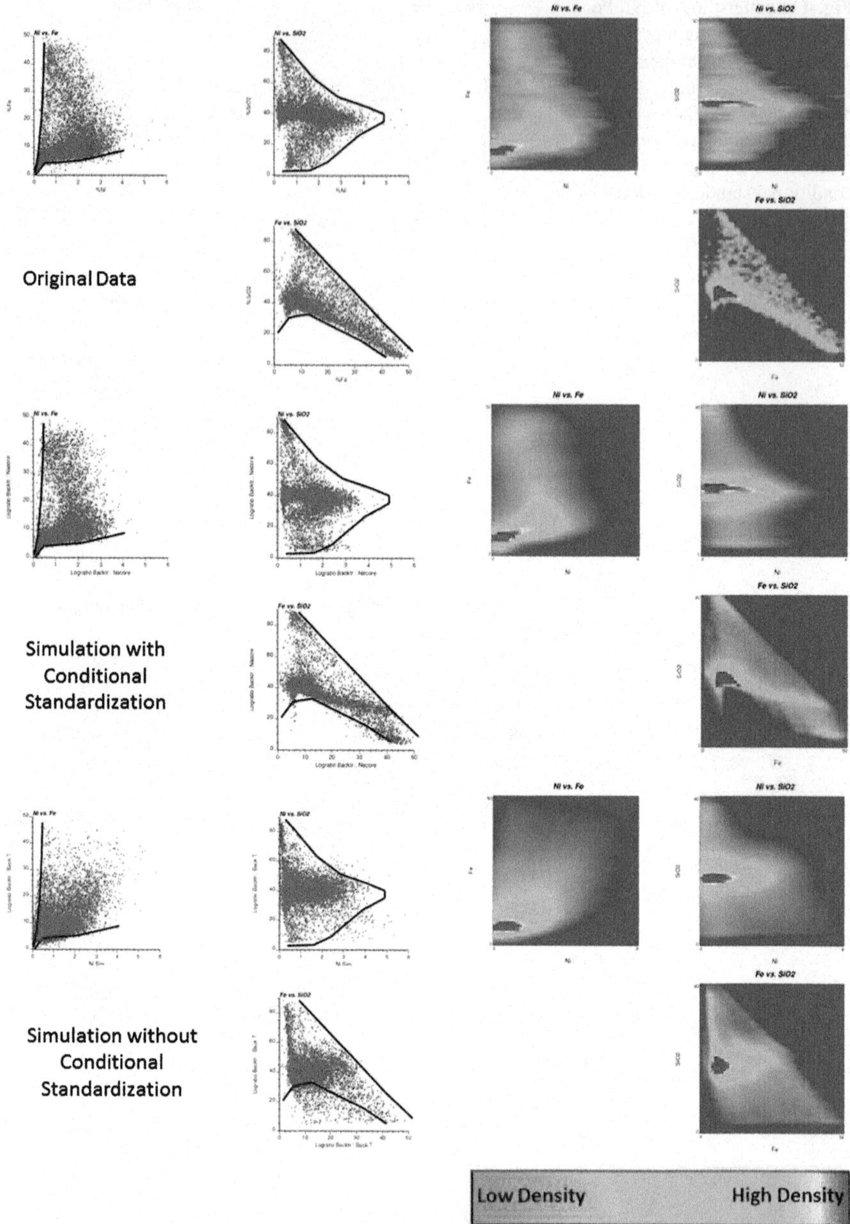

Fig. 5 Bivariate scatter and Gaussian kernel density plots for the original data (*top*), simulated values with (*middle*) and without (*bottom*) conditional standardization

semivariograms of the original data, and four arbitrary realizations from modeling workflows with and without the application of MAF. A significant improvement in the reproduction of the spatial correlation between variables is observed for the re-

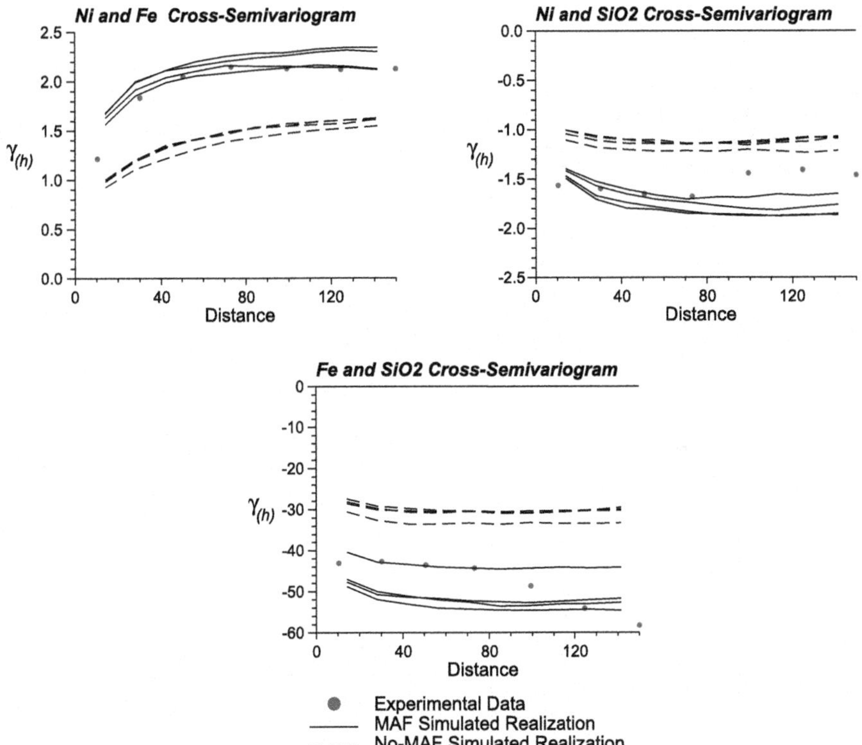

Fig. 6 Experimental cross-semivariogram of the original data (*dot*) and four arbitrary realizations from modeling workflows with (*solid*) and without (*dash*) the application of MAF

alizations where MAF has been applied. This is not surprising given that in addition to the correlation seen in Fig. 4, MAF also considers the correlation at a non-zero lag distance.

Conditional standardization will aid in the reproduction of constraints, potentially rendering logratios as a redundant transform. As logratios specifically target the compositional constraint, however, Fig. 7 displays histograms of the sum of three modeled components for the original assays, and simulated values with and without the use of logratios. It is observed from the maximum values, that the simulated locations without the use of logratios does not explicitly honor the compositional constraint. Although this issue is demonstrated with only three modeled variables, the results would be much more dramatic when considering the additional components that a typical Nickel Laterite mining model would require (such as MgO, Co, Al_2O_3, Cr_2O_3), which would serve to push the average summation of the components closer to the constraint.

Finally, to demonstrate the enforcement of histogram reproduction, Fig. 8 shows the quantile-quantile plots between declustered Ni data and the simulated distributions (corrected and uncorrected). Note that a slight deviation from the original distribution exists following correction of the simulated values, which is due to

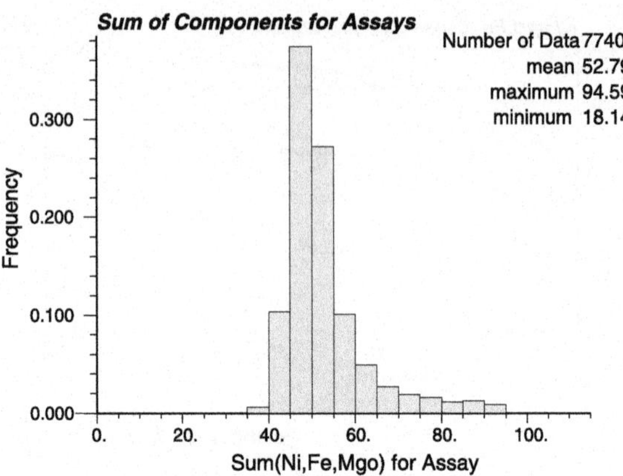

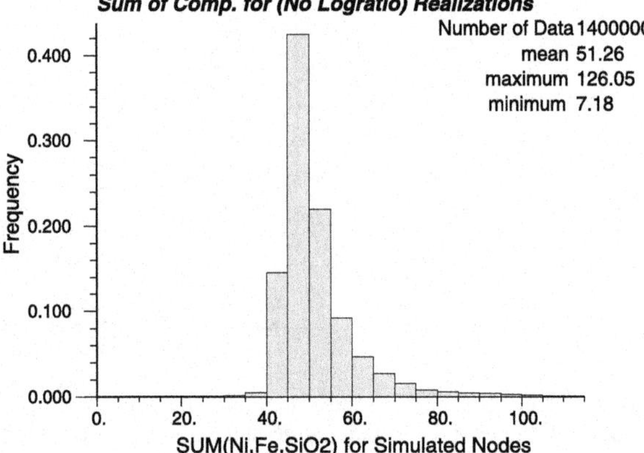

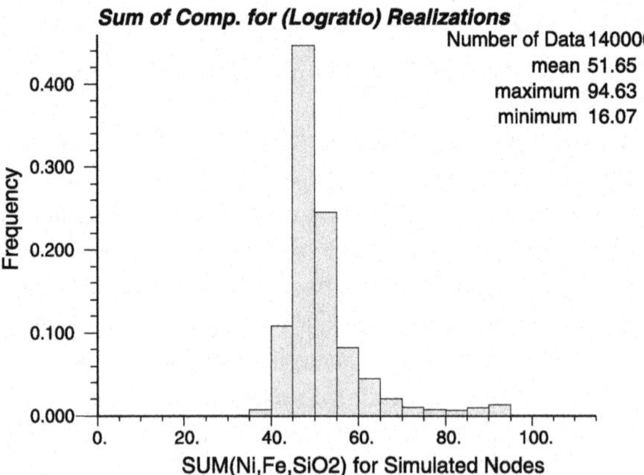

Fig. 7 Histograms of the sum of compositional components for the original data (*top*) and simulated realizations with (*middle*) and without (*bottom*) the application of logratios

weighting of the transform based on the kriging variance of each simulated location. Exact reproduction of the declustered distribution was therefore sacrificed in favor of reproduction at conditioning data locations.

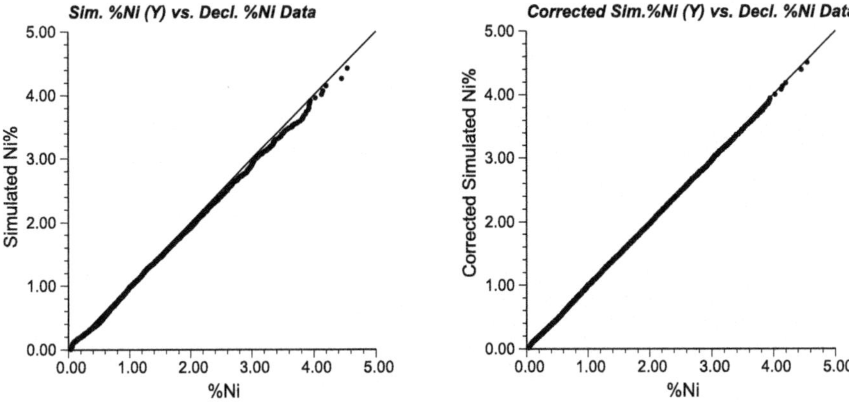

Fig. 8 Quantile-quantile plots between the original declustered Ni distribution and the simulated distribution before (*left*) and after (*right*) correction

6 Conclusion

Selecting and applying transformations to address multivariate complexities is a very challenging step when framing a geostatistical modeling workflow. Attractive features and practical limitations of the more popular transforms have been discussed, with a new technique named Conditional Standardization proposed to augment the established methods. To address multiple complexities, transforms may be applied in chained sequences, for which a general ordering has been proposed. A geometallurgical dataset was used to demonstrate a chained transformation modeling workflow, where a great deal of value was clearly extracted from the use of multiple techniques.

Many challenging problems remain to be addressed in the field of multivariate geostatistics. It is very difficult to attain the type of reproduction displayed in this paper for massively multivariate settings where 20 to 100 variables are being considered. Furthermore, every single transform that has been presented requires the presence of all variables at collocated locations. Transforms will therefore breakdown when variables are not homotopically sampled, or worse yet, where certain variables are never sampled at the same location. Along this same vein, integration of variables that are measured at varying scales and with varying quality control is not straight forward. Scaling of core measurements, channel samples and seismic sections to be considered together is an example of this commonly encountered issue. Such practical challenges will be the focus of future study.

References

1. Desbarats A, Dimitrakopoulos R (2000) Geostatistical simulation of regionalized pore-size distributions using min/max autocorrelation factors. Math Geol 32:919–942

2. Deutsch CV, Journel AG (1998) GSLIB: geostatistical software library and user's guide, 2nd edn. Oxford University Press, New York
3. Dunham S, Vann D (2007) Geometallurgy, geostatistics and project value—does your block model tell you what you need to know? In: Project evaluation conference, Melbourne, Victoria, 19–20 June 2007, pp 189–196
4. Isaaks E (1990) The application of Monte Carlo methods to the analysis of spatially correlated data. PhD thesis, Stanford University, USA
5. Johnson R, Wichern D (1988) Applied multivariate statistical analysis. Prentice Hall, Upper Saddle River
6. Leuangthong O, Deutsch C (2003) Stepwise conditional transformation for multivariate geostatistical simulation. Math Geol 35:155–172
7. Pawlowsky-Glahn V, Egozcue JJ (2006) Compositional data and their analysis: an introduction. In: Buccianti A, Mateu-Figueras G, Pawlowsky-Glahn V (eds) Compositional data analysis in the geosciences: from theory to practice. Geological Society special publication. Geological Society, London
8. Rosenblatt M (1952) Remarks on a multivariate transformation. Ann Math Stat 23:470–472
9. Switzer P, Green A (1984) Min/Max autocorrelation factors for multivariate spatial imaging. Technical report no. 6, Department of Statistics, Stanford University

Combined Use of Lithological and Grade Simulations for Risk Analysis in Iron Ore, Brazil

Debora Roldão, Diniz Ribeiro, Evandro Cunha, Ricardo Noronha, Amanda Madsen, and Lilian Masetti

Abstract The majority of economic feasibility studies of mineral projects use deterministic geological block models and grade estimation. Those deterministic block models do not allow project risk analysis. There is a smoothing of grade variability and the lithological dilution in each block is not considered. The uncertainty of geological block model can be evaluated by geostatistical simulation methods. The main objective of this project is to evaluate the impact of lithological and grade simulation in the economic studies of a world class iron ore deposit. An iron ore deposit from Carajás Province (Brazil), composed of low grade (jaspilite) and high grade ore (hematite), was selected as a case study. The hematite body is 9 km long, 3 km wide, and 300 m deep, with an average grade of 66 % iron. The uncertainty of the lithological contacts among hematite and waste/low grade rock is as important as the ore grade variability. For this study, drillhole database (hard data) and section interpretation (soft data) was used in order to improve the lithological conditioning. Different geostatistical simulations were combined to generate equiprobable realizations for lithologies (S_L), which are derived of superimposed different geological events; supergene and sedimentary/volcanic rocks. Two types of supergene events were simulated; the thickness of duricrust canga (2D TB simulation, S_C) and the transition of the weathering zone (3D SIS simulation, S_S). The levels of Banded Iron Formation and volcanic rock were simulated generating the primary

D. Roldão (✉) · D. Ribeiro · E. Cunha · R. Noronha · A. Madsen · L. Masetti
VALE-DIPF, Belo Horizonte, Brazil
e-mail: debora.roldao@vale.com

D. Ribeiro
e-mail: diniz.ribeiro@vale.com

E. Cunha
e-mail: evandro.cunha@vale.com

R. Noronha
e-mail: ricardo.noronha@vale.com

A. Madsen
e-mail: amanda.madsen@vale.com

L. Masetti
e-mail: lilian.masetti@vale.com

P. Abrahamsen et al. (eds.), *Geostatistics Oslo 2012*,
Quantitative Geology and Geostatistics 17,
DOI 10.1007/978-94-007-4153-9_34,

rock facies (3D SIS simulation, S_V), independent from weathering simulations. The grade variables were simulated in cascade, one grade simulation for each lithology simulation. The results of the simulations were compared with the official reserve calculated with deterministic block model in order to measure the project risk and the impact of waste dilutions in the quality of ore product.

1 Introduction

The economic previous feasibility studies of mineral projects in Vale Ferrous Department are conducted using deterministic geological models in which the chemical grades are estimated by ordinary kriging. The use of ordinary kriging smooths the grade variability of the block model. Geostatistical simulation is the most appropriate method to calculate the uncertainty of geological models. This technique is not routinely applied in the mining industry due to the demanding computational time and the lack of trained professionals. The main objective of this paper is to present the impact using simulated models instead of deterministic geological models for studying the risk in evaluation of resources and reserves.

A strategic project in the Ferrous Vale was chosen as case study. This project is located in Carajás Mineral Province, north of Brazil. The iron mineralization is formed by sedimentary deposits of Banded Iron Formation (BIF, jaspilites), from Paleoproterozoic, enriched by a supergenic process during the Cenozoic, that generates hematite bodies with high iron grade content. A geological characteristic of these deposits is the irregular shape of the contacts among hematite bodies and jaspilites, and the duricrust canga that covers the entire BIF sequence. Therefore, residual jaspilite in the basal contact of hematite bodies can be seen. In this type of deposit the risk due to uncertainty of geological contacts has a greater impact than grade variability.

This work proposes a combination of lithological and grade simulations. Three independent lithological simulations were combined in order to honor the stratigraphic, weathering and structural relationships between units.

The use of geostatistical simulation allows to measure, in a realistic way, the risk of a mineral project considering the uncertainty of geological interpretation and the grade distribution.

2 Local Geology

Iron ore deposits of Carajás belong to metavulcanosedimentary units of Grão Pará Supergroup. The mineralization is generated by weathering of banded iron formation unit (BIF) denominated Carajás Formation. BIF is composed of jaspilites enriched during supergenic weathering processes, which were responsible for the formation of hematite bodies. These hematites bodies have great extensions, high iron grade content (66 %), and are typically friable. The relation between contacts of high iron grade units and non-altered low iron grade unit (jaspilites) is sharp and irregular, having usual peaks of compact jaspilites in the middle of friable hematite ore. Those iron units are found interlayered in metavulcanic rocks of Parauapebas and Igarapé Cigarra formations and are covered by detritic soil and laterite units, denominated duricrust canga (CG), which are generated during the pedological evolution process.

3 Carajás Project

The previous feasibility study of Carajás Project was based on a deterministic geological model that was created using wireframes from the main lithological types. Those wireframes flagged a deterministic block model which unit cell has $25 \times 25 \times 15$ m. For this previous block model the chemical variables (iron and principal contaminants) were estimated using (co)kriging. The principal ore unit (hematite body) has the approximate dimensions of $9 \times 1 \times 0.3$ km, resulting in estimated resources of 5 Bt with 66.2 % of iron and reserves of 3.4 Bt with 66.7 % of iron.

An innovative mining method for iron ore, known as Truckless, was proposed due to the importance of the Carajás Project. This approach involves the use of load machines, mobile crushers and mining belts in mining ore and waste. There is a raw production prediction of 90 Mt per year. This method presents a series of advantages such as: reduction of particulate material emissions; reduction of mine infrastructure; reduction of environmental impact; greater and better automation. The great disadvantage of this methodology is the lower operational flexibility. In comparison with conventional methods it is more susceptible to geological changes such as the occurrence of non-interpreted peaks of jaspilite, which could happen in the deterministic model.

The mining operations will be composed of 4 operational mining faces with 7 crusher systems. The use of each system depends on the type of material to be mined. In the beginning, the mobile crushers will be used for friable materials and the jaw crushers for compact materials. Basically each system will work grouped every 3 benches, each one 15 m high. The projected equipment to work with the friable material has a production capacity of 8500 t/h, with a maximum continuous work time of 24 hours in compact material. After this period of work, the abrasion of the equipment will be elevated.

4 Database

Lithological simulation was implemented using two different types of information: (1) hard data—drilling database, an irregular grid of 200×200 m, on average and, 100×100 m in the center of the mineralized body; (2) soft data—interpreted geological sections, a grid of $200 \times 50 \times 30$ m was added to the database. The soft

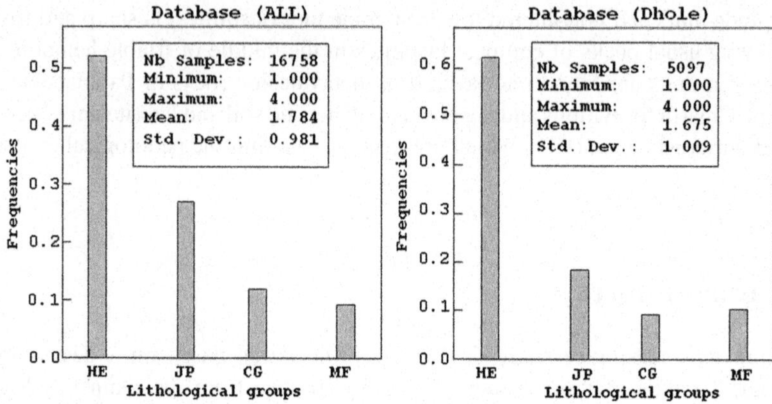

Fig. 1 Histogram of lithological proportions of drilling database (*right*) and drilling database + geological interpretation (*left*). HE—hematites, JP—jaspilites, CG—duricrust canga, MF—mafic rocks

Table 1 Descriptive statistics of variable Fegl of input database for simulation and deterministic block model

Variable	DataSet	Lithology	Count	Minimum	Maximum	Mean	StdDev	Variat.C
Fegl	Database	HE	2854	59.74	69.15	66.38	1.65	0.025
		JP	515	13.92	60.01	46.45	8.06	0.174
		CG	461	31.85	68.04	64.29	2.52	0.039
	Discrete block model	HE	172937	60.56	68.91	66.45	1.03	0.016
		JP	85165	26.01	65.13	42.18	4.08	0.097
		CG	33631	28.22	68.89	64.21	1.48	0.023

data was used only in lithological simulation, because lithological information from drilling was not sampled according to a regular grid. The majority of drilling was concentrated in the main ore body (hematite bodies), as usually done in mineral exploration deposits. Without the addition of soft data in simulation, the result will tend to simulate more ore, because simulation, independent of the method used, honors the lithological proportions. The use of soft data helped to decluster the samples to obtain the lithological proportions that were considered in the simulations (Fig. 1). The variographic analysis was made using only hard data.

The grade database is composed only by drilling information (hard data). This study will simulate the main chemical variable, iron grade (Fegl). Table 1 presents the iron grades from a declustered database and the previous deterministic block model. Those grades will be used to compare and validate the simulated block model.

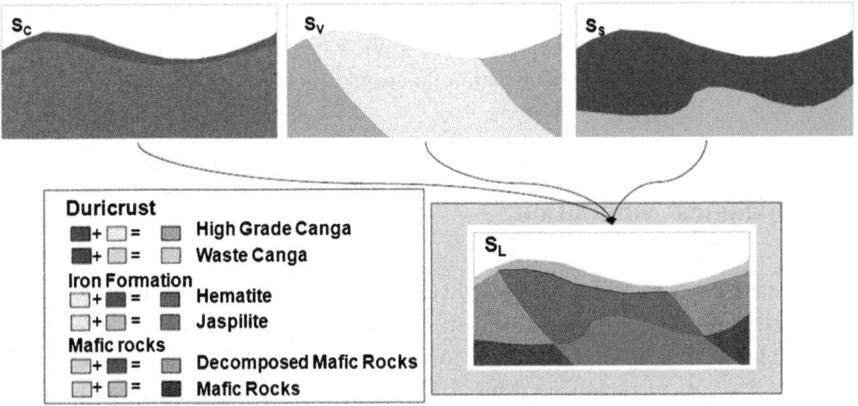

Fig. 2 Combination between 2D duricrust canga simulation (S_C), weathering simulation (S_S) and structural simulation (S_V), generating the final lithological simulation (S_L)

5 Lithological and Grade Simulations for Iron Deposits

The use of lithological and grade simulation is usually related to (co)simulation, where lithology and grade are simulated simultaneously [6]. This work will propose a methodology that combines simulation of categorical variables and concurrent and independent continuous variables, like "cascade" simulation [2]. In the beginning a lithological simulation (S_L) was done following a grade simulation (S_T) respecting the geological domains. The S_L is generated from the combination of three independent simulations that represents distinct geological events, duricrust canga formation, S_C, primary rock facies, S_V, and weathering facies, S_S (see Fig. 2):

$$S_L^n = (S_C^n \cap S_V^n) \cup ((S_V^n \cap S_S^n) \backslash S_C^n),$$

for each S_L^i, it is simulated one S_T^i. The final lithological simulations (S_L^n) are the result of the union of the tow group of simulation intersections: (i) the duricrust canga simulation (S_C^n) intersected by the primary rock facies (S_V^n); and (ii) the primary rock facies (S_V^n) intersected by the weathering facies (S_S^n), excluding the duricrust canga (S_C^n).

The Turning Bands (TB) methodology was used for simulation of 2D duricrust canga (S_C) and Sequential Indicator Simulation methodology (SIS) for categorical simulations of geological events.

Conditional simulation using TB is decomposed in two steps: appliance of a non-conditional simulation algorithm; condition normalized data using (co)kriging of error between estimated data using simple kriging and the simulated value. Initially it is generated n lines with random directions θ_i, where each line presents an independent stochastic process Y_i calculated from covariance $C_1(h)$ deduced of a known covariance $C(h)$ [5, 8].

SIS methodology was initially proposed by [1] and [7]. This is a methodology widely used for categorical simulations, because the implementation is very simple.

This method consists in generating random numbers that follows a Gaussian distribution, using Monte Carlo simulation [3, 4, 9]. After this, conditioning of data is done from local probability distribution function estimation (*lpdf*).

6 Lithological Simulation

The iron ore deposits of Carajás type are characterized by sharp geological contacts, with well defined stratigraphical and structural relations. To honor these relations, the simulation was divided according to geological events. Two 3D simulations of geological events were defined using SIS method and one 2D simulation for duricrust canga unit using TB method.

For the duricrust canga unit (CG) a TB simulation was adopted with a CG thickness in bidimensional domain, S_C. The SIS simulation was performed for the structural primary event, S_V, and another SIS for the weathering event, S_S. The S_V simulation separates the banded iron formation (FF) forming the hosting mafic rocks. The S_S simulation superimposed to S_V individualizes the decomposed and fresh rocks. It is important to notice that the high iron grade units (hematites) are formed from the alteration of fresh FF rocks (jaspilites).

The S_C simulation was generated from a 2D database, where the thickness variable for duricrust canga (CG) was normalized (*length_gauss*). A variographic model of variable *length_gauss* was adjusted: 1 Nugget Effect + 2 Spherical Structures. Afterwards, a 2D grid node was generated with grid spacing of $\frac{1}{4}$ in volume the deterministic model grid (12.5×12.5 m), where thickness data was estimated in a punctual support. Finally, the Gaussian data were back-transformed to the variable length.

The simulation process of two geological events (weathering and structural) was done in an independent way, but following the same methodology. Initially, binary variables were identified. The value 1 was used for the weathering event of decomposed rocks and 0 for fresh rocks. The value 1 was used for the structural event of banded iron formation (FF) and 0 for mafic rocks (MF). Then a variographic adjustment of indicator variables was performed, where 2 exponential structures were adjusted for each event. A 3D grid with cell dimensions of $12.5 \times 12.5 \times 7.5$ m, $\frac{1}{8}$ in volume of deterministic block cell size, was generated. The variables Ind_intemp (weathering event) and Ind_zero (structural event) were simulated using a punctual support. The final images resulting from the SIS process produced a good stratigraphic relation for weathering and structural facies.

The number of simulations used was 30. After 20 simulations the average of variance and the average of simulations were stabilized for this type of deposit.

After the simulation of those three distinct geological events, logical operations were used to compose the final simulated variable for lithology (lito_simul). The final variable was migrated to a grid of $25 \times 25 \times 15$ m block cell size.

7 Grade Simulation

Carajás Project is a deposit of high iron grade and high volume content, where the average grade of iron is 66 % and the variation coefficient is very low (0.020). The contaminants present low grades, where the benefit function is basically defined by the type of rock, in other words, the hematites are considered as ore and the other rocks are considered as waste.

A simplified method to simulate grades was defined taking into consideration the low variability of iron grade in the mineralized body and the contact relation with hosting rocks. The iron grade simulation is linked to the lithological simulation. Each lithological simulation (S_{L1}) has a grade simulation ($S_{T1'}$), simulated before and independently. Thirty simulations ($S_{T1'}$–$S_{T30'}$) of TB type were performed in 25 × 25 × 15 m block support.

The variographic analysis and simulations were done in the Gaussian domain, using a variographic model with 3 structures (1 Nugget Effect + 2 Spherical). The grade simulation respecting the geological simulated domains, where grades of each lithological domain, were treated separately.

8 Validation of Lithology and Grade Simulations

Initially a visual analysis of simulated lithology was conducted in order to verify if the S_L simulation honored the stratigraphical relations. The obtained results are good once the simulations honored the established stratigraphy and no unit extrapolation was observed. S_L simulations also fully honored database lithological proportions.

Regarding grade, it was verified if simulations reproduced the data histogram and the variography. Figure 3 presents the comparison between database and simulated scenarios; both histograms and variogram show that simulations fully respected the database used. It is important to notice that to evaluate if simulations reproduce the variography, a change of support was applied from point experimental database variogram to block ones.

Table 2 presents a descriptive statistics of mean of simulations, where it is possible to verify that the simulated model reproduced the database faithfully, including data variance (see Table 1).

Figure 4 presents the comparison between deterministic block models and some of the simulation scenarios. The uncertainty of lithological variable is more evident than the uncertainty of grade inside the same lithological unit, mainly for ore grade (hematite body), where the variance of simulated scenarios for ore is 1.7, considering simulated data variance without considering the lithological domains is 3.7.

9 Incorporating Uncertainty to the Project

The risk analysis presented in this work took into account the blocks flagged as measured or indicated resource in the deterministic block model (variable resource) and

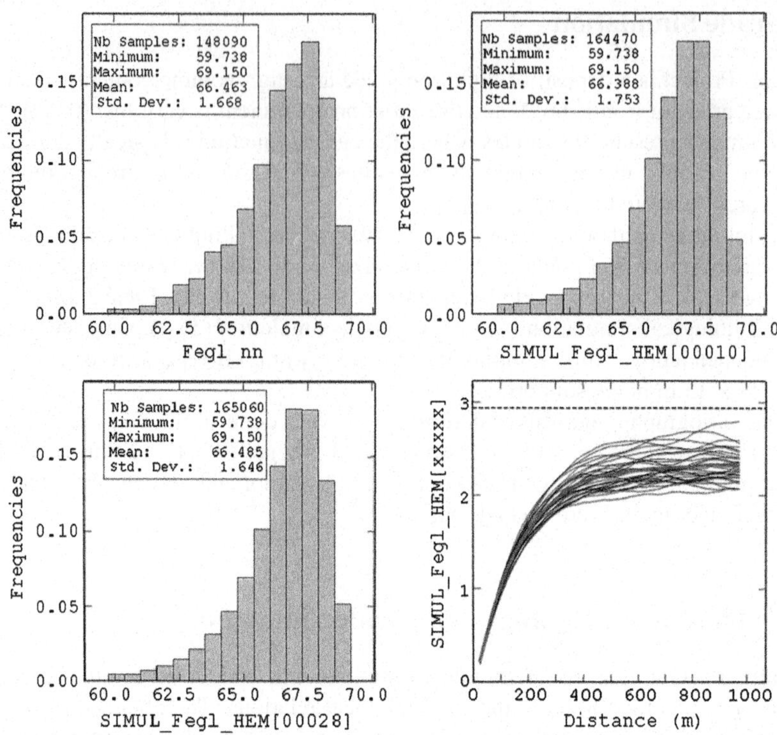

Fig. 3 Nearest neighbor database histogram (*upper left*) and two simulated scenarios (00010—*upper right*; 00028—*bottom left*). In the *bottom right* corner the experimental variogram is shown (*thin lines* represent the simulations) with the variographic model used in grade simulation (*thick line*)

Table 2 Descriptive statistics of mean of simulated scenarios

Variable	DataSet	Lithology	Count	Minimum	Maximum	Mean	StdDev	Variat.C
Simu_fegl	Mean of simulation models	HE	165243	59.74	69.15	66.38	1.71	0.026
		JP	60542	13.92	60.01	45.40	9.29	0.205
		CG	29131	31.85	68.04	64.05	3.09	0.048

the blocks also flagged as ore blocks (variable type). In deterministic block model the ore considered were those blocks flagged as measured or indicated resource of hematite and a use of 5 % of the duricrust canga cover (CM lithotype). After the parameters were fixed for resource variables and type, all simulated scenarios were evaluated in projected pits for sequential years (1, 2, 3, 4, 5, 7, 10, 15 and 20 years), with the objective of assessing the local and global variability of the block model.

For each realization it is possible to calculate the mass dilution and the ore loss considering the results of the simulations. The SMU simulation could be right or wrong compared to the deterministic model. The blocks assigned as ore or waste, in

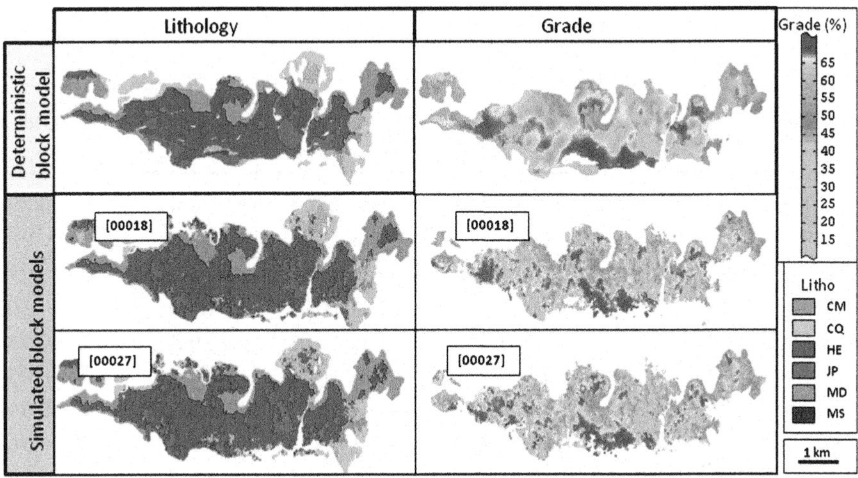

Fig. 4 Comparison between deterministic block model (*top*) and simulated block models (*middle* and *bottom*) of lithological simulation (*left*) and correspondent grade simulation (*right*). Where: CM and CQ are duricrust canga; HE is the hematites; JP is jaspilites; MD is decomposed mafic rocks; and MS is fresh mafic rocks

the deterministic model, could be misclassified; waste as ore (WO) and ore as waste (OW); or equally classified; ore as ore (OO) and waste as waste (WW).

In a general way the sequenced mining plans presented a low variability. Table 3 shows a comparison between assessed values in the deterministic model and the mean iron grade of simulated scenarios. The first comparison is related to all blocks with simulated grade (duricrust canga, jaspilite or hematite ore—OO + WO) and flagged as ore. The second comparison considers blocks flagged as ore, but only simulated and flagged blocks of hematite ore (OO).

In the first situation, which considers dilution of the waste, there is a mass loss of 4 % on average and a drop of iron grade from 66.63 % to 65.27 % (2 %), with the greater proportion of mass loss in the first year (16 % less). All scenarios have diluted iron grade lower than in the deterministic block model. Figure 5 shows the histogram of diluted iron grade in the mining plan for year 20, considering all scenarios. The mining plan already regarded a waste dilution of 5 %, which will be mined as ore. The projected plan does not consider the iron grade dilution, which will be in the best scenario of a 2 % decrease. This will affect the product standard, which must be beyond 66.7 % of total iron grade. This iron grade reduction is respected only for duricrust canga and jaspilite dilutions, but is not considered for mafic rocks.

If the exploitation could be more selective, considering the need to maintain the quality of the product with iron content of 66.7 %, the second scenario would be used to estimate ROM's grade, because in this case it only hematite blocks with low risk of dilution would be considered as ore. The mass loss of ore would be 14 % of total reserves calculated using the deterministic model.

Table 3 Comparison between deterministic block model and average grade of simulated blocks

	Deterministic Model		Mean of Simulations				Difference (%)			
			ALL Rock type		Hematite		ALL Rock type		Hematite	
Mining Plan	Mass (Mt)	Fegl (%)	Mass (Mt)	Fegl (%)	Mass (Mt)	Fegl (%)	Mass	Fegl	Mass	Fegl
Year 1	36.91	65.61	31.18	64.96	24.84	65.73	16 %	1.0 %	33 %	−0.2 %
Year 2	43.57	66.19	41.87	65.54	37.18	66.21	4 %	1.0 %	15 %	0.0 %
Year 3	65.55	66.48	62.83	65.23	56.43	66.48	4 %	1.9 %	14 %	0.0 %
Year 4	92.17	66.00	90.26	64.95	80.04	66.08	2 %	1.6 %	13 %	−0.1 %
Year 5	85.10	66.03	82.91	64.71	72.34	66.08	3 %	2.0 %	15 %	−0.1 %
Year 7	173.89	66.15	169.50	64.41	145.26	66.16	3 %	2.6 %	16 %	0.0 %
Year 10	258.49	66.71	248.46	65.30	221.01	66.69	4 %	2.1 %	14 %	0.0 %
Year 15	532.79	66.85	516.89	65.55	464.65	66.83	3 %	1.9 %	13 %	0.0 %
Year 20	441.55	66.90	421.81	65.44	385.73	66.94	4 %	2.2 %	13 %	−0.1 %
Total	1730.01	66.63	1665.72	65.27	1487.48	66.65	4 %	2.0 %	14 %	0.0 %

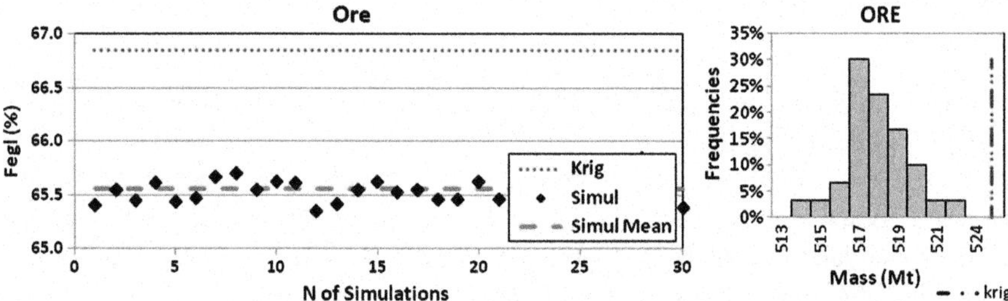

Fig. 5 Distribution of 30 simulations for iron grade (*left*) and total mass (*right*). The *solid line* represent the kriging values (grade/mass) of deterministic block model—mining plan 20

In the mining plan, the equipment was projected for friable material (mainly hematites) and compact material (waste), with low operational flexibility. Variability in terms of hardness (peaks of jaspilite, for example), will affect the operational life time of this equipment. The projected equipment to work with friable material has a maximum continuous work time of 24 hours in compact material, after this period of work the abrasion of the equipment is elevated. Considering the jaspilite density of 3.3 g/cm^3, the frequent occurrence of peaks of this rock with an overall dimension greater than 13 blocks will affect the considered OPEX (Operational Expenditure) index. A simulation is an excellent tool to indicate regions of greater probability of occurrence of those jaspilites peaks.

The product of geostatistical simulation could be a great tool to predict the occurrence of waste between the existing drilling grid, thus enabling the creation of an optimal grid for a region with a greater risk and also enabling mining plan revisions in order to smooth the dilution effect.

10 Conclusion

The uncertainty of the lithological contacts among hematite and waste/low grade rocks is as important as the ore grade variability. For this study, drillhole database (hard data) and section interpretation (soft data) were used, in order to improve the lithological conditioning. Different geostatistical simulations were combined to generate equiprobable realizations for lithologies which are derived from superimposing different geological events; supergene and sedimentary/volcanic rocks. The grade variables were simulated in cascade, one grade simulation for each lithology simulation. The results of the simulations were compared with the official reserve calculated using the deterministic block model in order to measure the project risk and the impact of waste dilutions in the quality of ore product.

The obtained results were good once the simulations honored the established stratigraphy and no extrapolation was observed. The product of geostatistical simulation could be a great tool to predict the occurrence of waste in between the existing drilling grid, thus enabling the creation of an optimal grid for regions with greater risk, and also enabling mining plan revisions in order to smooth the dilution effect.

The internal dilutions and ore loss were evaluated through the methodology proposed: combining lithological and grade simulations and evaluating the impact in ore reserve model for each scenario. The diluted ROM iron grade should be 2 % lower, on average, than the respective grade for the official ore reserve.

The ore reserve mass could be reduced in 14 % if the ore exploitation would be restricted to only hematite ore blocks with low geological risk. To ensure the iron grade of the official reserve (66.7 %), this option had low probabilities to be applied because the adopted mining methodology, truckless, had no operational flexibility to deal with geological uncertainty.

Future studies, such as considering other waste rock materials and optimizing the time consumption of the simulation processing, will contribute to enrich the level of application of the proposed methodology as a routine.

References

1. Alabert F (1987) Stochastic imaging of spatial distributions using hard and soft information. MSc thesis, Stanford University, Stanford, 197 pp
2. Cárceres A, Emery X (2010) Conditional co-simulation of copper grades and lithofacies in the Río Blanco—Los Bronces copper deposit. In: Proceedings of the 4th international conference on mining innovation—MININ 2010, Santiago, pp 312–320
3. Deutsch CV, Journel AG (1997) GSLIB: geostatistical software library and user's guide. Applied geostatistics series, 2nd edn. Oxford University Press, London, 369 pp
4. Dimitrakopoulos R (1998) Conditional simulation algorithms for modeling ore body uncertainty in open pit optimization. Int J Surf Min Reclam Environ 12:173–179
5. Emery X, Lantuéjoul C (2006) TBSIM: a computer program for conditional simulation of three-dimensional Gaussian random fields via the turning bands method. Comput Geosci 32(10):1615–1628
6. Freulon X, de Fouquet C, Rivorard J (1990) Simulation of the geometry and grades of a uranium deposit using a geological variable. In: Proceedings of the XXII international symposium on application of computers and operation research in the mineral industry, Berlin, pp 649–659

7. Journel AG (1989) Fundamentals of geostatistics in five lessons. Short course in geology, vol 8. American Geophysical Union, Washington, 91 pp
8. Lantuéjoul C (2002) Geostatistical simulation: models and algorithms. Springer, Berlin, 256 pp
9. Peroni RL (2002) Análise da sensibilidade do seqüenciamento de lavra em função da incerteza do modelo geológico. Tese de Doutorado, Universidade Federal do Rio Grande do Sul, Brasil, 126 pp

The Use of Geostatistical Simulation to Optimize the Homogenization and Blending Strategies

Diego Machado Marques and João Felipe Coimbra Leite Costa

Abstract Homogenization piles are largely used in the mining industry for variability reduction in the head grades feeding the processing plants. Various methods are found for homogenization piles design and most fail to incorporate the in situ grade variability intrinsic of a mineral deposit. The methodology proposed combines longitudinal piles and geostatistical simulation to emulate the in situ and the pile reclaimed grade variability. Variability reduction in large piles is based on the volume-variance relationship, i.e. the larger is the support the smaller is the variability. Based on a pre-defined mining sequence to select the blocks that will form each pile for each simulated block model, the statistical fluctuation of the grades derived from real piles can be simulated. These piles are characterized by their form, size (length and height) and number of layers. Using this methodology, one can evaluate within a certain time period the expected grade variability for various pile size and also the internal grade variability when a given pile is reclaimed. Results from a case study at two large iron mines operated by Vale proved the adequacy and functionality of the method. It is demonstrated the rate of variability decrease as the pile size increases and the internal grade variability to a given pile size, with different numbers of layers.

1 Introduction

ROM (Run of Mine) from the mine may vary in their characteristics due to the heterogeneity of the materials that compose it. Moreover, the heterogeneity of the material can be affected by the methods of loading and transporting ore and also on the procedures used in handling and preparation. According to [9], the steps of loading, transport, crushing, milling, storage and handling contribute somehow in

D.M. Marques (✉) · J.F.C.L. Costa
Federal University of Rio Grande do Sul, Av. Bento Gonçalves, 9500, Bloco IV, Prédio 75, Sala 101, Agronomia, Porto Alegre, RS 91501-970, Brazil
e-mail: diegommarques@yahoo.com.br

J.F.C.L. Costa
e-mail: jfelipe@ufrgs.br

P. Abrahamsen et al. (eds.), *Geostatistics Oslo 2012*,
Quantitative Geology and Geostatistics 17,
DOI 10.1007/978-94-007-4153-9_35,

the variability of the material, since they introduce a certain amount of restructuring of the spatial distribution of natural mineralization.

Blending piles play an important role on this task, as they substantially reduce the inherent variability of the deposit, even when mining follows a schedule defined along a stationary mining route [8]. Homogenization piles can be used as part of a system of quality control in order to reduce the variability of the material in the processing plant. For this system work properly, it is needed a measure on the quality and variability of the material forming the piles. This material is blended to deliver a product when reclaimed with less variability according to technical parameters required by the downstream processes. One of the simplest methods to reduce the variability of ore for a particular property is the construction of longitudinal piles, using elementary layers along the axis of the piles. Some previous works involving blending and homogenization piles and geostatistical simulations can be found in [1, 4, 7, 9]. The use of geostatistical simulation methods aim at reproducing in-situ variability, respecting the spatial continuity of a mineral property within an ore deposit. This study uses conditionally simulated model to reproduce the statistical moments of the original data set, i.e. their mean, variance and semi variograms. In a conditionally simulated model questions concerning the dispersion of the grades during mining or processing can be addressed, since the dispersion characteristics of the original data are maintained. As the spatial continuity and variability of the real deposit is properly determined, quality reliability in the numerically simulated model is obtained, helping to assess potential risks during the decision making process [8].

This work presents a study developed in two iron ore deposits, located at the Iron Quadrangle region in Brazil. The deposits are part of the Vargem Grande complex and are operated by Vale. The aim is to investigate the in situ grade variability obtained during mining and define an optimal pile dimension and geometry to attenuate this variability. The variable silica (SiO_2 %) was chosen for this analysis as it as constitutes the most critical contaminant for this iron ore.

2 Methodology

One of the most relevant parameters to be controlled in an iron mine is the silica content in the product. This parameter was chosen as it exhibits an erratic behavior (high variability) which need to be kept under a strict control below a maximum limit. The principles of geostatistical simulation were originally presented in [2]. In his original work Turning Bands Method (TBA) was presented and was used to obtain the grade simulations on this study. A 3D grade block model was built, where the algorithm generates multiple equally probably scenarios for the grade spatial distribution reproducing statistically their spatial continuity and histogram. Each of these equally probable 3D block models were submitted to a mine scheduling algorithm sequencing the order each block is mined. The steps involved are in this study are:

1. Generation of multiple equally probable 3D models using the Turning Bands Method (TBA) for both deposits.
2. Sequencing the simulated 3D block model according to the year 2004 production scheduled used at each mine.
3. Emulating the feeding of the homogenization piles (combination the blocks from the two mines).
4. Calculate the variability among piles for a given selected pile size for each 3D simulated block mode.
5. Calculate the variability during the reclamation of each pile formed (variance of the grades of each slice reclaimed).

To implement the algorithm which emulates homogenization piles, some considerations are necessary:

1. The model is based on configurations of longitudinal piles (chevron type piles).
2. The geometry of the cones at the ends of the piles is not considered.
3. Stacking of the material is performed along both directions of the moving stacker.
4. The stacker discharges a constant amount of material during its movement.
5. Recovery of the material is done perpendicular to the ground.
6. The stack recovers slices of one meter thick.
7. Segregation of the material was disregarded.
8. The particles that compose the block are infinitely small compared to the mass of the block.

The simplifications made in the settings of the algorithm do not significantly affect the calculation [4]. According to [3], two basic aspects should be considered in the size of the facilities of homogenization piles:

- Economic: it is desirable that equipment used for stacking has the lowest possible capital cost.
- Process: the material taken from the pile must have a standard deviation of the grades compatible with that required by the posteriors process.

Satisfying these two conditions, one can say that the system is satisfactory for homogenizing the ore. A detailed description on the algorithm used in this study is found in [4].

3 Case Study

3.1 Assessing Block Grade Uncertainty

The data used come from 400 vertical drill holes, which have their support regularized at 10 m composites, resulting in 2353 samples for mine 1 and 337 vertical drill holes, also with support regularized at 10 m, resulting in 2582 sample for mine 2. The conditional geostatistical simulations (TBA) had been applied for generating

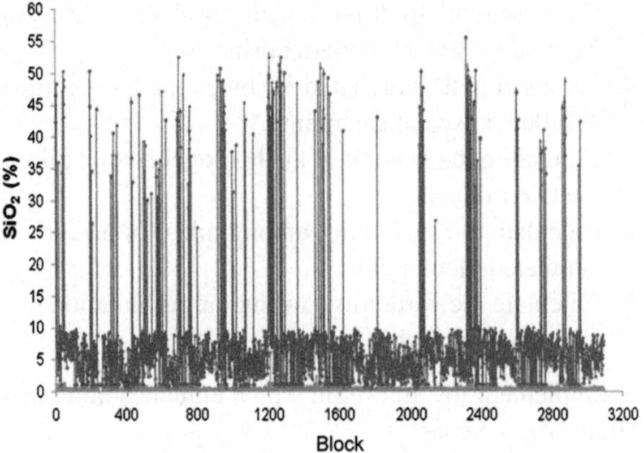

Fig. 1 Silica grades in a time series format for the blocks from mine 1. The *blue line* represents the lowest and *red* the highest simulated silica grades

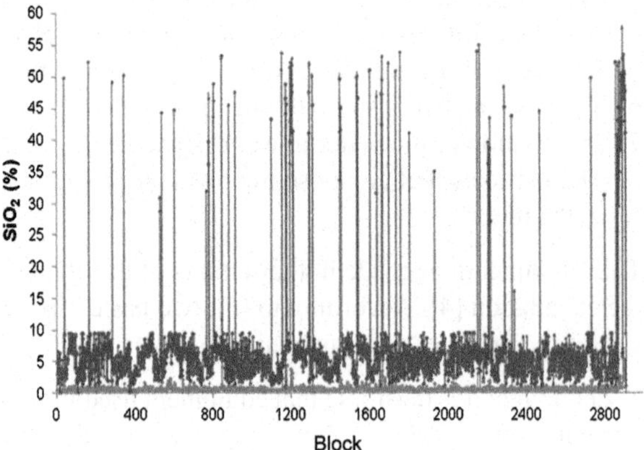

Fig. 2 Silica grades in a time series format for the blocks from mine 2. The *blue line* represents the lowest and *red* the highest simulated silica grades

of 50 equiprobable scenarios of SiO_2 spatial distribution at (10 m × 10 m × 10 m) 1000 m^3 blocks.

In Figs. 1 and 2 we display the time series grades for each block minded sequentially along a year (as the historical data of the company). The band of uncertainty associated with each block grade (max and min grades) is obtained through the simulated grades for mine 1 and mine 2, respectively. The maximum (red line) and minimum (blue line) grade values of silica (SiO_2) for each block is derived from the 50 equally probable scenarios ordered according the mine scheduling.

Combining the block sequence from the two mines provide the grades which will feed the blending piles, showed in a time series format at Fig. 3. Note the uncertainty in the grades (max-min) and their fluctuation along the year. This time series values are required as the input to the algorithm used to emulate the homogenization piles allowing an assessment on the reduction of variability.

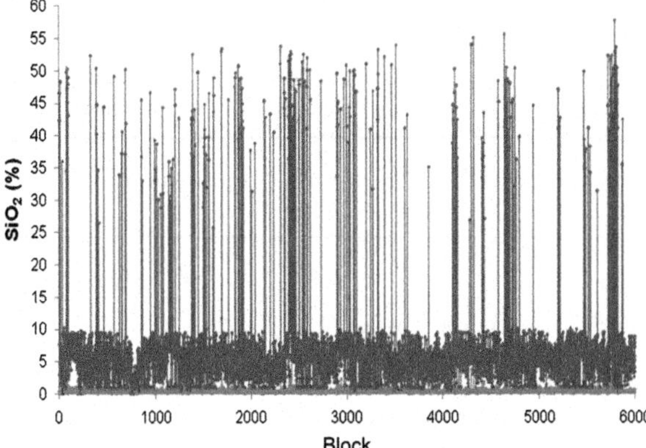

Fig. 3 Time series of the block grades feeding the homogenization piles. The *blue line* represents the lowest and the *red* the highest grade of silica

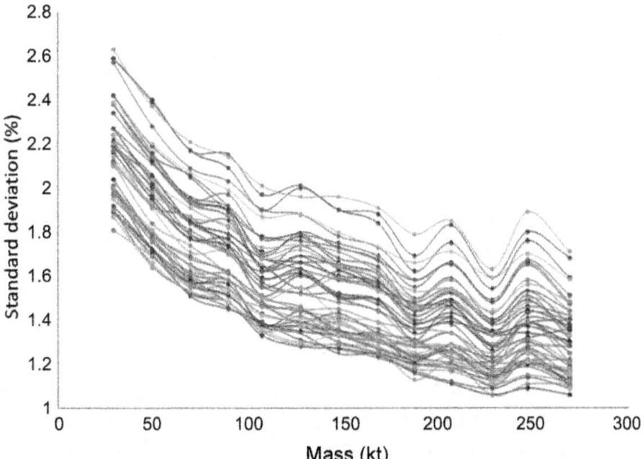

Fig. 4 Reduction of variability due to the increase of mass for 50 equiprobable scenarios

3.2 Inter Pile Grade Variability

Checking the fluctuation on the grades plotted at Fig. 3, one can understand why the need of homogenization piles. Note the high variability on silica grades from mine 1 and 2. By feeding directly the processing plant with this material would make the subsequent processes inefficient with high financial losses. To emulate how the homogenization piles could reduce the variability, a simulation process was defined with the following parameters:

- Initial mass of 30 kt (smallest pile).
- Final mass of 270 kt (largest pile).
- Increments of mass between pile to be tested 20 kt.

All 50 simulated block models were tested by the emulation algorithm. Figure 4 shows the graph of variability reduction as the size of the pile increases. Fifty curves were plotted each related to a simulated grade model.

Table 1 Reduction in grade variability for two scenarios: one the largest (26) and the one with the smallest reduction in variability (49) for 210 kt piles

Equiprobable scenarios	Standard deviation of input data (%)	Standard deviation of homogenization piles (%)	Reduction of variability (%)
26	4.56	1.12	75.6
49	4.90	1.71	65.1

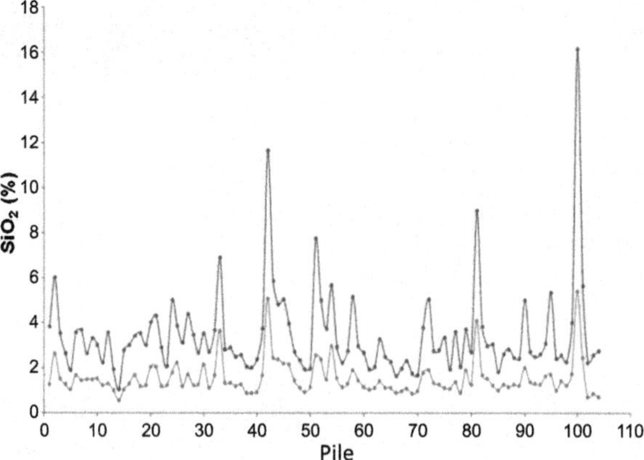

Fig. 5 Time sequence for the grades from 210 kt piles assembled along a year. The *blue line* represents the lowest possible grades and *red* the highest grade of silica, obtained from 50 realizations

Figure 4 summarizes the benefits by reducing the variability of the head grades obtained using different homogenization pile sizes, and simultaneously presents the expected variability of the head grades due to in situ grades uncertainty. For instance, a 50 kt pile will lead to head grades standard deviation in the 1.6–2.5 range (min-max variability). These results reproduce similarly a plot built for a different iron deposit [1]. The curve suggests an exponential decay in grade variability with an increase in pile size. This reproduces the volume-variance relationship discussed in [6]. Consequently, the grades measured at the reclaimed ore from the piles have a reduced variance in comparison to the grade variability that would be obtained from a block by block scheme feeding the processing plant.

Table 1 shows the reduction on the variability for two simulated scenarios of fifty (the ones with the highest and smallest reduction in grade variability) for a 210 kt (selected size for this mine site). These variability reduction forecasts obtained by combining in situ block grades simulation using geostatistical simulation and the chevron blending pile emulator reproduces production historical data [5]. This match between the model forecasts and real historical data demonstrates the efficiency of the methodology proposed.

Figure 5 shows a time series for the maximum and minimum grades for all 210 kt homogenization piles planned for one year (2004). Note a significant reduction on the width of possible grades interval compared to a block by block grade variability (Figs. 1 and 2).

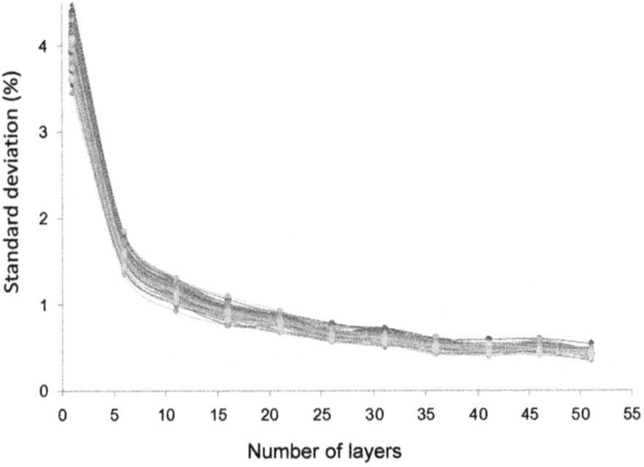

Fig. 6 The effect of the number on grade variability reduction for 210 kt piles

3.3 Intra Pile Variability

In this mine site, a homogenization pile of approximately 200 kt takes about three days to be formed. Because of the speed of this assembly process it is not possible to form piles with a big number of layers. For analyzing the internal variability of the grades within each one of the homogenization piles the process was emulated for various number of layers. The following parameters were used:

- Mass of the homogenization pile of 210 kt.
- Volume of material discharged by the stacker along each meter of the trajectory is 4.5 m^3 per meter.
- Bulking factor of the material: 1.3.
- Angle of repose for the material: 35 deg.
- Width of the pile base: 38 m.
- Initial number of layers: 1.
- Increasing steps in the number of layers: 5.
- Final number of layers: 150.

For this analysis the pile emulator mimics the heterogeneity distribution, i.e. the way the material is distributed within the pile. Each zone of the stack is different from another (in terms of grades), and are different from the average content of the stack. Organizing the lots (or blocks) along the pile in different layers leads to the reduction on grade variation for the materials reclaimed [1]. As the number of layers increases for a given pile the variability in the reclaimed material is reduced. This effect is due to fact that for a large amount of layers the final pile format tend to be formed by similar material along each vertical slice (Fig. 6).

Despite we had set the algorithm to calculate the reclaimed variability to up to 150 layers, it was noticed that at 51 layers the variability reaches its minimum asymptotically. Table 2 explains the geometric consequences of constructing 210 kt with different number of layers.

After 35 layers the gain in reduction of variability is not significant, however by increasing the number of layer the length of the stack can be reduced considerably.

Table 2 Average length of the homogenization piles for different numbers of layers

Number of layers	Average length of the piles (m)
1	17029.6
6	2837.8
11	1547.7
16	1064.1
21	810.4
26	654.5
31	548.8
36	472.6
41	414.7
46	369.7
51	333.4

Thus, an adequate solution would be the use of an equipment with higher flow discharge forming 210 kt piles with approximately 35 layers. This solution leads to suitable stock yard length dimensions with a minimized head grade variability feeding the processing plant.

4 Conclusions

The operation of a homogenization system brings different gains in reducing variability, depending on the pile size and number of layers selected. The parameters of pile formation (size and number of layers) should be carefully chosen in order to satisfy the maximum grade variation allowed in the downstream processes. The algorithm designed to predict the variability in the homogenization piles reflects in their results the uncertainty associated with the simulations. To obtain reliable results, should ensure the proper implementation of geostatistical simulation procedures. The largest amount in variability reduction from a homogenization system is associated with the increase in the mass of the pile (inter pile grade variability). However, the number of layers should be also considered in the formation of a pile homogenization, as a pile with a high mass but not properly mounted will not bring the expected reduction in variability (even the reduction of variability obtain in the inter pile could be lost). The effect of homogenization is best obtained using the greatest number of pile layers, having in mind that after a certain number of layers the reduction of variability is practically nil.

Acknowledgements CNPq (Brazilian research agency) is acknowledged for supporting scholarships to students associated with this project. Vale is acknowledged for supporting the research team and to provide the dataset and all information required for this study.

References

1. Costa JFLC, Koppe JC, Marques DM, Costa MSA, Batiston EL, Pilger GG, Ribeiro DT (2007) Incorporating in situ grade variability into blending piles design using geostatistical simulation. In: Proceedings of the third world conference on sampling and blending, pp 378–389
2. Journel AG (1974) Geostatistics for conditional simulation of ore bodies. Econ Geol 69(5):673–687
3. Ferreira FM, Chaves AP, Delboni H (1992) Conditional simulation method for design of blending piles. In: Proceedings, 23th international symposium on computer applications in the mineral industries, Phoenix, USA, pp 615–623
4. Marques DM (2010) Desenvolvimento de um algoritmo para simular a variabilidade do minério em pilhas de homogeneização. Dissertação de mestrado, PPGEM UFRGS
5. Marques DM, Costa JFCL, Ribeiro DT, Stangler RL, Castro EB, Koppe JC (2010) A comprovação da relação volume x variância na homogeneização da sílica em minério de ferro. Rev Esc Minas 63(2):355–361
6. Parker H (1979) The volume variance relationship: a useful tool for mine planning. Eng Min J 180:106–123
7. Ribeiro DT, Ezawa L, Moura MM, Pilger G (2007) PDFseq: mine scheduling simulation based on block probability distribution function: an iron ore application. In: Proceedings of the 33rd APCOM—international symposium on the application of computers and operations research in the mineral industry, Santiago, pp 369–374
8. Ribeiro DT, Roger LS, Vidigal M, Costa JFCL, Marques DM (2008) Conditional simulations to predict ore variability and homogenization pile optimal size: a case study of an iron deposit. In: Proceedings og the eighth international geostatistics congress, Santiago, pp 749–758
9. Schofield CG (1980) Homogenisation/blending systems design and control for minerals processing. TransTech Publications, Rockport

Plurigaussian Simulations Used to Analyze the Uncertainty in Resources Estimation from a Lateritic Nickel Deposit

Diego Machado Marques, Edgar Mario Müller, and João Felipe Coimbra Leite Costa

Abstract Mining industry increasingly uses simulations to predict fluctuations in ore grades. However, in some cases grade fluctuations analysis is not enough for risk assessment, as in some cases ore recovery at the processing plant depends on the rock type rather than the grades. Lateritic nickel weathering profiles are highly complex in its composition. The dataset used in this study is divided into five geological domains, which present in many locations changes in the sequence or lenses of different material trapped inside a major geological domain. It is common the absence of certain domains at some borehole intercepts. The use of a deterministic or interpreted model for defining the orebody contacts in this dataset is unsatisfactory. Instead to use deterministic models to define this orebody, based only in the interpretation of the drill holes (which does not take into account the uncertainty of lithological domains in areas with no information), this paper presents plurigaussian simulation as an alternative to generate equally probable scenarios from the orebody. Combining these realizations it is possible to map the uncertainty in the volumes and contacts. The simulation results showed good reproduction of the conditioning data, both laterally and vertically. Thus, the result of each simulation is also a possible geological model, presenting the spatial variability of the phenomenon, without the smoothing shown by other techniques. One advantage in using plurigaussian simulation for lateritic nickel profiles is the reduction of working time in a very complex geological model, using the simulations to mimic the reality and producing a set of different equally probable scenarios allowing to access the uncertainty in defining the geological domains.

D.M. Marques (✉) · E.M. Müller · J.F.C.L. Costa
Federal University of Rio Grande do Sul, Av. Bento Gonçalves, 9500, Bloco IV Prédio 75, Sala 101, Agronomia, Porto Alegre, RS 91501-970, Brazil
e-mail: diegommarques@yahoo.com.br

E.M. Müller
e-mail: edgarmmuller@gmail.com

J.F.C.L. Costa
e-mail: jfelipe@ufrgs.br

P. Abrahamsen et al. (eds.), *Geostatistics Oslo 2012*,
Quantitative Geology and Geostatistics 17,
DOI 10.1007/978-94-007-4153-9_36,

1 Introduction

Mine planning is traditionally based on a geological model of the deposit. As most of the mining decisions are made considering this model, it is necessary to be as accurate as possible in order to avoid unexpected surprises during mining. For this reason, it would be worthy to use all type of information which can improve the geological model. Traditionally, data collected from drill holes are preferentially used to estimate the volume and shape of the orebody. However, the information is limited to the sample location. The definition of the orebody in regions without information must be inferred, and in most cases underestimated the heterogeneity and variability of the geological contacts. Simulation techniques can be used as auxiliary tools for creating a geological model, each scenario is a realization of a possible representation from the orebody. Additionally, with an adequate number of realizations, it is possible to map the uncertainty associated with the geological model suggested, being possible to assess fluctuations on the volume and quality of ore that will be feed to the processing plant. Some methods have been developed over the years for simulation of facies and lithotypes, among them plurigaussian simulations [2]. As lateritic nickel profiles are highly complex, using only traditional techniques of interpreting the drill holes requires same geological simplifications, and in various cases smaller geobodies such as lenses of different material trapped inside a major geological domains sometimes are ignored.

2 Methodology

This study proposes plurigaussian simulations to simulate the geometry of the weathering domains along a complex orebody. The first step in a Plurigaussian simulation consists of calculating the proportions of each facies. But even before doing this, one has to make a series of preliminary choices. The lithotypes to be modeled are selected dividing the orebody into units if necessary, and choosing the reference levels to be used for the flattening each unit. Next, the parameters of the grid to be simulated are chosen [1]. The steps comprise:

- Define the lithotypes.
- Divide the orebody into units.
- Define the reference level.
- Choose the grid spacing.

The main steps in a plurigaussian simulation include:

- Estimating the parameter values: definition of the proportion of each facies (rock type), and determination of the thresholds, variogram modeling for the underlying Gaussian functions.
- Generating Gaussian values at wells/drill-holes.
- Simulating values at grid nodes given values at wells.

All these steps will be detailed below.

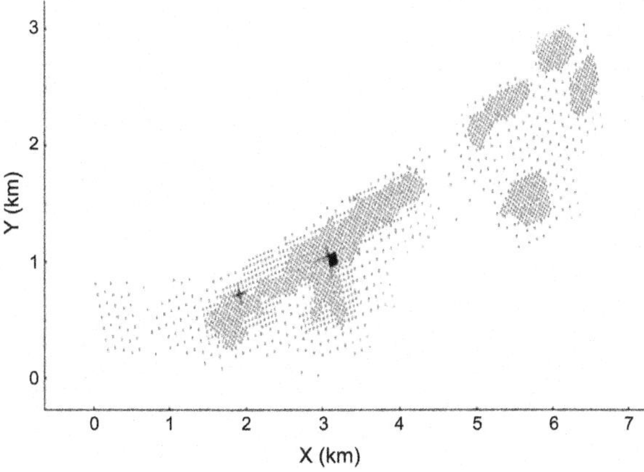

Fig. 1 Base map of the collars of the drill holes (2 132 data points) along the XY plan

3 Model Specification

The data set used in this study comprises 2 132 drill holes logs with 50 772 samples. The vertical drill holes are distributed in an area of approximately 11 km^2 (Fig. 1). The sampling grid spacing is approximately 35 m in the denser area and approximately 100 m in the remaining parts of the deposit. The lithologies were grouped according to their characteristics into five major groups (typical lateritic profile). These five geological domains will be used along this study and are:

- Limonitic Zone (LIM): primary textures absent with limonite (goethite, hematite, etc.) increase toward the top, becoming pisolitic. It becomes hard when exposed at the surface (yoke). Can show silicification. Retains some water in the voids.
- Transition Zone (TRN): textures disappear and the volume starts to get depressed.
- Saprolite (SAP): textures and structures are preserved. Presence of most clay minerals, but devoid of structures. It looks like a "house of cards" with many empty spaces, but it maintains the original volume. Retains water in the voids up to 40 percent (or more) weight.
- Boulders zone (BLD): mixture of partially weathered saprolite with original rock. All textures and structures are preserved.
- Bedrock (BRK): fresh rock weathered (lightly).

The definition of the lithotypes (domains) for this orebody and the numerical identification for the study are presented in Table 1. A typical lateritic profile should present the same disposal (sequence) as the shown in Table 1 (from top to bottom), but in this dataset occurs in many locations changes in the sequence or lenses of different material trapped inside a major geological domain. Frequently occurs the absence of certain domains at some borehole (heterotopic dataset). So, in this case, is not possible to define regular layers of the domain. The company uses a simpler model, using only three main layers (BRK, SAP, and LIM), obtained by combining a few layers, and avoid crossing them. This simplified model is used in the estimates and will be used for comparison with the results of the simulations.

Table 1 Numerical code and identification color for the samples in each domain of the dataset

Domain	Code
LIM	1
TRN	2
SAP	3
BLD	4
BRK	5

Table 2 Proportion of the data (samples) between the TOP and BRK surfaces

Domain	Proportions (%)
LIM	23.2
TRN	2.6
SAP	51.7
BLD	16.4
BRK	6.0

3.1 Reference Level

A key step in a study using Plurigaussian Simulation is the definition of the possible flattening surface (reference surface). The shape of the vertical proportion curve and the resulting simulations depend of the reference level [1]. Common choices for the reference level are the top or the bottom of a geological unit. For this work, it was tested three different surfaces and the results derived were compared. The surfaces used are top of the SAP domain (2 107 samples), top of BRK domain (2 107 samples) and the top of the orebody (TOP) (2 132 samples). For this, a 2D (along X and Y) dataset has been created, where the variable of interest is the highest value of elevation at each drill hole for the domains SAP, BRK and the top of the orebody (TOP), regardless of being a lense or a major domain. Since these surfaces were not defined within the entire area, it is required an estimation 2D grind of these surfaces using the data available.

3.2 Discretization and Flattening

A unit is a homogeneous stratigraphic formation possibly bounded by two surfaces top and bottom which can be flattened to work in the original stratigraphic system to get better correlations between the different lithofacies. Three discretizations were performed (one for each surface used for reference). The TOP surface will be used as the upper limit of the ore body and the surface of BRK will be the lower limit. The final proportion of the data selected between the surfaces can be seen in Table 2 (independent of the reference surface). After the process of simulation, the result of the simulation are back transformed in to 3D structural grid, constrained to the TOP and BRK surfaces.

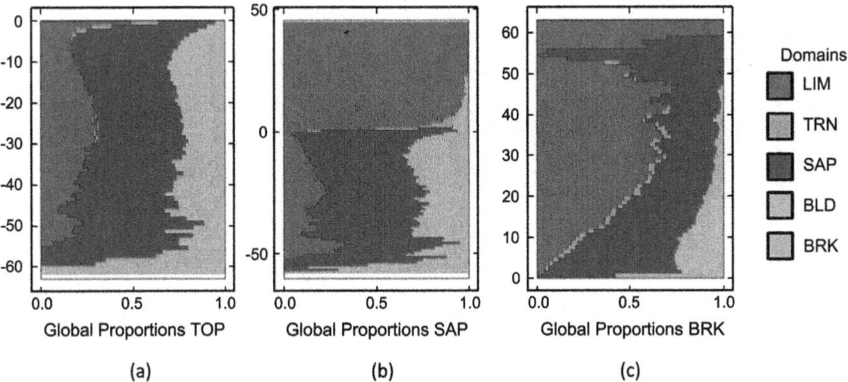

Fig. 2 Vertical proportion curves for (**a**) top of TOP, (**b**) top of SAP domain and (**c**) top of BRK domain

3.3 Vertical Proportion Curves

Vertical proportion curves (VPC) was first proposed by Matheron [3]. The VPC are a simple tool for quantifying the evolution in the amount of each facies or lithotype present as a function of depth. The vertical proportion curves for facies are computed along lines parallel to a given reference level, preferably a chronostratigraphic marker, not a unconformity or dispute, providing the average proportion of each facies at that level. The result is a graph showing the proportion of facies as a function of depth, which is very sensitive to the choice of reference level. For oil reservoirs this level is a specific geological marker which is used to restore the geometry of the reservoir at the time of deposition. This level must have been horizontal during the sedimentation, and, should, if possible, correspond to a time line. That is, it should be a chronostratigraphic marker not an erosional unconformity. The reservoir is then flattened using this as the reference level [1]. Common choices for the reference level are the top of the unit or the bottom. In this study, three different surfaces were tested to generate the global VPCs. Analyzing Fig. 2, it can be noted the different behavior of VPCs for the different reference surfaces tested. The best choice is the surface constructed from the top of BRK domain (Fig. 2c), as this surface shows a behavior similar to a typical lateritic profile, with LIM at the top, follow by the TRN, SAP, BLD and BRL at the bottom.

The analysis of stationarity on the VPC was performed. To investigate if orebody proportions are non-stationary (in a geological point of view); the study area was divided into six regions. Along each region it calculated the VPC, and compared against global VPC. Local proportions match the global.

3.4 Truncation, Thresholds and Variography of the Indicators

In this study, three different attempts of partition (lithotype rule) were tried. The first (Fig. 3a) follows the order of a theoretical disposal of lateritic nickel profile

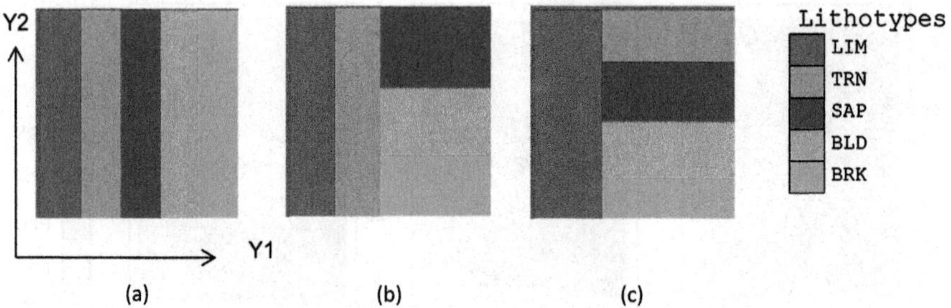

Fig. 3 Truncation of the facies correlated with the geology of the database. (**a**) The first attempt follows the order of arrangement of layers, (**b**) the second attempt consider the TRN as an extension of the LIM and (**c**) the third attempt considers a possibility of contact between LIM with all layers (which follow the theoretical disposal)

(LIM, TRN, SAP, BLD, BRK). In the second attempt (Fig. 3b), the transition zone (TRN) is considered as an extension of the limonite zone (LIM) and it is allowed possible contact with other domains (which present the theoretical disposition of the layers). The third attempt (Fig. 3c) considers a possibility of contact between the limonite zone (LIM) with all layers (which follow the theoretical disposal). For calculating the experimental indicator variograms, the main directions in the horizontal plane are aligned with the main directions of the orebody. The lag used was 70 m. The vertical variogram was calculated with a 1 m lag. Figure 4 shows the experimental variogram along the XY plane at N67E, at N157E and the experimental variogram along the vertical direction. The experimental vertical variogram shows a certain differentiation in the spatial structure in the limonitic domain if compared to the others, which suggests that one should use a Gaussian model unique to that domain. The other domains have a similar spatial structure (differing only in their *sill* relationship), which suggests that they can share a second Gaussian model. The three Lithotype Rules shown in Fig. 3 were tested to fitting a model to the experimental variograms. The best results were obtained with the Lithotype Rule present in Fig. 3c. The result for the variogram model can be seen in Fig. 4. Define the lag vector $\mathbf{h} = (h_1, h_2, h_3)$ where h_1 and h_2 are the distances measured along the direction N67 and N157 respectively, and h_3 is the vertical distance. The resulting parametric variogram for the first Gaussian models is

$$\gamma_1(\mathbf{h}) = 0.33 \times \mathrm{Sph}\left(\frac{h_1}{20\ \mathrm{m}}, \frac{h_2}{20\ \mathrm{m}}, \frac{h_3}{4\ \mathrm{m}}\right) + 0.67 \times \mathrm{Sph}\left(\frac{h_1}{175\ \mathrm{m}}, \frac{h_2}{80\ \mathrm{m}}, \frac{h_3}{50\ \mathrm{m}}\right), \tag{1}$$

and for the second Gaussian model the resulting parametric variogram is

$$\gamma_2(\mathbf{h}) = 0.6 \times \mathrm{Sph}\left(\frac{h_1}{20\ \mathrm{m}}, \frac{h_2}{20\ \mathrm{m}}, \frac{h_3}{5\ \mathrm{m}}\right) + 0.4 \times \mathrm{Sph}\left(\frac{h_1}{100\ \mathrm{m}}, \frac{h_2}{60\ \mathrm{m}}, \frac{h_3}{45\ \mathrm{m}}\right). \tag{2}$$

Here “Sph” denotes the standard spherical variogram. Now we have all the necessary information required, the process of simulation can be started.

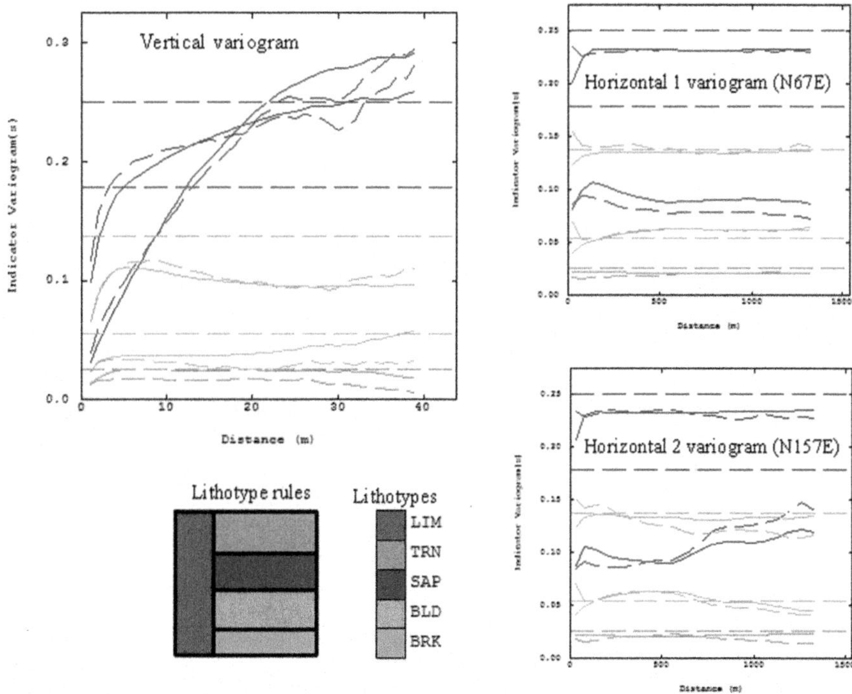

Fig. 4 Variograms of the indicators (for each domain). The *dashed lines* correspond to experimental variograms and the *continuous lines* corresponding to the variogram models

Table 3 Proportions of the facies simulated for realization 6, 23, 37 and 42 in %

Domains	Realization 6	Realization 23	Realization 37	Realization 42
LIM	25.12	23.15	24.27	24.11
TRN	2.30	2.26	2.21	2.35
SAP	47.87	49.12	48.16	47.98
BLD	16.23	16.76	16.53	16.65
BRK	8.48	8.71	8.83	8.91

4 Results

Four out fifth simulations were selected to be presented in Table 3 (realizations 6, 23, 37 and 42). The results of the simulation process are consistent with the proportions of the original data (Table 1). The simulation results in the block model (blocs with size of $10 \times 10 \times 1\ m^3$) can be seen in a vertical cross section, between the point AB (indicate in Fig. 5) in Fig. 6. The result of the simulations in Fig. 6 presents approximate the same stratification of the deposit, following the classical sequence of a laterite deposit but partially hidden by the complex shape of the geological model.

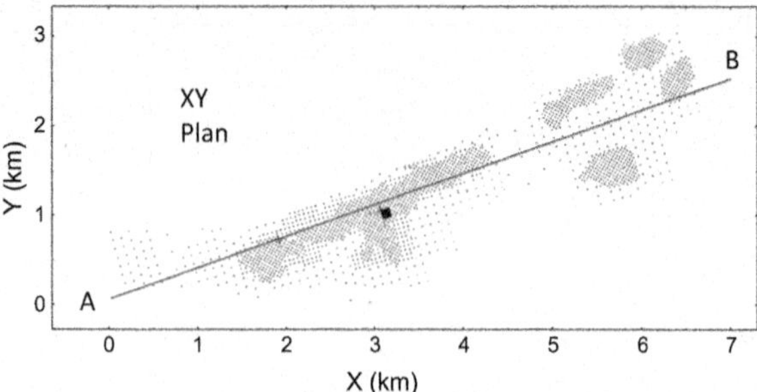

Fig. 5 Position of the cross section in the XY plan

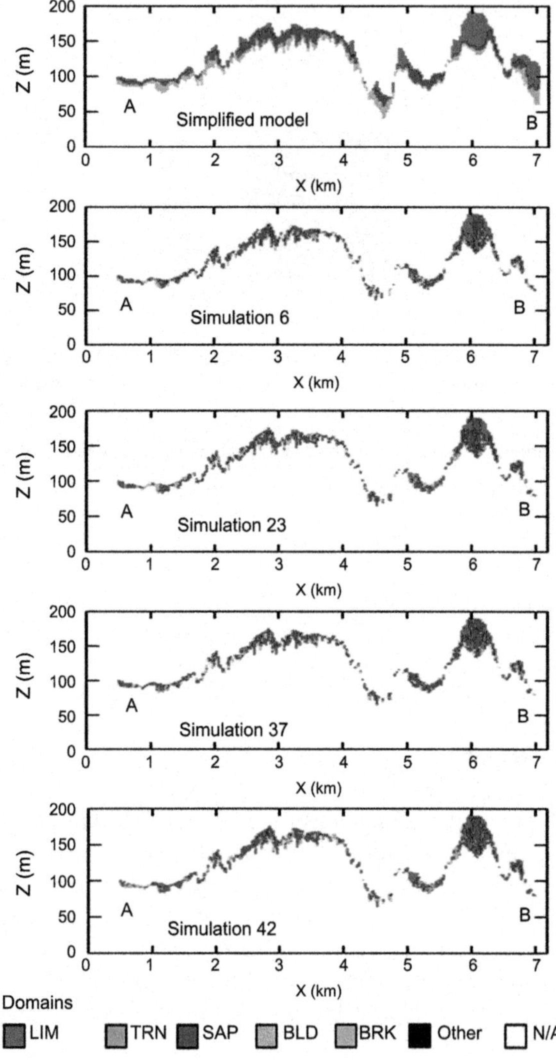

Fig. 6 Comparison between the simplified model used by the company for the estimatives and 4 different equiprobable scenarios obtained by for plurigaussian simulation

5 Conclusions

The generation of a 3D geological model as a possible representation of the orebody is required for all decision making process along the life of the mine. Each simulation provides an equally probable geological model, reproducing the data variability for the geological boundaries. These simulations aim at reproducing the spatial variability of the data and, in some cases, the connectivity features of the subdomains within the orebody. Using simulation models allows flexibility to evaluate multiple equiprobable scenarios and therefore analyzing the extension and continuity of the geological domains. Plurigaussian simulations are very sensitive to the model parameters and their choice impacts significantly the results. The calculation of the vertical proportion curves is extremely sensitive to the choice of reference surfaces.

Acknowledgements CFSG and Gaëlle Le Loc'h are acknowledged for help to development of this case study.

References

1. Armstrong M, Galli A, Le Loc'h G, Geffroy F, Eschard R (2003) Plurigaussian simulations in geosciences. Springer, Berlin, 144 pp
2. Galli A, Beucher H, Le Loc'h G, Doligez B, Heresim Group (1994) The pros and cons of the simulation gaussian method. In: Armstrong M, Dowd PA (eds) Geostatistical simulation: proceedings of the geostatistical simulation workshop, Fontainebleau-France, May 27–28, 1993. Kluwer Academic, Dordrecht, pp 217–233
3. Matheron G, Beucher H, de Fouquet C, Galli A, Ravenne C (1987) Conditional simulation of the geometry of fluvio-deltaic reservoirs. In: SPE annual technical conference and exhibition, Dallas, 27–30 September, pp 591–599

Domaining by Clustering Multivariate Geostatistical Data

Thomas Romary, Jacques Rivoirard, Jacques Deraisme, Cristian Quinones, and Xavier Freulon

Abstract Domaining is very often a complex and time-consuming process in mining assessment. Apart from the delineation of envelopes, a significant number of parameters (lithology, alteration, grades) are to be combined in order to characterize domains or subdomains within the envelopes. This rapidly leads to a huge combinatorial problem. Hopefully the number of domains should be limited, while ensuring their connectivity as well as the stationarity of the variables within each domain. In order to achieve this, different methods for the spatial clustering of multivariate data are explored and compared. A particular emphasis is placed on the ways to modify existing procedures of clustering in non spatial settings to enforce the spatial connectivity of the resulting clusters. K-means, hierarchical methods and model based algorithms are reviewed. The methods are illustrated on a simple example and on mining data.

T. Romary (✉) · J. Rivoirard
Geosciences/Geostatistics Center, Mines ParisTech, 35 rue Saint Honoré, 77305 Fontainebleau, France
e-mail: thomas.romary@mines-paristech.fr

J. Rivoirard
e-mail: jacques.rivoirard@mines-paristech.fr

J. Deraisme
Géovariances, 49bis avenue Franklin Roosevelt, BP 91, 77212 Avon Cedex, France
e-mail: deraisme@geovariances.com

C. Quinones · X. Freulon
Mining Business Unit, Direction Strategy—Resources and Reserves, AREVA NC, 1 place Jean Millier, 92084 Paris La Défense Cedex, France

C. Quinones
e-mail: cristian.quinones@areva.com

X. Freulon
e-mail: xavier.freulon@areva.com

P. Abrahamsen et al. (eds.), *Geostatistics Oslo 2012*,
Quantitative Geology and Geostatistics 17,
DOI 10.1007/978-94-007-4153-9_37,

1 Introduction

In mining assessment, once the delineation of mineralization envelopes has been performed, it is often necessary to partition the area inside this envelope into several homogeneous subdomains. This is particularly the case when the extracted materials have to be subsequently chemically processed. It is also helpful to assess the viability of a mining project for its planning optimization. A significant number of parameters (lithology, alteration, grades, ...) are to be combined in order to characterize domains or subdomains. This rapidly leads to a huge combinatorial problem. Methods to automatize this task are therefore necessary. Almost no method has been proposed in the literature expect from an univariate approach based on grade domaining [5]. Consequently, we focus here on a sensible approach that consists in adapting statistical clustering procedures.

Cluster analysis or clustering is the assignment of a set of observations into subsets (called clusters) so that observations in the same cluster are similar in some sense. Clustering is a method of unsupervised learning, and a common technique for statistical data analysis used in many fields, including machine learning, data mining, pattern recognition, image analysis, information retrieval and bioinformatics [7].

In the settings of independent observations, no particular structure is expected among the data. In a geostatistical context however, one expects to obtain a classification of the data that presents some spatial connexity.

Clustering in a spatial framework has been mainly studied in the image analysis and remote sensing context where the model is usually the following: the true but unknown scene, say t, is modeled as a Markov random field and the observed scene, say x, is interpreted as a degradation version of t, such that conditionally on t, the values x_i are independent to each other. In this model, label properties and pixel values need only to be conditioned on nearest neighbors instead of on all pixels of the map, see e.g. [6] for a review.

Clustering of irregularly spaced data (i.e. geostatistical data) has not been much studied. Oliver and Webster [8] proposed a method for clustering multivariate non-lattice data. They proposed to modify the dissimilarity matrix of the data by multiplying it by a variogram. Although this approach leads to a sensible algorithm, the method was not fully statistically grounded. Indeed, it terms to smooth the dissimilarity matrix for pairs of points at short distances but will not enforce the connexity of the resulting clusters. Contrarily, this tends to mitigate the borders between geologically different areas, making it difficult to differentiate between them.

Ambroise et al. [2] proposed a clustering algorithm for Markov random fields based on the expectation-maximization algorithm (EM, see [4]) that can be applied to irregular data using a neighborhood defined by the Delaunay graph of the data (i.e. the nearest-neighbor graph based on the Voronoi tessellation). However this neighborhood structure does not reflect a structure in the data, but rather the structure in the sampling scheme. A Gaussian Markov random field model, while adapted to lattice data, is not natural on such a graph. Furthermore, this method does not ensure the connexity of the resulting clusters either.

Finally, Allard and Guillot [1] proposed a clustering method based on an approximation of the EM algorithm for a mixture of Gaussian random functions model. However this method relies on strong assumptions that are not likely to be encountered in practice and particularly with mineral deposit data: the data are assumed to be Gaussian and data belonging to different clusters are assumed independent. Moreover, this last method is not suitable to large multivariate datasets as it computes the maximum likelihood estimator of the covariance matrix at each iteration of the EM algorithm. Thus, a single iteration requires several inversions of an $(N \times P) \times (N \times P)$ matrix, where N is the number of data and P is the number of variables. This becomes quickly intractable as N and P increase.

In this paper, we first review existing procedures in an independent context. In Sect. 2, we describe a novel geostatistical clustering algorithm that ensures the connexity of the resulting clusters. It is based on a slight modification of the hierarchical clustering algorithm. We compare its performances with other methods on a toy example. Finally, an application on mining data is exposed.

2 Review of Some Methods for Independent Observations

The goal of cluster analysis is to partition the observations into clusters such that those within each cluster are more closely related to one another than variables assigned to different clusters. A central notion for clustering is the degree of similarity (or dissimilarity) between the individual observations being clustered. A clustering method attempts generally to group the observations based on the definition of dissimilarity supplied to it.

2.1 Dissimilarity Matrix

Most of the clustering algorithms take a dissimilarity matrix as their input, the first step is to construct pairwise dissimilarities between the observations. For quantitative variables, one can choose among euclidean, squared euclidean, 1-norm (sum of absolute differences), ∞-norm (maximum over absolute differences). For ordinal variables, where the values are represented as contiguous integers (e.g. alteration degree), error measures are generally defined by replacing their N original values with

$$\frac{i - 1/2}{N}, \quad i = 1, \dots, N$$

in the prescribed order of their original values. They are then treated as quantitative variable on this scale. For unordered categorical variables however the degree of difference between pairs of values must be delineated explicitly (e.g. for geological factors). The most common choice is to take the distance between two observations to be 0 when they belong to different categories, 1 otherwise.

In a multivariate context, the next step is to define a procedure for combining the individual variable dissimilarities into a single overall measure of dissimilarity. This is done by means of a weighted average, where weights are assigned to regulate the relative influence of each variable. In general, setting the weight as the inverse of the average individual dissimilarity for all variables will cause each one of them to equally influence the overall dissimilarity between pairs of observations. Variable that are more relevant in separating the groups should be assigned a higher influence in defining object dissimilarity.

2.2 Partitioning Clustering

The most popular clustering algorithms directly assign each observation to a group or cluster without regard to a probability model describing the data. A prespecified number of clusters $K < N$ is postulated, and each one is labeled by an integer $k \in 1, \ldots, K$. Each observation is assigned to one and only one cluster. The individual cluster assignments for each of the N observations are adjusted so as to minimize a cost function that characterizes the degree to which the clustering goal is not met. A natural cost function is the sum over the clusters of the average distance between observations within each cluster. Cluster analysis by combinatorial optimization is straightforward in principle. As the amount of data increases however, one has to rely on algorithms that are able to examine only a very small fraction of all possible assignments. Such feasible strategies are based on iterative greedy descent. An initial partition is specified. At each iterative step, the cluster assignments are changed in such a way that the value of the criterion is improved from its previous value. The popular K-means algorithm and its variant K-medoids are built upon that principle. In order to apply K-means or K-medoids one must select the number of clusters K and an initialization, see [10] for a review. The number of clusters may be part of the problem. A solution for estimating K typically examine the within-cluster dissimilarity as a function of the number of clusters K, see [7], Chap. 14, for more details.

2.3 Hierarchical Clustering

In contrast to K-means or K-medoids clustering algorithms, (agglomerative) hierarchical clustering methods do not require the choice for the number of clusters to be searched and a starting configuration assignment. Instead, they require the user to specify a measure of dissimilarity between (disjoint) groups of observations, based on the pairwise dissimilarities among the observations in the two groups. As the name suggests, they produce hierarchical representations in which the clusters at each level of the hierarchy are created by merging clusters at the next lower level.

Agglomerative clustering algorithms begin with every observation representing a singleton cluster. At each of the $N-1$ steps the closest two (least dissimilar) clusters are merged into a single cluster, producing one less cluster at the next higher level. Therefore, a measure of dissimilarity between two clusters must be defined. *Single linkage* agglomerative clustering takes the intergroup dissimilarity to be that of the closest (least dissimilar pair). This is also often called the nearest-neighbor technique. *Complete linkage* agglomerative clustering (furthest-neighbor technique) takes the intergroup dissimilarity to be that of the furthest (most dissimilar) pair. *Group average* clustering uses the average dissimilarity between the groups. Although there have been many other proposals for defining intergroup dissimilarity in the context of agglomerative clustering (see e.g. [9]), the above three are the ones most commonly used.

2.4 Model-Based Clustering

Contrarily to the two previous methods, model-based clustering methods rely on the assumption that the data are drawn from a particular distribution. Generally, this distribution is a Gaussian mixture model, i.e. a weighted sum of Gaussian distributions each with a different mean (which corresponds to the centroid in K-means) and covariance. Each component of the mixture defines a cluster, that is, each observation will be considered to have been drawn from one particular component of the mixture.

The estimation of the parameters and the assignment of each observation to a cluster is conducted through an expectation-maximization (EM) algorithm [4]. The two steps of the alternating EM algorithm are very similar to the two steps in K-means. There exists a different version of the EM algorithm called classification EM (CEM, [3]) that may be more adapted to classification problems.

3 Geostatistical Hierarchical Clustering

In this section, we describe a novel geostatistical clustering algorithm that ensures the spatial connexity of resulting clusters. It is based on a slight modification of the hierarchical clustering algorithm described above. It practically consists of two steps: first, the data are structured on a graph according to their location; second, a hierarchical clustering algorithm is conducted where the merging of two clusters is conditioned by their connection in the graph structure. This enforce the spatial connexity of the clusters while respecting the dissimilarities between pairs of observation.

3.1 Algorithm

The first step of the proposed algorithm consists in building a graph over the data to structure them with respect to their proximity. In two dimensions, this task is straightforward as we can consider the Delaunay triangulation, associated to the sampling scheme of the data. Powerful algorithms exist to carry out this task efficiently. Figure 1(b) presents an example of a Delaunay triangulation associated to the sampling performed for the next section example. We can see that each point is connected to surrounding points, not necessarily to its nearest neighbors. Some branches of that graph may seem too long, particularly on the borders of the field. The graph can be post processed by pruning the longest edges so as to avoid spurious connections.

In a geological 3-dimensional context however, the Delaunay tetrahedralization, apart from being tricky to compute, may not be relevant for the purpose of domaining. As an example, consider a vein-type deposit with non horizontal veins. We would like one vein to belong to a unique cluster, which implies the samples belonging to the vein to be connected. Suppose that samples are located along parallel cores. The tetrahedralization will produce flat horizontal tetrahedra and the subsequent connections between points will be irrelevant with the geological configuration. Therefore, we propose to proceed in two steps to build the connections between sample points:

1. Compute the Delaunay graph for one or several 2D surrogates of the deposit (linking the cores), possibly post process it.
2. Extend the connections in the third dimension along the cores and between the cores by taking into account the geology (e.g. orientation), as far as possible.

Once the graph has been built, the second step of our method consists in running a slightly modified version of the hierarchical clustering algorithm (see Sect. 2.3), the trick being to authorize two clusters to merge only if they are connected (two clusters are considered connected if there exists a connected pair of points between the two clusters). This will ensure the connexity of the resulting clusters. We chose to perform complete linkage clustering upon numerical experiments results, as it tends to produce more compact clusters. Finally, the user can choose the hierarchical level of clustering to be considered in the final classification.

3.2 Example

Here, we describe a 2D example on which we evaluated the performances of some previously exposed methods. We consider a random function on the unit square which is made of a Gaussian random function with mean 2 and a cubic covariance with range 0.3 and sill 1 on the disk of radius 0.3 and center (0.5, 0.5) and a Gaussian random function with mean 0 and an exponential covariance with range 0.1 and sill 1. A realization is shown in Fig. 1(a) where increasing values take colors from

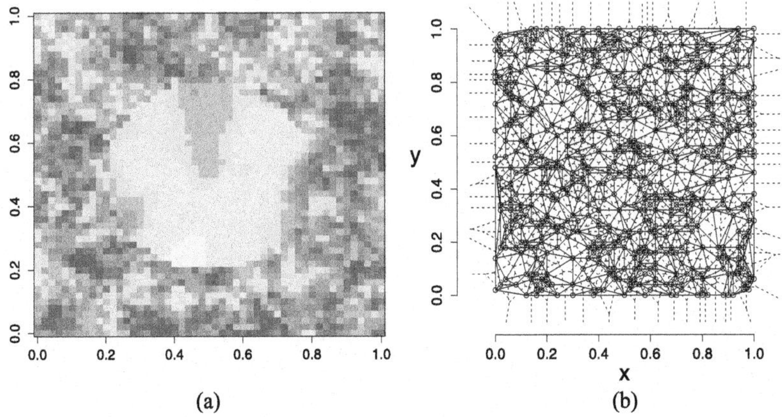

Fig. 1 Example dataset: full realization (**a**) and Delaunay triangulation corresponding to the sampling performed (**b**)

dark grey to white, while Fig. 1(b) corresponds to the Delaunay graph associated to the sampling performed by picking 650 points out of the 2601 points of the complete realization.

While we can clearly see a smooth surface with high values in the central disk in Fig. 1(a), it is much more difficult to distinguish between the two areas when the realization is sampled, which makes this example challenging. We now test the performances of four different methods for this task: the K-means algorithm, the complete linkage hierarchical clustering algorithm (HC), Oliver and Webster's method (O&W) and our geostatistical hierarchical clustering (GHC) algorithm.

Figure 2 shows the results obtained by each four methods. Each subpicture represents the dataset on scatterplots with respect to the coordinates (X and Y) and the sampled value (Z). K-means (a) identifies well the central area but the result lacks of connexity. It can be seen that the method discriminates between low and high values: the limiting value between the two clusters can be read as 0.5. HC (b) also discriminates between low and high value but the limiting value is lower. To sum up, those two classical methods in an independent observations context fail to produce spatially connected clusters. O & W's approach has been tested with various variograms and variogram parameter values but it never showed any structured result (c). The interpretation that we give is that multiplying the dissimilarity matrix by a variogram may erase some dissimilarities, inducing a loss in the structure of the data. The GHC algorithm succeeded in providing a clustering with spatial connexity (d). A part of the disk is misclassified however. If we turn back to the complete realization in Fig. 1(a), we can see that the misclassified area corresponds to the low values of the realization around the border of the disk that are very close to the values taken outside the disk and are thus difficult to classify well.

We applied each four algorithms to 100 realizations of the same geostatistical model each with a different uniform random sampling. Then we computed the mean,

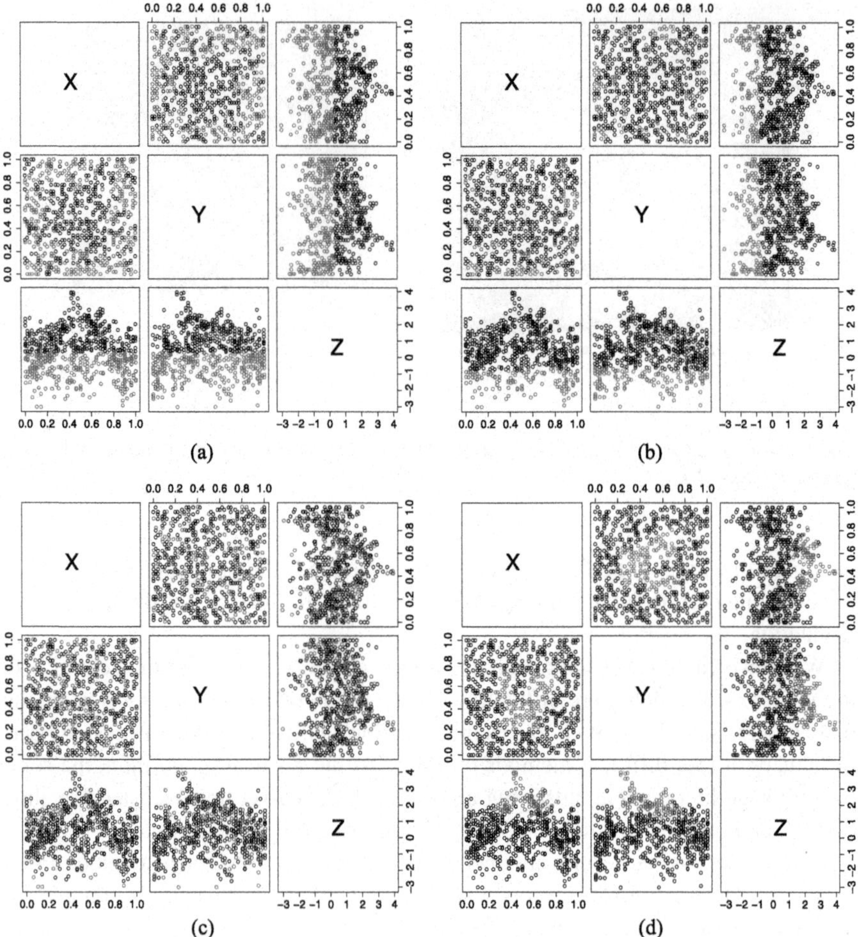

Fig. 2 Results of K-means (**a**), hierarchical clustering (**b**), Oliver and Webster's method (**c**) and geostatistical hierarchical clustering (**d**)

median and 10 % percentile of the rate of misclassified points. Results are summarized in Table 1.

GHC exhibits the best performances overall with 11 % misclassified points in average while K-means is not so far, O & W performing the worst with the HC in between. If we look at the median however, GHC has the lowest one with a larger margin. The 10 % percentile indicates that in the 10 % most favorable cases, GHC misclassified only 0.01 % of the points, while all the other algorithms performs a largely worse job. It can also be seen that the 10 % percentile are similar for the K-means and the HC. This can be explain by the fact that the HC, and GHC (its worse result in this task was a misclassification of almost 50 %), can sometimes perform really bad, whereas the K-means algorithm gives more stable results. In the

Table 1 Rates of misclassified points for the 4 algorithms

	K-means	HC	O & W	GHC
Mean	0.13	0.23	0.35	0.11
Median	0.12	0.20	0.34	0.09
10 % percentile	0.08	0.08	0.28	0.01

favorable cases however, this algorithm works as well as the K-means. Concerning GHC, it performed worse than the K-means in less than 10 % of the cases.

4 Application to an Ore Deposit

In this section, we present a preliminary study for the application of statistical clustering methods on an ore deposit. We describe the different steps and exhibit some results.

The first step has been to select the data that will be used for the domaining. The following variables have been chosen:

- Coordinates, X, Y and Z.
- Uranium grades.
- A geological factor describing the socle.
- Hematization degree.

This choice has been made upon an exploratory analysis of the data and discussion with geologists. Some transformations of the data have been performed:

- The coordinates have been normalized.
- Uranium grades have been log-transformed and normalized.
- The degree of hematization has been transformed into a continuous variable, then normalized.

Then, the connections between close samples have been designed. As the mineralization envelop of the deposit exhibits a horizontal shape, we chose to build a 2D Delaunay triangulation first. To do that, we selected for each core the observation closest to the median altitude of the whole sample. Then we performed the Delaunay triangulation on that subsample. Some branches of the graph were very long and did not correspond to vicinity. Consequently, the longest branches were pruned. We obtained the graph pictured in Fig. 3(a). Then, for each core, the adjacent observations are connected along the core. Close observations from two connected cores are connected as well. In this way, we built the adjacency matrix, an $n \times n$ lower diagonal matrix whose entries are one if the points are connected and zero otherwise. This matrix is plotted in Fig. 3(b). Note that the black points near the diagonal line corresponds to the connections along the cores. The diagonal of this matrix contains only zero.

The next step has been to build the dissimilarity matrix. This has been done using all the variables listed above and considering a particular distance for the geological

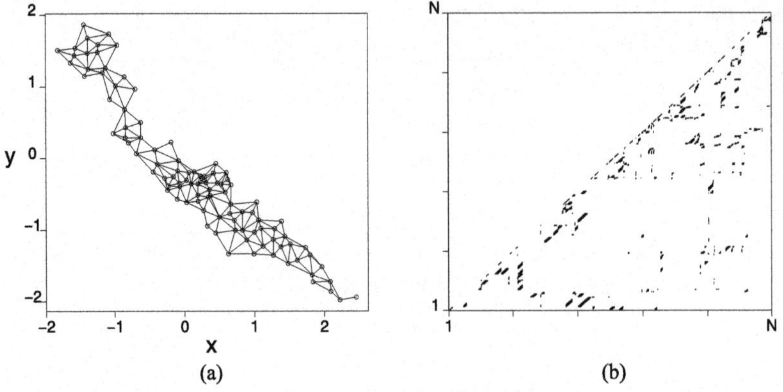

Fig. 3 Pruned Delaunay triangulation at median altitude (**a**) and adjacency matrix (**b**)

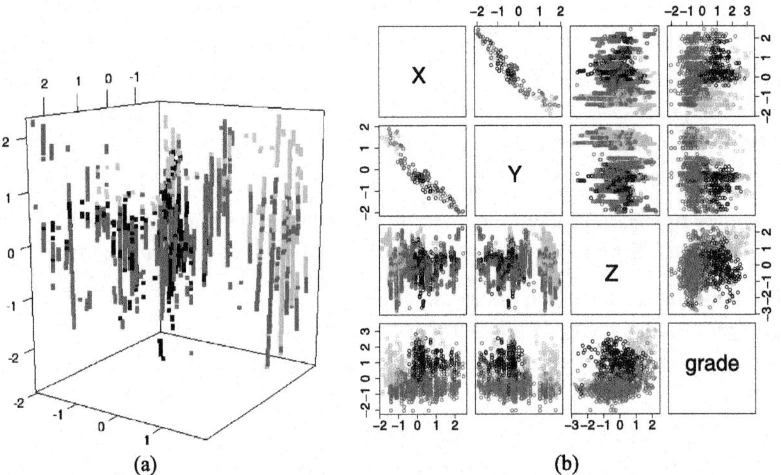

Fig. 4 Results of the algorithm with 6 clusters: in 3D (**a**) and scatterplots (**b**)

factor: it has been chosen to be 1 when the samples have different factor values and 0 otherwise. Weights have been set by trial and error: we finally set the weights to 1 for the coordinates, 4 for the grade, 2 for the hematization degree and 10 for the geological factor.

Finally, we were able to run the geostatistical hierarchical algorithm described in Sect. 3.1. We retained six clusters after the visualization of different hierarchical configurations. The results are depicted in Fig. 4 and described in Table 2.

Table 2 Description of the resulting clustering

	Lighter grey	Dark grey	Darker grey	Black	Medium grey	Light grey
(X, Y)	middle	S-E	middle to S-E		N-W	
Z	top	top	middle to bottom		all along	
Grade	high	middle	middle	high	middle	high
Geol.	sand	sand	socle		sand & socle	
Hemat.	middle to large	low	no role		no role	

5 Conclusions

In this paper, we presented an insight towards geostatistically adapted clustering procedures. We presented an hierarchical algorithm conditioned to a connexity structure imposed on the data. Two applications have been provided, the first one on a toy example and the second on the deposit data.

The results shown on the toy example clearly assess the superiority of our method over tested ones as it is able to produce compact, connected clusters. The results obtained for the application where also satisfactory as they depicted a synthesized description of the deposit. Moreover, thanks to the sequential nature of the algorithm, our method generates a whole ensemble of clusterings that can be useful to the user: he can visualize the results at different hierarchical levels which leads to different interpretation levels. Furthermore, the user can also play with the weights of each variable to produce different clusterings, according to its knowledge of the geology.

Still, some improvements can be done. The first point is the way how we connect the observations in 3D. Performing the Delaunay to a 2D surrogate can be extended to non horizontal deposits by e.g. transforming the mineralization envelop into a flat manifold. Then the observations between connected cores should be connected if and only if they are not to distant away. Second, ways to define properly the weights associated to each variable according to the desired results should be investigated. Then, we could think of a more adapted linkage criterion than the complete linkage in the hierarchical algorithm. This new criterion would account for instance for the homogeneity of the cluster.

Finally, implementing a K-medoids algorithm based on the connection relations may be an interesting perspective, as it presents more appealing theoretical properties than hierarchical algorithms and is much faster.

References

1. Allard D, Guillot G (2000) Clustering geostatistical data. In: Proceedings of the sixth geostatistical conference.
2. Ambroise C, Dang M, Govaert G (1995) Clustering of spatial data by the EM algorithm. In: Soares, A et al (eds) geoENV I—geostatistics for environmental applications. Kluwer Academic, Norwell, pp 493–504

3. Celeux G, Govaert G (1992) A classification EM algorithm for clustering and two stochastic versions. Comput Stat Data Anal 14:315–332
4. Dempster AP, Laird NM, Rubin DB (1977) Maximum likelihood from incomplete data via EM algorithm (with discussion). J R Stat Soc, Ser B, Stat Methodol 39:1–38
5. Emery X, Ortiz J (2004) Defining geological units by grade domaining. Tech. rep., Universidad de Chile
6. Guyon X (1995) Random fields on a network. Springer, Berlin
7. Hastie T, Tibshirani R, Friedman J (2009) The elements of statistical learning, 2nd edn. Springer, Berlin
8. Oliver M, Webster R (1989) A geostatistical basis for spatial weighting in multivariate classification. Math Geol 21:15–35
9. Saporta G (2006) Probabilités, analyses des données et statistiques, 2nd edn. Edition Technip, Paris
10. Steinley D, Brusco MJ (2007) Initializing k-means batch clustering: a critical evaluation of several techniques. J Classif 24:99–121

The Influence of Geostatistical Techniques on Geological Uncertainty

Joanna K. Marshall and Hylke J. Glass

Abstract Geological uncertainty is a recognition that our understanding of geology is based on a limited set of observations that capture only part of reality. The necessary spatial interpolation of geological observations, expressed in terms of indicators, can utilize the same geostatistical techniques otherwise applied to drillcore grades. This study investigates the relationship between geological uncertainty and geostatistical techniques. Four distinct types of geostatistical techniques are considered: indicator kriging, sequential indicator simulation, plurigaussian simulation, and multiple-point simulation. Each technique is applied to a dataset featuring the geology of an iron ore deposit in the Pilbara, Western Australia. Within plurigaussian simulation and multiple-point simulation, the effect of variation of the input parameters is also investigated. Overall, PGS and MPS give the most consideration to the geological interpretation of the deposit, but these require significant knowledge of the deposit. Post-processing indicates that the complex techniques (PGS and MPS) generally appear more precise than the simpler techniques (SIS and IK), with more of the nodes reoccurring as the same domain across all simulations. If a comparison to the block model is considered for accuracy, the simpler technique SIS relates most closely to the block model, and generates less geological uncertainty during pre-simulation processing and simulation. This study supports the notion that geostatistical simulation can assist in the modeling of geological uncertainty.

1 Introduction

Geological uncertainty is a recognition that our understanding of geology is based on a limited set of observations that capture only part of reality. When aiming, for resource assessment purposes, to recreate a virtual reality using geostatistical techniques, geological observations routinely assist in the definition of domains.

J.K. Marshall (✉) · H.J. Glass
University of Exeter, Cornwall Campus, Penryn, TR10 9EZ, UK
e-mail: Joanna.Marshall@riotinto.com

H.J. Glass
e-mail: H.J.Glass@ex.ac.uk

P. Abrahamsen et al. (eds.), *Geostatistics Oslo 2012*,
Quantitative Geology and Geostatistics 17,
DOI 10.1007/978-94-007-4153-9_38,

Table 1 Comparative work program

	Estimation process	Variation		
IK	1 krige × 4 domains	Indicator datasets		
SIS	40 simulations × 4 domains	Indicator datasets		
PGS	40 simulations	Lithotype rule 1	Lithotype rule 2	Lithotype rule 3
MPS	40 simulations × 4 domains × 3 grids	30 × 30 × 10 m^3 over 10 × 10 × 10 grid	30 × 30 × 10 m^3 over 20 × 20 × 20 grid	

This is followed by more or less independent consideration of the drillcore grades within each domain. Consequently, geological uncertainty may influence the global resource assessment and its interpretation.

Geological observations may be translated into indicators which are interpolated using the same geostatistical techniques applied to drillcore grades. It is instructive to examine the effect of geostatistical techniques on the geological uncertainty. In this study, the geology of an iron ore deposit serves as a case study to compare the following four geostatistical techniques:

1. Indicator Kriging (IK), a geostatistical technique which minimizes local error variance,
2. Sequential Indicator Simulation (SIS), a two-point simulation technique with reproduce the global statistics such as histograms and variograms,
3. Plurigaussian Simulation (PGS), another two-point simulation techniques using lithotype rules to reproduce lithology relationships within the deposit,
4. Multiple Point Simulation (MPS), which reproduces spatial continuity, with the use of a training image instead of variograms.

Further details on these techniques are found in [2, 4, 6, 8, 9, 11, 12].

2 Methodology

2.1 Dataset and Indicators

The dataset relates to an iron ore deposit in the Pilbara whose geology is classified into four geological domains: (1) Detritals and Miscellaneous, (2) Banded Iron Formation (BIF)/Waste, (3) Hydrated Mineralized domain, and (4) Mineralized domain. For each geological domain, indicator datasets using binary values are created: value 1 indicates the presence of a selected geological domain and 0 indicates the presence of a different geological domain. Each indicator dataset is modeled with a separate set of variograms. It should be noted that all data processing in this study is performed with Isatis software.

Table 2 Downward and upward probability matrix. L1 = Detritals and miscellaneous, L2 = BIF/waste, L3 = Hydrated, L4 = Mineralized

Downward probability					Upward probability				
From/to	L1	L2	L3	L4	From/to	L1	L2	L3	L4
L1	0.663	0.189	0.057	0.091	L1	0.700	0.088	0.088	0.124
L2	0.004	0.962	0.006	0.028	L1	0.007	0.903	0.011	0.079
L3	0.013	0.045	0.715	0.227	L3	0.012	0.031	0.957	0.000
L4	0.006	0.102	0.000	0.892	L4	0.05	0.034	0.072	0.8

2.2 Geostatistical Techniques

Variation of the parameters for the four different geostatistical techniques (IK, SIS, PGS and MPS) are shown in Table 1.

2.2.1 Plurigaussian Simulation

Plurigaussian simulation consists of a number of steps. As a starting point, vertical proportion curves are established. These capture the proportions of lithotypes observed at a selected level in the drillhole data. In parallel, a lithotype rule is developed based on characterization of the contacts between lithotypes, typically through interpretation of a matrix of transition probabilities. Table 2 shows the transition probabilities applicable to this case study. It suggests that L4 (mineralized) will not occur directly above or below L3 (hydrated) and that L1 (detritals and misc) will hardly occur directly below the L4 (mineralized) or L2 (BIF/waste). This supports the visual impression that detritals and miscellaneous and the hydrated domain are confined to the upper layers of the deposit.

A suitable lithotype rule should reflect the contacts and the sequencing of the lithotypes. The most simple case is when sequencing of the different lithotypes is ordered throughout the deposit. In this case, the truncated Gaussian method applies and definition of a single Gaussian random function. When more complicated sequencing is observed, more than one Gaussian random function may be configured, leading to the Plurigaussian simulation method. Representing the lithotype rule by a square, the horizontal edge shows the value for the first Gaussian random function while the vertical edge shows the value for a second Gaussian random function. Inside the square, partitions are created to accommodate the principal spatial relationships between the lithologies. For this deposit, the use of multiple Gaussian random functions is deemed necessary. Three different lithotype rules are constructed, as depicted in Fig. 1.

Lithotype rule 1 is established as a base lithotype rule, which considers the main domain contacts within the deposit. Rules 2 and 3 are designed to test variation in the geological interpretation. Rule 2 differs from rule 1 by a change in contact: the

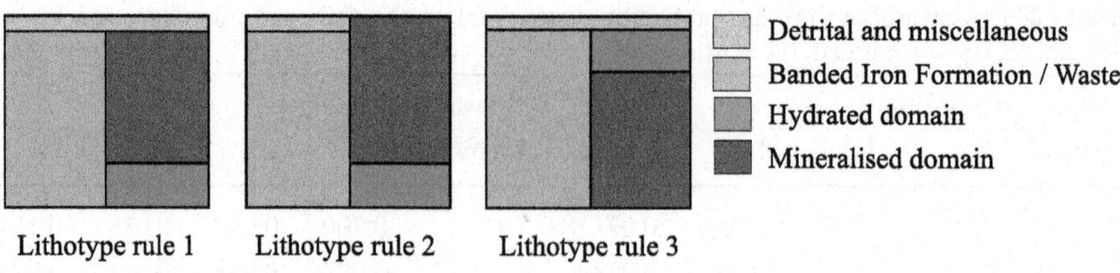

Fig. 1 Selected lithotype rules for PGS

Table 3 Grids used in this study

	Grid size	Block size	Number of blocks
Deposit grid:	$3000 \times 2100 \times 855$ m^3	$15 \times 15 \times 15$ m^3	1 596 000
Small meso grid:	$300 \times 300 \times 100$ m^3	$30 \times 30 \times 10$ m^3	1 000
Large meso grid:	$600 \times 600 \times 100$ m^3	$30 \times 30 \times 10$ m^3	4 000

Detritals and Miscellaneous domain does not follow the Mineralized domain. Comparison of simulations with rules 1 and 2 should provide insight into the direction of contact. It should be noted that the change in rule 2 does not affect the proportion of the Detritals and Miscellaneous and the Mineralized domains, which are calculated independently of the lithotype rule. Rule 3 differs from rule 1 by a domain reversal: the Hydrated and Mineralized domains are switched, changing the nature of the contact with the Detritals and Miscellaneous domain. Comparison of simulations with rules 1 and 3 should reveal the effect of domain reversal.

Besides the vertical proportion curves and the lithotype rule, PGS requires modeling of the variograms associated with the Gaussian variables as well as determination of the correlation between the Gaussian variables.

2.2.2 Multiple-Point Simulation

The multiple-point simulation used the block model developed by the company mining the deposit as a training image and the sneism simulation technique.

2.2.3 Grids

Simulations with IK, SIS and PGS are performed on a grid across the entire deposit (Table 3). Due to the large amount of memory required, MPS simulations are restricted to meso grids of two different sizes (Fig. 2).

Three 300×300 m^2 grids cover an area dominated by waste domain (Grid A), an area of dominated by mineralized domain (Grid B) and an area of a boundary between the waste and the mineralized domains (Grid A/B). Three 600×600 m^2 grids are also selected across the area. To enable a comparison, simulations are also performed over the meso grids with SIS.

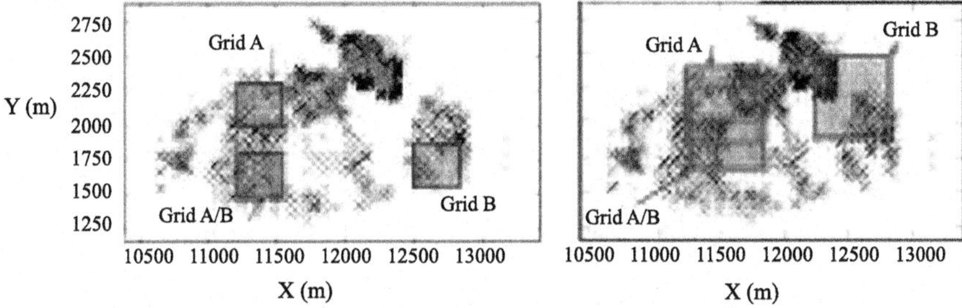

Fig. 2 Locations of study areas across the deposit

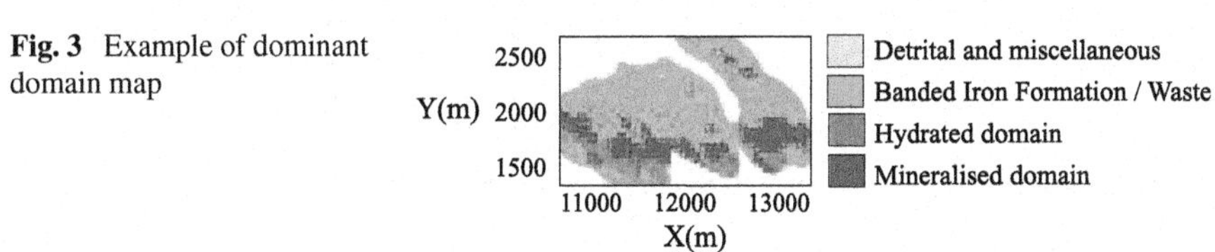

Fig. 3 Example of dominant domain map

2.2.4 Post-processing

Post-processing is required for all techniques. With IK, a probability map is produced for each domain, allowing for comparison of the probabilities at each node. The highest probability at a node determines the domain which is assigned to the node. For SIS, PGS and MPS, 40 simulations are created for each domain. After combining the 40 simulations, the average probability of encountering a domain at each node is determined. The highest probability determines which domain is assigned to the node, creating a map of dominant domains. For all techniques, the average probability associated with the dominant domain is used to produce repeatability maps. Results of the different techniques are compared with each other and with a given block model of the deposit.

3 Analysis

3.1 Dominant Domain Maps

The dominant domain maps obtained with IK, SIS and PGS are broadly similar, with a band of mineralization and a hydrated zone near the southern periphery. Figure 3 shows a typical dominant domain map across the entire deposit.

For all geostatistical techniques, good agreement between the dominant domain maps is generally observed within the more homogeneous, waste-dominated grid A. More significant differences between the location and structures of the mineralized and hydrated domains appear in the mixed domain grid (A/B) and the ore grid (B).

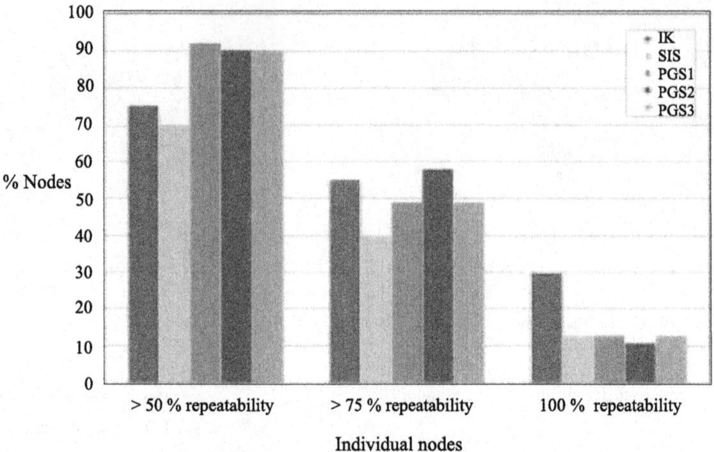

Fig. 4 Repeatability of results across the entire deposit with different geostatistical techniques

3.1.1 Repeatability

Repeatability is defined here as the average percentage of the occurrence of the dominant domain at the nodes simulated across the deposit. A greater repeatability points to a higher confidence that a domain is actually encountered at a particular node and suggests a lower geological uncertainty. Figure 4 shows that IK has the greatest number of nodes with 100 % repeatability of the dominant domain. However, noting the generally high levels of repeatability of PGS, it appears that PGS outperforms IK and SIS in terms of precision. In the meso scale grids (Fig. 5), MPS appears to outperform SIS.

To enable a comparison of MPS to IK and PGS, results for IK and PGS are cut to the areas of the $600 \times 600\ m^2$ grids. Overall, MPS appears to have the greatest repeatability and precision (Fig. 6).

3.1.2 Accuracy

All results are compared to the block model developed by the company mining the deposit. When results for the entire deposit are considered, Fig. 7 shows that SIS has the greatest similarity to the block model in the deposit scale maps.

When comparing all techniques on the meso scale grid, Fig. 8 indicates that MPS and SIS generally produce the greatest correspondence to the block model. However, it must be considered that the training image of the MPS is the block model.

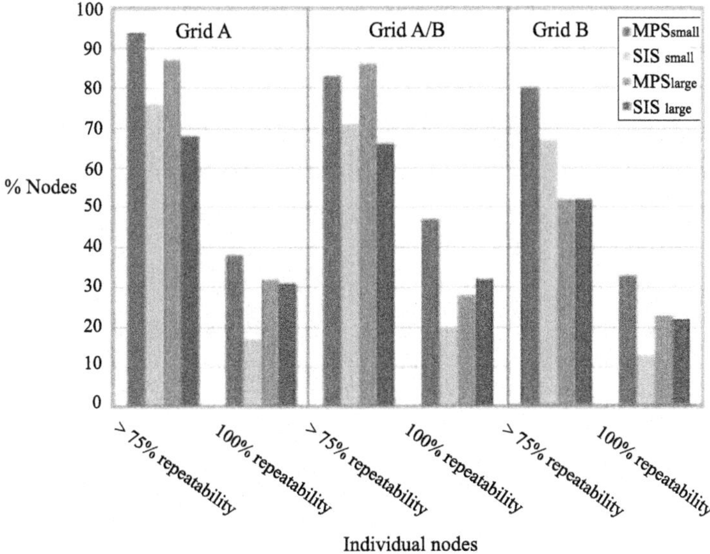

Fig. 5 Repeatability of results across 300 × 300 m^2 grids

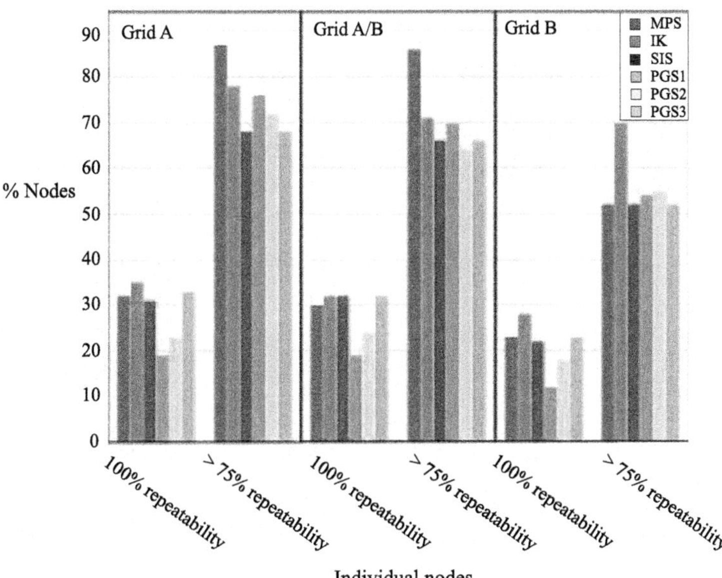

Fig. 6 Repeatability of results across 600 × 600 m^2 grids

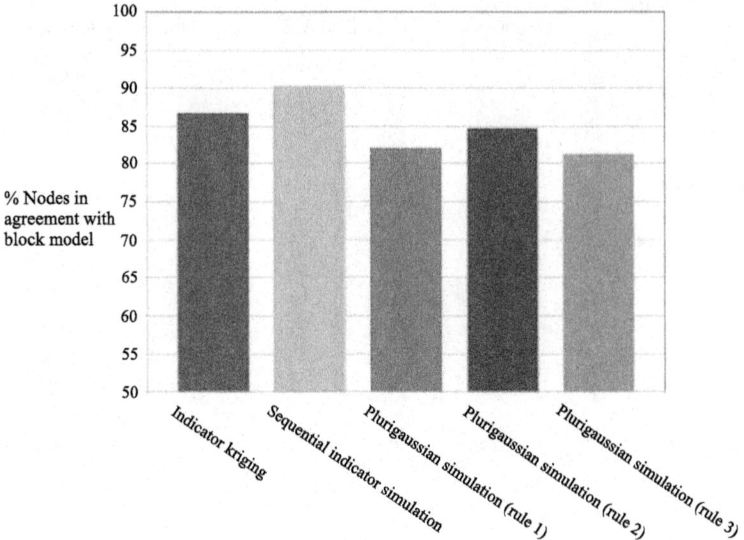

Fig. 7 Agreement of results across the entire deposit with the given block model

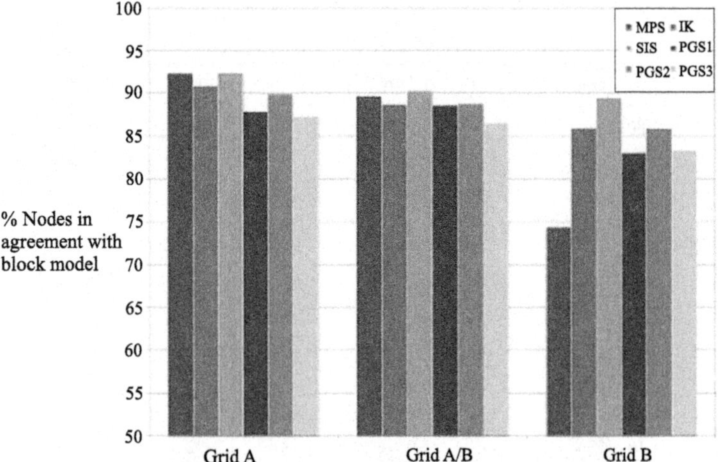

Fig. 8 Agreement of results across 600×600 m^2 grids with the given block model

4 Discussion

4.1 Plurigaussian Simulation

Setting up PGS consists of a large number of steps and requires a large amount of prior information. This may build uncertainty into the subsequent simulation process. For example, when creating vertical proportion curves in Isatis, incomplete 'raw' curves must extend over the full vertical dimension. This entails extrapolation of the domains from the upper and lower surfaces of the 'raw' curves. Completed curves can be edited, but this process is intricate and potentially confusing. It requires movement of domain boundaries through placement of control points on the domain to be edited while freezing all other domains. However, editing a domain affects the previously edited domains, which then need to be revisited. Noting that there are over 60 raw curves in this study, editing was only performed in the selected cases where extrapolation was clearly unrealistic.

The modelling of PGS variograms is also challenging. PGS is strongly influenced by the variogram models, as defined by the choice of a preliminary set of variogram parameters [7]. Initial variogram parameters are entered prior to calculation of the Gaussian functions for every line of the lithotype rule. When the variograms are calculated for the different Gaussian functions, the effect of changes in the variogram parameters on the variogram is either negligible for subtle changes or unpredictable for large changes. While these affect the underlying multi-Gaussian functions through an iterative process, selection of the early sets of variogram parameters is an issue which could determine the overall result. Selection of the variogram parameters is not obvious and especially challenging when involved with early stages of deposit evaluation [3]. Hence, it is important to conduct sensitivity studies to assess the appropriateness of the chosen parameters.

Another complication with PGS is definition of the lithotype rule. The lithotype rule is particularly applicable for ordered sequences of lithotypes. In practice, however, these geological domains all have some degree of contact with each other and aspects of multiple lithotype rules may be appropriate. In this study, three different lithotype rules were created. Comparison of simulations obtained with lithotype rules 1 and 3 reveals few differences. This suggests that the effect of domain reversal in the lithotype rule is limited. More significant differences are observed between simulations with rules 1 and 3 and simulations with rule 2, suggesting that variation of the type of contact has a significant effect here. Visually, simulations with rules 1 and 3 show areas of fine mineralization around the edge of the deposit which are largely absent in simulations based on rule 2. With the latter, the majority of the mineralization appears in a distinct band of mineralization, with an outer band of hydrated mineralization.

4.2 Multiple-Point Simulation

A distinguishing feature of Multiple-Point Simulation (MPS) is the use of a training image instead of a variogram. In this study the block model is used as the training image and its accuracy should be scrutinized: lack of accuracy in the block model would propagate during MPS. The block model itself is based on the use of estimation techniques, which may require knowledge of variograms. The latter would undermine the advantage of the MPS in being essentially non-reliant on variograms. For effective application of MPS, design of the training image selection is vital and must be representative of the structural characteristics of the deposit [1]. However, representativity is not guaranteed because an image of the target spatial phenomena is created with another two-point simulation or kriging technique. Training images could be constructed from geological interpretation and data gained in previously mined areas of the deposit [8]. This effectively limits the use of MPS to mining areas which have a history of mining, which generally implies that a detailed understanding of the geology has been gained. This may not be available with new mines or for exploration stage deposits. In addition, it is unlikely that an accurate image of a deposit can be formed when the deposit has a large vertical dimension.

A further consideration relating to MPS is the very large amount of RAM required to perform the simulations. Due to RAM requirements and the associated computation time required to perform MPS, a number of smaller areas were analyzed with MPS. Comparisons between MPS and the other techniques are hence restricted to selected subareas.

4.3 Techniques

It is generally recognised that the smoothing or loss of variance in kriging is detrimental for subsequent interpretation. However, simulation techniques have their own drawbacks. For example, SIS suffers from possible order relationship problems, which can produce unreliable results. Contrary to kriging and SIS, PGS and MPS give more consideration to the geology of the area: PGS achieves this through the use of the lithotype rule and MPS through the use of the training image. It should be noted that both the lithotype rule and the training image contain intrinsic uncertainty.

Various authors state that the modern PGS and MPS techniques are more sophisticated that established techniques such as SIS and IK [5, 8, 10]. In practice, however, PGS and MPS techniques require relatively detailed knowledge of the deposit area in order to represent the geological complexity of the deposit. This may limit the application of PGS and MPS to areas where substantial information is already available and makes these techniques unsuitable for application in exploration or early mine development.

5 Conclusions

Based on a case study of an iron ore deposit, it emerges that geostatistics can assist in the modeling of geological uncertainty. PGS and MPS are relatively complex techniques which appear more precise than SIS and IK. When confidence/repeatability maps are considered, both MPS and PGS outperform SIS and IK. PGS and MPS also give the most consideration to the geology of the area, but they require significant knowledge of the area. When the accuracy of the results are considered, the simpler technique of SIS relates most closely to the block model, and produces less uncertainty during pre-simulation processing and simulation. Due to this, SIS appears to outperform the more complex techniques. Overall, the complex techniques of PGS and MPS appear more precise than the more simple techniques of IK and SIS. However, the more simple techniques appear more accurate. Further development of the modeling approaches is expected to improve the way in which geological heterogeneity of deposits is taken into account.

Acknowledgements The Rio Tinto Chair at Camborne School of Mines (H.J.G.) wishes to thank Rio Tinto for supporting this research. Rio Tinto Iron Ore is thanked for providing the dataset.

References

1. Boisvert JB, Pyrcz MJ, Deutsch CV (2010) Multiple point metrics to assess categorical variable models. Nat Resour Res 19:165–175
2. Caers J (2000) Adding local accuracy to direct sequential simulation. Math Geol 32:815–850
3. Deraisme J, Field M (2006) Geostatistical simulation of kimberlite orebodies and application to sampling optimisation. In: 6th international mining geology conference, Darwin, pp 194–203
4. Deutsch CV (2006) A sequential indicator simulation program for categorical variables with point and block data: BlockSIS. Comput Geosci 32:1669–1681
5. Dowd PA, Pardo-Iguzquiza E, Xu C (2003) Plurigau: a computer program for simulating spatial facies using the truncated plurigaussian method. Comput Geosci 29:123–141
6. Liu YH, Journel A (2004) Improving sequential simulation with a structured path guided by information content. Math Geol 36:945–964
7. Mariethoz G, Renard P, Cornaton F, Jaquet O (2009) Truncated plurigaussian simulations to characterize aquifer heterogeneity. Ground Water 47:13–24
8. Osterholt V, Dimitrakopoulos R (2007) Resource model uncertainty at the Yandi channel iron deposits, Western Australia–an application of multiple-point simulations for orebody geology. In: Iron ore conference, Perth, WA, pp 147–151
9. Sinclair AJ, Blackwell B (2002) Applied mineral inventory estimation. Cambridge University Press, Cambridge, 382 pp
10. Strebelle S (2002) Conditional simulation of complex geological structures using multiple-point statistics. Math Geol 34:1–21
11. Wang XL, Tao S, Dawson RW, Wang XJ (2004) Uncertainty analysis of parameters for modeling the transfer and fate of benzo(a)pyrene in Tianjin wastewater irrigated areas. Chemosphere 55:525–531
12. Webster R, Oliver MA (2007) Geostatistics for environmental scientists. Wiley, New York

Part IV
Environmental, Climate and Hydrology

A Study on How Top-Surface Morphology Influences the Storage Capacity of CO_2 in Saline Aquifers

Anne Randi Syversveen, Halvor Møll Nilsen, Knut-Andreas Lie, Jan Tveranger, and Petter Abrahamsen

Abstract The primary trapping mechanism in CO_2 storage is structural trapping, which means accumulation of a CO_2 column under a deformation in the caprock. We present a study on how different top-seal morphologies will influence the CO_2 storage capacity and migration patterns. Alternative top-surface morphologies are created stochastically by combining different stratigraphic scenarios with different structural scenarios. Stratigraphic surfaces are generated by Gaussian random fields, while faults are generated by marked point processes. The storage capacity is calculated by a simple and fast spill-point analysis, and by a more extensive method including fluid flow simulation in which parameters such as pressure and injection rate are taken into account. Results from the two approaches are compared. Moreover, by generating multiple realizations, we quantify how uncertainty in the top-surface morphology impacts the primary storage capacity. The study shows that the morphology of the top seal is of great importance both for the primary storage capacity and for migration patterns.

A.R. Syversveen (✉) · P. Abrahamsen
Norwegian Computing Center, PO Box 114, Blindern, 0314 Oslo, Norway
e-mail: Anne.Randi.Syversveen@nr.no

P. Abrahamsen
e-mail: Petter.Abrahamsen@nr.no

H.M. Nilsen · K.-A. Lie
SINTEF ICT, PO Box 124, Blindern, 0314 Oslo, Norway

H.M. Nilsen
e-mail: hnil@sintef.no

K.-A. Lie
e-mail: Knut-Andreas.Lie@sintef.no

J. Tveranger
Center for Integrated Petroleum Research, Uni Research, PO Box 7800, 5020 Bergen, Norway
e-mail: Jan.Tveranger@uni.no

P. Abrahamsen et al. (eds.), *Geostatistics Oslo 2012*,
Quantitative Geology and Geostatistics 17,
DOI 10.1007/978-94-007-4153-9_39,

1 Introduction

Sequestration of CO_2 in subsurface rock formations is commonly advocated as a necessary means to reduce the rate of anthropogenic carbon emission into the atmosphere. Whether CO_2 can be stored in saline aquifers, un-mineable coal seams, or abandoned petroleum reservoirs is mainly a question of costs and the risk associated with the storage operation; most of the technology required to inject CO_2 is already available from the petroleum and mining industry.

A prospective storage site must have sufficient storage capacity, one must be able to inject CO_2 at a sufficient rate without creating a pressure increase that may threaten the caprock integrity, and the overall probability for leakage during and after the injection must be acceptable. Comprehensive numerical simulation capabilities (see e.g., [2, 3]) have been developed to provide confident assessments of prospective storage sites, which, in turn, has nurtured an often lopsided emphasis on numerical- and modeling-based uncertainties in this area of research [3, 7]. Uncertainties from formation properties have received less attention: Academic studies of CO_2 injection frequently employ simplified or conceptualized reservoir descriptions in which the storage formation has highly idealized geometry and is considered nearly homogeneous. Geological knowledge and experience from petroleum production, on the other hand, show that typical rock formations can be expected to be heterogeneous on the relevant physical scales, regardless of whether the target formation is an abandoned petroleum reservoir or a pristine aquifer.

Complex geology introduces tortuous subsurface flow paths, baffles, and barriers, which in turn influence reservoir behavior during injection. It is important that the effect of geological heterogeneity is quantified by the research community. A key challenge when assessing the impact of heterogeneity is to quantify the uncertainty associated with the precise spatial structure of formation properties. To provide a statistically sound frame of reference, several sedimentological scenarios need to be evaluated. Likewise, multiple geostatistical realizations of each sedimentological scenario are required to quantify the relative effect of uncertainties associated with depositional and structural architecture and their associated petrophysical properties. This will facilitate both improved understanding of subsurface flow at operational CO_2 injection sites, and allow comparison with simulated flow in ideal homogeneous models and upscaled versions of these.

In this paper, we will consider a very favorable, synthetic storage scenario in which the reservoir consists of a huge body of good sand, buried underneath an impermeable caprock that dips slightly in one direction. The scenario is inspired by the Utsira formation in the North Sea, into which CO_2 has been injected from the Sleipner B platform since 1996. Because of density differences between the supercritical CO_2 and the resident brine, the injected CO_2 will move upward to form a plume that spreads out underneath the caprock, implying that top-surface morphology will be the main driver for uncertainty. Herein, we will therefore study how reliefs in this morphology, occurring at length scales close to the seismic resolution, may impact estimates of primary storage capacity as well as migration patterns.

2 Geological Modeling

Geological heterogeneity influences fluid movement in the subsurface and occurs at all scales. Most geological parameters used for describing reservoir-type rocks (e.g., mineralogy, grain-size, grain shape, sorting, cementation, sedimentary structures, bed-thickness) express heterogeneity at often very fine scales. Physical properties of the rock are linked to these descriptive geological parameters and their scale through direct measurement, empirical databases, or established physical relationships (e.g., pore-size, pore shape, pore throat diameter, connectivity, elasticity, shear strength).

In reservoir models, heterogeneity is commonly expressed as spatial distributions of porosity and permeability on the scale dictated by model resolution. Heterogeneity at scales below model resolution is treated either implicitly, by considering the cell sizes of the model as common representative elementary volumes (REV) for all features in the model, or alternatively derived from more detailed models through upscaling. In either case, one must be careful to correctly incorporate effects of unresolved heterogeneity when formulating simplified models. Mapping out the impact of geological features on CO_2 sequestration would therefore ideally require a study covering all possible types of geological features, how to upscale them and charting their effect on fluid flow through a series of simulation studies. The scale of such an effort is beyond the scope of any single project or the capacity of any single group, but it should certainly form a clear goal and ambition for a collective research effort.

Being a buoyant fluid, CO_2 moves up from the point it is being injected until encountering a barrier that prevents further upward or lateral movement. At this point, the fluid will move laterally upslope along the barrier until it either reaches the end of the barrier or a closure/trap where accumulation can take place. As the trap fills, it may either overspill, causing further lateral migration of CO_2, or experience pressure build-up to a point where trap integrity may be compromised and seal bypass occurs, at which point CO_2 will intrude into the seal and leak into the overlying formation. It follows that the morphology of the interface between the reservoir and overlying seal will affect CO_2 migration pathways, shape, and size of local accumulations as well as the evaluation of seal integrity on reservoir scale. To investigate these effects, a series of generic scenarios were defined which included depositional and structural features affecting top-reservoir morphology in a synthetic reservoir model measuring 30×60 km^2. The scale of the features included was constrained by the $100 \times 100 \times 5$ m^3 resolution of the model.

For depositional features, two scenarios were chosen for which it was considered likely that a depositional/erosional topography could be preserved under a thick regional seal; the latter commonly formed by marine shale. The two scenarios reflect situations were sand deposition in an area similar to the model size is succeeded by deposition of fines as a result of marine transgression:

1. Offshore sand ridges covered by thick marine shale (OSS)
2. Preserved beach ridges under marine shale (FMM)

Generic input for the scales and geometries of the two scenarios was compiled from published literature on recent and ancient offshore sandbanks and drowned beach ridges. Input data are summarized in Table 1.

Table 1 Morphometric data for preserved topographic features (offshore sand ridges, OSS and flooded marginal marine, FMM) capped by the top seal

Scenario label	OSS	FMM
Amplitude	< 20 m	1–10 m
Width	2–4 km	10–300 m
Length	10–60 km	< 15 km
Spacing	2–4 km	40–300 m

In addition to the preserved depositional topography, a series of conceptual structural scenarios were generated with fault patterns extending over the entire model area. The scenario details are listed in Table 2.

3 Stochastic Modeling

Once geological base-case scenarios are defined, their intrinsic uncertainty can be explored using geostatistical methods. Factors that affect the behavior of the injected CO_2 in the reservoir include the morphology of the overlying seal, presence of barriers like sealing faults and reservoir porosity, and permeability distribution. To explore the impact of top-seal morphology, a set of plausible top-reservoir surfaces were generated by superimposing different sinusoidal structures, reflecting the geological features to be modeled, onto a base-case surface. The shape of the base-case surface was chosen so as to keep the injected CO_2 plume within the model area. It has the shape of an inverse half-pipe parallel to the longest axis of the model, with a 500 m height difference over a distance of 60 km. This corresponds to a gradient of 0.48 degrees, which is low, but not unrealistic, for basins of this size.

The top surfaces of the OSS and FMM models were created by Gaussian random fields. A sinusoidal covariance of form $\sin x/x$ was used for both models. For OSS, the range was 1000 m along the long axis and 7000 m along the short axis with standard deviation equal 13 m. For the FMM case, the range along the long axis was 200 m and 7000 m along the short axis with standard deviation equal 5 m. One realization of each model is shown in Fig. 1. Faults were generated by the fault modeling tool HAVANA [4], which is based on a marked-point model. For UP1 and UP2, we use strong repulsion to get uniformly distributed faults and a small variance

Table 2 Geometric definitions of fault populations for the four faulted reservoir scenarios

Scenario label	UP1	NP1	UP2	NP2
Displacement	uniform; 100 m	random; 20–150 m	uniform; 100 m	random; 20–150 m
Length	uniform; 4000 m	random; 300–6000 m	uniform; 4000 m	random; 300–6000 m
Strike	uniform; 90°	uniform; 90°	30° and 90°	30° and 90°

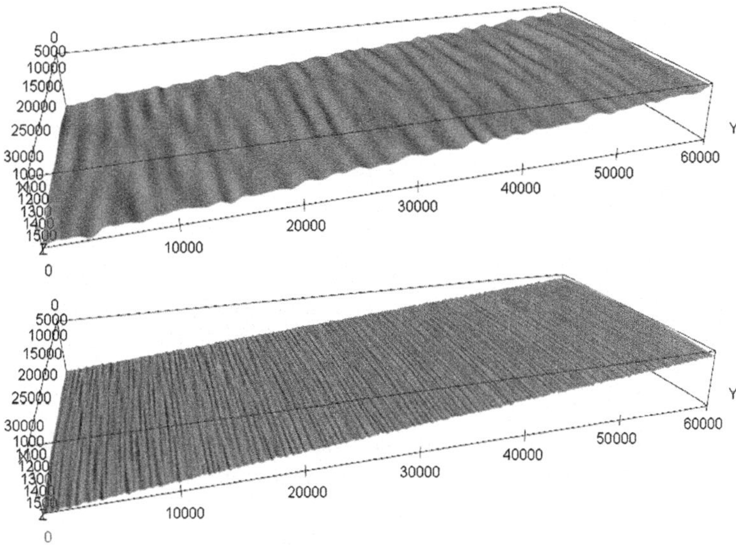

Fig. 1 Top surfaces: the *top plot* shows offshore sand ridges (OSS) and the *bottom plot* shows a flooded marginal marine (FMM) deposition

for the fault length to have constant length. In NP1 and NP2, the fault length has a larger variance, and the repulsion between faults is weaker.

Variations in reservoir porosity and permeability have so far not been considered, but are important heterogeneity factors that should be investigated at a later stage.

4 Flow Simulation and Estimation of Structural Trapping

To quantify the impact of changing geological parameters, we will consider a simple scenario in which CO_2 is injected from a single point (injection well). Chief among the immediate concerns during CO_2 injection are primarily pressure build-up during injection, and secondly storage capacity and migration of injected CO_2. The topography of the top surface is unlikely to have significant impact on pressure build-up during injection, and thus our attention will be focused towards storage capacity and CO_2 migration.

As a simple estimate of fluid migration, we will use a spill-point calculation [1] in which fluid is injected at an infinitesimal rate and the buoyant forces dictate flow. Such calculations are extremely fast, and allow the full model suite to be quickly assessed in terms of maximum upslope migration distance for a given injection volume from a specific injection point. A slightly more advanced analysis is achieved by identifying the cascade of all structural traps associated with a given top-surface morphology, defined so that primary traps contain a single local peak, secondary traps contain one saddle-point and more than one peak, tertiary traps contain two

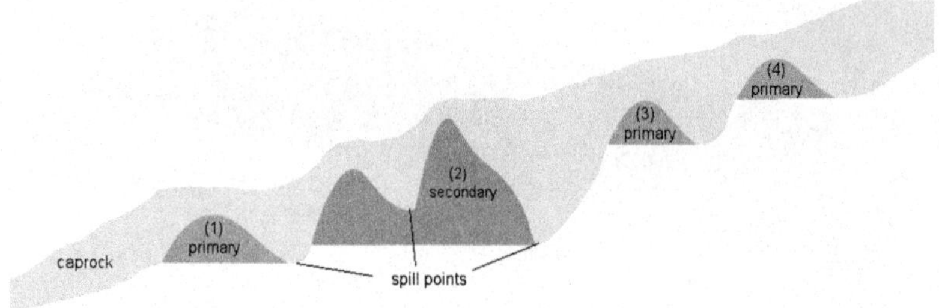

Fig. 2 Illustration of the calculation of spill-points upslope of an injection point. In the cascade of structural traps, a primary trap has no interior spill-point, secondary trap has one interior spill-point, and so on

saddle-points, etc., see Fig. 2. Using this cascade of traps, one can bracket the potential for structural trapping, estimate structural trapping for finite injection rates, optimize placement of injection points, etc.

When CO_2 is injected at a finite rate, it will form a plume that may move too fast in the upslope direction to be able to fill all structural traps predicted by the spill-point analysis. Likewise, the plume will also spread in the transverse direction and possibly contact other structural traps that cannot be reached by the spill-point algorithm. A full 3D simulation of the high-resolution stochastic models is not possible, at least not with the commercial and in-house simulators we have available. Instead, we will use a reduced model based upon vertical integration of the two-phase flow equations, see e.g., [5, 6]. Since we are mainly concerned with the long-term migration, the implied time-scales make the assumption of vertical segregation and equilibrium a robust choice from the perspective of upscaling.

5 Results

Using the setup presented in the previous sections, we have conducted a study of how top-surface morphology affects structural and residual trapping and to what extent the CO_2 plume is retarded during its upslope migration.

Example 1 (Structural trapping) In the first example, we will study how the top-surface morphology affects capacity estimates for structural trapping. To study the impact of geological uncertainty, we generate one hundred realizations of each of the fifteen model scenarios (except for the flat, unfaulted scenario, for which only a single realization is needed). For each realization, we computed the total volume available to residual trapping based upon the cascade of structural traps described in the previous section. Table 3 reports the mean volumes for a uniform porosity of 0.25 with uncertainty specified in terms of one standard deviation. In the table, the structural complexity increases from left to right and the complexity of the sedimentary topography increases from top to bottom.

Table 3 The total volume available for structural trapping for the fifteen different types of top-surface morphologies. The table reports mean volumes in 10^6 m^3 and one standard deviation estimated from one hundred realizations for each scenario assuming a porosity of 0.25

	Unfaulted	UP1	NP1	UP2	NP2
Flat	0 ± 0	96 ± 5	74 ± 23	79 ± 5	50 ± 14
OSS	608 ± 122	648 ± 99	715 ± 120	639 ± 115	629 ± 118
FMM	227 ± 2	278 ± 21	314 ± 38	260 ± 20	259 ± 27

For the flat depositional topography, all structural traps are fault traps. Here, the fault patterns with all faults normal to the flow direction (UP1 and NP1) give larger volumes than the cases that have additional faults with a strike angle of 30° relative to the flow direction (UP2 and NP2). This reduction in volume depends critically on how effective the faults that are not parallel to the trapping structure are at limiting their trapping volume. As expected, we also observe a larger uncertainty in each fault pattern when introducing a random length and displacement.

For the unfaulted cases, all structural traps are fold traps induced by the depositional topography. Here, the case with offshore sand ridges (OSS) has significantly larger storage capacity, mainly because the fold traps have lobes with larger amplitude, width, and length. Compared with the flat cases, we see that the volumes in the fold traps are (almost) one order of magnitude larger than the volumes in the fault traps.

For the cases having a combination of fold and fault traps, faults normal to the flow direction increase the storage capacity, in particular for the flooded marginal marine (FMM) cases. On the other hand, faults having a strike angle of 30° relative to the upslope direction will open some of the fold traps and hence lead to a (slightly) lower structural trapping capacity. Because the OSS scenarios have fewer and larger lobes, the variation between different realizations is larger than for the FMM cases. This is illustrated in Fig. 3, which shows the cascade of traps and how they are connected through spill paths for two specific cases; traps that are not connected will form different trees. We see that the OSS case gives trees with more branches, while the FMM case has more nodes in the biggest tree.

The study reported in Table 3 only considers potentially favorable injection scenarios in which lobes in the depositional topography are orthogonal to the upslope direction. Rotating the lobes (and the fault strikes) ninety degrees resulted in very small structural trapping capacity.

Example 2 (Single injector) In practice, it will be very difficult to utilize all the potential storage volume that lies in the structural traps unless one is willing to drill and operate a large number of injection wells. To get a more realistic estimate of actual trapping in a plausible injection scenario, we perform a spill-point calculation with a fixed injection point at coordinates (15, 15) km. Table 4 reports the corresponding mean trapped volumes with an added uncertainty of one standard deviation. Figure 4 shows structural traps computed for one realization of each of the fifteen scenarios.

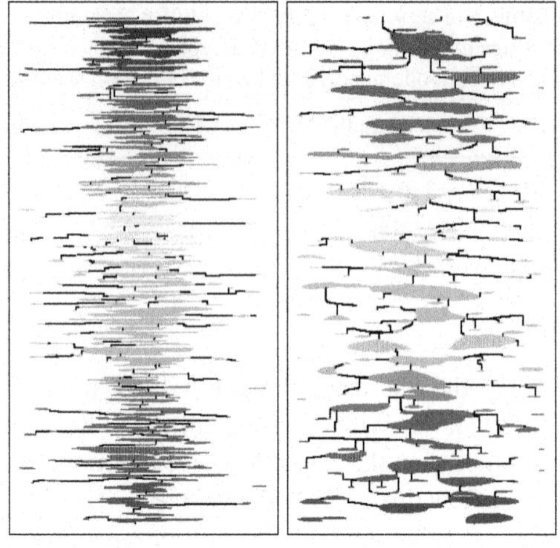

Fig. 3 The cascade of structural traps for FMM UP1 (*left*) and OSS UP1 (*right*). In the plots, the traps are presented in a tree structure with *black lines* denoting spill paths connecting traps in the upslope direction. *Colors* are used to distinguish different traps (numbered in the upslope direction)

As expected, the spill-point analysis predicts that only a fraction of the structural traps will be filled with CO_2. For the preserved beach ridges (FMM), the spill-point and total volumes are almost the same for the unfaulted cases and the cases with no crossing fault. Here, the lobes in the top-surface morphology are narrow, tightly spaced, and relatively long in the transverse direction, which means that the CO_2 will spread out laterally before migrating upslope. This observation is confirmed by the FMM-unfaulted and FMM-UP1 cases in Fig. 4, where we see that the spill path connects with almost all traps in the middle of the formation. In some of the FMM-NP1 cases—e.g., the one shown in the figure—leakage over the edges prevents the injected CO_2 from reaching the top. The actual trapped volume will therefore be much smaller than the total capacity of the whole top surface. Leakage over the edges also explains why the variation in Table 4 has increased significantly compared with Table 3 for some of the OSS and FMM scenarios. If cases with leakage are disregarded, the variation in volumes becomes more similar to the variation seen in Table 3. For crossing faults, the faults having a strike angle of 30° will accelerate the upslope migration of CO_2 and reduce the lateral filling, see the FMM-NP2 case in Fig. 4. As a result, the spill-point analysis predicts that approximately 67 % and 71 % of the available volume will be filled. For offshore sand ridges, the spill-point analysis predicts a filling degree of 60 to 69 %. Because the fold traps are much

Table 4 Trapped volumes in units of 10^6 m^3 computed by a spill-point analysis with a single source at coordinates (15, 15) km

	Unfaulted	UP1	NP1	UP2	NP2
Flat	0 ± 0	20 ± 5	30 ± 19	13 ± 3	15 ± 12
OSS	419 ± 123	431 ± 153	441 ± 180	404 ± 153	379 ± 141
FMM	239 ± 24	268 ± 24	278 ± 94	175 ± 25	184 ± 45

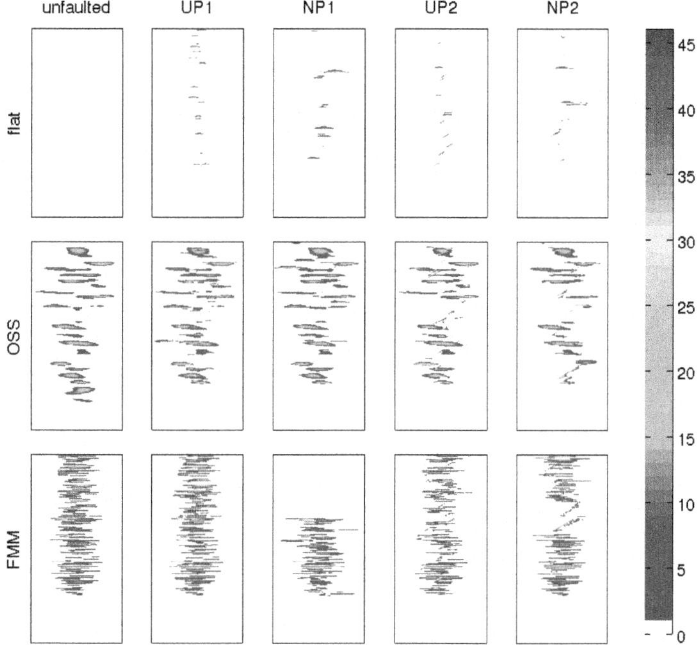

Fig. 4 Height in meters inside structural traps computed by a spill-point analysis. The *columns* show different structural scenarios and the *rows* different depositional scenarios

larger than for the FMM cases, spill paths may miss large traps on their way to the top.

For the flat cases, the spill path resulting from a single injection point will only contact a few of the available fault traps, and hence the analysis predicts that only 16–40 % of the available volume is filled for the particular injection point chosen herein. For cases having only fault traps, a much better injection strategy would thus be to use an array of injectors to improve the utilization of structural trapping.

Next, we will study how our estimates of structural trapping depend upon the placement of the injection point. To this end, we pick one specific realization for each scenario as shown in Fig. 4 and consider fifteen different injection points placed on a regular mesh with nodes at $x = 5, 10, \ldots, 25$ km and $y = 10, 15, 20$ km. Table 5 reports the corresponding trapping volumes calculated by a spill-point analysis. As observed in the previous example, the variation in volumes is significantly larger for the cases with offshore sand ridges than for the flooded marginal marine cases.

Example 3 (Prediction by flow simulation) We continue with the set of fifteen realizations used in Example 2 and a single well at (15, 15) km injecting at a rate of one million cubic meters per year. For simplicity, the reservoir is assumed to be homogeneous, with an isotropic permeability of 500 mD. Table 6 shows the free and residually trapped volumes computed by a flow simulation using a vertically-

Table 5 Volumes in units 10^6 m^3 from spill-point calculations with varying placement of the injection point

	Unfaulted	UP1	NP1	UP2	NP2
Flat	0 ± 0	21 ± 2	32 ± 3	11 ± 2	19 ± 11
OSS	311 ± 94	366 ± 121	340 ± 106	317 ± 48	250 ± 75
FMM	272 ± 21	276 ± 20	182 ± 24	208 ± 17	196 ± 22

Table 6 Free and residually trapped volumes in units of 10^6 m^3 computed by a flow simulation with a vertically-integrated model and a single injection point at (15, 15) km

	Unfaulted		UP1		NP1		UP2		NP2	
Flat	214	297	248	263	248	263	238	273	263	248
OSS	355	156	365	146	357	153	357	154	360	151
FMM	300	211	304	207	315	196	305	206	305	206

integrated model. The free volume is defined as the volume that is not residually trapped and includes volumes confined in fold and fault traps. At a first glance, the volumes for all the different scenarios may seem surprisingly similar and it may appear counter-intuitive that the residual trapping is largest for the flat cases. However, for cases with a flat deposition there is (almost) no relief in the top-surface morphology that will retard the plume migration. Hence, the plume will either reach or come very close to the top of the model within a migration period of 5000 years (see Fig. 5), and in the process sweep a relatively large volume, which results in large volumes of residually trapped CO_2. For the scenarios with offshore sand ridges, on the other hand, the large lobes in the top surface will force the upslope migration of the injected CO_2 to predominantly follow ridges in the morphology, which retard the plume migration and reduce the residual trapping compared with the flat scenarios. The FMM cases are somewhere in between. We also observe that having faults of varying length (NP1 and NP2) retards the plume migration slightly compared with cases having uniform faults (UP1 and UP2).

With regard to structural trapping, the situation is completely different. Here, OSS has the largest volumes, as seen in Table 7, while the flat scenarios have almost no structural trapping, even in the faulted cases. Comparing FMM and OSS, we see that the OSS scenarios have approximately 50 % more structurally trapped CO_2. To compare structural trapping predicted by flow simulation and by spill-point analysis, we note that in the flow simulation, the CO_2 plume has not reached the top of the structure after 5000 years for the OSS and FMM cases. The spill-point calculations, on the other hand, are run until the top is reached and hence overestimate the structurally trapped volumes. Conversely, for the flat depositional scenario, the spill-point analysis predicts a thin CO_2 trail that only contacts a few fault traps, whereas in the flow simulations the injected CO_2 plume spreads laterally and therefore contacts more fault traps, and hence gives higher trapped volumes.

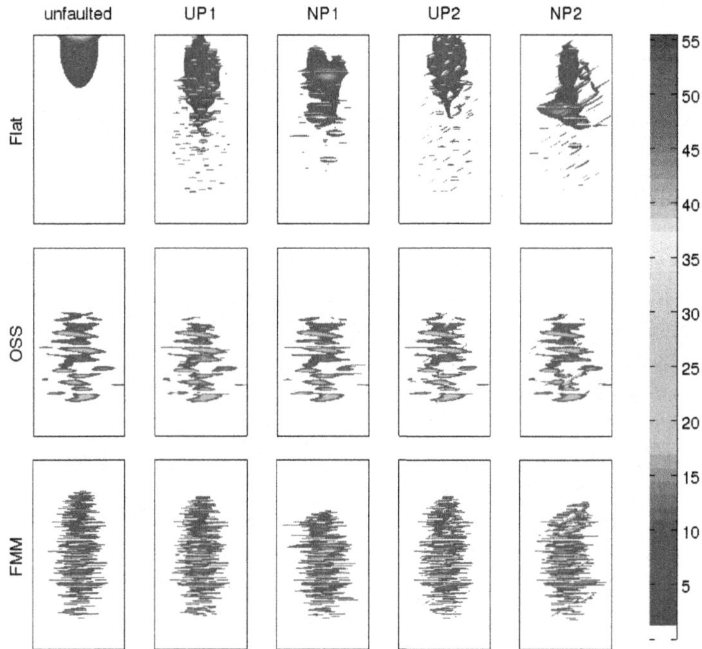

Fig. 5 Height in meters for the plumes of free CO_2. The *columns* show different structural scenarios and the *rows* different depositional scenarios

Table 7 Comparison of structurally trapped volumes in units of 10^6 m^3 computed by spill-point analysis and by simulation with a vertically-integrated model

	Unfaulted		UP1		NP1		UP2		NP2	
Flat	0	0	22	34	31	20	11	24	16	27
OSS	324	150	370	164	352	150	315	158	250	141
FMM	276	111	289	101	183	115	209	93	205	95

The volume of movable CO_2 is obtained by subtracting the structurally trapped volume from the free volume. A surprising result is that the movable CO_2 volume is approximately the same for all of the models and represents between 39 % and 45 % of the total injected volume. In terms of risk, however, quoting only the movable percentage is quite misleading. For the flat deposition, almost all the movable CO_2 volume has accumulated close to the upslope boundary. The OSS and FMM cases, on the other hand, show different degrees of retention and hence the probability of leakage will also vary a lot. For the OSS cases, most of the movable CO_2 is still far from the boundary and will gradually become structurally and residually trapped. Here, the CO_2 plume also moves very slowly since it is forced to cross the spill-points decided by the top-surface morphology. The FMM cases appear somewhere between the OSS and the flat cases, and here the plume moves a bit further because it 'feels' less effects from the top-surface morphology in the beginning of the migration when the height of the plume is larger than the height of the sand ridges.

6 Concluding Remarks

The CO_2 trapping is affected by top-reservoir morphology. However, the spread of the CO_2 plume is only inhibited if the height of the plume is of the same scale as the amplitudes of the relief. For low injection rates, a modest relief may slow migration, whereas for high injection rates, creating a thick plume, the effect is negligible. Spill-point calculations are fast, but capture only structurally trapped CO_2. A large number of realizations can be run quickly using flow simulations based upon a vertically-integrated formulation, thereby allowing a fast way of measuring volumetric uncertainty of CO_2 retention.

Our study highlights the necessity of including geological detail to models forecasting realistic CO_2 migration. The manner in which these elements affect migration must be understood, which calls for new modeling initiatives in which more detailed geology is considered.

References

1. Abrahamsen P, Hauge R, Heggland K, Mostad P (1998) Uncertain cap rock geometry, spillpoint and gross rock volume. In: SPE49286 proceedings from SPE annual technical conference and exhibition, New Orleans, Louisiana
2. Celia MA, Nordbotten JM, Bachu S, Kavetski D, Gasda S (2010) Summary of Princeton workshop on geological storage of CO_2. Princeton-Bergen series on carbon storage. http://arks.princeton.edu/ark:/88435/dsp01jw827b657
3. Class H, Ebigbo A, Helmig R, Dahle HK, Nordbotten JM, Celia MA, Audigane P, Darcis M, Ennis-King J, Fan Y, Flemisch B, Gasda S, Jin M, Krug S, Labregere D, Beni AN, Pawar RJ, Sbai A, Thomas SG, Trenty L, Wei L (2009) A benchmark study on problems related to CO_2 storage in geologic formations. Comput Geosci 13(4):409–434. doi:10.1007/s10596-009-9146-x
4. Hollund K, Mostad P, Nielsen BF, Holden L, Gjerde J, Contursi MG, McCann AJ, Townsend C, Sverdrup E (2002) Havana—a fault modeling tool. In: Koestler AG, Hunsdale R (eds) Hydrocarbon seal quantification. Norwegian Petroleum society conference, 16–18 October 2002, Stavanger, Norway. NPF special publication, vol 11. Elsevier, Amsterdam
5. Ligaarden IS, Møll Nilsen H (2010) Numerical aspects of using vertical equilibrium models for simulating CO_2 sequestration. In: Proceedings of ECMOR XII—12th European conference on the mathematics of oil recovery. EAGE, Oxford
6. Nordbotten JM, Celia MA (2012) Geological storage of CO_2: modeling approaches for large-scale simulation. Wiley, New York
7. Pruess K, Garca J, Kovscek T, Oldenburg C, Rutqvist J, Steefel C, Xu T (2004) Code intercomparison builds confidence in numerical simulation models for geologic disposal of CO_2. Energy 29(9–10):1431–1444. doi:10.1016/j.energy.2004.03.077

Modeling and Analysis of Daily Rainfall Data

Bjørn H. Auestad, Andreas Henriksen, and Hans A. Karlsen

Abstract In this work we have tried to develop a statistical model for the daily rainfall measurements. Data series from Bergen the last 107 years and from Sviland, Rogaland, the last 115 years, obtained by The Norwegian Meteorological Institute, has specifically been studied. This statistical approach is based only on measured precipitation data. The high frequency of daily data may contain important information of the underlying meteorological process and we investigate possibilities to extract this information. Simple parametric modeling with components for seasonal variation is used to represent the data. A parsimonious parametric model may contribute to increased statistical power of analyses and hypothesis testing of possible changes in the meteorological process. Our parametric model with quite few parameters can used for a detailed study of the properties of trends and changes over time. Properties such as occurrence of wet days, expected amount of rain, spell lengths and extreme events are of special interest. Similar modeling of daily rainfall data has been used by others, but not for the same purposes as we suggest. We have used generalized linear models as statistical tool for fitting the data to the model. Based on simulation studies, comparisons of model estimated quantities and corresponding data quantities, the model seems to fit series of daily rainfall data very well. We also combine this modeling with techniques for statistical process control to detect changes in the rainfall process. For two particular series considered, we see indications of development in jumps between levels rather than slowly evolving trends.

B.H. Auestad (✉)
Faculty of Science and Technology, University of Stavanger, 4036 Stavanger, Norway
e-mail: bjorn.auestad@uis.no

A. Henriksen
Kavli Senter, forskningssenter for aldring og demens, Haraldsplass Hospital, Ulriksdal 8C, 5009 Bergen, Norway
e-mail: andreas.h.henriksen@gmail.com

H.A. Karlsen
Dep. of Math., University of Bergen, PB 7800, 5020 Bergen, Norway
e-mail: hans.karlsen@uib.no

P. Abrahamsen et al. (eds.), *Geostatistics Oslo 2012*,
Quantitative Geology and Geostatistics 17,
DOI 10.1007/978-94-007-4153-9_40,

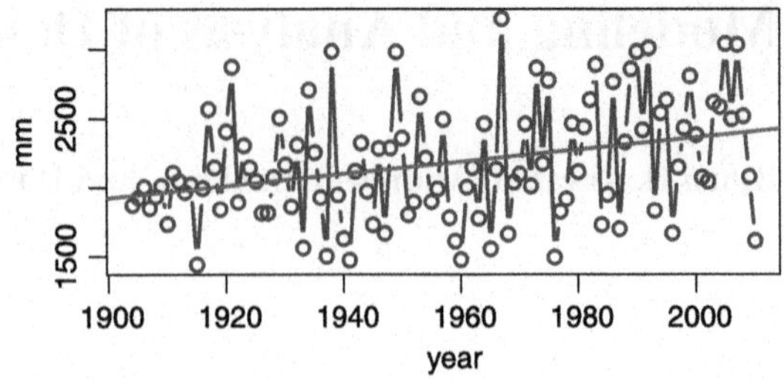

Fig. 1 Total yearly rain in Bergen during the years 1904–2010. *Red line*: fitted linear regression with slope estimate: 4.375 (significant p-value: 0.0008)

1 Introduction

Climate evolvement attains much attention. Changes in e.g. temperature and precipitation are analyzed and sought predicted. Often the focus is on aggregated quantities such as e.g. yearly amounts of rain. In Fig. 1 the total yearly precipitation in Bergen is shown for the years 1904–2010. An increasing trend seems fairly clear. Such an increase in yearly amounts could be composed of more expected rain evenly spread over the seasons of the year. Alternatively it might be caused by more heavy rain concentrated to e.g. autumn and winter. Changes in distribution of spell lengths may also contribute to changes in year totals and other possibilities obviously exist. Daily data may contain useful information on such structural properties of rainfall. Adequate modeling and establishing of methods for analyses and inference would be useful tools for studies of changes in the structure over time.

In this paper we discuss a model for daily rainfall. Based on this model we suggest analyses that may give answers to questions as those posed above. Similar modeling have been used earlier (see e.g. [8]) but to our knowledge not for the purposes we suggest. In addition we clarify the theoretical basis for statistical inference based on this model as well as methods of estimation. Related works are [2, 3, 9, 10].

Section 2 of the paper describes the model and possible analyses of daily rainfall data. In Sect. 3 examples of applications are shown.

2 The Model

The theoretical basis for the modeling is described in the first sub section. Estimation and sampling properties of the estimators are then showed and interpretations of the model are discussed in the third sub section.

2.1 Description

We model daily rain data as a bivariate time series, $\{(Y_t, U_t)\}$, where $\{Y_t\}$ is the occurrence process and $\{U_t\}$ is the amount of rain on day t. The occurrence process consists of binary variables where, $Y_t = 1$ if day t is wet and $Y_t = 0$ if day t is defined as dry. In our model $\{Y_t\}$ is a logistic INAR(p) process, i.e., an integer valued autoregressive time series (cf. [5, Chap. 2]).

In this setting the occurrence process is controlling the other process. That is the conditional distribution of U_t given the past and $Y_t = 1$, is exponential with time dependent mean μ_t and likewise if $Y_t = 0$ then $U_t = 0$. The observed daily rainfall is $X_t = Y_t U_t$. The model implies the conditional independence property which means that all the dependence in the time series goes through the occurrence process and given this process up to time t, all the U's are independent up to time t. We denote this the conditional independence assumption. The assumption may appear to be a limitation of the model but our data seem to fulfill this restriction quite well. This finding is also in accordance studies (cf. e.g. [8]). This will be further discussed in the data section. However, the observed daily rain, $\{X_t\}$, is in general a time series with a dependence structure.

For the occurrence process we model the probability of rain on day t conditional on the past as

$$P\left(Y_t = 1 \middle| \mathcal{F}_{t-1}^{Y}\right) = \frac{\exp(\theta_{t0} + \sum_{j=1}^{p} \theta_{tj} Y_{t-j})}{1 + \exp(\theta_{t0} + \sum_{j=1}^{p} \theta_{tj} Y_{t-j})}, \tag{1}$$

where $\mathcal{F}_{t-1}^{Y}$ represents all the information about $\{Y_t\}$ prior to time t. i.e., this is a binary autoregression with inverse link $F(z) = \exp(z)/(1 + \exp(z))$, the standard logistic cumulative distribution function. The linear predictor, $\theta_{t0} + \sum_{j=1}^{p} \theta_{tj} Y_{t-j}$, depends on p lagged values of Y_t. The dependence of the probability of rain at day t on the p preceding days is modeled by the autoregressive parameters $\theta_{t1}, \ldots, \theta_{tp}$. The coefficient θ_{t0} expresses the non stochastic level of the probability of rain on day t.

To handle the seasonal variation and obtain moderate number of unknown parameters the coefficients $\theta_{t0}, \theta_{t1}, \ldots, \theta_{tp}$ are further parameterized as sums of harmonics

$$\theta_{tj} = \alpha_{j0} + \sum_{k=1}^{q} \left\{ \alpha_{jk} \cos\left(2\pi k \frac{t}{\tau}\right) + \beta_{jk} \sin\left(2\pi k \frac{t}{\tau}\right) \right\}, \quad j = 0, 1, \ldots, p, \tag{2}$$

where $\alpha_{j0}, \alpha_{j1}, \ldots, \alpha_{jq}$ and $\beta_{j1}, \ldots, \beta_{jq}$, $j = 1, \ldots, p$ are unknown parameters and $\tau = 365$ is the number of days in a year without leap year. In practice $q = 2$ harmonics may be sufficient. The autoregressive order, p, is in our data quite small, i.e., one or two.

For $\{U_t\}$, on a wet day the expectation $\mu_t = \mathrm{E}(U_t)$ is modeled with seasonal variation as

$$\mu_t = \exp\left[a_0 + \sum_{k=1}^{q} \left\{ a_k \cos\left(2\pi k \frac{t}{\tau}\right) + b_k \sin\left(2\pi k \frac{t}{\tau}\right) \right\} \right]. \tag{3}$$

2.2 Estimation and Distributional Properties of the Estimators

We can express the log likelihood in terms of the bivariate process $\{(Y_t, U_t)\}$:

$$l = \sum_{t=1}^{T} \log P_{t,\boldsymbol{\alpha},\boldsymbol{\beta}}(Y_t \mid Y_{t-1}, \ldots, Y_{t-p+1}) + \sum_{t=1}^{T} Y_t \log g_{t,\boldsymbol{a},\boldsymbol{b}}(U_t) = l_1 + l_2,$$

where $g_{t,\boldsymbol{a},\boldsymbol{b}}$ is the exponential density with rate μ_t given by (3), and $P_{t,\boldsymbol{\alpha},\boldsymbol{\beta}}$ is (1) parameterized by (2). The $\boldsymbol{\alpha}$, $\boldsymbol{\beta}$, $\boldsymbol{a}$ and $\boldsymbol{b}$ are vectors of the parameters $\alpha_{j0}, \alpha_{j1}, \ldots, \alpha_{jq}; \beta_{j1}, \ldots, \beta_{jq}; a_0, a_1, \ldots, a_q$ and $b_1, \ldots, b_q$, respectively.

Since the decomposition of the log likelihood function also corresponds to a division of the parameters in two separate sets, the log likelihood equations are given by

$$\nabla_{\boldsymbol{\alpha},\boldsymbol{\beta}} l_1(\widehat{\boldsymbol{\alpha}}, \widehat{\boldsymbol{\beta}}, Y_t, \ldots, Y_1) = 0 \quad \text{and} \quad \nabla_{\boldsymbol{a},\boldsymbol{b}} l_2(\widehat{\boldsymbol{a}}, \widehat{\boldsymbol{b}}, U_t, \ldots, U_1) = 0.$$

This means that the parameters of the $\{U_t\}$ and $\{Y_t\}$ processes can be estimated independently. This is a considerable advantage for of the estimation procedure.

The variables $\{U_t\}$ follow a generalized linear model. Estimation procedures are available in standard software packages, and the distributional properties of the estimators are well established. The estimators are asymptotically consistent and multivariate normally distributed,

$$\sqrt{T}\left(\begin{bmatrix} \widehat{\boldsymbol{a}} \\ \widehat{\boldsymbol{b}} \end{bmatrix} - \begin{bmatrix} \boldsymbol{a} \\ \boldsymbol{b} \end{bmatrix}\right) \xrightarrow{d} N\left(0, \boldsymbol{I}^{-1}\right),$$

where $\boldsymbol{I}$ is the information matrix.

Taking advantage of the Markovian structure, $\{Y_t\}$, estimation can be done using ordinary maximum likelihood and an explicit and computable expression for the information matrix is obtainable. However, here we will use the approach of [5]. To see that our model fits into the framework of [5], define the three covariates

$$Z_{t,0} = Y_t, \qquad Z_{t,1,k} = \cos\left(2\pi k \tau^{-1}\right) Y_t,$$
$$Z_{t,2,k} = \sin\left(2\pi k \tau^{-1}\right) Y_t; \quad k = 1, \ldots, q,$$

the parameter vector,

$$\boldsymbol{\gamma}' = \left[(\alpha_{j,0},\ \alpha_{j,k};\ k = 1, \ldots, q,\ \beta_{j,k};\ k = 1, \ldots, q)\ j = 1, \ldots, p\right],$$

the vector of the random variables,

$$\boldsymbol{Z}_t' = [Z_{t,0}, Z_{t,1,1}, \ldots, Z_{t,1,q}, \quad Z_{t,2,1}, \ldots, Z_{t,2,q}],$$

and the inverse link function:

$$F(z) = \frac{\exp(z)}{1 + \exp(z)}.$$

Then the logistic autoregression (1) can then be written as $P(Y_t = 1 | \mathcal{F}_{t-1}^{Y}) = F(\boldsymbol{\gamma}'\boldsymbol{Z}_t)$, as in [5].

Note that the dimension of γ is $(1 + 2q) \times (p + 1)$. When $q = p = 1$ or $q = p = 2$, this dimension is 6 and 15, respectively. These are quite moderate numbers compared to data series of approximately 40 000 observations.

2.3 Interpretations of the Model

For the amounts model, $\{U_t\}$, the parameter a_0 represents the main level and the $a_1, \ldots, a_q$ and $b_1, \ldots, b_q$ represent seasonal adjustments to the main level. The number of sine and cosine terms, q, needed for a satisfactory description of the data, may be judged based on criteria as AIC or BIC. Also the likelihood ratio test may be employed to evaluate need for higher q.

Estimates of $a_0, a_1, \ldots, a_q$ and $b_1, \ldots, b_q$ from different time periods may be compared to investigate possible differences in the level and seasonal distribution of expected daily rain. Since the estimators are approximately multivariate normal, such tests may be carried out using standard tests for multinormal vectors, (see e.g. [4, Chap. 6]).

The parameters of the expected daily rain, μ_t, may also be time dependent to allow for time trends in the parameter values. A linear trend may be modeled by replacing a_k with $a_{k0} + a_{k1}t$ and b_k with $b_{k0} + b_{k1}t$, $k = 0, 1, \ldots, q$. Existence of trends may then be judged by hypothesis tests of the slope parameters a_{k1} and b_{k1}, $k = 0, 1, \ldots, q$. Results for such tests are parts of standard output of a generalized linear models analysis.

For the occurrence model given by (1) and (2) the parameter θ_{t0} represents the level of probability of rain at day t. For θ_{t0} and the autoregressive parameters, $\theta_{t1}, \ldots, \theta_{tp}$, the interpretations of seasonal effects represented by α_{j0}, $\alpha_{j1}, \ldots, ts, \alpha_{jq}$ and $\beta_{j1}, \ldots, \beta_{jq}$, $j = 1, \ldots, p$, are similar to that of the amounts model.

With an autoregressive order $p = 0$ in (1), we are left with an ordinary logistic regression model, and similar kinds of analyses may be performed of the $\{Y_t\}$ process as was indicated for the $\{U_t\}$ process. As will be demonstrated below (Sect. 3.4), data strongly support autoregressive dependence in the occurrence process which means that the order p is greater than zero. With lagged values of Y_t in the predictor of $P(Y_t = 1|\mathcal{F}^Y_{t-1})$, the sampling results are based on assumptions that do not permit inclusion of a linear time trend in the seasonal parameters, (cf. [5, Chap. 2.2]). As an alternative we suggest using techniques from statistical process control to study time evolving changes in the parameters of the occurrence process. We have chosen to consider averaged parameter estimates in non overlapping five year periods. The sequence of estimates for each parameter may be plotted in a Shewart diagram, (cf. [6, Chap. 6]). The use of such diagrams is a well established method used to detect changes e.g. in factors contributing to the output of industrial processes. Other statistical process control techniques, e.g. exponential weighted moving average, may have better power for detecting small changes in parameter values than the Shewart diagram and may enhance results (cf. [6]).

The distribution of lengths of periods with consecutive wet/dry days, spell lengths, are of interest. For spells of dry days, let L_0 be the number of consecutive dry days from day t, conditional on day t being dry, $Y_t = 0$. The probability of l consecutive dry days is given by $P(L_0 = l) = P(Y_{t+l} = 1, Y_{t+l-1} = 0, \ldots, Y_{t+1} = 0|Y_t = 0)$. For autoregressive order $p = 1$ and using the Markov property of $\{Y_t\}$, we have $P(L_0 = l) = P_{t+1,00} \cdots P_{t+l-1,00} P_{t+l,01}$, $l = 1, 2, 3, \ldots$,

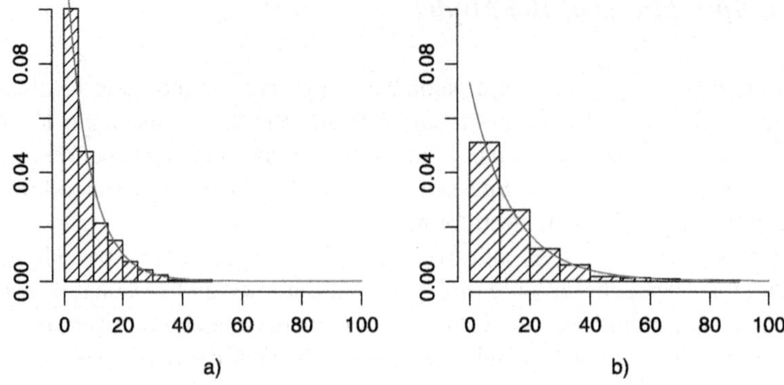

Fig. 2 Distribution of daily rainfall data compared with fitted exponential distribution. Data from the Bergen series 1982–2010, (**a**) daily rainfall in May months and (**b**) daily rainfall in November months. *Red line*: fitted exponential densities

where $P_{t,ij} = P(Y_t = j|Y_{t-1} = i)$, $i, j = 0, 1$. For $l = 1$ we define $P_{t+l,00} = 1$ in this expression. For spells of wet days, L_1, we have similarly $P(L_1 = l) = P_{t+1,11} \cdots P_{t+l-1,11} P_{t+l,10}$, $l = 1, 2, 3, \ldots$. In our model for the occurrence process we have with $p = 1$

$$P_{t,ij} = P(Y_t = j|Y_{t-1} = i) = \frac{(1 - j) + j \exp(\beta_{t0} + \beta_{t1} i)}{1 + \exp(\beta_{t0} + \beta_{t1} i)}, \quad i, j = 0, 1.$$

Thus changes in spell length distributions may be studied through changes over time in estimates of β_{t0} and β_{t1}.

It may be reasonable to assume that $P_{t,ij}$ varies slowly with time, and it seems fair to use the approximation: $P_{t,00} = P_{t+1,00} = \cdots = P_{t+l-1,00}$ and $P_{t+l,01} = P_{t,01}$ if l is not unreasonably large. Therefore the spell length probabilities for L_0 may be approximated by $P(L_0 = l) = P_{t,00}^{l-1} P_{t,01}, l = 1, 2, 3, \ldots$. This is the geometric distribution having mean $\mathrm{E}(L_0) = 1/P_{t,01}$ and variance $\mathrm{Var}(L_0) = (1/P_{t,01})^2$. The validity of this approximation can be examined using the estimated model. In the same manner the mean and variance of spell lengths of wet days, L_1, may be approximated by mean $\mathrm{E}(L_1) = 1/P_{t,10}$ and variance $\mathrm{Var}(L_1) = (1/P_{t,10})^2$.

3 Applications

The model has been fitted to several data sets. We present some results for two of the series. One is from Bergen, Norway, containing measurements of daily rainfall from January 1, 1904 to December 31, 2010. This series is composed from measurements from two locations; calibrated using simple regression. The other series is from Sviland in Rogaland county, Norway, with daily data from January 1, 1896 to December 31, 2010. To simplify February 29 in leap years are disregarded. We have used a threshold of 0.7 mm rain for definition of a wet day. The program package [7] is used in the analyses.

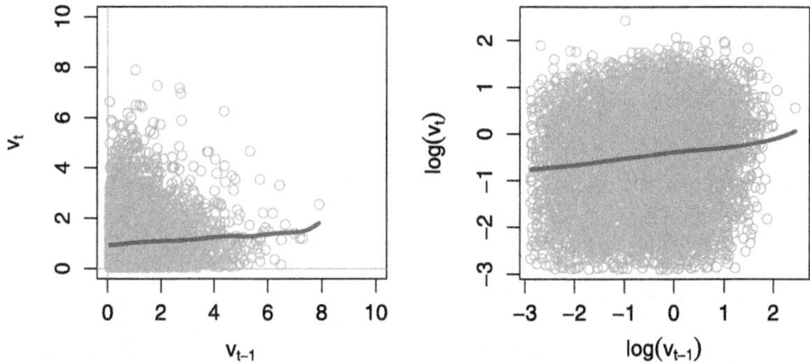

Fig. 3 Scatter plot of consecutive standardized amounts of daily rain, $v_t = u_t/\widehat{\mu}_t$. *Left figure* shows (v_{t-1}, v_t) and *right figure* shows log transformed quantities, $(\log(v_{t-1}), \log(v_t))$. Data from Bergen series 1982–2010. The expectation, μ_t, is estimated with the data from the same period using (3) with $q = 2$. *Red line*: kernel smoother estimating the conditional expectation $\mathrm{E}(V_t | V_{t-1} = v)$ and the corresponding for log transformed quantities

3.1 Distribution and Conditional Independence of the U_t's

The assumption of independence and exponential distribution for the U_t's have been used in similar modeling (see [1, 8]). Simple graphical analysis gives strong support to these assumptions. In Fig. 2 we have shown histograms of daily rainfall in May and November months from the Bergen series 1982–2010. The exponential densities are fitted with the average rainfall as mean. The estimated distribution seems to fit the data very well. Using other months and periods of the series also show similar good fits.

The validity of the conditional independence assumption, may be judged by looking at scatter plots of pairs, (u_{t-1}, u_t), of rainfall on succeeding wet days. Scatter plots of rainfall amounts are shown in Fig. 3 and log transformed amounts on succeeding wet days. Standardized amounts, U_t/μ_t, are used so that for all t the mean and variance is 1 when $U_t \sim \exp(\mu_t)$. None of the plots indicate strong dependencies between U_{t-1} and U_t. Estimated correlations are 0.099 with approximate 95 % confidence interval $(0.073, 0.124)$ and 0.142 with interval $(0.117, 0.167)$ for the original (and standardized) data and the log transformed data, respectively. The very weak positive correlations are significant, but this is as expected with moderately large data set, 5828 data pairs, and it seems reasonable to model these quantities as independent. We are aware of that there may by stronger dependencies if precipitation data are sampled at higher rate than daily.

3.2 The $\{U_t\}$ Process Fitted to Data

Data series from Sviland in the periods 1931–1980 and 1981–2010 have been fitted to the amounts model $\{U_t\}$. The reason for choosing these periods will be explained

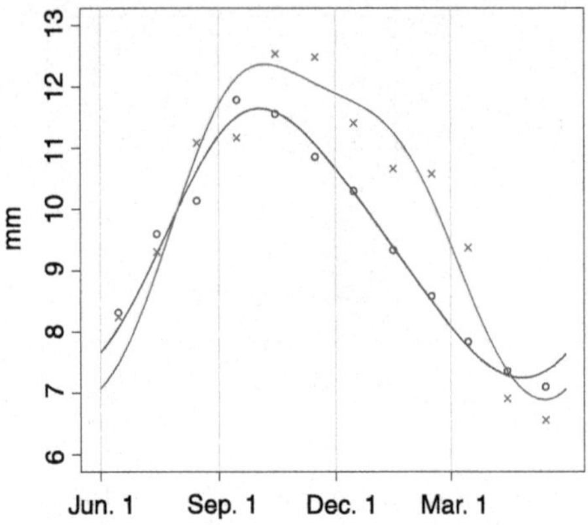

Fig. 4 Estimated seasonal profile of expected daily rain on wet days from two different periods of the Sviland series, *blue*: 1931–1981 and *red*: 1982–2010. *Solid lines* are model estimated amount comparable with the *crosses/circles* which are data averages of daily rain on wet days in the corresponding periods

below when use of statistical process control techniques is discussed. Figure 4 shows the estimated seasonal profile of expected daily rain on wet days, i.e., estimates of μ_t through a whole year. The model (3) with $q = 2$ is used. The values of the estimated parameters together with standard errors are shown in Table 1. Together with each of the curves from the two periods are also the corresponding averages of daily rainfall. The averages are made for data in each month to reveal the seasonality of the rainfall. It is seen that the estimated model produces a fairly good fit to the observed averages in both periods. Also there is a marked difference in the two profiles. This is most apparent during autumn and winter where estimated expected amounts are highest in the 1981–2010 period. The significance of the difference of the two profiles can be investigated by means of e.g. confidence bands or point wise confidence intervals.

3.3 Extreme Values

Extreme events are of special interest and we have shortly examined the ability of the model to reveal occurrences of extreme precipitation. In the Bergen series the event of at least 100 mm rain have occurred 8 times during the 150 years from 1861 to 2010. We have conducted a small simulation experiment to study the capability of the model to reproduce a realistically number of such extreme events. The model with $p = 1$ for the $\{Y_t\}$ and $q = 2$ for both $\{U_t\}$ and $\{Y_t\}$ is fitted to the Bergen series data in the two equally long periods 1904–1932 and 1982–2010. Based on these models, 100-years realizations of daily rain are simulated and the number of days with at least 100 mm rain is registered. With the model from the first period we find an average of 4.1 occurrences per 100 years with a standard error equal to 0.2 and from the second period the result shows an average of 8.3 occurrences per 100 years with standard error equal to 0.24. From each model one 100 100-years realizations were simulated. These numbers seem to be comparable with the observed number of such events in the Bergen series, which is 5.3 per 100 years. This indicates that

Table 1 Amounts model (3) with $q = 2$ fitted to the Sviland series in the periods 1931–1981 (left part) and 1982–2010 (right part). Est.: parameter estimate and S.E.: estimated standard error

Parameter	1931–1981		1982–2010	
	Est.	S.E.	Est.	S.E.
a_0	2.2266	0.0107	2.2825	0.0133
a_1	0.0617	0.0151	0.1697	0.0188
b_1	−0.2243	0.0152	−0.2252	0.0187
a_2	−0.0039	0.0149	−0.0058	0.0186
b_2	0.0262	0.0152	0.0734	0.0188

our models have capability to reflect some extreme event properties of the true daily rainfall process and this may be somewhat surprising since model is mainly based on ordinary values and not extreme values.

3.4 AR Order of $\{Y_t\}$

The results of fitting the occurrence model (1) to the Sviland series are shown in Table 2. Models with AR order $p = 1$ and $p = 2$ are used and $q = 2$ in both cases. The rightmost columns in both sub tables show the standardized value of the parameter estimates. These quantities are approximately standard normally distributed under the null hypothesis of zero parameter value. For the AR(1) model it is clearly seen that both α_{10} and α_{11} are significant showing strongly dependencies at lag 1. The results of fitting the AR(2) model also show strong indications of dependencies at lag 2. Also the AIC and BIC values are lower for the second order model. In both cases some of the seasonal parameters are significant thus showing that level of probability of rain at day t and autoregressive dependence structure changes during the season. When interpreting significance of the parameters it should be borne in mind that the data sets are very large and small deviations from the null are also likely to be detected. Multiple testings problem might also pose a challenge.

3.5 Shewart Diagram for Parameters of $\{U_t\}$

Changes in parameter values governing the processes $\{Y_t\}$ or $\{U_t\}$ may be studied using techniques from statistical process control (SPC). Such methods are aimed at detecting time points at which the system under consideration changes from a defined normal state to an out-of-control state. The out-of-control state is characterized by a change in parameter values of the statistical model of the process considered. See e.g. [6] for more details.

One possibility to apply SPC to daily rain processes is to fit the model in shorter periods, e.g. five years, and plot the estimated parameter values in Shewart diagram. This is done for the Sviland series where the amounts model (3) with $q = 2$ is fit in

Table 2 Occurrence model (1) fitted to the Sviland series, 1896–2010. Number of data: $T =$ 41 975; Left table: results when model with $p = 1$ and right table: results when model with $p = 2$. In each table the first column is the parameter estimate, second column shows the estimated standard error and the third column shows the statistic $z =$ estimate/standard error. First order model: $-2\log$L: 50001.39; AIC $=$ 50 021, BIC $=$ 50 108. Second order model: $-2\log$L: 49 661.68; AIC $=$ 49 692, BIC $=$ 49 821

First order model				Second order model			
Parameter	Est.	S.E.	z	Parameter	Est.	S.E.	z
α_{00}	−0.9249	0.0153	−60.4510	α_{00}	−1.0552	0.0171	−61.7076
α_{01}	0.0047	0.0216	0.2176	α_{01}	−0.0264	0.0241	−1.0954
β_{01}	−0.1889	0.0216	−8.7454	β_{01}	−0.1817	0.0241	−7.5394
α_{02}	−0.0006	0.0216	−0.0278	α_{02}	0.0000	0.0242	0.0000
β_{02}	0.0509	0.0215	2.3674	β_{02}	0.0513	0.0240	2.1375
α_{10}	1.7731	0.0217	81.7097	α_{10}	1.6089	0.0233	69.0515
α_{11}	0.3028	0.0307	9.8632	α_{11}	0.2481	0.0329	7.5410
β_{11}	−0.0651	0.0307	−2.1205	β_{11}	−0.0719	0.0329	−2.1854
α_{12}	0.0087	0.0307	0.2834	α_{12}	0.0053	0.0329	0.1611
β_{12}	−0.0273	0.0307	−0.8893	β_{12}	−0.0235	0.0329	−0.7143
				α_{20}	0.4303	0.0233	18.4678
				α_{21}	0.0815	0.0330	2.4697
				β_{21}	0.0230	0.0330	0.6970
				α_{22}	0.0006	0.0329	0.0182
				β_{22}	−0.0096	0.0329	−0.2918

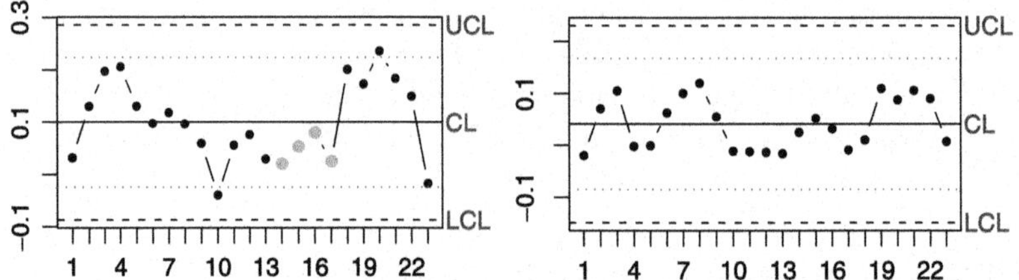

Fig. 5 Control charts for the parameters a_1 (*left*) and b_2 (*right*) of (3) when fitted in the periods 1896–1900, 1901–1905, ..., 2006–2010. The x-axis shows 1 for first five year period, 2 for second and so on. *Dashed lines* show lower and upper control limit

the 23 five-year periods 1896–1900, 1901–1905, ..., 2006–2010. Control chart for the parameters a_1 and b_2 are shown in Fig. 5. Especially diagram for a_1 shows a period between the 9'th and 17'th five year period with parameter value at a lower level and periods with higher values earlier and later. This pattern is also to a certain extent visible in the chart for b_2. For the amounts process this may be seen as indications of development in jumps between levels rather than slowly evolving trends. The charts for the other parameters showed no clear patterns. The indication of change at the 9'th and 17'th five year period is reason for dividing the data in the periods 1931–1981 and 1982–2010 and studying possible differences.

References

1. Buishand TA (1978) Some remarks on the use of daily rainfall models. J Hydrol 36(3–4):295–308
2. Buishand TA, Klein Tank AMG (1996) Regression model for generating time series of daily precipitation amounts for climate change impact studies. Stoch Hydrol Hydraul 10:87–106
3. Chandler RE, Wheater HS (2002) Analysis of rainfall variability using generalized linear models: a case study from the west of Ireland. Water Resour Res 38(10):10
4. Johnson RA, Wichern DW (2002) Applied multivariate statistical analysis, 5th edn. Wiley, New York
5. Kedem B, Fokianos K (2002) Regression models for time series analysis. Wiley-Interscience/Wiley, New York
6. Montgomery DC (2009) Statistical quality control. A modern introduction, 6th edn. Wiley, New York
7. R Development Core Team (2011) A language and environment for statistical computing. R Foundation for Statistical Computing, Vienna, Austria. ISBN:3-900051-07-0. url:http://www.R-project.org/
8. Stern RD, Coe R (1984) A model fitting analysis of daily rainfall data. J R Stat Soc, A (General) 147(1):1–34
9. Villarini G, Smith JA, Napolitano F (2010) Nonstationary modeling of a long record of rainfall and temperature over Rome. Adv Water Resour 33(10):1256–1267
10. Yang C, Chandler RE, Isham VS, Wheater HS (2006) Quality control for daily observational rainfall series in the UK. Water Environ J 20(3):185–193

A Stochastic Model in Space and Time for Monthly Maximum Significant Wave Height

Erik Vanem, Arne Bang Huseby, and Bent Natvig

Abstract It is important to take severe sea state conditions into account in ship design and operation and there is a need for stochastic models describing the variability of sea states. These should also incorporate realistic projections of future return levels of extreme sea states, taking into account possible long-term trends related to climate change. The stochastic model presented herein is such a model. The model has been fitted by significant wave height data for an area in the North Atlantic ocean and it aims at describing the temporal and spatial variability of significant wave height in this area. The model will be outlined and the results obtained by using monthly maximum data will be discussed.

1 Introduction and Background

A correct and thorough understanding of meteorological and oceanographic conditions such as the significant wave height is of paramount importance to maritime safety. Currently, metocean data for the last 50+ years are available and wave parameters obtained from hindcast of these data are commonly used as basis for design of marine and offshore structures. However, ships and other marine structures are designed for a lifetime of several decades and design codes and standards should take this into account. In particular, considering potential impacts of climate change, there is a need for statistical models that can take long-term trends properly into account.

A thorough literature survey on stochastic models for ocean waves, previously presented in [10], identified Bayesian hierarchical space-time models as promising candidates for modeling the spatial and temporal variability of wave climate

E. Vanem (✉) · A.B. Huseby · B. Natvig
Department of Mathematics, University of Oslo, Oslo, Norway
e-mail: erikvan@math.uio.no

A.B. Huseby
e-mail: arne@math.uio.no

B. Natvig
e-mail: bent@math.uio.no

P. Abrahamsen et al. (eds.), *Geostatistics Oslo 2012*, Quantitative Geology and Geostatistics 17, DOI 10.1007/978-94-007-4153-9_41,

[5, 18, 19]. Thus, such a model has been developed for data of significant wave height for an area in the North Atlantic ocean. The impact of climate change on extreme wave conditions, and the uncertainties involved, have been investigated in several previous studies such as [4, 6–8, 14–16], see e.g. [10] for a more comprehensive review.

The model for significant wave height will be outlined and the structure of the paper is as follows: The first part of this paper consists of an initial data description. The main part will outline the model and simulation results for various model alternatives. Previous results from this model have been reported for daily and monthly data for both the original data [11] and the log-transformed data [13], but not for monthly maxima.

1.1 Area Description

The ocean area in the North Atlantic between 51°–63° North and 12°–36° West will be the focus of this study. This was chosen due to the fact that North Atlantic conditions are often used as basis for design of ships and offshore structures. The spatial resolution of the data is $1.5° \times 1.5°$, hence a grid of $9 \times 17 = 153$ data-points covers the area.

1.2 Data Description and Initial Data Inspection

In this study, the modeling has been based on the corrected ERA-40 data of significant wave height, which covers a 45-year period from September 1957 to August 2002 [2, 3].[1] It includes fields of significant wave height sampled every 6th hour covering the period from January 1958 to February 2002.

Before developing the spatio-temporal model for the significant wave height data, an initial data inspection will be carried out. The density of the monthly maxima for the complete period and area is shown in Fig. 1 and two distinct modes can be identified, one around 5 m and another at about 8 m. It is believed that these correspond to different characteristics during calm and rough seasons. For the whole dataset, the mean monthly maximum is 7.5 m, but the average monthly maxima for each month varies from 4.42 (July) to 9.87 (January). The density plots for the months January to March and October to December have peaks around 8 m and for the months May to August around 5 m. The remaining months seem to be rather flat with most probability mass between 5 and 8 m. Hence, the two modes in the density may be explained by typical monthly maximum at 5 or 8 m for calm and rough seasons respectively.

[1] Provided by Dr. Andreas Sterl at the Royal Netherlands Meteorological Institute (KNMI).

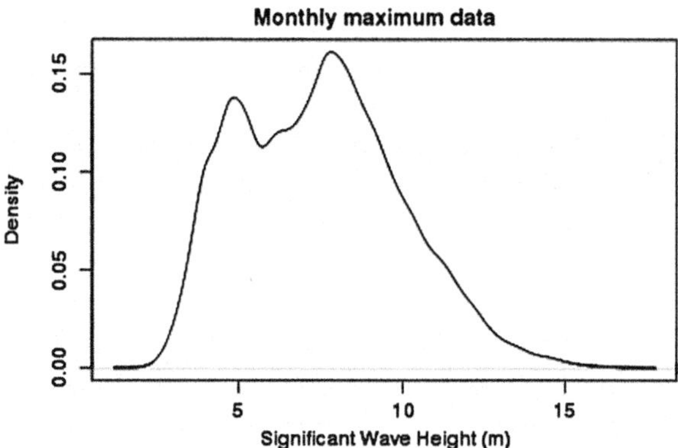

Fig. 1 The density of the monthly maximum data

It is observed that higher wave heights are associated with higher uncertainty, and therefore the model was also run for log-transformed data in order to account for this. Normal probability plots of the original and log-transformed data show that the data are neither Gaussian nor log-normal. Investigation of time-series of monthly maxima for each month at arbitrary locations revealed that notable increasing trends can be detected for some months, most notably February and December, but not for others.

2 Model Description

Hierarchical models are known to model spatio-temporal processes with complex dependence structures at different scales [17]. Therefore, a Bayesian hierarchical space-time model, along the lines drawn out by e.g. [18] will be attempted to model the significant wave height data in space and time. In the following, the spatio-temporal data will be unambiguously identified by two indices; an index x to denote spatial location with $x = 1, 2, \ldots, X = 153$ and an index t to denote a point in time with $t = 1, 2, \ldots, T = 530$.

The structure of the model will be outlined below; first the main model will be outlined, and then various model alternatives are suggested for comparison. The model will also be used together with a logarithmic transformation of the data.

2.1 Main Model

At the first level, the observations, Z, i.e. the monthly maxima at location x and month t, are modeled in the observation model as latent (or hidden) variables, H, corresponding to the underlying maximum significant wave height process, and some random noise, ε_Z:

$$Z(x,t) = H(x,t) + \varepsilon_Z(x,t) \quad \forall x \geq 1, t \geq 1. \tag{1}$$

It is noted that all noise terms to be included in the model are assumed to be independent of all other noise terms in the model, and the generic notation $\varepsilon_\zeta(x,t) \overset{\text{i.i.d}}{\sim} N(0,\sigma_\zeta^2)$ is introduced.

The underlying process for the significant wave height at location x and time t is modeled by the state model, which is split into a time-independent component, $\mu(x)$, a time-space component $\theta(x,t)$ and spatially independent seasonal and long-term trend components, $M(t)$ and $\mathrm{T}(t)$:

$$H(x,t) = \mu(x) + \theta(x,t) + M(t) + \mathrm{T}(t) \quad \forall x,t. \tag{2}$$

$\mathrm{T}(t)$ is, in fact, the component of most interest in this particular case. In line with the model presented in [9], no noise terms are introduced in the model at this level. For the remainder of this paper, the following notation is used for neighboring locations of x in space: x^D = the location of the nearest gridpoint in direction D from x, where $D \in \{N, S, W, E\}$ and N = North, S = South, W = West and E = East. If x is at the border of the area, the value at the corresponding neighboring gridpoint outside the data area is taken to be zero. The time-independent part is modeled as a first-order Markov Random Field (MRF) with different dependence parameters in lateral and longitudinal direction:

$$\begin{aligned}\mu(x) = \mu_0(x) &+ a_\phi\{\mu(x^N) - \mu_0(x^N) + \mu(x^S) - \mu_0(x^S)\}\\ &+ a_\lambda\{\mu(x^E) - \mu_0(x^E) + \mu(x^W) - \mu_0(x^W)\} + \varepsilon_\mu(x) \quad \forall x.\end{aligned} \tag{3}$$

The spatially specific mean, $\mu_0(x)$, is modeled as having a quadratic form with an interaction term in latitude and longitude, letting $m(x)$ and $n(x)$ denote the longitude and latitude of location x respectively:

$$\begin{aligned}\mu_0(x) = \mu_{0,1} &+ \mu_{0,2}m(x) + \mu_{0,3}n(x) + \mu_{0,4}m(x)^2 + \mu_{0,5}n(x)^2\\ &+ \mu_{0,6}m(x)n(x) \quad \forall x.\end{aligned} \tag{4}$$

The spatio-temporal dynamic term $\theta(x,t)$ is modeled as a vector autoregressive model of order one, conditionally specified on its nearest neighbors in all cardinal directions:

$$\begin{aligned}\theta(x,t) = b_0\theta(x,t-1) &+ b_N\theta(x^N,t-1) + b_E\theta(x^E,t-1)\\ &+ b_S\theta(x^S,t-1) + b_W\theta(x^W,t-1) + \varepsilon_\theta(x,t), \quad \forall x,t.\end{aligned} \tag{5}$$

The $b.$ parameters are assumed invariant in space and to have interpretations related to the underlying wave dynamics.

The seasonal part is modeled as an annual cyclic contribution where the seasonal contribution is assumed independent of space (6). The long-term trend is modeled as a simple Gaussian process with a quadratic trend:

$$M(t) = c\cos\omega t + d\sin\omega t + \varepsilon_m(t) \quad \forall t \geq 1, \tag{6}$$

$$\mathrm{T}(t) = \gamma t + \eta t^2 + \varepsilon_t(t) \quad \forall t \geq 1. \tag{7}$$

Table 1 Specification of prior distributions

Parameters:	$\sigma_{\cdot}^2$	$a_{\cdot}, b_{\cdot}$	$\mu_{0,1}$	$\mu_{0,i}\ i \neq 1$
Prior distribution:	$IG(3,2)$	$N(0.2, 0.25)$	$N(3.5, 2)$	$N(0, 2)$

Parameters:	$\theta(x, 0)$	c	d	γ, η
Prior distribution:	$N(0, 15)$	$N(2, 0.5)$	$N(0, 0.2)$	$N(0, 0.1)$

When the model is used with the log-transformed data, setting $Y(x,t) = \ln Z(x,t)$, the observation equation (1) becomes $Y(x,t) = H(x,t) + \varepsilon_Y(x,t)$ and the significant wave height is modeled according to

$$Z(x,t) = \exp H(x,t) \times \exp \varepsilon_Y(x,t) \quad \forall x, t. \tag{8}$$

This yields a fundamentally different interpretation of the various components, i.a. the long-term trend, which become multiplicative rather than additive factors and hence different for extreme and moderate sea states.

2.1.1 Prior Distributions

Specification of prior distributions for the parameters together with initial values for θ_0 is needed, and then the full conditional distributions can be derived. All prior distributions are assumed independent. Conditionally conjugate priors will be employed and for the purpose of this study, the same priors as in [11, 13] will be adopted (Table 1). Reference is made to [11] for the rationale behind choice of priors. The results are found not to be very sensitive to the actual choice of hyperparameters, and in [13] it is argued that it is appropriate to use the same priors for the log-transformed data. Slightly modified priors for monthly maxima were used in [12], but end results were not affected. This is not unreasonable since the amount of data is large.

2.2 Model Alternatives

Simulations on both the original and the log-transformed data are run for five model alternatives, i.e. with a quadratic, linear and no temporal trend respectively and for both one and two temporal noise terms:

Model alternative 1: $\mathrm{T}(t) = \gamma t + \eta t^2 + \varepsilon_t(t)$.
Model alternative 2: $\mathrm{T}(t) = \gamma t + \varepsilon_t(t)$.
Model alternative 3: $\mathrm{T}(t) = 0$.
Model alternative 4: $M(t) = c \cos \omega t + d \sin \omega t + \gamma t + \eta t^2 + \varepsilon_m(t)$.
Model alternative 5: $M(t) = c \cos \omega t + d \sin \omega t + \gamma t + \varepsilon_m(t)$.

In the first three models above, the seasonal component $M(t)$ remains unchanged as in (6). It is noted that including two temporal noise terms in the model might give identifiability issues, and this is further discussed in [11].

2.2.1 Model Comparison

Two loss functions based on predictive power will be considered where one-step predictions of the spatial field at the final time point are compared to data. The standard loss function is

$$L_s = \left[\frac{1}{Xn}\sum_{x=1}^{X}\sum_{j=1}^{n}\left(Z(x) - Z(x)_j^*\right)^2\right]^{\frac{1}{2}}, \tag{9}$$

where, $Z(x)$ denotes the data at location x and $Z(x)_j^*$ denotes the predicted value at location x in iteration j. The weighted loss function, where the squared prediction errors have been weighted according to the actual observed significant wave height, is defined as

$$L_w = \left[\frac{1}{n\sum_x Z(x)}\sum_{x=1}^{X}\sum_{j=1}^{n} Z(x)\left(Z(x) - Z(x)_j^*\right)^2\right]^{\frac{1}{2}}. \tag{10}$$

When the model is run with a logarithmic transform, it is the prediction $e^{Y^*(x,j)}$ that should be compared with the data. That is, in order to compare, one must calculate the loss associated with the significant wave height, and not its log-transform.

3 Simulation Results

The results from running the model on monthly maxima are reported below.

3.1 Implementation and MCMC Simulations

Markov Chain Monte Carlo techniques (Gibbs sampler with Metropolis-Hastings steps) were employed in order to simulate the model. This requires the full conditional distributions of the model parameters which may be derived from the model specification [11]. A total of 45,000 iterations were run, with an initial burn-in period of 20,000 and a batch size of 25. Thus, a total of 1000 samples of the multi-dimensional parameter vector were obtained from each simulation.

No formal tests on convergence has been carried out, but visual inspection indicates that convergence occurs relatively quickly. A few simulations have also been performed with different starting values for the parameter set and the results indicate that the Gibbs sampler has indeed converged. In addition, a few control simulations with much longer burn-in were run, and these showed nearly identical results, indicating that convergence did indeed occur within the burn-in period.

3.2 Results for Non-transformed Data of Monthly Maxima

A visual check of the residuals indicates that the Gaussian model assumptions in (1) might be reasonable. Most of the marginal posterior distributions are symmetric and resemble Gaussian distributions, and the mean and standard deviation of the posterior distributions for some of the parameters of the different model alternatives are given in Table 2. Apart from the temporal trend part, most of the model parameters do not vary significantly between model alternatives. It is observed that b_N is smaller than the other $b.$ parameters, but no particular explanations for this have been sought.

The posterior time independent part $\mu(x)$ looks reasonable with values in the order of 6.1 to 7.3 m. The posteriors of the space-time dynamic part take values between -1.1 and 1.8 m and a notable part of the variation in the data is explained by this component. The long-term effect of this component should ideally be zero, but no constraints are imposed to ensure this. Nevertheless, the long term effect of this component is found to be negligible upon inspection of the results. The seasonal contribution corresponds to an annual cyclic variation of about ± 2.5 m.

The component of perhaps most interest, the long-term trend part, is illustrated in Fig. 2. The sampled mean trend is shown together with the mean, 5- and 95-percentiles of the trend contribution obtained by calculating $\gamma t + \eta t^2$ for (γ, η) pairs from the posterior distributions and an increasing trend can be extracted from the data. For the quadratic model (model 1) the mean yearly trend corresponds to nearly 70 cm over the period. The 90 % credible interval embraces overall trends between 44 and 93 cm. The corresponding trend of the linear model (Model 2) corresponds to an increase of about 69 cm over the period with 90 % credible interval ranging from 45–94 cm. For the models with one temporal noise term, the seasonal and trend components were not sampled individually, but the joint temporal components is shown along with the mean seasonal part and the mean, 5- and 95-percentiles of the trend contribution. The quadratic trend model (Model 4) estimates a mean increase over the period of 68 cm, with the 90 % credible interval ranging from 39 to 96 cm. The linear trend model (Model 5) estimates a mean increase of 69 cm during the period, with a 90 % credible interval ranging from 40 cm to 96 cm. It is noteworthy that all trends are comparable to the others and agrees well.

3.3 Results for Log-Transformed Monthly Maxima

The normal probability plot of the residuals for the log-transformed data looks much better than for non-transformed data which indicates that the log-transform might be an improvement. The mean and standard deviation of some marginal posterior distributions for different parameters are given in Table 3. It is emphasized that these parameters pertain to the log-transformed data, and the values are therefore not comparable to the values obtained for the non-transformed data. However, dependence parameters in space and time are presumably similar.

Table 2 Posterior marginal distributions: mean and standard deviation

	Model 1	Model 2	Model 3	Model 4	Model 5
σ_Z^2	(1.2, 0.0085)	(1.2, 0.0081)	(1.2, 0.0080)	(1.2, 0.0079)	(1.2, 0.0081)
σ_μ^2	(0.035, 0.0042)	(0.035, 0.0042)	(0.035, 0.0040)	(0.035, 0.0042)	(0.035, 0.0044)
σ_θ^2	(0.074, 0.0059)	(0.074, 0.0056)	(0.076, 0.0054)	(0.074, 0.0052)	(0.074, 0.0051)
σ_m^2	(0.63, 0.16)	(0.56, 0.13)	(1.1, 0.073)	(1.1, 0.068)	(1.1, 0.069)
b_0	(0.23, 0.0094)	(0.23, 0.0095)	(0.23, 0.0094)	(0.23, 0.0095)	(0.23, 0.0096)
b_N	(0.0068, 0.011)	(0.0064, 0.012)	(0.0090, 0.011)	(0.0055, 0.011)	(0.0059, 0.011)
b_E	(0.22, 0.010)	(0.22, 0.0097)	(0.22, 0.0098)	(0.22, 0.0098)	(0.22, 0.0098)
b_S	(0.35, 0.026)	(0.35, 0.023)	(0.34, 0.023)	(0.35, 0.023)	(0.35, 0.022)
b_W	(0.23, 0.011)	(0.23, 0.010)	(0.23, 0.0099)	(0.23, 0.010)	(0.23, 0.010)
c	(2.4, 0.062)	(2.4, 0.065)	(2.4, 0.067)	(2.4, 0.064)	(2.4, 0.067)
d	(1.0, 0.067)	(1.0, 0.063)	(1.0, 0.067)	(1.0, 0.066)	(1.0, 0.065)
$\mu_{0,1}$	(3.4, 1.5)	(3.4, 1.3)	(3.4, 1.4)	(3.4, 1.4)	(3.3, 1.3)
$\mu_{0,2}$	(−0.28, 0.068)	(−0.28, 0.068)	(−0.28, 0.067)	(−0.29, 0.066)	(−0.29, 0.064)
$\mu_{0,3}$	(1.9, 0.39)	(1.8, 0.40)	(1.9, 0.39)	(1.9, 0.39)	(1.9, 0.38)
$\mu_{0,4}$	(0.00049, 0.00017)	(0.00048, 0.00017)	(0.00049, 0.00017)	(0.00050, 0.00017)	(0.00050, 0.00016)
$\mu_{0,5}$	(−0.013, 0.0024)	(−0.13, 0.0023)	(−0.013, 0.0022)	(−0.013, 0.0023)	(−0.013, 0.0023)
$\mu_{0,6}$	(−0.0010, 0.00095)	(−0.0010, 0.00097)	(−0.0011, 0.00098)	(−0.0011, 0.00097)	(−0.0011, 0.00096)
σ_t^2	(0.50, 0.16)	(0.56, 0.13)	–	–	–
γ	(0.0022, 0.0011)	(0.0013, 0.00028)	–	(0.0018, 0.0014)	(0.0013, 0.00033)
η	$(-1.7 \times 10^{-6}, 2.1 \times 10^{-6})$	–	–	$(-9.1 \times 10^{-7}, 2.5 \times 10^{-6})$	–

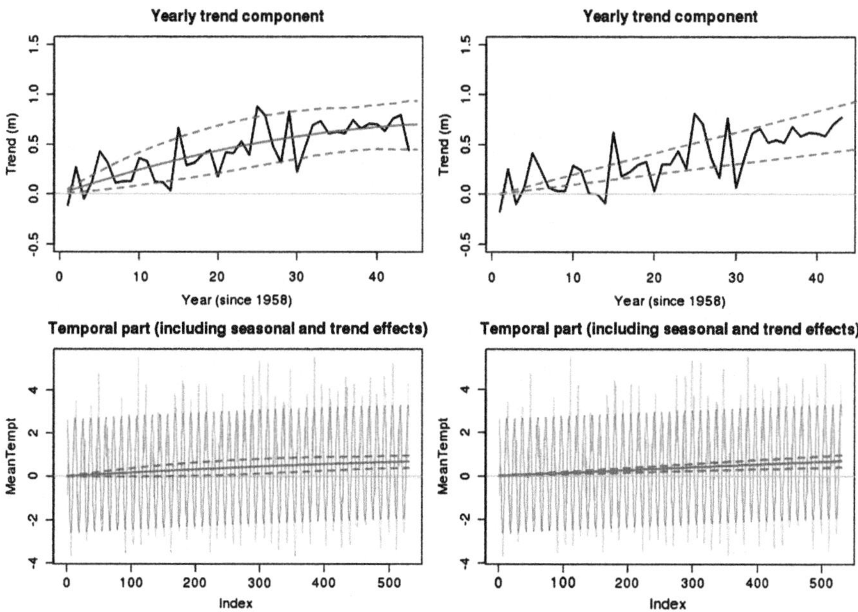

Fig. 2 Estimated temporal trends of the 45-year period; Model 1 (*top left*), Model 2 (*top right*), Model 4 (*bottom left*) and Model 5 (*bottom right*)

The contribution from $\mu(x)$ is in the order of 1.76–1.95 but the interpretation is different; $e^{\mu(x)}$ is now a multiplicative factor for the significant wave height at location x varying between 5.8 and 7.0. The mean contributions from the space-time dynamic part, $\theta(x,t)$ correspond to factors between 0.70 and 1.4 for different space and times. Hence, this space-time dynamic part, contributes with a factor from -30 to $+40$ %. The seasonal component corresponds to a factor between 0.68 for calm seasons and 1.5 in rough seasons (i.e. the seasonal component $M(t)$ varies cyclically between ± 0.38).

Model 1 estimates a mean yearly trend corresponding to a factor of about 1.09 over the period, with 90 % credible interval ranging from 1.03 to 1.14. For monthly maximum significant wave heights of 5 and 8 m this corresponds to a mean increase of 43 and 68 cm, respectively. The mean estimate of the long-term trend from model 2 corresponds to an increase of about 1.07 over the period. The 90 % credible interval ranges from 1.03 to 1.12. For monthly maximum significant wave heights of 5 and 8 m, this corresponds to an expected increase of about 36 and 57 cm respectively. The trends estimated from the models with one temporal noise term are very similar to the ones estimated with two noise terms. Model 4 estimates a mean increase over the period of about 1.05, but with the 90 % credible interval ranging from 1.00 to 1.10 whereas Model 5 estimates a mean increase of 1.05 during the period, with a 90 % credible interval ranging from 1.00 to 1.09. Assuming average monthly maximum significant wave heights of 5 and 8 m, these factors correspond to an expected trend of about 26–27 cm and 41–43 cm respectively. Overall, these

Table 3 Posterior marginal distributions (log-transformed data)

	Model 1	Model 2	Model 3	Model 4	Model 5
σ_Z^2	(0.013, 0.00051)	(0.013, 0.00051)	(0.013, 0.00047)	(0.013, 0.00052)	(0.013, 0.00050)
σ_μ^2	(0.027, 0.0031)	(0.027, 0.0029)	(0.027, 0.0031)	(0.027, 0.0030)	(0.027, 0.0031)
σ_θ^2	(0.012, 0.00051)	(0.012, 0.00051)	(0.012, 0.00048)	(0.012, 0.00052)	(0.012, 0.00050)
σ_m^2	(0.026, 0.0021)	(0.025, 0.0021)	(0.029, 0.0017)	(0.029, 0.0018)	(0.029, 0.0018)
b_0	(0.012, 0.0037)	(0.012, 0.0037)	(0.013, 0.0037)	(0.012, 0.0038)	(0.012, 0.0036)
b_N	(0.011, 0.0041)	(0.011, 0.0038)	(0.011, 0.0040)	(0.011, 0.0040)	(0.011, 0.0040)
b_E	(0.0090, 0.0036)	(0.0092, 0.0037)	(0.0091, 0.0037)	(0.0093, 0.0037)	(0.0091, 0.0035)
b_S	(0.0043, 0.0038)	(0.0043, 0.0039)	(0.0043, 0.0039)	(0.0045, 0.0038)	(0.0041, 0.0039)
b_W	(0.012, 0.0035)	(0.012, 0.0036)	(0.013, 0.0036)	(0.013, 0.0037)	(0.012, 0.0039)
c	(0.35, 0.014)	(0.35, 0.014)	(0.35, 0.011)	(0.35, 0.010)	(0.35, 0.011)
d	(0.15, 0.014)	(0.15, 0.013)	(0.15, 0.010)	(0.15, 0.010)	(0.15, 0.010)
$\mu_{0,1}$	(3.5, 1.4)	(3.4, 1.4)	(3.5, 1.4)	(3.4, 1.5)	(3.5, 1.4)
$\mu_{0,2}$	(−0.056, 0.038)	(−0.055, 0.038)	(−0.054, 0.037)	(−0.056, 0.037)	(−0.058, 0.037)
$\mu_{0,3}$	(0.28, 0.22)	(0.28, 0.21)	(0.27, 0.21)	(0.29, 0.21)	(0.30, 0.22)
$\mu_{0,4}$	$(9.0 \times 10^{-5}, 9.7 \times 10^{-5})$	$(8.8 \times 10^{-5}, 9.4 \times 10^{-5})$	$(8.7 \times 10^{-5}, 9.3 \times 10^{-5})$	$(9.1 \times 10^{-5}, 9.3 \times 10^{-5})$	$(9.7 \times 10^{-5}, 9.4 \times 10^{-5})$
$\mu_{0,5}$	(−0.0021, 0.0012)	(−0.0021, 0.0012)	(−0.0020, 0.0012)	(−0.0022, 0.0011)	(−0.0022, 0.0012)
$\mu_{0,6}$	(−0.00012, 0.00054)	(−0.00010, 0.00051)	(−0.00011, 0.00052)	(−0.00012, 0.00052)	(−0.00014, 0.00052)
σ_t^2	(0.026, 0.0023)	(0.025, 0.0021)	–	–	–
γ	(0.00036, 0.00024)	$(0.00011, 5.2 \times 10^{-5})$	–	$(-6.4 \times 10^{-5}, 0.00023)$	$(7.1 \times 10^{-5}, 5.1 \times 10^{-5})$
η	$(-4.4 \times 10^{-7}, 4.7 \times 10^{-7})$	–	–	$(2.6 \times 10^{-7}, 4.1 \times 10^{-7})$	–

trends are somewhat smaller but still in fairly good agreement with the trends estimated from the non-transformed data. A QMLE-estimate for bias correction due to re-transformation have been adopted [13].

3.4 Future Projections

It is tempting to extend the linear trends into the future. Assuming that estimated linear trends will continue, the results from the original data correspond to an expected increase in monthly maximum significant wave height of 1.6 m over 100 years, with a 95 % credibility of an increase of at least around 1.0 m. The two linear trend estimates, with credibility bands, are nearly identical. The estimates from the log-transformed data correspond to an expected increasing factor of 1.15 over 100 years, with a 95 % credibility of a trend-factor larger than 1.04. (An expected increase of 1.10 with 94 % credibility of a positive trend is obtained from the linear model with one temporal noise term.) Assuming such trends valid for an average monthly maximum sea states of 5 and 8 m in calm and rough seasons, respectively, the expected increases would be about 75 cm and 1.2 m respectively (51 cm and 82 cm for one noise term) over 100 years.

This seems to be in reasonable agreement with other projections made for significant wave height in the North Atlantic and other areas [4, 8, 14, 15], and also with the trends obtained for non-maximum daily and monthly data [11, 13], even though the trends in the maxima are somewhat stronger than the trends in the data overall. It is emphasized that even though the models detect trends in the data, they are not necessarily related to anthropogenic climate change; they might be a result of decadal natural variability, as discussed in e.g. [1]. Furthermore, the validity of extrapolating the trend into the future is questionable and the model might need to be extended with regression terms in order to yield reliable projections. Hence, regression of future trends on projections of atmospheric CO_2 levels has been added to the model and will be presented in future publications [12].

3.5 Model Comparison and Selection

The losses corresponding to the two loss functions were estimated for the various model alternatives and are given in Table 4. Preference is given to the linear trend models for the original data but the quadratic model with one temporal noise term is better for the log-transformed data. It is also interesting to note that, in spite the fact that the normal probability plot for the log-transformed data is improved compared to the original data, the models for the original data seem to perform consistently better with regards to short-term prediction. Notwithstanding, it is difficult to assess the models in terms of long-term prediction and it is still an open question which model alternative would be better in this respect. It is also seen that the loss functions are not really able to distinguish between the various alternatives, which all seem to describe the data similarly well.

Table 4 Model selection: Estimated losses

Model alternative	Original data		Log-transformed data	
	L_s	L_w	L_s	L_w
Model 1	2.630	2.748	3.383	3.456
Model 2	2.576	2.691	3.346	3.412
Model 3	2.775	2.902	3.035	3.127
Model 4	2.609	2.726	2.970	3.045
Model 5	2.580	2.695	2.987	3.070

4 Summary and Conclusions

This paper has presented a Bayesian hierarchical space-time model for significant wave height and the results obtained from running the model on original and log-transformed data for an area in the North Atlantic ocean. Overall, the model seems to perform reasonable well, and captures various dependence structures in space and time at different scales. The model also includes a long-term temporal part and a positive long-term trend is extracted from the data. This trend has been extrapolated to yield future projections of the wave climate which seem to be in reasonable agreement with previous studies. Notwithstanding, it is believed that further work is needed and that some reasonable covariates should be included in the model for reliable future projections.

References

1. Caires S, Sterl A (2005) 100-year return value estimates for ocean wind speed and significant wave height from the ERA-40 data. J Climate 18:1032–1048
2. Caires S, Sterl A (2005) A new nonparametric method to correct model data: application to significant wave height from ERA-40 re-analysis. J Atmos Ocean Technol 22:443–459
3. Caires S, Swail V (2004) Global wave climate trend and variability analysis. In: Preprints of 8th international workshop on wave hindcasting and forecasting
4. Caires S, Swail VR, Wang XL (2006) Projection and analysis of extreme wave climate. J Climate 19:5581–5605
5. Cressie N, Wikle CK (2011) Statistics for spatio-temporal data. Wiley, New York
6. Debernard J, Sætra Ø, Røed LP (2002) Future wind, wave and storm surge climate in the northern North Atlantic. Clim Res 23:39–49
7. Debernard JB, Røed LP (2008) Future wind, wave and storm surge climate in the Northern Seas: a revisit. Tellus A 60:427–438
8. Grabemann I, Weisse R (2008) Climate change impact on extreme wave conditions in the North Sea: an ensemble study. Ocean Dyn 58:199–212
9. Natvig B, Tvete IF (2007) Bayesian hierarchical space-time modeling of earthquake data. Methodol Comput Appl Probab 9:89–114
10. Vanem E (2011) Long-term time-dependent stochastic modelling of extreme waves. Stoch Environ Res Risk Assess 25:185–209
11. Vanem E, Huseby AB, Natvig B (2011) A Bayesian hierarchical spatio-temporal model for significant wave height in the North Atlantic. Stoch Environ Res Risk Assess. doi:10.1007/s00477-011-0522-4

12. Vanem E, Huseby AB, Natvig B (2012) Bayesian hierarchical spatio-temporal modelling of trends and future projections in the ocean wave climate with a CO_2 regression component. (submitted)
13. Vanem E, Huseby AB, Natvig B (2011) Modeling ocean wave climate with a Bayesian hierarchical space-time model and a log-transform of the data. Ocean Dyn 62:355–375
14. Wang XJ, Zwiers FW, Swail VR (2004) North Atlantic ocean wave climate change scenarios for the twenty-first century. J Climate 17:2368–2383
15. Wang XL, Swail VR (2006) Climate change signal and uncertainty in projections of ocean wave heights. Clim Dyn 26:109–126
16. Wang XL, Swail VR (2006) Historical and possible future changes of wave heights in northern hemisphere ocean. In: Perrie W (ed) Atmosphere-ocean interactions, vol 2. Advances in fluid mechanics, vol 39. WIT Press, Ashurst, pp 185–218 (Chap. 8)
17. Wikle CK (2003) Hierarchical models in environmental science. Int Stat Rev 71:181–199
18. Wikle CK, Berliner LM, Cressie N (1998) Hierarchical Bayesian space-time models. Environ Ecol Stat 5:117–154
19. Wikle CK, Milliff RF, Nychka D, Berliner LM (2001) Spatiotemporal hierarchical Bayesian modeling: tropical ocean surface winds. J Am Stat Assoc 96:382–397

Interpolation of Concentration Measurements by Kriging Using Flow Coordinates

Martine Rivest, Denis Marcotte, and Philippe Pasquier

Abstract Groundwater contaminant plumes frequently display a curvilinear anisotropy, which conventional kriging and geostatistical simulation approaches fail to reproduce properly. In many applications, physically relevant coordinate transformations are used to modify the relationships between data points and simplify the specification of nonlinear anisotropy. In this paper, we present a kriging approach that uses a coordinate transformation to improve the interpolation of contaminant concentrations. The proposed alternative flow coordinates (AFC) consist in the hydraulic head and one (2D) or two (3D) streamline-based coordinates. In 2D, the mapping obtained using AFC is similar to that yielded by the natural coordinates of flow (i.e. hydraulic head and stream function). AFC can be generalized to 3D flow and to the presence of wells, which is not the case with the natural coordinates. The performance of the approach is investigated using a simple 3D synthetic case. Kriged concentration maps obtained with the AFC reproduce the curvilinear features found in the reference plume. Performance statistics suggest AFC improves plume delineation compared to conventional kriging on a Cartesian grid. However, the performance of the AFC transformation is limited by the fact that, while it accounts for advection, it does not consider the effects of dispersion on the shape of the plume. This aspect and further testing on complex cases is the subject of ongoing research.

1 Introduction

Conventional kriging approaches often yield poor results when mapping groundwater contaminants, in particular with plumes displaying a curvilinear shape that

M. Rivest (✉) · D. Marcotte · P. Pasquier
École Polytechnique de Montréal, C.P. 6079 Succ. Centre-Ville, Montréal, QC, Canada, H3C 3A7
e-mail: martine.rivest@polymtl.ca

D. Marcotte
e-mail: denis.marcotte@polymtl.ca

P. Pasquier
e-mail: philippe.pasquier@polymtl.ca

P. Abrahamsen et al. (eds.), *Geostatistics Oslo 2012*,
Quantitative Geology and Geostatistics 17,
DOI 10.1007/978-94-007-4153-9_42,

cannot be accounted for using a global definition of anisotropy. Several solutions exist to cope with curvilinear anisotropy, such as using locally defined directions of anisotropy [26, 29], calculating non-Euclidean distances along features which outline the studied phenomenon continuity [2, 9, 10, 14, 23, 28] or else, applying a coordinate transformation that simplifies the description of anisotropy [3, 6, 11–13, 18]. In some cases, geographic features (river, city map) may indicate the local direction of anisotropy. If not, a dense sampling of the studied variable or the availability of directional data, such as dip, azimuth or gradient, is generally required to capture the spatial continuity. In addition, using non-Euclidean distances raises an issue relative to the positive definiteness of covariance functions [9, 10]. To get around this issue, multidimensional scaling (MDS) maps the points into a higher dimensional space where inter-point Euclidean distances correspond approximately to the computed non-Euclidean distances in "real" space [1, 4, 5, 19, 21, 25]. In the same way, coordinate transformations based on geographic features or physical processes also have the advantage of precluding problems related to the admissibility of the covariance function. Furthermore, they provide a clear interpretation of the transformed space [18].

The principal direction technique [15] uses the natural flow coordinates (NFC), which correspond to the hydraulic head and stream function, to enhance numerical solution of contaminant transport. In that approach, finite element equations are reformulated in terms of the NFC; thus, mass fluxes are oriented in the principal direction of the grid and the dispersion tensor is a diagonal matrix [8]. In recent work, [24] proposed a coordinate transformation similar to that used for the principal direction technique, but adapted to the kriging of groundwater contaminants in 2D. In 3D, however, construction of natural coordinate grids, although feasible in simple situations [20], is not straightforward [7]. Furthermore, Cauchy-Riemann conditions, which express the coupling between potential and stream function, are restricted to 2D [22].

This paper presents an alternative coordinate transformation, which is based on streamlines, to provide an approximate substitute for the natural coordinates concept in 3D. The proposed coordinate system consists of the hydraulic head h, and two streamline-based coordinates, u and v. In the following, we detail the computation of the alternative flow coordinates (AFC). Then, a simple 3D example is used to show how a coordinate transformation based on AFC can be used to enhance the kriging of groundwater contaminant concentrations. Results and limitations are discussed in Sect. 4.

2 Methodology

2.1 Alternative Flow Coordinates

In this section, we propose streamline-based coordinates that can be seen as approximate NFC. Keeping hydraulic head as one of the flow coordinates, the proposed

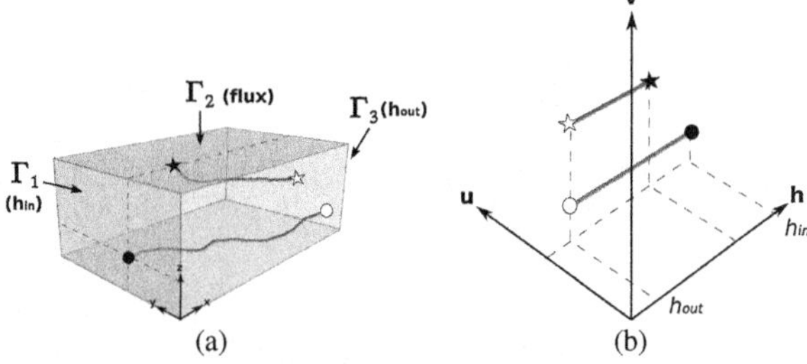

Fig. 1 Illustration of the alternative flow coordinate transformation. The rational behind the AFC is that each streamline can be identified using its start point. (**a**) Streamlines within a flow domain. Boundary Γ_1 is a constant head boundary, whereas Γ_2 is a flux boundary. The *black symbols* are on the inflow boundaries Γ_1 and Γ_2, while the *white* ones are on the outflow boundary Γ_3 (on the *right*, opposite to Γ_1). All other boundaries are no flow. (**b**) Transformed space. Streamlines depicted in (**a**) become *straight lines* in the huv space

alternative flow coordinates (AFC) replace the stream function by one (2D) or two (3D) boundary-conforming coordinates related to the position of streamline start points on the inflow boundaries. To alleviate the presentation, we only consider the 3D case in the following.

Let Γ be an inflow boundary belonging to a 3D domain Ω of a finite element model, we define u and v as two boundary-conforming coordinates on the surface Γ. Suppose a calibrated hydraulic head solution is available. Streamline 1 start point (x_0^1, y_0^1, z_0^1) located on Γ is mapped into AFC (h_0^1, u^1, v^1) using the calibrated hydraulic head solution and numerical surface grid generation techniques [27] (for u and v). AFC u^1 and v^1 values are then propagated to points inside the domain using particle tracking algorithm. Assigning u and v values to every data and grid points this way would be somewhat cumbersome. Instead, we consider tracking a limited number of streamlines and using linear interpolation, which preserve monotonicity, to estimate the coordinates where needed. As an example, consider the simple box-shaped model shown in Fig. 1(a), with inflow boundaries Γ_1 and Γ_2. AFC requires both inflow boundaries to be mapped into a single surface so they share continuous u and v coordinates. In this case, the geometry of the model allows a simple mapping of start points located on inflow boundaries into u and v:

$$u(x_0, y_0, z_0) = y_0 \quad \text{if } (x_0, y_0, z_0) \in \Gamma_1 \cup \Gamma_2 \tag{1}$$

$$v(x_0, y_0, z_0) = \begin{cases} z_0 & \text{if } (x_0, y_0, z_0) \in \Gamma_1, \\ z_{\max} + \alpha x & \text{if } (x_0, y_0, z_0) \in \Gamma_2, \end{cases} \tag{2}$$

where $z_{\max}$ is the value of z on the top boundary, Γ_2, and α is a proportionality constant to account for a change in flux magnitude on Γ_2, compared to Γ_1. These relations are used to assign u and v to a grid of k start points located on the inflow boundaries. The h coordinate at the start points is given by the numerical hydraulic head solution. In the transformed space, streamlines appear parallel to the h coordinate. The following algorithm is used to allow the computation of (h, u, v) coordinates everywhere in the domain.

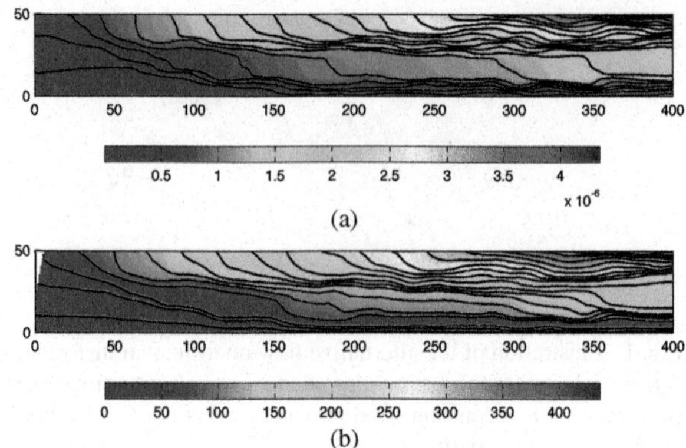

Fig. 2 Comparison between the AFC and the stream function for a 2D flow. (**a**) Stream function, (**b**) AFC

1. Given a start point grid, compute streamlines and ensure they reach the outflow boundary Γ_3.
2. For each streamline, use mapping relations in (1) and (2) to compute (u, v) at the start point and assign it to the set of points discretizing that streamline.
3. Interpolate u and v to data and grid points, using the values defined in 2.
4. Use the numerical solution to compute coordinate h to data and grid points.
5. Check that the mapping is one-to-one i.e. that no points have been given the same AFC.

The result is a table which associates each point in the xyz coordinates to its corresponding coordinates in the huv system. Note that, in general, the mapping relations between the xyz values on the inflow boundaries and the uv coordinates are likely to be more complex. In this case, numerical surface grid generation techniques as presented in [27] should be used to compute the boundary-conforming coordinates, u and v. These approaches treat the determination of the curvilinear grid on a surface as a boundary-value problem; therefore, one must specify u and v on the edges of the inflow boundaries.

As an example, Fig. 2 illustrates how AFC compares with the stream function for a 2D flow. Despite local discrepancies, the alternative flow coordinate shown in Fig. 2(a) yields a reasonable approximation of the stream function main features. Note the scale of the coordinates is not important, as it will be accounted for by the range parameter in the kriging process.

2.2 Concentration Estimation

Once the correspondence between xyz and huv coordinates as been computed, contaminant concentrations are interpolated. No secondary information or drift is considered; thus, ordinary kriging (OK) is used with kriging weights λ_i obtained by solving the OK system:

$$\begin{bmatrix} \boldsymbol{\Sigma} & \mathbf{1} \\ \mathbf{1}^{\mathrm{T}} & 0 \end{bmatrix} \begin{bmatrix} \boldsymbol{\lambda} \\ \mu \end{bmatrix} = \begin{bmatrix} \boldsymbol{\sigma}_0 \\ 1 \end{bmatrix}, \tag{3}$$

where $\boldsymbol{\Sigma}$ is the covariance matrix between concentration observations, $\mathbf{1}$ is a vector of ones, $\boldsymbol{\lambda}$ is the vector of kriging weights, μ is the Lagrange multiplier, and $\boldsymbol{\sigma}_0$ represent the covariance between the estimation point and the observation points. The only difference with the usual OK procedure is in $\boldsymbol{\Sigma}$ and $\boldsymbol{\sigma}_0$, where covariance is computed using Euclidean distances in the huv space instead of the usual Cartesian space. The effect of this coordinate change was seen in Fig. 1(b): in the transformed space, groundwater flows parallel to the h axis. This enables kriging to be performed using a single global anisotropy. A global kriging neighborhood is used.

2.3 Performance Assessment

This section presents the different statistics used to compare the AFC approach with the conventional approach i.e. Cartesian coordinates kriging. First, mean absolute error is obtained by direct comparison between reference and estimated concentration maps. It is computed using

$$MAE = \frac{1}{V_T} \sum_{i=1}^{n_e} \left| c_i^R - c_i^* \right| V_i, \tag{4}$$

where V_T is the total volume of the domain, V_i is the volume of the ith element, c_i^R and c_i^* are, respectively, the reference and estimated concentration values interpolated at the center of the ith element and n_e is the number of mesh elements. We also look at the error on contaminant mass, RAE_{mass}. For each concentration map, mass and the relative error, are computed using

$$M = \sum_{i=1}^{n_e} c_i V_i \quad \text{and} \quad RAE_{\text{mass}} = \frac{|M^R - M^*|}{M^R}, \tag{5}$$

where c_i is the concentration interpolated at the center of the ith element, V_i is the volume of the ith element, M^R and M^* are respectively the masses computed from the reference and kriged concentration maps. In addition, Pearson's r is used to compare the correlation between the kriged and the reference concentrations.

Furthermore, we also use statistics that depend on the use of a contamination threshold value to delineate the plume. Among those is the relative error between the estimated and the reference contaminated volumes $RAE_{\text{vol.}}$. To achieve this, an indicator variable is used to classify which elements exceed the threshold. The contaminated volume and $RAE_{\text{vol.}}$ are computed by

$$V_s = \sum_{i=1}^{n_e} I_i V_i \quad \text{and} \quad RAE_{\text{vol.}} = \frac{|V_s^R - V_s^*|}{V_s^R}, \tag{6}$$

where V_s is the area above threshold s and I_i is the indicator value for the ith element, V_s^R is the reference volume and V_s^* is the kriged volume. In addition, we use a confusion matrix [16, 17] to further assess if the estimated plume is properly

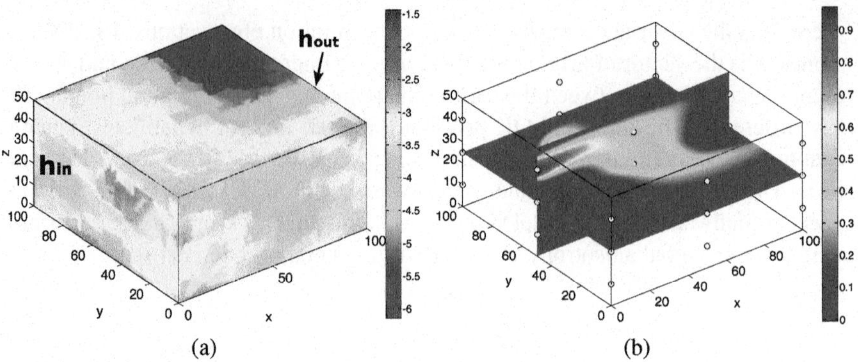

Fig. 3 Reference model. (**a**) Hydraulic conductivities and boundary conditions, (**b**) Reference plume and data locations

located, as this is not indicated by $RAE_{\text{vol.}}$. The confusion matrix allows one to evaluate the performance of a classification algorithm. In the present case, we use it to see if the kriged plumes produce a reliable division between contaminated and uncontaminated volumes. In this study, the three statistics computed from a confusion matrices are accuracy, precision and Pearson's ϕ:

$$ACC_s = \frac{a+d}{a+b+c+d} \tag{7}$$

$$PREC_s = \frac{d}{b+d} \tag{8}$$

$$\phi_s = \frac{ad-bc}{\sqrt{(a+b)(c+d)(a+c)(b+d)}}, \tag{9}$$

where a, b, c and d corresponds to summed volumes of elements where:

a is the concentration correctly estimated as under the threshold (true negative),
b is the concentration incorrectly estimated as above the threshold (false positive),
c is the concentration incorrectly estimated as under the threshold (false negative),
d is the concentration correctly estimated as above the threshold (true positive).

Accuracy represents the fraction of the domain where concentrations are correctly estimated with respect to a given threshold s. Note that accuracy should be used when the two classes are equally represented on the domain [17]; therefore, we use it only with a threshold corresponding to the median of the concentrations. Precision is the proportion of the estimated plume that is correctly classified with respect to threshold s. Finally, Pearson's ϕ measures the correlation between two binary variables.

3 Example

In this section, a simple example illustrates the approach and its potential. A finite element solver (COMSOL Multiphysics) is used to produce a steady-state flow field and contaminant plume (Fig. 3(b)), from the synthetic model shown in Fig. 3(a). The

Table 1 Synthetic model parameters

Parameter	Value
Geometry	
Dimensions	100 m × 100 m × 50 m
Max. element size	7 m
Flow parameters	
mean K	6.1×10^{-4} m/s
Variance of K	7.6×10^{-7} $(\mathrm{m/s})^2$
Mean hydraulic gradient	0.0032
Transport parameters	
porosity	0.35
D_m	1.34×10^{-9} m/s
$\alpha_1, \alpha_2, \alpha_3$	5 m, 0.15 m, 0.15 m

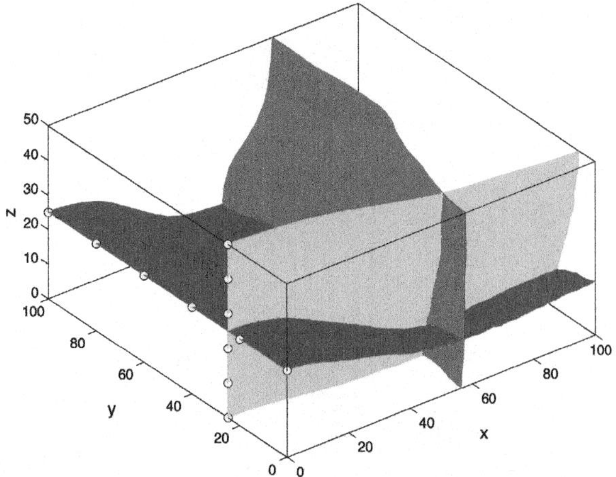

Fig. 4 Isosurfaces of h (*red*), u (*green*) and v (*blue*). u and v are the result of the combined streamline tracking and linear interpolation presented in Sect. 2, h is obtained from the numerical flow model. *Yellow dots* represent some streamline start points

model parameters are presented in Table 1. The plume is generated by the constant release of a contaminant on the left hand side boundary from an area bounded by $35\ \mathrm{m} \le y \le 90\ \mathrm{m}$ and $35\ \mathrm{m} \le z \le 45\ \mathrm{m}$, and it is sampled on a regular grid of 27 points. The procedure detailed in Sect. 2 is used to compute the alternative flow coordinates. A 5 m grid-spacing is used for the streamline start points grid, which is located on the left hand side boundary as this is the only inflow boundary for this model. Three surfaces, representing constant values of the resulting h, u and v coordinates, are presented in Fig. 4; the intersection of the u (green) and v (blue) surfaces is the streamline corresponding to that particular pair of uv coordinates.

Kriging is performed on the Cartesian and alternative flow grids yielding the plumes shown in Figs. 5(a) to 5(d), which display the concentrations above a threshold corresponding to the median value (0.0476 kg/mm^3). The proposed coordinate transformation generates a curvilinear anisotropy when the kriging estimate is mapped back in the Cartesian space. Indeed, the plume in Figs. 5(c) and 5(d) curves

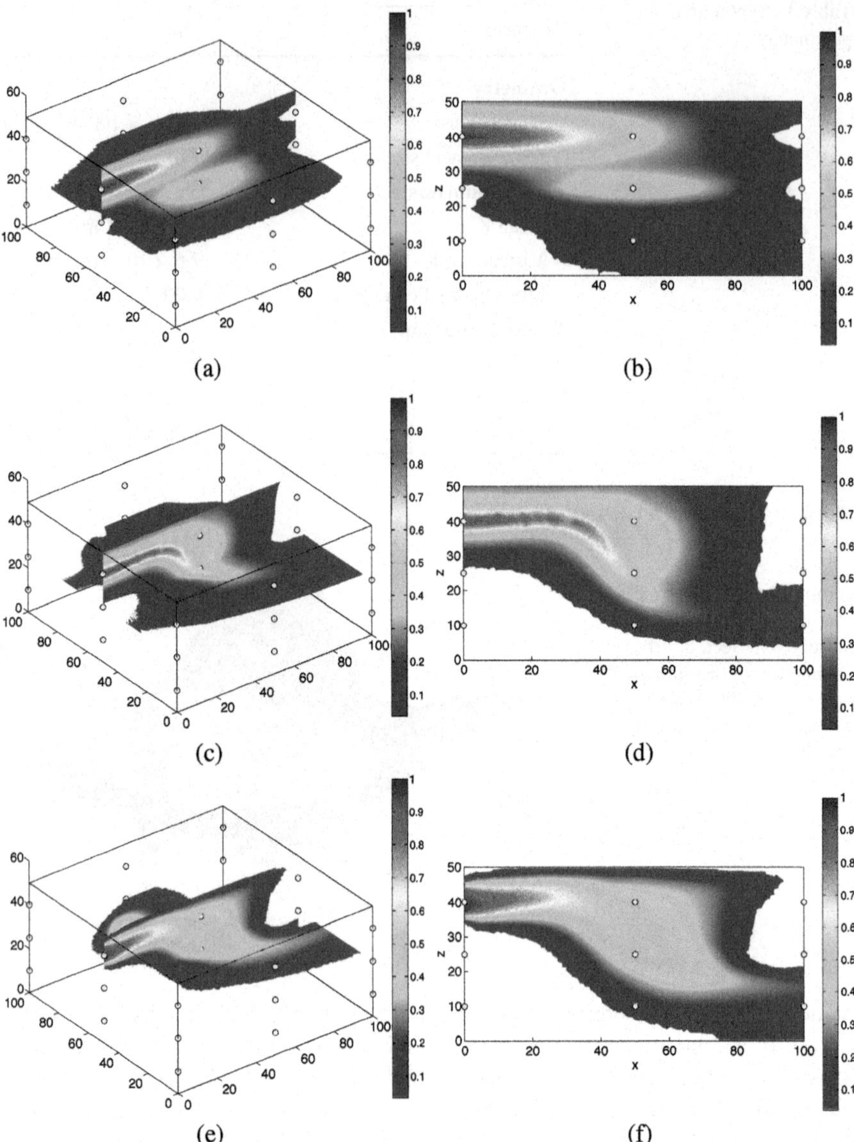

Fig. 5 Kriged and reference concentration maps. (3D view of $y = 50$ m and $z = 25$ m and plane view of $y = 50$ m) (**a**) and (**b**) CCK, (**c**) and (**d**) AFCK, (**e**) and (**f**) Reference map

towards a high concentration value located in the center of the domain. This is not the case for the Cartesian coordinate kriging (CCK) plume in Figs. 5(a) and 5(b), for which the anisotropy directions parallels the xyz axes. Figure 5(d), however, illustrates one drawback of the AFC approach: the plume stretches excessively the

Table 2 Error statistics on estimated concentrations and % Improvement over CCK

	MAE	RAE_{mass}	r	$RAE_{vol.}$[a]	ACC[a]	$PREC$[a]	ϕ[a]
CCK	0.0843	0.0081	0.71	0.4508	0.7228	0.6514	0.5002
AFCK	0.0693	0.0582	0.76	0.2400	0.8006	0.7406	0.6198
% Imp.	17.8	−614.1[b]	7.5	46.8	10.8	13.7	23.9

[a]Threshold used corresponds to concentration median (0.0476 kg/m^3)

[b]Negative values indicate a better performance by CCK

high value found in (0, 50, 40). Issues regarding the choice of the range are further discussed in Sect. 4.

Performance statistics presented in Table 2 suggest global improvement over CCK, except for RAE_{mass}. Accuracy and precision indicate the concentration values estimated with AFCK reflect more correctly the location of the actual plume than those obtained with CCK (confusion table results are presented in Table 3). Seen in this light, the low RAE_{mass} of CCK potentially is less meaningful if estimated concentration values are not located in the actual plume. Figure 6 shows the evolution of the ϕ statistic for threshold values corresponding to concentration percentiles 10th to 90th.

In this example, the range parameters where first estimated using a maximum likelihood procedure. Then, a sensitivity analysis was performed to assess the robustness of the results to range parameters. The results presented here are those that yielded the lowest MAE for each kriging approach. While individual performance statistics are influenced by the change in range parameters, the sensitivity analysis indicates a better overall performance by AFCK. Providing a different anisotropy angle for CCK did not yield much improvement, both for visual and statistical assessment.

4 Discussion and Conclusion

The simple example presented above illustrates the capability of the proposed coordinate transformation to produce curvilinear plume estimates and to account for groundwater flow when kriging concentration data in 3D. Although alternative flow coordinates only approximate the concept of natural flow coordinates, and involve several numerical procedures (particle tracking, interpolation), they can capture useful flow patterns from the reference groundwater flow model. Results obtained from the different performance statistics are encouraging, suggesting AFCK yields better plume delineation compared to CCK. It was noticed during the sensitivity analysis that range parameters which gave the best results (both for CCK and AFCK) were often a trade-off between accurate estimation of either the high concentrations values found near the source or the low concentrations values downstream. Figures 5(c) and 5(d) illustrate this: high concentrations near the source stretch unrealistically,

Table 3 Results from confusion matrices for a threshold corresponding to concentration median 0.0476 kg/m^3 (values are in m^3)

	CCK	AFCK
a—true negative	127766	173271
b—false positive	125023	79518
c—false negative	13581	20194
d—true positive	233630	227017

despite it was the best cases in term of *MAE*. The proposed coordinate transformation includes the effect of advection in the interpolation process; however, contaminant plumes are shaped by dispersion processes as well. Intuitively, dispersion suggests that the range parameters should be smaller close to the source and increase away from it. Similar results were observed with the 2D cases presented in [24]. The proposed coordinate transformation is relevant in a context where flow displays curvilinear features; otherwise, it obviously looses its advantage over CCK. Moreover, the advantage of AFCK over CCK will likely decrease as more concentration data are available. The approach appears especially interesting in a sparse data context as it allows the interpolation of concentration in a meaningful direction. The alternative flow coordinates can also be used in 2D. In this case, the streamlines are related to a single coordinate. The computation of the streamline-based coordinates u and v requires that every streamline enters the domain by the inflow boundaries. Therefore, the approach, as it is presented here, could be used in presence of a pumping well, but not with sources inside the domain. However, it could be modified to use the streamline end points instead and compute the coordinates using the outflow boundaries, thus allowing the presence of sources (i.e. streamline starting inside the domain). For complex cases involving the presence of both sources and wells, splitting the domain into sub-domains using dividing streamlines similarly to [7] should be considered. The idea behind this is to make sure each streamline receives a unique pair of uv coordinates, so the mapping is one-to-one.

In conclusion, the proposed alternative flow coordinates enable kriging to account for a plume curvilinear anisotropy and to incorporate secondary information, groundwater flow, into the interpolation of concentration data. This is generally not possible with conventional interpolation approaches. Results obtained from a synthetic 3D case show that concentration values are mapped more accurately with the coordinate transformation than with Cartesian coordinates kriging. The possibility of using the approach for 3D cases and in the presence of wells makes it more general than the previously studied natural coordinate transformation [24]. The present study did not consider the effect of uncertainty regarding the flow model; however, the results are clearly dependent on the accuracy of the hydrogeological conceptual model at hand. Further work is required to account for the effect of dispersion. Moreover, the alternative flow coordinates analogy with Lagrangian coordinates suggest that time could be used as a coordinate instead of the hydraulic head. These topics, as well as the assessment of flow uncertainty on the resulting maps and the application of the approach to a real test case, are the subject of ongoing research.

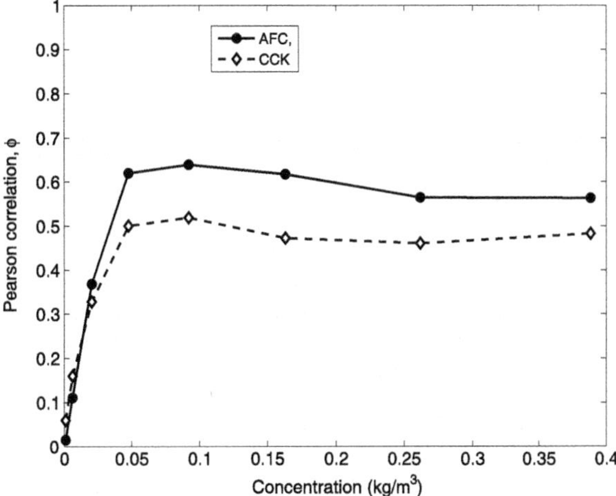

Fig. 6 Pearson's ϕ between the reference concentration map and the AFCK map, taking different concentration percentiles (10th to 90th) as threshold values. AFC yielded higher correlation for the 40th percentile and above

Acknowledgements Financial support for this research was provided by a scholarship from the Natural Sciences and Engineering Research Council of Canada (NSERC).

References

1. Almendral A, Abrahamsen P, Hauge R (2008) Multidimensional scaling and anisotropic covariance functions. In: Proceedings of the eight international geostatistics congress, Santiago, Chile
2. Bailly J, Monestiez P, Lagacherie P (2006) Modelling spatial variability along drainage networks with geostatistics. Math Geol 38(5):515–539
3. Barabas N, Goovaerts P, Adriaens P (2001) Geostatistical assessment and validation of uncertainty for three-dimensional dioxin data from sediments in an estuarine river. Environ Sci Technol 35(16):3294–3301
4. Boisvert JB, Deutsch CV (2011) Programs for kriging and sequential Gaussian simulation with locally varying anisotropy using non-Euclidean distances. Comput Geosci 37:495–510
5. Boisvert JB, Manchuk JG, Deutsch CV (2009) Kriging in the presence of locally varying anisotropy using non-Euclidean distances. Math Geosci 41:585–601
6. Christakos G, Hristopulos DT, Bogaert P (2000) On the physical geometry concept at the basis of space/time geostatistical hydrology. Adv Water Resour 23:799–810
7. Cirpka OA, Frind EO, Helmig R (1999) Numerical methods for reactive transport on rectangular and streamline-oriented grids. Adv Water Resour 22(7):711–728
8. Cirpka OA, Frind EO, Helmig R (1999) Streamline-oriented grid generation for transport modelling in two-dimensional domains including wells. Adv Water Resour 22(7):697–710
9. Curriero FC (1996) The use of non-Euclidean distances in geostatistics. PhD thesis, Kansas State University
10. Curriero FC (2006) On the use of non-Euclidean distance measures in geostatistics. Math Geol 38(8):907–926
11. Dagbert M, David M, Crozel D, Desbarats A (1984) Computing variograms in folded strata-controlled deposits. In: Vergy G, David M, Journel A, Marchal A (eds) Geostatistics for natural resources characterization. Reidel, Dordrecht, pp 71–89
12. Deutsch C, Wang L (1996) Hierarchical object-based stochastic modeling of fluvial reservoirs. Math Geol 28(7):857–879

13. Deutsch CV (2002) Geostatistical reservoir modeling. Oxford University Press, London
14. de Fouquet C, Bernard-Michel C (2006) Modèles géostatistiques de concentrations ou de débits le long des cours d'eau. C R Géosci 338:307–318
15. Frind EO (1982) The principal direction technique: a new approach to groundwater contaminant transport modeling. In: Finite elements in water resources. Proceedings of the 4th international conference, Hanover, Germany, June 1982
16. Kohavi R, Provost F (1998) Glossary of terms. Mach Learn 30:271–274
17. Kubat M, Holte RC, Matwin S (1998) Machine learning for the detection of oil spills in satellite radar images. Mach Learn 30:195–215
18. Legleiter CJ, Kyriakidis PC (2006) Forward and inverse transformation between Cartesian and channel-fitted coordinate systems for meandering rivers. Math Geol 38(8):927–958
19. Loland A, Host G (2003) Spatial covariance modelling in a complex coastal domain by multidimensional scaling. Environmetrics 14:307–321
20. Matanga GB (1993) Stream functions in three-dimensional groundwater flow. Water Resour Res 29(9):3125–3133
21. Monestiez P, Sampson P, Guttorp P (1993) Modeling of heterogeneous spatial correlation structure by spatial deformation. Cah Géostat 3:35–46
22. Narasimhan TN (2008) Laplace equation and Faraday's lines of force. Water Resour Res 44:W09412
23. Rathbun SL (1998) Spatial modelling in irregularly shaped regions: kriging estuaries. Environmetrics 9(2):109–129
24. Rivest M, Marcotte D, Pasquier P (2012) Sparse data integration for the interpolation of concentration measurements using kriging in natural coordinates. J Hydrol 416–417:72–82. doi:10.1016/j.jhydrol.2011.11.043
25. Sampson PD, Guttorp P (1992) Nonparametric-estimation of nonstationary spatial covariance structure. J Am Stat Assoc 87(417):108–119
26. te Stroet CBM, Snepvangers JJJC (2005) Mapping curvilinear structures with local anisotropy kriging. Math Geol 36:635–649
27. Thompson JF, Warsi ZUA, Mastin CW (1985) Numerical grid generation: foundations and applications. North-Holland, Amsterdam
28. Ver Hoef JM, Peterson E, Theobald D (2006) Spatial statistical models that use flow and stream distance. Environ Ecol Stat 13:449–464
29. Xu W (1996) Conditional curvilinear stochastic simulation using pixel-based algorithms. Math Geol 28:937–949

A Comparison of Methods for Solving the Sensor Location Problem

Rodolfo García-Flores, Peter Toscas, Dae-Jin Lee, Olena Gavriliouk, and Geoff Robinson

Abstract A problem that frequently arises in environmental surveillance is where to place a set of sensors in order to maximize collected information. In this article we compare four methods for solving this problem: a discrete approach based on the classical k-median location model, a continuous approach based on the minimization of the prediction error variance, an entropy-based algorithm, and simulated annealing. The methods are tested on artificial data and data collected from a network of sensors installed in the Springbrook National Park in Queensland, Australia, for the purpose of tracking the restoration of biodiversity. We present an overview of these methods and a comparison of results.

1 Introduction

The importance of sensor networks is ever increasing, as the developments summarized in the review [15] testify. The problem of optimizing the locations of a set of sampling points over a region A, which we address in this paper, is very common in spatial statistics, particularly in environmental monitoring networks. An optimal *design* (i.e., a set of sensor locations) is commonly determined using measurements Z_i: $i = 1, \dots, n$ of the values on a set of sampling points $s_i \in A$: $i = 1, \dots, n$. These measurements are the product of a process $\{\mathbf{Z}(\mathbf{s})\colon \mathbf{s} \in D \subset \mathbb{R}^2\}$, $\mathbf{Z}(\mathbf{s}) = \mathbf{Y}(\mathbf{s}) + \varepsilon(\mathbf{s})$, where ε is assumed to be independent Normal with zero mean and variance τ^2, and $\mathbf{Y}(\mathbf{s}) = \mu(\mathbf{s}) + \eta(\mathbf{s})$. In this expression, $\mathbf{Y}(\mathbf{s})$ is usually assumed to be a stationary Gaussian process with $E[\mathbf{Y}(\mathbf{s})] = \mu(\mathbf{s})$, and $\eta \sim N(0, \Sigma)$, so that $\mathrm{Var}[\mathbf{Y}(\mathbf{s})] = \Sigma_{ii}$, where Σ is the spatial covariance matrix [4, Chap. 3].

A natural approach to solve this problem is to minimize the prediction error variance. Brus and Heuvelink [3] solve the problem by minimizing the spatially-averaged universal kriging variance, which incorporates trend estimation error as well as spatial interpolation. Weng and Hong [14] model the covariance spatial

R. García-Flores (✉) · P. Toscas · D.-J. Lee · O. Gavriliouk · G. Robinson
CSIRO Mathematics, Informatics and Statistics, Private Bag 33, Clayton South, VIC 3168, Australia
e-mail: Rodolfo.Garcia-Flores@csiro.au

P. Abrahamsen et al. (eds.), *Geostatistics Oslo 2012*, Quantitative Geology and Geostatistics 17,
DOI 10.1007/978-94-007-4153-9_43,

structure using random fields and propose a utility-maximization algorithm that considers the sensors' battery life. One of the methods used in this paper is based on triangular integration to minimize prediction error variance as introduced by Robinson [11].

Unfortunately, a covariance function may not always be available. Space filling designs are geometric methods based on the idea that sampling points are placed so as to minimize a criterion based solely on the distance between sensor locations and all other locations. This problem is also known as the *k-median location problem*. An example of this approach is [2], who discretize the area of interest in a grid and minimize the mean squared shortest distance by k-means clustering. The method used here is due to [12], whose criterion measures the coverage of a design with respect to the area of interest.

If, on the other hand, the covariance function is available, the calculation of a relevant optimization criterion may be computationally expensive, as it needs to be calculated for each subset of locations. For this reason, meta-heuristic search using simulated annealing is an attractive alternative (see for example [13]). Finally, another family of methods commonly used is based on the *entropy* criterion. The rationale behind these methods is that the optimal design is made up of the sensors that are most informative with respect to the entire design space. A good design would therefore minimize the conditional entropy of the measurements in all possible sensor locations, given the selected locations.

Many of the examples of applications of sensor networks in environmental monitoring are dedicated to the measurement of pollutant concentrations in air or water. Murray et al. [8] present a methodology to design, test and deploy contamination warning systems for drinking water. Nunes at al. [9] propose a method to optimize a monitoring network for detecting and delimiting underwater contamination plumes. Nunes et al. [10] present a comparison of approaches for the design of a water monitoring network. The final network layout is obtained using simulated annealing.

In this paper we compare four methods for solving the sensor location problem. After an introduction to the Springbrook National Park Wireless Sensor Network (SWSN), an overview of the methods is presented. Simulated annealing (SA) and cover design make no assumption about spatial correlation structure, and minimum entropy and minimization of prediction error variance assume a linear Gaussian model and use an exponential variogram model. The results presented consist of, first, a comparison of the methods using randomly-generated linear Gaussian measures in a unit square, and second, a prospective analysis of the SWSN data. The paper ends with a conclusion.

2 The Springbrook Sensor Network

The Springbrook National Park is located 96 km south of Brisbane and covers an area of 61.97 km^2 at between 600 and 1000 m above sea level. SWSN (Fig. 1) is a joint research effort of Queensland's Department of Environment and Resource Management, CSIRO and other government agencies, which aims to enhance

Fig. 1 Springbrook sensors with known location

knowledge of rain forest restoration and to monitor biodiversity. At present, the measured variables include air and soil temperature, air humidity and soil moisture; air temperature is the variable used throughout this paper to compare methods. The network will expand in the near future to account for direct indicators of wildlife presence. At the moment, however, the data collected is being used to monitor the re-growth of local vegetation.

Section 4 includes results on *prospective* sensor design of artificial data, as well as on *retrospective* design on the SWSN data to determine the best locations of future sensors. The former consists on defining a new network, whereas the latter is the addition (or removal) of sensors to an existing network in order to increase (or sustain) measurement accuracy.

3 Methods

The first three methods select among a set of potential locations; only prediction error variance (PEV) minimization is free to select any location in the unit square. All algorithms except cover design use a fitted variogram model.

Algorithm 1 Obtaining a design with minimum average PEV

Input: Desired number of sensors $n = |D|$,
Sill c_0 and range a_0 of fitted exponential variogram $\gamma(\cdot)$.
Output: An optimal design D

1 **foreach** *point* **g** *of a design* G **do**
2 $\quad g_x := U(0,1)$, $g_y := U(0,1)$
3 **end**
4 **while** *number of iterations* $\leq$ *max number of iterations* **do**
5 $\quad$ **foreach** *point* **g** *in* G **do**
6 $\quad\quad B := G, C := G$
7 $\quad\quad$ **if b == g then**
8 $\quad\quad\quad b_x := b_x + \varepsilon$
9 $\quad\quad$ **end**
10 $\quad\quad$ **if c == g then**
11 $\quad\quad\quad c_y := c_y + \varepsilon$
12 $\quad\quad$ **end**
13 $\quad\quad \delta_{g_x} := (PEV(B) - PEV(G))/\varepsilon$, $\delta_{g_y} := (PEV(C) - PEV(G))/\varepsilon$
14 $\quad$ **end**
15 $\quad$ objective value $h^* := \arg\min_h PEV(G + h\delta_G)$
16 $\quad G := G + \phi h^* \delta_G$, where ϕ is a relaxation factor.
17 **end**
18 $D := G$

3.1 Minimization of Prediction Error Variance

Let D denote a design of n points in the unit square and **d** a point in D. Algorithm 1 produces a design with minimum average prediction error variance in the unit square, assuming that the spatial dependence is known in the form of a semi-variogram model. For the experiments, ordinary kriging was used with an exponential covariance fitted to the empirical variograms. The parameters were estimated using maximum likelihood.

The algorithm starts with a random design G and repeats the following procedure for a fixed number of iterations. For each coordinate of each point in G it calculates the gradient δ_G of the average of the prediction error variance *PEV* (lines 5 to 14), as explained below. A search is made on the distance h in the direction of steepest descent to reduce the value of the objective function (line 15). The gradient and h^* and are used to correct G (line 16).

Function *PEV* calculates the average of the prediction error variance in two steps. First, it obtains a Delaunay triangulation of the region defined by G, using the sensor locations as vertices. The algorithm checks that the region is convex; if not, the point locations are adjusted accordingly. The second step is integration over the triangles of the prediction error variance of the residual,

$$\sigma^2(\mathbf{s}) = \mathbf{c}(0) - \mathbf{c}_0' \mathbf{C}^{-1} \mathbf{c}_0, \tag{1}$$

where $\mathbf{C}$ is the $n \times n$ variance-covariance matrix of the n residuals and $\mathbf{c}_0$ is the vector of covariances between the residuals at the observation and prediction locations. Integration is done using a Gaussian quadrature rule

$$\int_A f(\alpha, \beta, \gamma)\,\mathrm{d}A = A \sum_{i=1}^{ng} w_i f(\alpha_i, \beta_i, \gamma_i), \tag{2}$$

for a triangle of area A, where ng is the number of points, w_i is the Gaussian weight, and $(\alpha_i, \beta_i, \gamma_i)$ are natural coordinates such that $0 = \alpha_i + \beta_i + \gamma_i - 1$. A detailed discussion of quadrature rules is beyond the scope of this paper, but the interested reader may refer to [5]. Additional costs treated in the same optimization problem may include those associated with linking the sensors; this is done using Kruskal's [7] minimum spanning tree algorithm.

3.2 Minimization of a Space-Filling Criterion

Unlike covariance-based criteria, the approach of minimizing a space-filling criterion proposed by [12] does not require the knowledge of the covariance function of the underlying process. The problem consists of selecting the optimal design D from a larger set of candidate points, C. Let n be the size of D and N the size of C. Given a point $\mathbf{s}$,

$$d_p(\mathbf{s}, D) = \left(\sum_{\mathbf{u}\in D} \|\mathbf{s} - \mathbf{u}\|^p\right)^{1/p} \tag{3}$$

is a measure of the distance between this point and a particular design D. It is easy to see that, when $p < 0$, $d_p(\mathbf{s}, D) \to 0$ as $\mathbf{s}$ converges to a member of D. The most commonly used metric to evaluate the goodness of a space covering design is an average of (3) over all candidate points,

$$\Omega_{p,q}(D) = \left(\sum_{\mathbf{s}\in C} d_p(\mathbf{s}, D)^q\right)^{1/q}. \tag{4}$$

The optimal coverage design is the subset of n elements in D from the N elements in C which minimize the criterion $\Omega_{p,q}(D)$.

The basic idea for finding the optimum design is to start with random configurations and then decrease the coverage criterion by swapping a candidate point by a design point. For each point in the current design, replace this point by members of the candidate set. If a particular swap reduces the coverage criterion over the initial design, then this new point is included in the design and the old point is moved in the candidate set. This process is repeated until there are no more productive swaps.

Royle and Nychka [12] note that the efficiency of the algorithm is increased if, instead of calculating all distance matrices $\mathbf{D}$ in every swap, only the column and row elements of D are changed. Hence, if $\mathbf{D}$ is an $N \times n$ matrix such that $D_{ij} = \|s_i - y_i\|^p$ and r_i the row sum $r_i = \sum_{j=1}^{n} D_{ij}$, then (4) becomes

$$\Omega_{p,q}(D) = \left(\sum_{i=1}^{N} (r_i)^{q/p} \right)^{1/q}. \tag{5}$$

If the new design is $\mathbf{D}^*$, and $y_{i'}$ is replaced by candidate element $s_{j'}$, to compute the resulting criterion for $\mathbf{D}^*$ one needs to only update the i'th row and j'th column of $\mathbf{D}$. This reduces the number of calculations from $N \times n$ to $N + n - 1$. With r_i^* the new row sums, if $i \neq i'$,

$$r_i^* = r_i - \|y_{i'} - s_i\|^p + \|s_{j'} - s_i\|^p , \tag{6}$$

and if $i = i'$,

$$r_i^* = \sum_{j \neq j'} \|y_j - y_{i'}\|^p + \|s_{j'} - y_{i'}\|^p. \tag{7}$$

3.3 Entropy

A sensible request for a good design is that it is made up of the sensors that are most informative with respect to the entire design space. A good design would therefore minimize the conditional entropy of the measurements in all possible sensor locations, given the selected locations. Using the notation of Sect. 3.2,

$$H(X_{C \setminus D} | X_D) = - \iint p(\mathbf{x}_{C \setminus D}, \mathbf{x}_D) \log p(\mathbf{x}_{C \setminus D} | \mathbf{x}_D) \, d\mathbf{x}_{C \setminus D} \, d\mathbf{x}_D, \tag{8}$$

where $X_{C \setminus D}$ and X_D refer to sets of random variables at locations $C \setminus D$ and D respectively. A multivariate Gaussian distribution $P(X_C = \mathbf{x}_C) = N(\mu_C, \Sigma_{CC})$ is assumed. We want to find

$$D^* = \underset{D \subset C:\, |D|=k}{\arg\min} \; H(X_{C \setminus D} | X_D) = \underset{D \subset C:\, |D|=k}{\arg\max} \; H(X_D) \tag{9}$$

since $H(X_{C \setminus D} | X_D) = H(X_C) - H(X_D)$. To determine the optimal design, the following heuristic was used: start from a set of empty locations D_0 and greedily add placements until $|D| = k$. At each iteration, add to D_i the location $y_H^* \in C \setminus D$ that has the highest conditional entropy,

$$y_H^* = \underset{y}{\arg\max} \, H(X_y | X_{D_i}), \tag{10}$$

which is the location we are most uncertain about given the sensors placed thus far [6, 16]. The same source reports that mutual information has better performance than entropy alone, but this is beyond the scope of this paper.

3.4 Simulated Annealing

The central idea of simulated annealing is to accept (with a reasonable probability) solutions that are far from the optimal at the beginning of the search, in order to explore the solution space. This probability decreases as the search progresses. The optimization criterion used in the experiments is the total sum of the variances at the potential sensor locations. The logarithmic cooling schedule is as given in [1, p. 890]. The starting temperature parameter was set to 1000 and the number of evaluations at each temperature was set to 1000.

4 Results and Discussion

Entropy minimization and PEV minimization were coded in R. The cover design algorithm used is the one in the R library "fields", and Simulated Annealing was called from method `optim` using option "SANN". All results were obtained using a 64-bit Intel Xeon CPU with 2 processors of eight cores (2.27 GHz) each and 48 GB of RAM.

4.1 Data Sets

The algorithms were tested using, first, artificial data consisting of ten test sets of 100 points each, and second, actual sensor locations and air temperature measurements from SWSN.

The (isotropic) underlying random process used for the measurements at each location in the artificial data sets was simulated from the true function given by $-(2x-1)^2-(2y-1)^2+5$, where x and y are drawn from a random Uniform distribution in $[0,1]$. We also add an additive white noise. For the models that need a covariance function, an exponential function and its corresponding variogram were fitted from the empirical variogram of each test data set. The results of kriging are shown in Figs. 2 and 3 for the test and SWSN data, respectively.

4.2 A Comparison

Figure 4 shows the results of selecting 30 sensors from the 100 potential sites in one of the test data sets. Potential unselected sites are marked with a small dot and selected locations are marked with a full circle. Results for PEV minimization do not show potential sites because this algorithm considers the search space as a continuum and any point in the unit square may be chosen.

The cover design and entropy algorithms provide a good coverage of the total area. The particular artificial data set shown did not produce many points in the

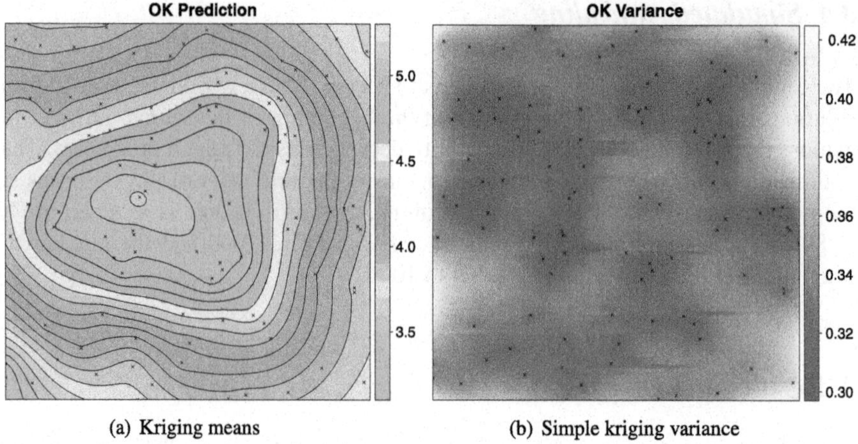

(a) Kriging means (b) Simple kriging variance

Fig. 2 Simple kriging of one of the test data sets

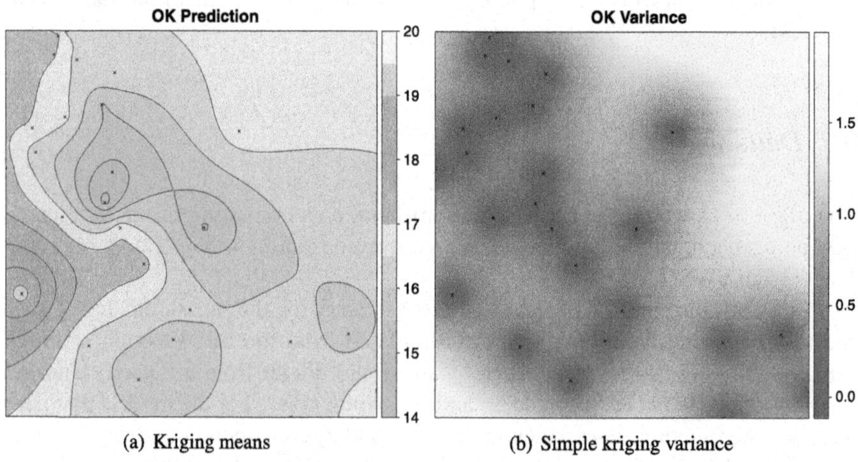

(a) Kriging means (b) Simple kriging variance

Fig. 3 Simple kriging for Springbrook air temperature data

lower middle of the unit square and this area goes noticeably uncovered by all the algorithms except PEV minimization. In contrast, PEV minimization seems to leave uncovered the central part of the square. This is probably due to the "flatness" of the variable value and the relatively constant variance in the central part of the area. This algorithm places sensors only in the vertices of the integration triangles, and the algorithm places only few triangles covering the central area. Simulated annealing tends to place more sensors on the right of the unit square, which happens to be the area with a higher density of potential locations. This happens because the criterion SA uses in the experiments is to minimize the total sum of the variances at the potential sensor locations. Kriging variance increases at areas with a low density

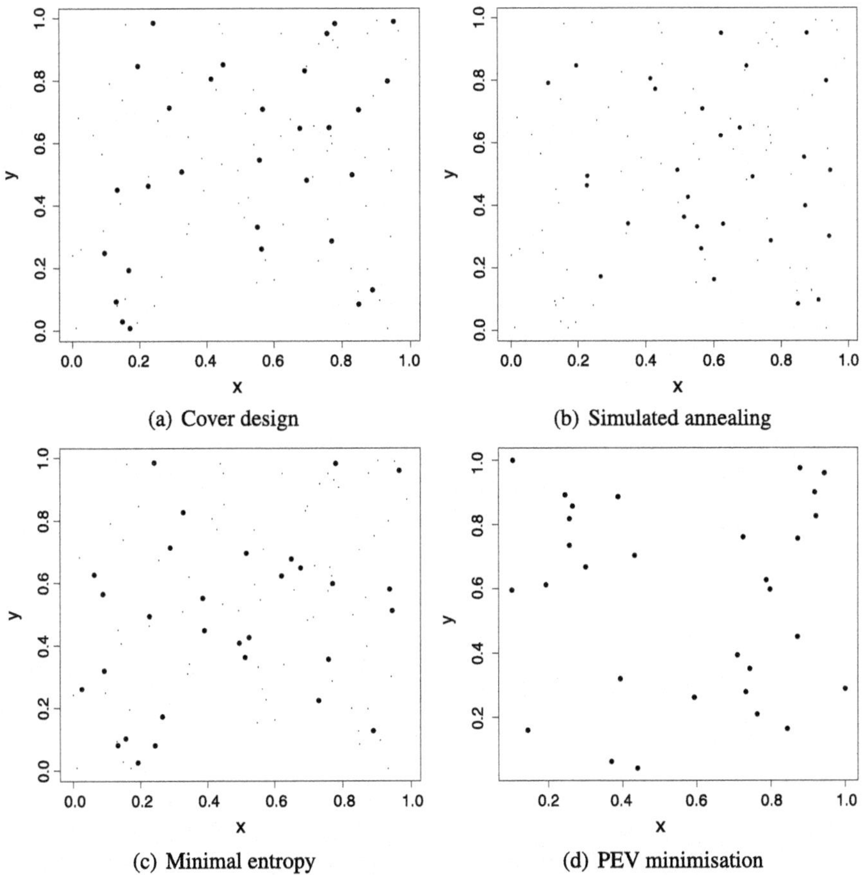

(a) Cover design

(b) Simulated annealing

(c) Minimal entropy

(d) PEV minimisation

Fig. 4 Typical results of the sensor location algorithms. Selected sites are marked with a *full circle* and potential unselected sites are marked with a *dot*

Table 1 Execution times in seconds for the sensor location algorithms

	Mean	SD
Cover design	0.61	0.29
Simulated annealing	92.82	0.43
Entropy	20.30	0.55
PEV Minimization	4135.85	3546.32

of potential locations, and SA tends to place sensors in sites with low prediction variances.

Execution times are shown in Table 1. Not surprisingly, PEV minimization is by far the most time-consuming algorithm, but also the one that produces the most consistent pattern in all data sets. These are not shown due to space limitations.

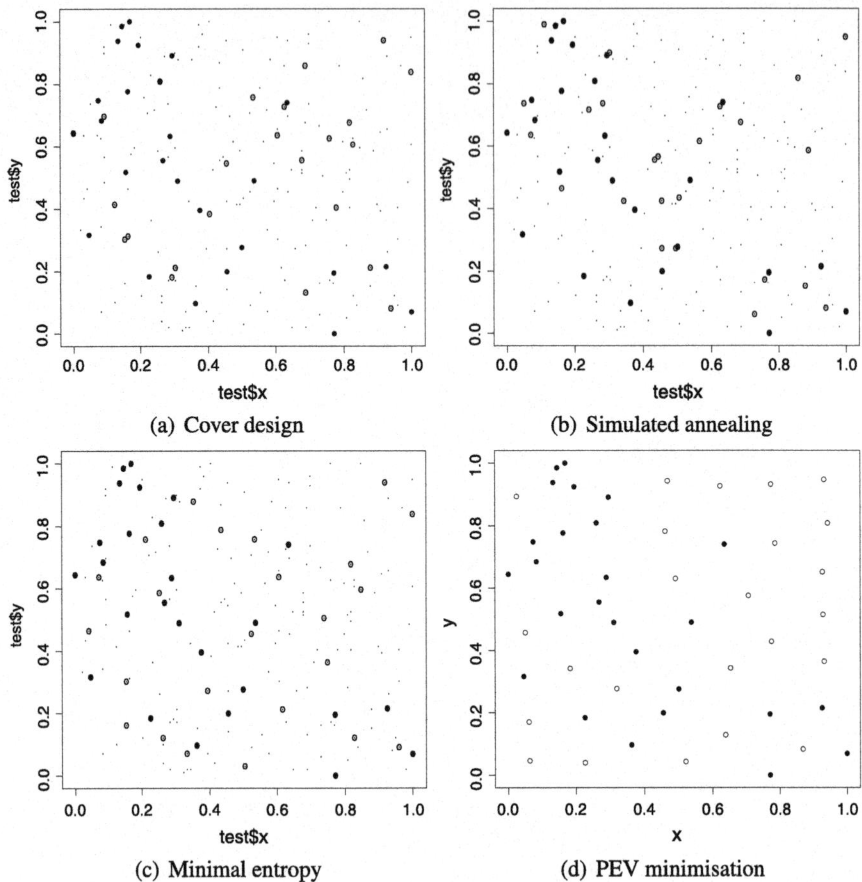

(a) Cover design
(b) Simulated annealing
(c) Minimal entropy
(d) PEV minimisation

Fig. 5 Prospective designs for the Springbrook National Park Sensor Network. *Full circles* represent the original network, *empty circles* represent additional sensors

Figure 5 shows the results of a prospective design study using the data from SWSN. Full circles represent the original network, empty circles represent additional sensors selected, and dots represent the potential unselected sites, when applicable. The results show again that preferred additional points in SA tend to be near the existing sensor locations because the variance near these points is lower. Cover design gives a more even distribution, but not as uniform as PEV minimization. Entropy gives a more interspersed design, although there is no great difference in the designs obtained by entropy and cover design methods.

It is worth noting that prospective design is much more computationally expensive for PEV minimization than free selection of all nodes. The run time for free selection is 4136 s, whereas a run constrained by existing selected locations increases to 47925 s. The reason is that fixing the original set of points slows down the speed at which the whole design converges because the gradient is sub-optimal (step 13 in Algorithm 1).

5 Conclusions and Directions for Research

A problem that frequently arises in environmental surveillance is where to place a set of sensors in order to maximize collected information. In this paper we have compared four methods for solving this problem, and we applied them to a prospective design study for the Springbrook National Park Sensor Network. The minimum entropy method and space filling approaches give comparable results for the isotropic process assumed. Simulated annealing tends to add points in regions of high density of potential locations. Minimization of prediction error variance is the most computationally expensive method, but due to its freedom to select any point in the area of interest and its prior knowledge on the variable, it is also the best performing method. PEV minimization takes a different approach by computing the variance of average prediction error by high-order numerical integration, rather than by discretization values at selected locations. Future work includes, after expanding the network in coverage and measured variables, trying alternate criteria for SA, exploring the use of mutual information instead of entropy, and improving PEV minimization to increase the speed of the method when doing prospective designs.

References

1. Belisle C (1992) Convergence theorems for a class of simulated annealing algorithms on R(D). J Appl Probab 29(4):885–895
2. Brus D, de Gruijter J, van Groningen J (2006) Designing spatial coverage samples using the k-means clustering algorithm. In: Digital soil mapping: an introductory perspective. Elsevier, Amsterdam, pp 183–192
3. Brus D, Heuvelink G (2007) Optimization of sample patterns for universal kriging of environmental variables. Geoderma 138(1–2):86–95
4. Cressie N (1993) Statistics for spatial data. Wiley series in probability and mathematical statistics. Wiley–Interscience, New York
5. Dunavant D (1985) High degree efficient symmetrical Gaussian quadrature rules for the triangle. Int J Numer Methods Eng 21:1129–1148
6. Krause A, Singh A, Guestrin C (2008) Near-optimal sensor placements in Gaussian processes: theory, efficient algorithms and empirical studies. J Mach Learn Res 9:235–284
7. Kruskal J (1956) On the shortest spanning subtree of a graph and the travelling salesman problem. Proc Am Math Soc 7(1):48–50
8. Murray R, Hart W, Phillips C, Berry J, Boman E, Carr R, Riesen L, Watson J, Haxton T, Herrmann J, Janke R, Gray G, Taxon T, Uber J, Morley K (2009) US environmental protection agency uses operations research to reduce contamination risks in drinking water. Interfaces 39(1):57–68
9. Nunes L, Cunha M, Ribeiro L, Azevedo J (2007) New method for groundwater plume detection under uncertainty. In: Ferreira LL, Viera J (eds) Water in Celtic countries: quantity, quality and climate variability. IAHS publication, vol 310. International Association of Hydrological Sciences, Institute of Hydrology, Wallingford, pp 191–198

10. Nunes L, Paralta E, Ribeiro L (2007) Comparison of variance-reduction and space-filling approaches for the design of environmental monitoring networks. Comput-Aided Civ Infrastruct Eng 22:489–498
11. Robinson G (2010) Computing strategies for choosing sensor locations to minimize prediction error variance. Tech. Rep. EP-101092, CSIRO Mathematics, Informatics and Statistics
12. Royle J, Nychka D (1998) An algorithm for the construction of spatial coverage designs with implementation in SPLUS. Comput Geosci 24(5):479–488
13. Su Z, Shang F, Wang R (2009) A wireless sensor network location algorithm based on simulated annealing. In: Proceedings of the 2nd international conference on biomedical engineering and informatics, vols 1–4. IEEE Press, New York, pp 1248–1252
14. Weng Y, Hong W (2009) Local optimal sensor deployment for random field reconstruction. In: Proceedings of 4th international conference on computer science and education. IEEE Press, New York, pp 419–424
15. Yang C, Raskin R, Goodchild M, Gahegan M (2010) Geospatial cyberinfrastructure: past, present and future. Comput Environ Urban Syst 34:264–277
16. Zidek J, Sun W, Le N (2000) Designing and integrating composite networks for monitoring multivariate Gaussian pollution fields. Appl Stat 49:63–79

Comparing Geostatistical Models for River Networks

Gregor Laaha, Jon Olav Skøien, and Günter Blöschl

Abstract Geostatistical methods have become popular in various fields of hydrology, and typical applications include the prediction of precipitation events, the simulation of aquifer properties and the estimation of groundwater levels and quality. Until recently, surprisingly little effort has been undertaken to apply geostatistics to stream flow variables. This is most likely because of the tree-like structure of river networks, which poses specific challenges for geostatistical regionalization. Notably, the shape of catchments (irregular block support), the nestedness of catchments along the river network (overlapping support), and the definition of a relevant distance measure between catchments pose specific challenges. This paper attempts an annotated survey of models proposed in the literature, stating contributions and pinpointing merits and shortcomings. Two conceptual viewpoints are distinguished: one-dimensional models which use covariances along a river network based on stream distance, and two-dimensional models where stream flow is conceptualized as the integral of the spatially continuous local runoff process over the catchment area. Both geostatistical concepts are evaluated relative to geostatistical standard methods based on Euclidean distances. It is shown how the methods perform in various examples including spatial prediction of environmental variables, stream flows and stream temperatures.

G. Laaha (✉)
Institute of Applied Statistics and Computing, University of Natural Resources and Life Sciences Vienna (BOKU), Vienna, Austria
e-mail: gregor.laaha@boku.ac.at

J.O. Skøien
Joint Research Centre of the European Commission, Institute for Environment and Sustainability, Ispra, Italy
e-mail: jon.skoien@jrc.ec.europa.eu

G. Blöschl
Institute for Hydraulic and Water Resources Engineering, Vienna University of Technology, Vienna, Austria
e-mail: bloeschl@hydro.tuwien.ac.at

P. Abrahamsen et al. (eds.), *Geostatistics Oslo 2012*,
Quantitative Geology and Geostatistics 17,
DOI 10.1007/978-94-007-4153-9_44,

1 Introduction

The estimation of stream flow and stream flow related variables is a fundamental problem in water resources management. Water uses, such as hydropower operation, agricultural irrigation, drinking water supply, and navigation depend on available stream flow quantity. River ecosystems are controlled by the water quality, variables such as oxygen concentrations, stream temperature, and the concentration of pollutants. For locations where appropriate measurements are missing these variables can be estimated from available measurements in the region using statistical regionalization techniques. Estimates can be obtained by a range of methods [1], including statistical, conceptual and mechanistic methods. Statistical methods are usually based on some kind of correlation of the stream flow variable of interest: Multivariate methods exploit the information of auxiliary variables representing physical or geographical properties of river locations, while geostatistical methods are based on the spatial autocorrelation of the stream variable. Geostatistical methods have evolved in the mining industry. The main problem consisted of estimating the expected ore grade (and its uncertainty) of a block using point samples of the ore grade in the area. The problem in predicting on a river network is quite different. The main difference is that catchments related to points of the river network are organized into subcatchments, i.e. they are nested. It is therefore clear that upstream and downstream catchments would have to be treated differently from neighboring catchments that do not share a subcatchment. Therefore Euclidean distances between catchments are not the natural way of measuring the spatial distance of catchments. Estimation of variables on stream networks needs to use a topology that is different from the usual Euclidean topology. Most applications of geostatistics to river network, so far, have indeed used Euclidean distance between catchments, usually measured as the Euclidean distance between the gauges or the catchment centroids [17]. Given the obvious nested structure of catchments it is surprising that very little research has been done on extending geostatistical concepts to catchments. Recently, a few approaches have been proposed which can be divided into two groups, one-dimensional and two-dimensional river network conceptualizations. These major conceptual viewpoints are reviewed hereafter.

2 Geostatistical Models for River Networks

2.1 1D Models

The first approach treats river network variables as a one-dimensional problem. In this approach, discharge is a spatial variable which is defined for any point along the river network. Then, spatial prediction on a river location x_0 can be performed using the ordinary kriging predictor:

$$\hat{z}(x_0) = \sum_{i=1}^{n} \lambda_i z(x_i), \tag{1}$$

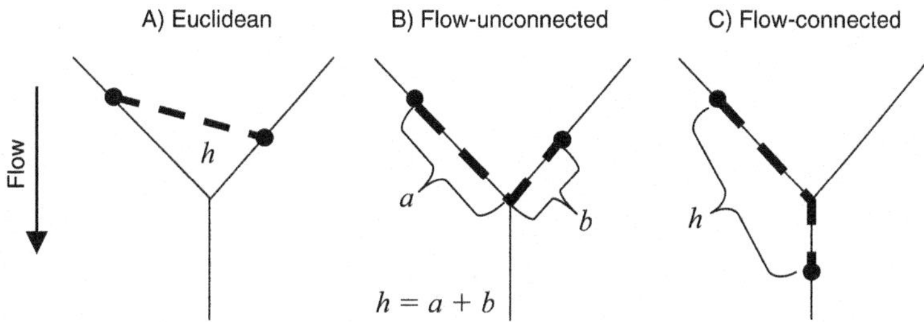

Fig. 1 Distance metrics for river networks: Euclidean distance (**A**) and stream distance of flow-unconnected (**B**) and flow-connected (**C**) locations (*black circles*). Schematic, from [11]

where λ_i are the optimal kriging weights obtained by minimizing estimation variance subject to unbiasedness constraints. The method requires a meaningful distance metric together with a valid covariance function for river networks. Stream distance (i.e. distance measured along the river network) is an obvious choice of the metric and has been used by several authors [2, 6, 15]. Figure 1 shows how stream distance can be defined for flow-unconnected (panel B) and flow-connected (panel C) river branches. As it can be seen from the figure, stream distance may differ considerably from classical Euclidean distance (panel A).

Kriging based on stream distance requires a valid covariance function, and [5] was probably the first to develop a method for calculating covariance along a river network based on stream distance. As kriging based on stream distance is not generally consistent with the equation of continuity, particularly at river junctions, [6] proposed to include water balance constraints in the kriging system. The constraints guarantee that the interpolated lateral inflow of a river branch between two gauges is the difference between up- and downstream observations. References [15] and [2] proposed an alternative method based on a moving average representation of the latent process. Their approach is analogue to ordinary kriging and employs unilateral convolution kernels which go either upstream (Tail-up model) or downstream (Tail-down model) to derive covariance functions for river networks. It uses a representation of the random process by smoothing a white noise Brownian process by a square integrable kernel. The resulting random function is a Gaussian process with a spatial covariance function which is the self convolution of the smoothing kernel. A schematic of the typical kernel shapes of both models are shown in Fig. 2. The Tail-up model either uses catchment area [15] or stream order [2] to weight the spatial covariance of the branches of a stream junction. The Tail-down model does not require additional variables. To combine the advantages of the two models, [4] proposed a hybrid model. It allows for the possibility of stronger autocorrelation among flow-connected pairs of sites. There are further extensions of this model to account for correlations with auxiliary variables [11, 14], but all of them refer to one of the two kernel models to account for spatial dependences along the river network.

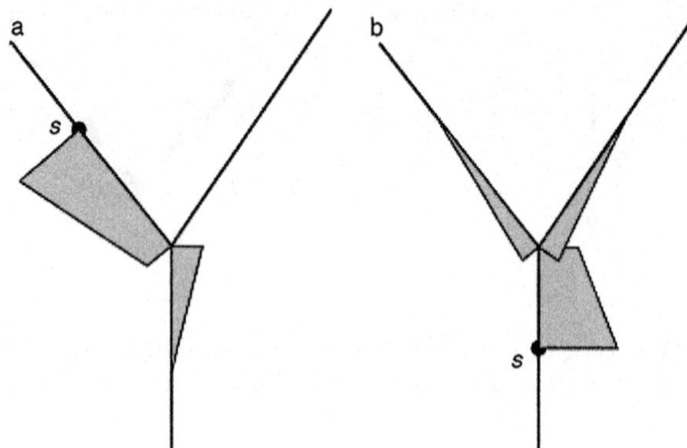

Fig. 2 Kernel shapes for (**a**) Tail-down model and (**b**) Tail-up model (schematic, from [4])

2.2 2D Models

The second approach treats the river network as a two-dimensional problem. In this approach, the runoff process is conceptualized as a spatially continuous process which exists at any point in the landscape. Instead of observing a pointwise realization of this process, data $z(A_1), z(A_2), \ldots, z(A_n)$ are observed at stream gauges, where

$$z(A_i) = \frac{1}{|A_i|} \int_{A_i} z(x)\,\mathrm{d}x \tag{2}$$

and A_i denotes the spatial support of $z(A_i)$. For stream flow variables, A_i is the catchment which drains into a river location x_i, and is its surface area. In this context, the transfer of information between river locations adds up to the area-to-area change of support problem in geostatistics [8], based on the classical Euclidean distance metric. For spatial prediction on a river location x_0 with catchment area A_0 from non-point samples $z(A_1), z(A_2), \ldots, z(A_n)$, the linear block-kriging predictor given by

$$\hat{z}(A_0) = \sum_{i=1}^{n} \lambda_i Z(A_i) \tag{3}$$

is used, where λ_i are the optimal kriging weights, again obtained by minimizing estimation variance subject to unbiasedness constraints. The calculation of λ requires aggregated values of spatial covariance or semivariance, which are calculated from a theoretical point-variogram by means of regularization (Fig. 3). Due to their different scale, the point variogram cannot be estimated directly from aggregate data. Instead, a back-calculation approach according to [9, 10] is used to identify a point variogram whose aggregates yield best fit to observed covariance or semivariance, as compared to other point-variograms.

The application of the 2D geostatistical models on river networks is rather novel. References [7, 12] were the first who proposed a 2D concept to disaggregate mean and variance of runoff observed at stream gauges to a partition of this basin. They

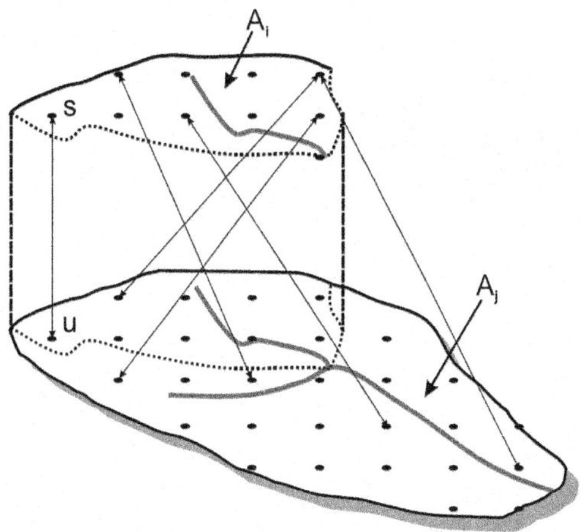

Fig. 3 Calculation of regularized variogram for two nested catchments (from [13]). Catchment i is nested within catchment j, A_i and A_j are the respective catchment areas (support), and s and u are the discretization points within each catchment

found it superior to the nested approach where disaggregation of discharges is performed between nearest neighbors only. Reference [13] extends the work of [12] to account for the stronger spatial correlation between nested basins than between un-nested basins. They showed that their method, known as topological kriging or Top-kriging, can be used, in an approximate way, for a range of stream flow related variables including variables that do not aggregate linearly and are non-stationary. Also, they apply a Kriging with uncertain data estimator [3] to account for local uncertainties of the observations. An example of the variogram model used in Top-Kriging (point variogram together with a number of regularized variograms for different catchment areas) is shown in Fig. 4. The model defines similarity according to distance, support (catchment size), and nestedness of catchments.

3 Comparison of Concepts

Both geostatistical concepts for river networks use kriging approaches to perform interpolation by a weighted average of observed data. To this end, the interpolation depends only on the kriging weights, which reflect the combined effect of distance metric and variogram model, conditional to a data structure. It is therefore straightforward to characterize the various concepts by their kriging weights, and how they are distributed in space. For the calculation of kriging weights, we assume the variogram model of Fig. 4, but we underline that the relative effects of size and topology would be similar for other variograms.

Secondly, we argue that water follows a river network, and the crucial aspect is how kriging weights are distributed between flow-connected and flow-unconnected neighbors. We analyze this aspect for two limiting situations, (i) equally distant neighbors, and (ii) a much more distant flow-connected neighbor. The first situation corresponds to source areas in the same geographic environment. The second situation corresponds to predicting at source areas, either using information from a

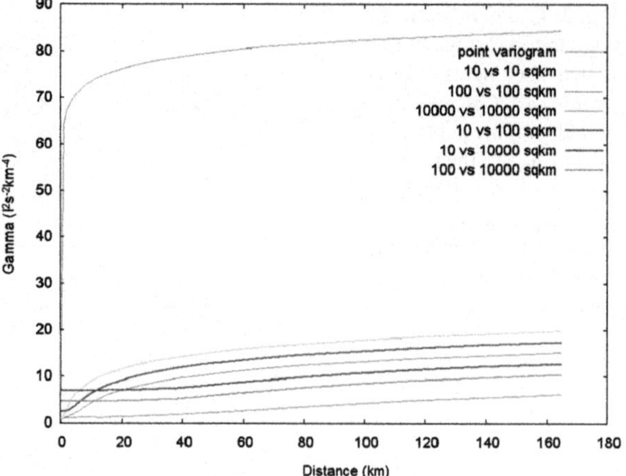

Fig. 4 Variogram model of Top-kriging, example of low streamflows q95 ($\mathrm{l\,s^{-1}\,km^{-2}}$), i.e. daily flow exceeded 95 % of the time, divided by catchment area

neighbor source in a similar environment, or from a much downstream river location in a likely different environment.

We first analyze the performance of ordinary kriging based on Euclidean distance (OK-Euclid). We do this because OK-Euclid is still the most frequently used kriging approach for the river network problem and constitutes a benchmark for any innovative model. OK-Euclid distributes kriging weights according to distance only (Fig. 5a). It gives equal weights for equally distant neighbors, regardless if they are flow-connected or flow-unconnected. This means that OK-Euclid clearly ignores the topology of the river network. If the flow-connected neighbor is much further away, OK-Euclid tends to give much more weight to the closer flow-unconnected neighbor. For the example shown in the right panel of Fig. 5a, the connected site receives a weight $\lambda_i = 0.1$ whereas the unconnected site receives almost all the weight with $\lambda_i = 0.9$, but this indeed depends on the structure of the variogram model. The behavior of OK-Euclid at large distance is rather realistic and reflects the decreasing difference of Euclidean and stream distance metric with increasing distance. Overall, however, OK-Euclid seems to give too much weight according to distance in geographic space and too little weight according to river topology.

Kriging based on stream distance (OK-Stream) distributes kriging weights in a different way (Fig. 5b). Stream distance takes finite values only for connected river systems. All weight ($\lambda_i = 1$ in the case of two neighbors) is therefore given to flow-connected neighbors and no weight ($\lambda_i = 0$) is given to the flow-unconnected neighbor. Even when the flow-connected site is far away, let's say at the river mouth, the estimation at the source would use the distant information exclusively, rather then using the more relevant information of the adjacent, but unconnected source area. This situation occurs illustratively when estimating at major water divides, e.g. at the source of the River Danube in the Black Forest, by a gauge situated close to the Black Sea, rather then by an adjacent and geographically similar source of the River Rhine. Similar effects occur along the entire river network, but often in a less pronounced and less transparent way. Notably, if there is much information available at flow-connected parts of the river system, the model will tend to give

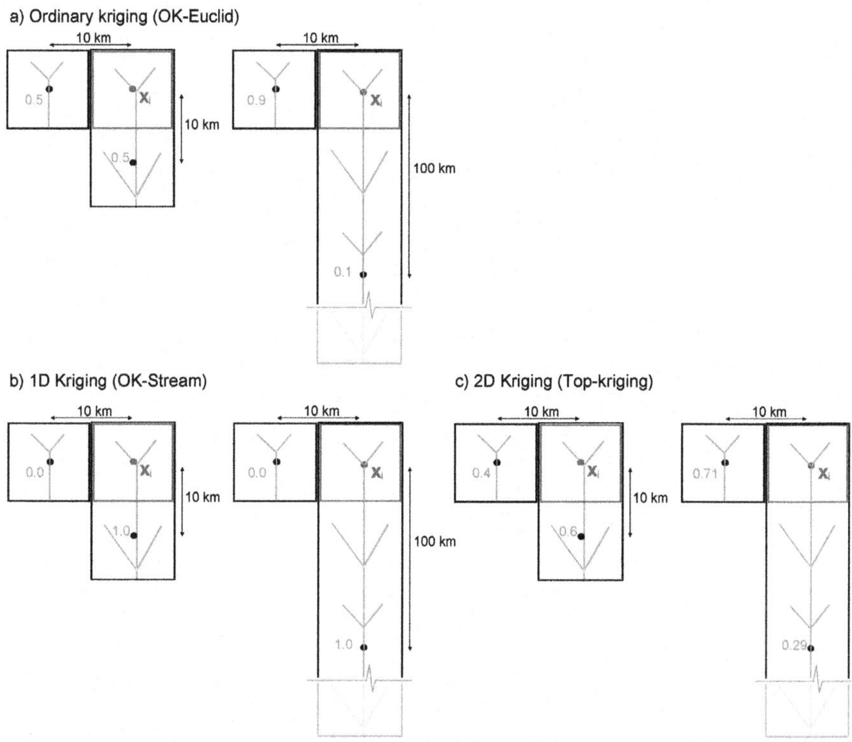

Fig. 5 Effect of stream network topology on the kriging weights λ_i (*green numbers*) of Top-kriging. (**a**) Ordinary kriging based on Euclidean distance, (**b**) ordinary kriging based on stream distance, and (**c**) Top-kriging. x_i indicates the center of the target catchment (*red boxes*), *black points* indicate the centers of the measured catchments (*black boxes*)

good results. But overall, OK-Stream seems to give too much weight according to river topology and too little weight according to distance in geographic space.

In Top-kriging, as opposed to ordinary kriging and kriging based on stream distance, the kriging weights are distributed in a natural way, depending on both, distance and river network topology. When a close-by site at the same river system is available, Top-kriging gives less weight to an adjacent river system. For the example shown in the left panel of Fig. 5c, the connected site receives a weight $\lambda_i = 0.60$ whereas the unconnected site receives a weight of only $\lambda_i = 0.40$. However, when the next site at the same river system is much further away, much weight is given to a close-by site at an adjacent river system. For the example shown in the right panel of Fig. 5c, the distant connected site receives a weight of only $\lambda_i = 0.29$ whereas the unconnected site receives a weight of $\lambda_i = 0.71$. This distribution of kriging weights is very realistic and corresponds well with physical understanding of stream flow processes behavior in space [17].

4 Review of Case Studies

4.1 1D Modeling of Environmental Variables

1D models have been applied to various environmental stream flow related variables, including environmental exceedances [2]. Reference [4] performed a comparative assessment of Tail-up and Tail-down models in the context of summer stream temperature and nitrate concentration. Annual stream temperature is a flow related variable which depends on local climate (continuous processes) and heat transfer in the river network. Nitrate loads depend on diffuse (agriculture) and local (waste water) input, and dilution related to stream flow quantity. Analyses were based on data from 141 nitrate and 187 temperature monitoring stations situated at the Meuse and Moselle basin in north-eastern France. Reference [4] found that for summer temperature, the Tail-up model performed better then the Tail-down model. For nitrate, the inverse was true. A hybrid model which is a combination of the Tail-up and the Tail-down model performed significantly better than each of the models separately. A closer analysis of the hybrid model yielded a stepwise pattern of prediction errors. Segments without observation have significantly higher estimation errors than segments with observations. For nitrate rates, the corresponding relative errors amount to about 60 % and 10 %, respectively. The pattern suggests that prediction is unreliable in the extrapolation case, but reliable in the interpolation case, with an abrupt change of performance in between. Pure Tail-up or Tail-down models, in contrast, appear less appropriate for environmental variables due to the fact that these variables are generally influenced by several natural processes, such as continuous processes at the catchment scale.

4.2 2D Modeling of Stream Flow

The second case study shows the performance of the 2D-model Top-kriging for streamflow estimation. The focus is on low flow q95, i.e. the daily flow exceeded 95 % of the time, divided by catchment area. There are two main groups of processes that control stream flow. The first group consists of runoff generating processes acting over the catchment area. These processes are continuous in space. The second group is related to runoff aggregation and routing. These processes are related to the river network topology. An extensive Austrian data set (8000 catchments, 490 of them are gauged) is used in the analysis. The study area is characterized by a large range of landscapes and flow regimes, and permits to assess how the model performs in various data settings. On average over the Austrian study area, Top-kriging explains 75 % of the variance of the low flows. The performance of Top-kriging mainly depends on the (intrinsic) homogeneity of the observations and the density of the gauging network in the region. One would expect the performance of Top-kriging to increase with increasing density of the network and increasing catchment size. The latter is because runoff is an integrating process, so low flows tend to vary

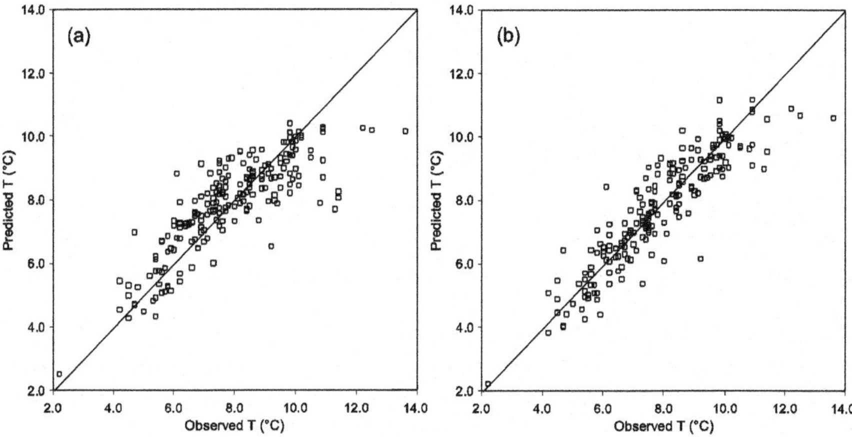

Fig. 6 Scatter plots of predicted vs. observed mean annual stream temperature (°C). (**a**) exponential regression and (**b**) Top-kriging with external drift

more smoothly along the stream for large catchments. The analyses of the data set in this paper indicate that this is indeed the case. Top-kriging explains 91 % and 59 % of the variances in the non-headwater catchments (with up-stream gauge) and headwater catchments (without up-stream gauge) respectively. The median catchment sizes of these two groups of catchments are 252 and 62 km^2. Top-kriging integrates continuous processes in the geostatistical concept and appears well suited to perform stream flow estimation at river networks.

4.3 2D Modeling of Stream Temperature

In the third case study we analyze the performance of Top-kriging for average annual stream temperature in Austrian. Annual stream temperature, equally to summer stream temperature, depends on local climate and heat transfer in the river network. For the alpine study area, the observed stream temperature patterns clearly depend on catchment altitude, and can be assumed as an output from random and deterministic processes. Hence, some adaptation of the model is in order, and we propose an approach similar to external drift kriging [16] to model non-stationarity. We term the extended model Top-kriging with external drift, because of this analogy. We tested the method for a comprehensive Austrian stream temperature dataset. The drift was modeled by exponential regression with catchment altitude. Leave-one-out cross-validation analysis reveals that the model performs well for the study domain (Fig. 6). When using Top-kriging in addition to the regression model the root mean squared error decreases by 20 % from 1.01 to 0.80 °C. For regions where the observed stream temperatures deviate from the expected value of the drift model, Top-kriging corrects regional biases. Top-kriging is able to improve the local adjustment of the drift model for the main streams and the tributaries, by exploiting the topological information of the stream network. The additional performance increases with increasing density of the gauging network.

5 Conclusions and Outlook

Most applications of geostatistics to river network, so far, have used ordinary kriging based on Euclidean distance (OK-Euclid) between catchments. This is surprising, giving the obvious nested structure of catchments. Recently, geostatistical models for river networks have been developed. The models are based either on a one-dimensional or two-dimensional conceptualization of the river network. For 1D-concepts, spatial similarity is modeled by covariances along a river network which are based on stream distance. In this context, prediction on a river location is performed using the ordinary kriging predictor based on stream distance. For 2D models, stream flow is conceptualized as the integral of spatially continuous local runoff over the catchment area. In this context, the prediction on river locations adds up to the area-to-area change of support problem. We compared OK-Euclid with 1D and 2D geostatistical river network models. OK-Euclid distributes kriging weights according to distance only. It may give equal weights for equally distant neighbors, regardless if they are flow-connected or flow-unconnected. OK-Euclid seems to give too much weight according to distance in geographic space and too little weight according to river topology. 1D-models, in contrast, distribute kriging weights according to river network topology. This yields more plausible estimates for situations when most information is available at flow-connected parts of the river system. Otherwise, OK-Stream gives all weight to flow-connected gauges at the same river, while close-by sites situated at a flow-unconnected river are not taken into account. Overall, OK-Stream seems to give too much weight according to river topology and too little weight according to distance in geographic space. 2D-models distribute the kriging weights in a natural way, depending on both, distance and river network topology. When a close-by site at the same river system is available, the 2D model Top-kriging gives less weight to an adjacent river system. However, when the next site at the same river system is much further away, much weight is given to a close-by site at an adjacent river system. This distribution of kriging weights is very realistic and corresponds well with physical understanding of stream flow processes behavior in space. The behavior of 1D and 2D models in different data settings was illustrated in case studies of stream flow and stream flow related variables. It would be interesting to extend these studies to other variables, such as conservative and non-conservative solutants. It would also be interesting to perform a direct comparison of 1D and 2D models on a common data set. This is, however, ongoing work and results will be reported in near future.

References

1. Blöschl G (2005) Rainfall-runoff modeling of ungauged catchments. In: Encyclopedia of hydrological sciences, pp 2061–2080

2. Cressie N, Frey J, Harch B, Smith M (2006) Spatial prediction on a river network. J Agric Biol Environ Stat 11(2):127–150
3. De Marsily G (1986) Quantitative hydrogeology. Academic Press, London
4. Garreta V, Monestiez P, Ver Hoef J (2010) Spatial modeling and prediction on river networks: up model, down model or hybrid? Environmetrics 21(5):439–456. doi:10.1002/env.995. url:http://doi.wiley.com/10.1002/env.995
5. Gottschalk L (1993) Correlation and covariance of runoff. Stoch Hydrol Hydraul 7(2):85–101
6. Gottschalk L (1993) Interpolation of runoff applying objective methods. Stoch Hydrol Hydraul 7(4):269–281
7. Gottschalk L, Krasovskaia I, Leblois E, Sauquet E (2006) Mapping mean and variance of runoff in a river basin. Hydrol Earth Syst Sci 10(4):469–484. doi:10.5194/hess-10-469-2006. url:http://www.hydrol-earth-syst-sci.net/10/469/2006/
8. Gotway C, Young L (2002) Combining incompatible spatial data. J Am Stat Assoc 97(458):632–648
9. Kyriakidis P (2004) A geostatistical framework for area-to-point spatial interpolation. Geogr Anal 36(3):259–289
10. Mockus A (1998) Estimating dependencies from spatial averages. J Comput Graph Stat 7(4):501–513
11. Peterson E, Ver Hoef J (2010) A mixed-model moving-average approach to geostatistical modeling in stream networks. Ecology 91(3):644–651
12. Sauquet E, Gottschalk L, Leblois E (2000) Mapping average annual runoff: a hierarchical approach applying a stochastic interpolation scheme. Hydrol Sci J 45:799–815. doi:10.1080/02626660009492385. url:http://www.tandfonline.com/doi/abs/10.1080/02626660009492385
13. Skøien J, Merz R, Blöschl G (2006) Top-kriging-geostatistics on stream networks. Hydrol Earth Syst Sci 10(2):277–287
14. Ver Hoef J, Peterson E (2010) A moving average approach for spatial statistical models of stream networks. J Am Stat Assoc 105(489):6–18
15. Ver Hoef J, Peterson E, Theobald D (2006) Spatial statistical models that use flow and stream distance. Environ Ecol Stat 13:449–464. doi:10.1007/s10651-006-0022-8. url:http://www.springerlink.com/index/10.1007/s10651-006-0022-8
16. Wackernagel H (1995) Multivariate geostatistics. Springer, Berlin
17. Wehrly K, Brenden T, Wang L (2009) A comparison of statistical approaches for predicting stream temperatures across heterogeneous landscapes 1. J Am Water Resour Assoc 45(4):986–997

Author Index

P. Abrahamsen et al. (eds.), *Geostatistics Oslo 2012*,
Quantitative Geology and Geostatistics 17,
DOI 10.1007/978-94-007-4153-9, © Springer Science+Business Media Dordrecht 2012

Subject Index

P. Abrahamsen et al. (eds.), *Geostatistics Oslo 2012*,
Quantitative Geology and Geostatistics 17,
DOI 10.1007/978-94-007-4153-9,

CPSIA information can be obtained
at www.ICGtesting.com
Printed in the USA
LVOW02*2226050516
486884LV00001B/9/P